WJEC GCSE GEOGRAPHY

Includes coverage of WJEC Eduqas GCSE (9–1) Geography A

■ Andy Owen (SERIES EDITOR) ■ Gregg Coleman ■ Val Davis
■ Bob Digby ■ Andy Leeder ■ Glyn Owen

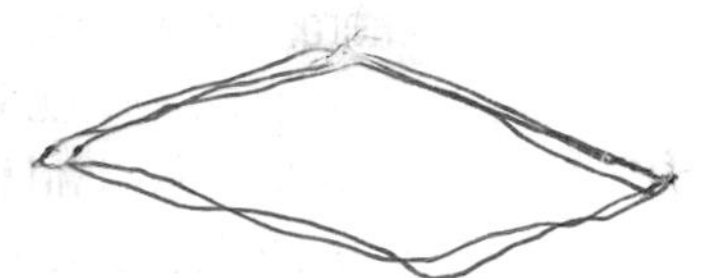

HODDER EDUCATION
AN HACHETTE UK COMPANY

Orders: please contact Bookpoint Ltd, 130 Milton Park, Abingdon, Oxon OX14 4SE. Telephone: +44 (0)1235 827720. Fax: +44 (0)1235 400454. Email: education@bookpoint.co.uk. Lines are open from 9 a.m. to 5 p.m., Monday to Saturday, with a 24-hour message answering service. You can also order through our website: www.hoddereducation.co.uk

ISBN: 978 1 5104 7755 1

First edition published in 2016. Second edition published in 2020 by

Hodder Education,

An Hachette UK Company

Carmelite House

50 Victoria Embankment

London EC4Y 0DZ

www.hoddereducation.co.uk

Impression number 10 9 8 7 6 5 4 3 2 1

Year 2024 2023 2022 2021 2020

Cover photo © Aurora Photos/Alamy

Illustrations by Aptara, Inc. and Barking Dog Art

Typeset by Aptara Inc.

Printed in Slovenia

A catalogue record for this title is available from the British Library.

CONTENTS

INTRODUCTION

How to use this book

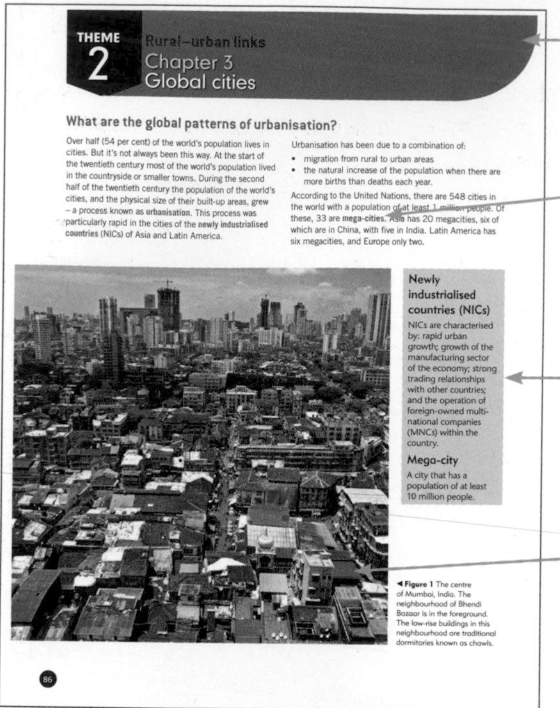

THEME 2 Rural–urban links

Chapter 3 Global cities

What are the global patterns of urbanisation?

Over half (54 per cent) of the world's population lives in cities. But it's not always been this way. At the start of the twentieth century most of the world's population lived in the countryside or smaller towns. During the second half of the twentieth century the population of the world's cities, and the physical size of their built-up areas, grew – a process known as **urbanisation**. This process was particularly rapid in the cities of the **newly industrialised countries** (NICs) of Asia and Latin America.

Urbanisation has been due to a combination of:

- migration from rural to urban areas
- the natural increase of the population when there are more births than deaths each year.

According to the United Nations, there are 548 cities in the world with a population of at least 1 million people. Of these, 33 are **mega-cities**. Asia has 20 megacities, six of which are in China, with five in India. Latin America has six megacities, and Europe only two.

Newly industrialised countries (NICs)

NICs are characterised by: rapid urban growth; growth of the manufacturing sector of the economy; strong trading relationships with other countries; and the operation of foreign-owned multinational companies (MNCs) within the country.

Mega-city

A city that has a population of at least 10 million people.

◄ **Figure 1** The centre of Mumbai, India. The neighbourhood of Bhendi Bazaar is in the foreground. The low-rise buildings in this neighbourhood are traditional dormitories known as chawls.

86

Each theme has a different colour so you can find your way around the book easily.

Important geographical terms are shown in **bold red font**. You can check the meaning of these words in the glossary at the back of the book and expand your geographical vocabulary.

The most important geographical terms are explained in a coloured box.

Photographs show what real places look like. This photograph shows Mumbai – a unique place. But the photograph shows some features that are common in urban landscapes in many NIC cities. You should study the photographs carefully. What can you learn about NIC cities from this image?

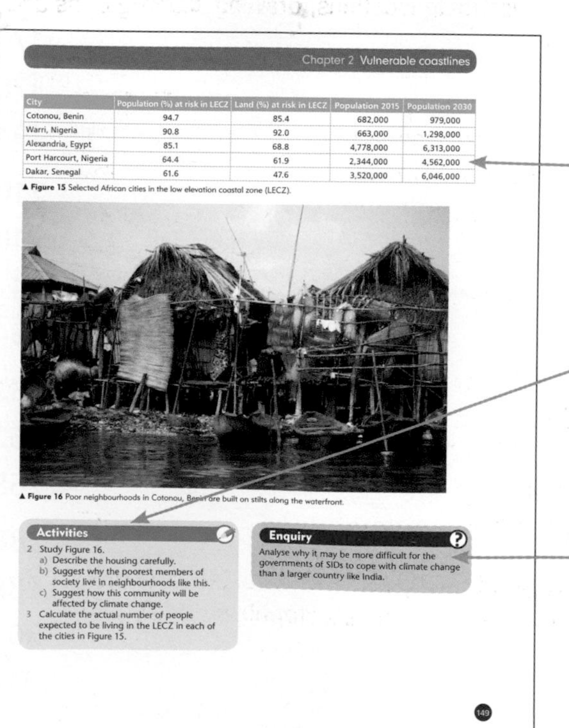

Chapter 2 Vulnerable coastlines

City	Population (%) at risk in LECZ	Land (%) at risk in LECZ	Population 2015	Population 2030
Cotonou, Benin	94.7	85.4	682,000	979,000
Warri, Nigeria	90.8	92.0	663,000	1,298,000
Alexandria, Egypt	85.1	68.8	4,778,000	6,313,000
Port Harcourt, Nigeria	64.4	61.9	2,344,000	4,562,000
Dakar, Senegal	61.6	47.6	3,520,000	6,046,000

▲ **Figure 15** Selected African cities in the low elevation coastal zone (LECZ).

▲ **Figure 16** Poor neighbourhoods in Cotonou, Benin are built on stilts along the waterfront.

Activities

2 Study Figure 16.
   a) Describe the housing carefully.
   b) Suggest why the poorest members of society live in neighbourhoods like this.
   c) Suggest how this community will be affected by climate change.

3 Calculate the actual number of people expected to be living in the LECZ in each of the cities in Figure 15.

Enquiry

Analyse why it may be more difficult for the governments of SIDs to cope with climate change than a larger country like India.

149

Maps, graphs and tables of data like this one provide us with evidence about the state of the world. Thinking like a geographer means that you need to look for patterns and trends in this evidence.

The activities will make you think carefully about the geographical information that has been presented in the photographs, maps, graphs and tables on the page. Doing them will help to build your geographical confidence and your ability to describe features, spot patterns and trends and explain why things happen.

Enquiries are longer activities. Some of them require further research, debate or discussion. Many will ask for your opinion and build your skills of analysis, evaluation and decision making. These tasks will encourage you to take an enquiry approach to learning – helping you to think like a geographer.

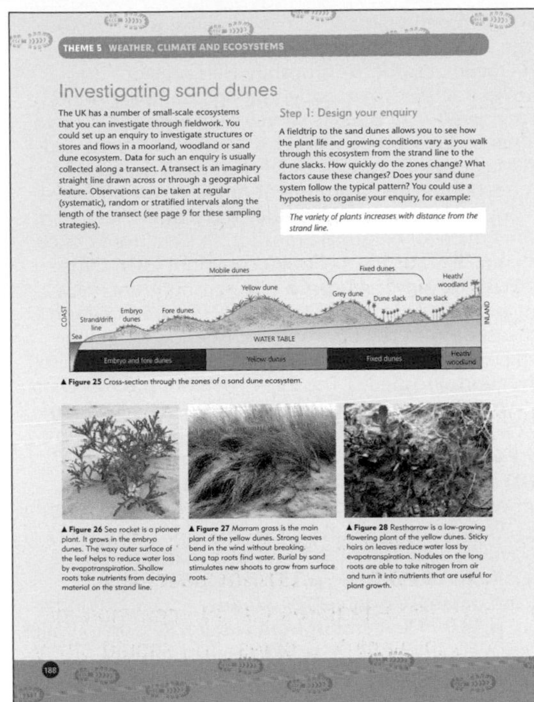

THEME 5 WEATHER, CLIMATE AND ECOSYSTEMS

Investigating sand dunes

The UK has a number of small-scale ecosystems that you can investigate through fieldwork. You could set up an enquiry to investigate structures or stores and flows in a moorland, woodland or sand dune ecosystem. Data for such an enquiry is usually collected along a transect. A transect is an imaginary straight line drawn across or through a geographical feature. Observations can be taken at regular (systematic), random or stratified intervals along the length of the transect (see page 9 for these sampling strategies).

Step 1: Design your enquiry

A fieldtrip to the sand dunes allows you to see how the plant life and growing conditions vary as you walk through this ecosystem from the strand line to the dune slacks. How quickly do the zones change? What factors cause these changes? Does your sand dune system follow the typical pattern? You could use a hypothesis to organise your enquiry, for example:

The variety of plants increases with distance from the strand line.

▲ Figure 25 Cross-section through the zones of a sand dune ecosystem.

▲ Figure 26 Sea rocket is a pioneer plant. It grows in the embryo dunes. The waxy outer surface of the leaf helps to reduce water loss by evapotranspiration. Shallow roots take nutrients from decaying material on the strand line.

▲ Figure 27 Marram grass is the main plant of the yellow dunes. Strong leaves bend in the wind without breaking. Long tap roots find water. Burial by sand stimulates new shoots to grow from surface roots.

▲ Figure 28 Restharrow is a low-growing flowering plant of the yellow dunes. Sticky hairs on leaves reduce water loss by evapotranspiration. Nodules on the long roots are able to take nitrogen from air and turn it into nutrients that are useful for plant growth.

188

Fieldwork enquiries are described on pages which have a coloured background with footprints at the top and bottom of the page. Some pages focus on how to use specific fieldwork methods for collecting or representing data – like this page which describes the use of a transect to collect data in a sand dune ecosystem. Other fieldwork pages provide advice on sampling strategies, measuring flows and investigating concepts such as place.

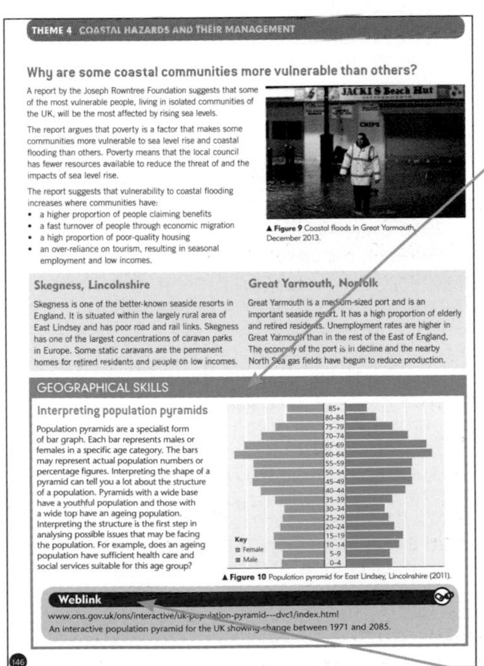

THEME 4 COASTAL HAZARDS AND THEIR MANAGEMENT

Why are some coastal communities more vulnerable than others?

A report by the Joseph Rowntree Foundation suggests that some of the most vulnerable people, living in isolated communities of the UK, will be the most affected by rising sea levels.

The report argues that poverty is a factor that makes some communities more vulnerable to sea level rise and coastal flooding than others. Poverty means that the local council has fewer resources available to reduce the threat of and the impacts of sea level rise.

The report suggests that vulnerability to coastal flooding increases where communities have:

- a higher proportion of people claiming benefits
- a fast turnover of people through economic migration
- a high proportion of poor-quality housing
- an over-reliance on tourism, resulting in seasonal employment and low incomes.

▲ Figure 9 Coastal floods in Great Yarmouth, December 2013.

Skegness, Lincolnshire

Skegness is one of the better-known seaside resorts in England. It is situated within the largely rural area of East Lindsey and has poor road and rail links. Skegness has one of the largest concentrations of caravan parks in Europe. Some static caravans are the permanent homes for retired residents and people on low incomes.

Great Yarmouth, Norfolk

Great Yarmouth is a medium-sized port and is an important seaside resort. It has a high proportion of elderly and retired residents. Unemployment rates are higher in Great Yarmouth than in the rest of the East of England. The economy of the port is in decline and the nearby North Sea gas fields have begun to reduce production.

GEOGRAPHICAL SKILLS

Interpreting population pyramids

Population pyramids are a specialist form of bar graph. Each bar represents males or females in a specific age category. The bars may represent actual population numbers or percentage figures. Interpreting the shape of a pyramid can tell you a lot about the structure of a population. Pyramids with a wide base have a youthful population and those with a wide top have an ageing population. Interpreting the structure is the first step in analysing possible issues that may be facing the population. For example, does an ageing population have sufficient health care and social services suitable for this age group?

▲ Figure 10 Population pyramid for East Lindsey, Lincolnshire (2011).

Weblink

www.ons.gov.uk/ons/interactive/uk-population-pyramid---dvc1/index.html
An interactive population pyramid for the UK showing change between 1971 and 2085.

146

Geographical Skills panels describe how to carry out some important skills that are needed by geographers. These panels cover subjects like describing locations, drawing scattergraphs and reading hydrographs.

Weblinks will allow you to carry out further research or explore interactive maps on sites that use Geographical Information Systems (GIS).

ACKNOWLEDGEMENTS

Text acknowledgements: p.4 Data from www.nationalparks.gov.uk/learningabout/whatisnationalpark/factsandfigures; **p.7** *t and b* Extracts from Clwydian Range AONB: Sustainable Tourism Strategy and Action Plan, the Tourism Company 2009–2014; *m* Data from www.clwydianrangeaonb.org.uk/files/1906706421-Tourism%20Strategy%20Final%20Document.pdf; *b* Shropshire Hills Area of Outstanding Natural Beauty Management Plan 2014–2019; **p.15** Ordnance survey map extract © Crown copyright and database rights 2020. Hodder Education under licence to OS.; **p. 23** Data from British Geological Survey; **p.30** *m* Ordnance survey map extract © Crown copyright and database rights 2020. Hodder Education under licence to OS; **p.32** Ordnance survey map extract © Crown copyright and database rights 2020. Hodder Education under licence to OS; **p.34** *br* Data from Gerd Masselink, Plymouth University; **p.41** Screenshot from Met Office, © Crown copyright 2016 Met Office; **p. 42** *m* Graph data © Crown copyright 2016 Environment Agency; **p.46** *b* Based on data provided by the UK National River Flow Archive, hosted by the Centre for Ecology and Hydrology, Natural Environment Research Council. http://nrfa.ceh.ac.uk; **p.55** Data from ww.ukcensusdata.com; **p.57** *t* Data from Office for National Statistics; *m* Data from Office for National Statistics; **p.59** Data from Welsh Government; **p.61** Map from Office of National Statistics, © Crown copyright 2016 Office of National Statistics; **p.62** *r* Map from Welsh Government, © Crown copyright 2015 Welsh Government; **p.63** *t* Screenshot from www.neighbourhood.statistics.gov.uk, © Crown copyright 2015 Office of National Statistics; **p.65** *Daily Mail*, 15 October 2019; **p.66** r Graph © Crown copyright 2014 Office for National Statistics; **p.74** *tr* Data from ONS; **p.75** Screenshot from the Environment Agency, © Crown copyright 2016 Environment Agency; **p.78** Data from ONS; **p.80** Data from Google *m* Data from Trussell Trust; **p.87** *all* Data from United Nations Department of Economic and Social Affairs, Population Division, World Urbanizations Prospects: The 2014 Revision; **p.89** *all* Data from United Nations Department of Economic and Social Affairs, Population Division, World Urbanizations Prospects: The 2014 Revision; **p.92** *tr* Data from Census of India; **p.93** *t* Data adapted from National Sample Survey Organization (NSSO), http://timesofindia.indiatimes.com/; *m* Data from Census of India and www.devinfolive.info/censusinfodashboard; **p.94** *m* Data from http://blogs-images.forbes.com/niallmccarthy/files/2014/09/Bollywood_2.jpg; **p.99** *tl* Data from National Census; **p.101** *r* Data from National Census; **p.103** Used by permission of *The Guardian*; **p.111** *tr* Diagram courtesy of the USGS; **p.115** *t* Data from World Bank; **p.119** *t* Data from UNSD Demographic Statistics; **p.123** Data from United Nations, Department of Economic and Social Affairs, Population division (2014), World Urbanization Prospects: The 2014 Revision, CD-ROM Edition; **p.124** *tr* Diagram based on Seismic Waves Radiate from the Focus of an Earthquake, University of Waikato; *m* Data from US Geological Survey and World Bank; **p.129** *b* Data from World Bank; **p.131** *b* Graph © Crown copyright 2014 Met Office; **p.137** *l* Ordnance survey map extract © Crown copyright and database rights 2020. Hodder Education under licence to OS.; **p.139** Extract from 'A Simple Life of Luxury' © 2016. All Rights Reserved; **p.140** *m* Screenshot from Natural Resources Wales, © Crown copyright 2016 Natural Resources Wales; *b* Screenshot from Data Shine © Crown copyright 2015; **p.141** *t* Data from Neighbourhood Statistics, 2011; **p.144** *r* Data from Environment Agency; **p.145** Map from 'Safecoast – trends in flood risk', www.safecoast.org (July 2008), reproduced by permission of Rijkswaterstaat-Centre for Water Management; **p.146** Graph from LRO 2011 Census Population Estimates: East Lindsey; **p.147** *l* Data from Office for National Statistics; r Extract from Joseph Rowntree Foundation Report, Summary of 'Impacts of climate change on disadvantaged UK coastal communities' (6 March 2011); **p.148** Map copyright: James Morgan / Panos published by SciDev.Net; **p.149** Data from UN Habitat, State of the World's Cities 2008/2009; **p.150** *bl* Map copyright © 2008, United Nations Environment Programme; **p.151** *r* Maps from 'Impact of sea level rise on the Nile delta' © GRID-Arenday. Used with permission; **p.163** *t* Data from Pakistan Meteorological Department; **p.164** Data from www.seatemperature.org; **p.165** Satellite images © NASA; **p.166** Map copyright 2015 National Drought Mitigation Center; **p.167** Graph copyright 2015 National Drought Mitigation Center; **p.169** Graph data from https://www.ncdc.noaa.gov/cag/; **p.171** *m* Satellite image © NASA; **p.174** Weather map © Crown copyright 2014 Met Office; **p.187** Quotation from http://www.ceh.ac.uk/news-and-media/blogs/farmland-ditch-management-hydrologist-perspective © NERC Centre for Ecology & Hydrology 2016; **p.196** Data from www.obt.inpe.br/degrad; **p.197** *all* Data from American Soybean Association; **p.200** Data from World Bank; **p.207** Maps © NASA; **p.217** *l* Data from Nike.com; *r* Estimates based on *Washington Post*, 1995; **p.221** *l* Text extract from http://qz.com/389741/the-thing-that-makes-bangladeshs-garment-industry-such-a-huge-success-also-makes-it-deadly/; *r* Text extract © Institute for Global Labour and Human Rights; **p.224** Data from World Bank; **p.226** *m* Data from World Bank; **p.227** *tr* OECD used under a CC-BY-SA 3.0 licence; **p.228** Data from UNWTO Tourism Highlights 2015 Edition; **p.230** Data from World Bank; **p.236** Infographic copyright © 2018 United Nations. **p.237** *tl* Infographic copyright © 2016 United Nations. Reprinted with permission of the United Nations; **p.242** Data from Aquastat; **p.243** Text extract from *Rural Women in the Sahel and Their Access to Agricultural Extension: Overview of Five Country Studies*. 1994. Report No. 13532 AFR. AF5AE. Washington, D.C.: World Bank; **p.245** *t* Map from Food and Agriculture Organization of the United Nations,2016, Proportion of total water withdrawal withdrawn for agriculture, http://www.fao.org/nr/water/aquastat/maps/WithA.WithT_eng.pdf. Reproduced with permission; *bl* Graph © UNESCO 2015; *br* Data from Aquastat; **p.252** *l* Screenshot from World Resources Institute, © World Resources Institute (http://creativecommons.org/licenses/by/4.0/); **p.256** *bl* Screenshot from Annual Survey of Hours and Earnings, © Crown copyright 2015 Office of National Statistics; **p.259** Ordnance survey map extract © Crown copyright and database rights 2020. Hodder Education under licence to OS; **p.263** Screenshots from Statistics South-Africa; **p.264** *l and r* Data from World Bank; **p.265** Map © www.worldmapper.org; **p.269** All graph data from https://www.populationpyramid.net/; **p.270** *b* Extract from Breaking Free from Child Labour © UNICEF India; **p.271** Graph from Making Progress Against Child Labour: Global Estimates and Trends 2000–2012, © 1996–2013 International Labour Organization (ILO); **p.272** *b* © UNICEF India; **p.273** Data from ILO; **p.274** Data © Amnesty International; Reprinted with permission of the United Nations; **p.278** *t* data from UNICEF, *b* map © Worldmapper; **p.279** *b* Data from National Statistical Office, Malawi; Prevalence Rate, 2014 The Global HIV/AIDS Epidemic, Kaiser Family Foundation; **p.282** © Avert, 2019; **p.286** Data from UN Water; **p.288** *m* Data from IMO GHG study, 2009; *b* Data from Eurostat; **p.289** *m* Quotation courtesy of Guardian News & Media; **p.291** Map data from www.oafrica.com; **p.291** Map data from: http://standardgraphs.ices.dk/ViewCharts.aspx?key=4121_as © ICES – All Rights Reserved; **p.295** Data from FAO Yearbook 2014, Fishery and Aquaculture Statistics; **p.295** Data from www.worldpalmoilproduction.com/; **p.296** *b* Data from www.palmoilextractionmachine.com/FAQ/what-is-palm-oil-used-for.html; **p.298** Screenshot from Future Flooding Report, © Crown copyright 2012 Department for Business Information & Skills; **p.306** Data from www.thetravelfoundation.org.uk; **p.308** *b* Artwork © 1996–2016 Australian Institute of Marine Science; **p.311** This article was first published by the Asian Development Bank (www.adb.org). **p.312** Ordnance Survey map extract © Crown copyright and database rights 2020. Hodder Education under licence to OS; **p.314** Data from UN; **p.315** Infographic © UN Women; **p.316** Data from http://www.wetterzentrale.de/; **p.320** Data from https://www.parallel.co.uk/; **p.322** Data from World Bank; **p.323** Data from National Institute for Space Research; **p.324** Data from World Bank; **p.326** Data for pyramids from https://www.populationpyramid.net/; **p.327** Data from World Bank; **p.328** Met Office © Crown Copyright 2020; **p.329** Data from World Economic Forum.

Photo credits can be found on page 346.

THEME 1 Landscapes and physical processes

Chapter 1 Distinctive landscapes

Upland landscapes of the UK

The **upland** landscapes of the UK are areas of mountain or moorland. The highest upland regions are open landscapes with few, if any, field boundaries. At lower altitudes, the UK's upland landscapes contain river valleys. These lower areas are more affected by people, with farms, field boundaries and settlements. Each regional upland landscape is slightly different, affected by factors such as:

- geology
- the history of settlement and farming that have affected land use
- natural processes in the past (erosion by ice during the last ice age) and now (river erosion)
- type of vegetation.

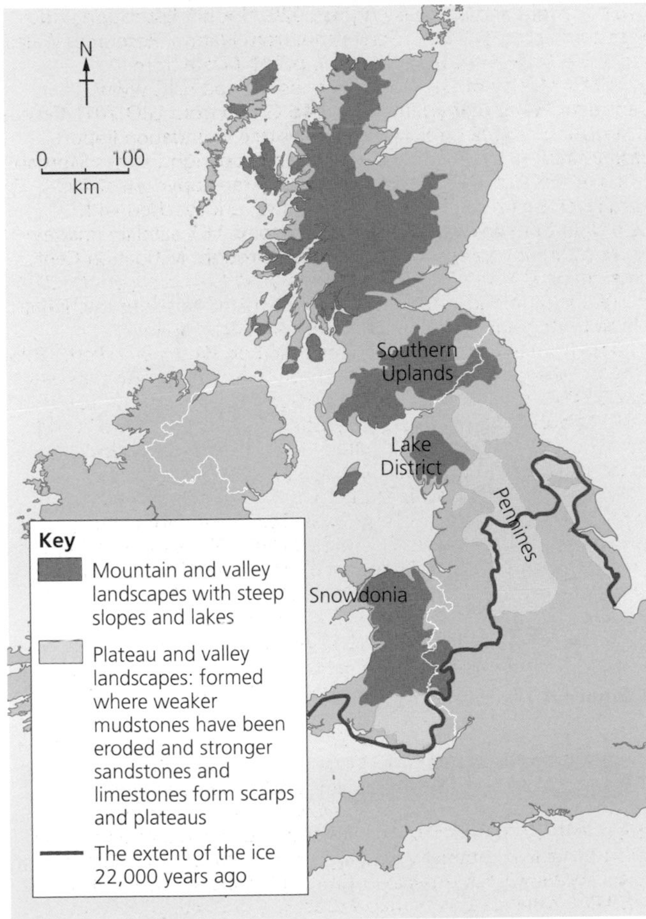

▲ **Figure 1** The upland landscape types of the UK and the extent of the ice during the last ice advance 22,000 years ago.

Upland

A landscape that is high above sea level. Upland areas are hilly and sometimes mountainous.

▲ **Figure 2** Mountain and valley landscape in the Lake District.

Activities

1 a) Use Figure 1 to describe the distribution of:
   i) mountain and valley landscapes
   ii) plateau and valley landscapes.
  b) What do you notice about the relationship of these landscape types and the extent of the ice?

2 a) Describe the landscape in Figure 2.
  b) What features make this landscape special?

Enquiry

How do upland landscapes in the UK compare?

- Research some photos of Snowdonia National Park.
- What are the similarities and differences between Figure 2 and the photos you have found?

Understanding complex landscapes

Every landscape contains many separate features and landforms. Rather like the pieces of a jigsaw, it is the combination of these different features that creates a **place** identity that we see (and experience) when we view or visit a landscape in the UK.

Study Figure 3. It shows the sand dunes at Ynyslas on the west coast of Wales. Sand dunes themselves are not uncommon. Figure 4 shows that a large number of coastal locations have similar landscape features.

▲ **Figure 3** The sand dunes at Ynyslas at the mouth of the River Dyfi.

Study Figure 5. The sand dunes can be seen at point 6. They are just one small feature in a complex landscape. The sand dunes cover an area of less than one square kilometre, but the river estuary behind is of a much larger **scale**. This area includes a number of river and coastal landforms and a variety of ecosystems. This landscape looks natural but it includes human influences. Much of the boggy land in the estuary has been drained with ditches and is grazed. The landscape is popular with tourists and includes a number of caravan sites and small towns. It is the combination of these different human and physical features that give the estuary of the River Dyfi its unique landscape qualities.

Scale

A geographical concept used to describe the relative size of something. A landscape contains features, like the sand dunes, that are relatively small scale. They are described as being local in scale. Other features, like the hills, are much larger. They cover large parts of Wales, so are described as regional in scale.

Place

A geographical concept that is used to describe what makes somewhere special, unique or distinct. Each place includes many different features of the human and physical environment, such as landscape features and landmarks, local styles of building, ecosystems and habitats, and local historical and cultural features. Each of these features may be relatively common across the UK; it is the unique combination of these geographical features, however, that creates an identity for any one place.

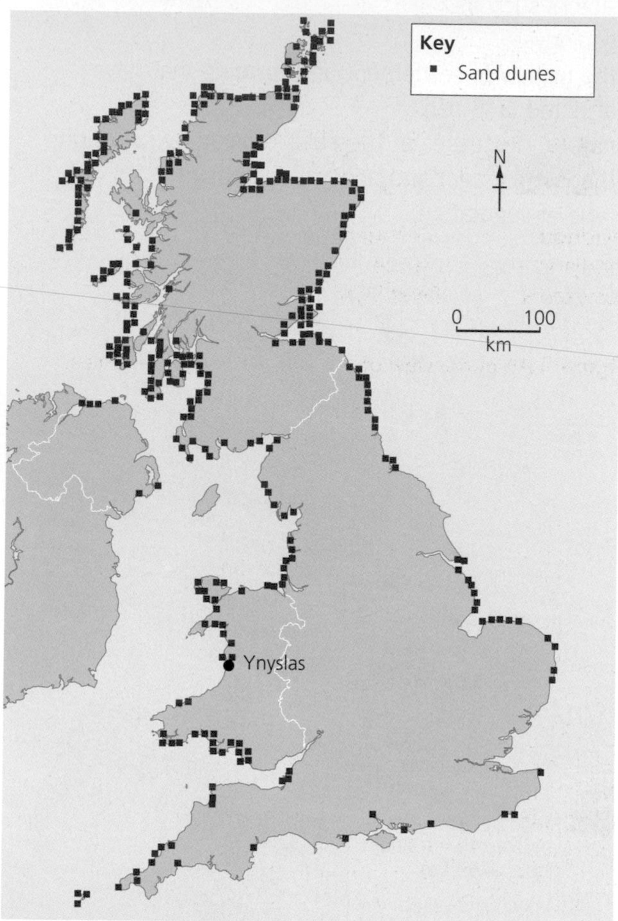

▲ **Figure 4** Distribution of sand dunes in the UK.

Activities

1. Study Figure 3. Describe the features of this landscape that can be seen at A and B.
2. Describe the distribution of sand dunes shown in Figure 4.
3. Match each label to the correct numbered feature of Figure 5.
4. Use evidence in Figures 3 and 5 to explain the concepts of scale and place identity.

1. Deciduous woodland ecosystems
2. Meanders on the River Dyfi
3. Sand dunes at Ynyslas
4. Upland grazing and moorland habitat
5. Features of the estuary including salt marsh and sheep pasture
6. Small towns such as Aberdyfi
7. Coastal landforms – beaches and bars

▲ **Figure 5** An aerial view of the Dyfi estuary, Mid Wales.

▲ **Figure 6** The small market town of Machynlleth, about 9 km from the mouth of the River Dyfi.

▲ **Figure 7** Ponds in the valley of the River Dyfi at Ynyshir, about 3 km from the mouth of the river.

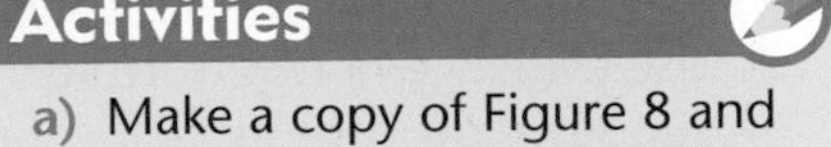

Activities

5 a) Make a copy of Figure 8 and use it to calculate a bi-polar score for each photograph.

b) Share your scores with at least four members of your class. Calculate a mean score for each photograph.

	+5	+4	+3	+2	+1	−1	−2	−3	−4	−5	
Attractive											Unattractive
Varied											Monotonous
Rare											Common
Natural											Human

▲ **Figure 8** Bi-polar statements to assess landscape.

Honeypot sites

Millions of people enjoy leisure activities in the countryside. Walking, jogging and cycling are all popular activities that are good for your health. Leisure activities are also good for the rural economy: day visitors spend money at local attractions, in shops and on food and drink; other visitors stay overnight in guest houses. This allows **rural diversification** – where the rural economy branches out from farming and invests in tourism-related service industries.

However, too many visitors can cause problems. Litter, parking and footpath erosion are all issues that need careful management. These issues become acute when the number of visitors exceeds the **carrying capacity** of the location and an activity begins to damage the landscape or ecosystem. Carrying capacity is most likely to be exceeded at the UK's **honeypot sites**. Like bees around a pot of honey, these sites attract the largest numbers of people because they are:

- exceptionally beautiful or interesting
- accessible by road and within easy reach of people living in larger towns or cities.

The majority of the UK's natural honeypot sites occur in either one of the UK's Areas of Outstanding Natural Beauty (AONBs) or one of our National Parks.

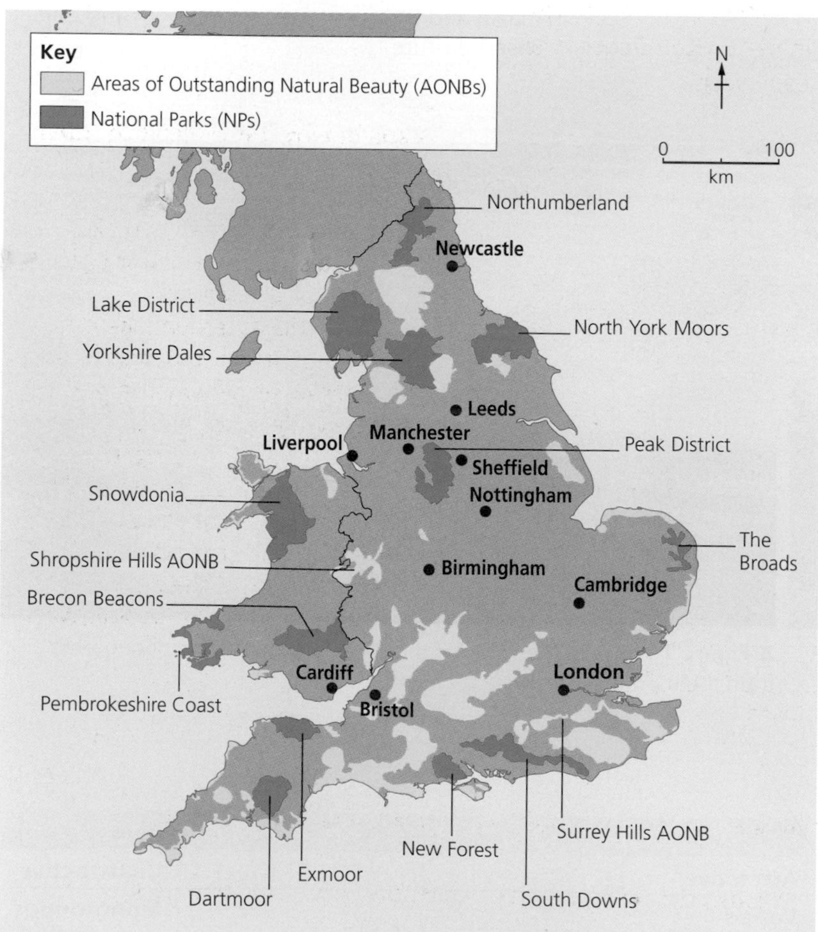

▲ **Figure 9** The location and distribution of Areas of Outstanding Natural Beauty and National Parks in England and Wales; AONBs are selected because of their exceptional scenic qualities.

Rural diversification

The development of new businesses in the countryside. These new businesses are often in leisure or tourism rather than in farming.

Carrying capacity

The number of people that a landscape can support before any lasting damage is done.

590,000 visitors stayed overnight in hotels, bed and breakfasts or campsites.

Visitors spent over **2.1 million** days in the Park.

Tourism creates **4,325** full time equivalent (FTE) jobs in direct employment and another **5,188** full time equivalent (FTE) jobs in indirect employment.

Visitors to the Park spent a total of **£239 million.**

▲ **Figure 10** The importance of tourism to Yorkshire Dales National Park (2017) in numbers.

Activities

1. Study Figure 9.
   a) Describe the location of Shropshire Hills AONB.
   b) List five urban areas that fall within 100 km of the Peak District National Park.
   c) Which National Park is furthest from any large city?
2. Use evidence in Figure 10 to suggest how tourism helps the rural economy to diversify.

Repairing footpath erosion in the Brecon Beacons

The Brecon Beacons is a mountainous region of South Wales. It is a popular area for walkers and mountain bikers. Some of the most-used paths are on land owned by public bodies such as the National Park Authority and the National Trust. During the period 2005 to 2015, the National Trust improved 15 km of paths. This work has allowed bare soil to become re-vegetated and areas of stunted vegetation to recover – in all, over 75,000 m^2 of vegetation has been restored. The National Trust aims to repair a further 10 km of footpaths where erosion is greater than 4 m in width.

▲ **Figure 11** Volunteers repairing a footpath close to Pen Y Fan, Brecon Beacons, Wales.

The restoration work relies heavily on donations and volunteer labour. During the summer of 2015, a section of path at Cefn Cwm Llwch, close to Pen y Fan, was restored. This is a very remote area and the stone needed for the path was delivered by helicopter in 1 tonne bags. In all, 70 bags of stone were flown in from a nearby quarry. The stones are placed upright in the ground and packed tight to make a hard-wearing and natural-looking path. Once the path is built, the area on either side can be re-vegetated. The cost of footpath management in the Brecon Beacons to the National Trust alone is around £100,000 each year.

Activities

3 Explain why it is difficult to repair environments like the one in Figure 11.

4 a) Make a copy of Figure 12.
   b) Add a suitable label for each stage of the diagram.
   c) Copy the statements below the diagram and put them in order to show why the path gets wider over time.

5 Suggest why it is important to manage the effects of footpath erosion. Think about the impact on both the physical environment and the rural economy if footpath erosion was not managed.

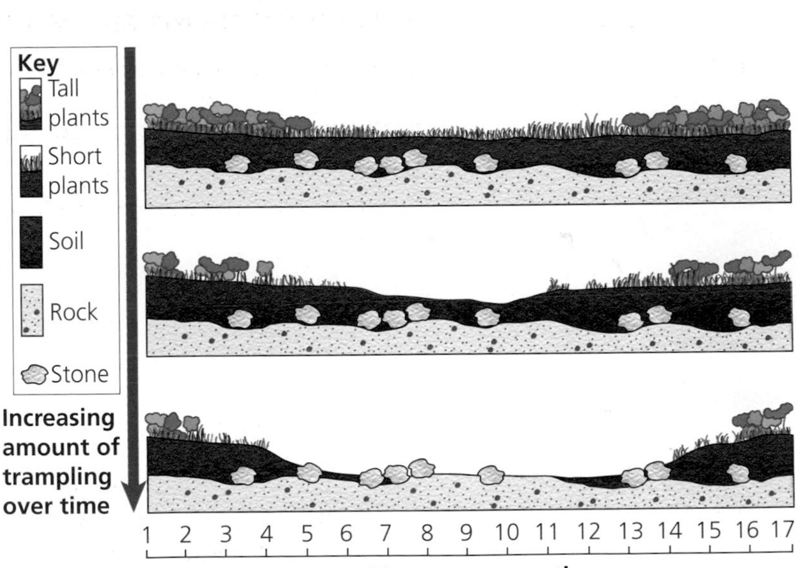

Plants die | Soil is exposed to rainwater | Soil is eroded by rain splash and gulley erosion

The path becomes wider and wider | Plants are short and stunted where they have been trampled

Walkers avoid the central muddy section of path so walk at the edge | Stones are exposed as soil is eroded

▲ **Figure 12** Causes of footpath erosion.

Managing distinctive landscapes

The UK countryside is managed by landowners. Farmers manage their land to produce food and conserve habitats and wildlife. The UK's National Parks and AONBs are managed carefully to ensure that special and unique qualities of the landscape, wildlife and cultural heritage are recognised and conserved. The work is done by a small team of full-time staff and a large number of volunteers. Each of the UK's AONBs produces a five-year management plan and identifies action points that need to be met.

Managing the Clwydian Range and Dee Valley AONB

The Clwydian Range is a limestone upland area in northeast Wales. It is a popular area for visitors with an estimated 4.5 million people living within a 90-minute drive of it. Its distinctive landscape, seen in Figure 14, shows the importance of its geology. The limestone is resistant to erosion, forming steep escarpments with long scree slopes below them. Above the escarpment, the limestone soils on the plateau are thin and plants are specialised. Limestone landscapes are relatively uncommon in the UK. Their distribution is shown in Figure 15.

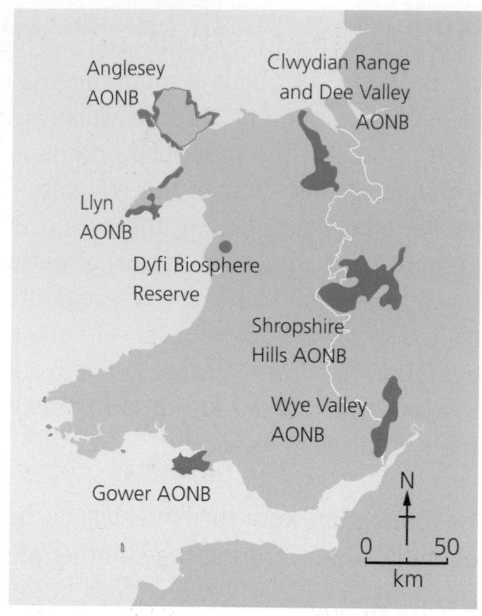

▲ **Figure 13** The location of AONBs and biosphere reserves in, and close to, Wales.

▲ **Figure 14** The limestone landscape of Eglwyseg in the Clwydian Range AONB, Wales.

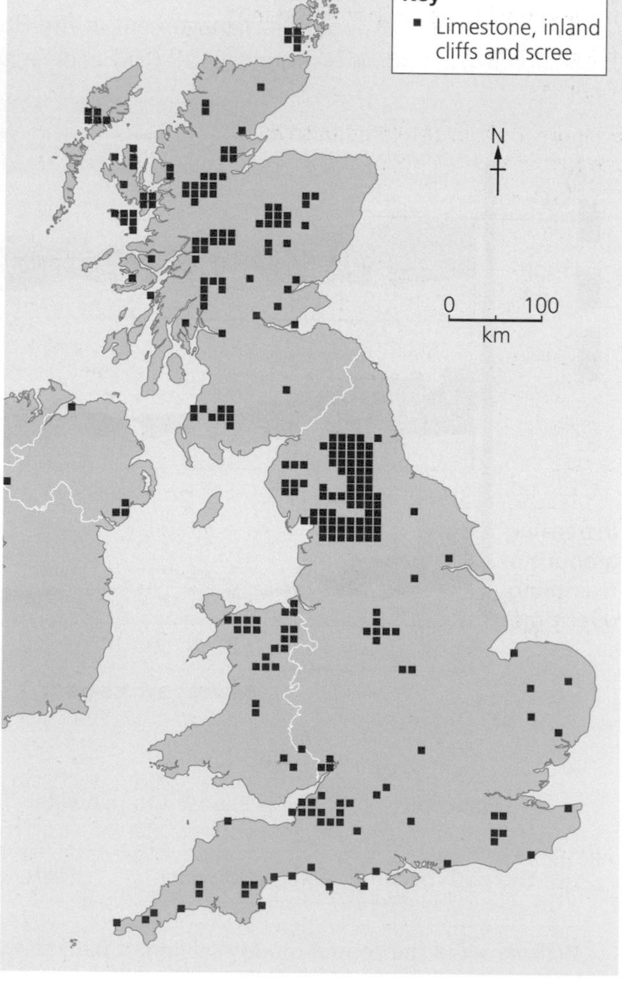

▲ **Figure 15** The distribution of limestone pavements in the UK.

Activities

1 Write an appropriate label for each of the features, 1 to 6, in Figure 14.

2 a) Describe the distribution of limestone pavements in the UK.

b) Explain why, although the limestone pavement in Figure 14 occurs in other places, this feature helps to create a sense of place identity for Eglwyseg.

The unique features that attract visitors to the Clwydian Range AONB

The richness of the landscape comes from a combination of undulating upland, including open heather clad moors and rocky outcrops, gently rolling farmland and wooded valleys. The area is blessed with extensive views in all directions. Attractive hedgerows, stone walls and a number of picturesque small villages add to the striking visual appeal.

The rich landscape of the Clwydian Range is the result of geological processes acting over 500 million years. The landscape of heather-clad uplands, ice-carved valleys, rocky limestone, rolling farmland and wooded valleys are a unique and spectacular resource.

Much of the landscape pattern visible today of villages and isolated farmsteads dates from the medieval period. Offa's Dyke is an important heritage feature of the borderland between England and Wales, and Offa's Dyke Path, a designated National Trail, is an important recreational resource.

▲ **Figure 16** Extract from the Clwydian Range AONB Action Plan.

Activity	Mean score
Driving around in attractive countryside	2.41
Taking longer walks (over 2 hours)	2.57
Taking shorter walks (up to 2 hours)	2.61
Exploring small towns and villages	2.64
Visits to attractions	2.83
Cycling (including mountain biking)	3.07
Horse riding	3.53
Fishing	3.61

▲ **Figure 17** Tourist-related businesses were asked to score each activity on a scale of one to five, where one is 'important to most of my visitors' and five is 'not important to any of my visitors'.

Landscape quality and special character	Establish fixed-point photography monitoring of key viewpoints	Work with power companies to put overhead power lines underground in the most sensitive locations	Organise one day per year when local communities are asked to share their wealth of knowledge about the area
Access and recreation	Improve signage by renewing and replacing finger posts	Improve access by reducing the number of stiles or replacing stiles with kissing gates	Monitor footpath erosion in key areas, e.g. around hill forts, and carry out one erosion project per year
Culture and people	Object to development proposals that involve the loss of community facilities, such as local shops and public houses	Explore options to secure consistent mobile phone coverage across the AONB and to overcome existing 'black spots' without harming the character and appearance of the area	Work with local planning and housing authorities to promote affordable housing schemes for local people

▲ **Figure 18** Selected action points in the Shropshire Hills AONB management plan for 2014–19.

Activities

3 Use Figure 16 to describe how geological, geographical and cultural features of the landscape attract visitors to this AONB.

4 Choose an appropriate technique to represent the data in Figure 17.

5 a) Discuss the nine action points in Figure 18.
b) Make a diamond nine diagram (like the one on page 85) and place the action points in the diagram, putting those that you think are essential at the top of the diagram.
c) Explain why you have chosen your top three action points. You should explain how your chosen action points will help to conserve or recognise landscapes, wildlife or local communities/heritage.

Investigating visitor use of a honeypot site

The Stiperstones is a rocky ridge in the Shropshire Hills AONB. This AONB is in easy driving distance of the towns and cities of the West Midlands. The hills are not large: the summits of most are around 400 m above sea level. The slopes are not too steep so they are accessible to walkers and cyclists of all abilities and ages.

The soils on the Stiperstones are particularly thin. In fact, rocky outcrops called tors stick out along the backbone of the ridge. People walk up to the top to climb over the tors and admire the view. However, trampling along the same routes has damaged the vegetation and led to erosion of the soil. You can see the scars made by footpath erosion in Figure 20.

Sampling strategies

A sample is a set of data that provides us with a good understanding of what is happening without having to record everything. The sample must be representative – a fair reflection of the whole. It is essential that the sample is not biased. There are three main sampling strategies and these are shown in Figure 19. In addition, sampling can sometimes be 'opportunistic'. Opportunity sampling means that the sample has been chosen because it is convenient and easily or safely available. For example, sample points along a river are usually chosen because access into the river is permitted (the landowner has agreed) and safe. The problem with this type of sampling is that it isn't fair or unbiased, so your sample may not be representative.

What's the difference between quantitative and qualitative data?

Geographers can collect all sorts of different kinds of data during a fieldwork enquiry but it will always fall into one of two categories:

Quantitative data is information that can be measured and recorded as numbers. Counting traffic, measuring the width and velocity of a river, or measuring the size of pebbles on a beach are all examples of quantitative data.

Qualitative data is information that is not numerical. You can collect qualitative data in a wide variety of ways such as by taking photographs, field sketches, videos or audio recordings. Interviewing people to find their opinions or perceptions about an issue is also an example of qualitative data collection. The bi-polar technique (page 104) is a way to try to make the collection of opinions more quantitative.

	Random sampling	Systematic sampling	Stratified sampling
What is it?	Every sample point has an equal probability of being sampled.	Data is collected at regular intervals. These intervals can be in time or space (distance).	Proportional samples are taken from separate groups or strata.
How might you use it?	To investigate how busy a country park is you could select a range of locations across the park and carry out pedestrian counts at each of these sites. Put the names of all locations in a bag and pull out five.	To investigate how tourist impact changes as you move further away from a visitor centre, you might carry out litter counts and environmental impact surveys every 100 m away from the centre.	When carrying out a questionnaire, make sure that you sample appropriate numbers from different age groups based on data from the census, i.e., 20% aged 0–18, 25% aged 19–45, 35% aged 45–65 and 20% aged 66+.
Advantages	Removes bias completely in the selection of sites.	Covers a range of locations and provides an even spread over the whole survey area.	Provides a true representative sample.
Disadvantages	Survey sites may be clustered together, so you do not get an even spread of locations.	If samples are too far apart you may miss important variations.	You will need information on the size of each group before you start.

▲ **Figure 19** Sampling strategies.

Using a control to investigate the impact of trampling

Figure 20 clearly shows the difference in vegetation between areas that have been heavily trampled by visitors and areas that have not been walked across. However, as Figure 12 (page 5) suggests, people walk through the vegetation on either side of the path to avoid walking on the uneven stones. This leads to an enquiry question: are some species of plants more vulnerable to damage than others?

Are some species of plants more vulnerable to damage than others?

If so, you might find that there is a lower percentage of some plants growing immediately next to the path because they have been damaged by trampling. To find out you would need to set up a control – an experiment to find the average amount of each species of plant. This is what you would do:

Step 1: Use a **quadrat** to measure the percentage of various plants in an area at a distance away from the path that appears to be unaffected by trampling. A quadrat is a metal or plastic grid – usually about half a metre across. You can estimate the percentage of each plant in each quadrat. Take at least five control measurements in different places and then calculate the mean percentage of each plant in these areas.

Step 2: Set up a **transect** (see page 188) across the footpath. You will need to start and end the transect several metres away from the eroded section of path so that you sample the plants on the edge of the trampled area. Each quadrat will need to be 2 m apart along the transect.

	Quadrat				
	1	2	3	4	5
Bilberry	30	10	40	25	55
Heather	60	80	60	60	40
Other plants	10	10	0	15	5
Bare soil/rock	0	0	0	0	0

▲ **Figure 21** Percentage of each type of vegetation in the control quadrats.

	Quadrat									
	1	2	3	4	5	6	7	8	9	10
Bilberry	30	10	0	0	0	0	0	10	20	35
Heather	60	70	70	0	20	0	70	80	80	65
Other plants	10	20	10	0	0	0	30	10	0	0
Bare soil/rock	0	0	20	100	80	100	0	0	0	0

▲ **Figure 22** Percentage of each type of vegetation in each quadrat; each quadrat was 2 m apart.

Activities

1 How does trampling affect different species of plant?
   a) Use the data in Figure 21 to calculate the mean percentage of each plant in the control.
   b) Choose a suitable technique to represent the data in Figure 22. See page 189 to see how to draw a kite diagram.
   c) What conclusions can you draw about the effect of trampling on bilberry and heather plants at the edge of the path?

Enquiry

How would you go about designing an enquiry into what gives your local area a sense of identity? What quantitative and qualitative data could you collect?

◀ **Figure 20** Footpath erosion is not a widespread problem in the Shropshire Hills AONB but it is a localised problem on the Stiperstones ridge.

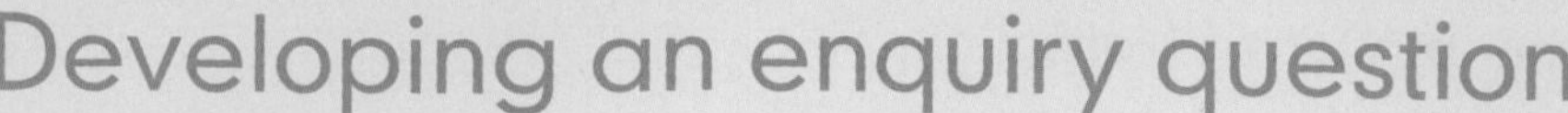

Developing an enquiry question

Imagine you are about to go on a fieldtrip to the Stiperstones. Your enquiry will need an enquiry question. First you need to get a feel for the place and the issues it faces. Using the internet to read about the Shropshire Hills AONB or using Google Maps or Google Street View are all good places to research the location of your fieldwork before you visit. Then you can pose possible enquiry questions. For example:

What impact do visitors have on the environment at the Stiperstones?

This question can be broken down into smaller sub-questions to help to give it structure:

1 *What do we know about the visitors to the Stiperstones?*

2 *What is the environment of the Stiperstones?*

3 *How have the visitors affected the environment of the Stiperstones?*

Using questionnaires

Questionnaires have both strengths and limitations. They collect subjective information, which is very useful if you are interested in people's opinions. However, a limitation is that you cannot guarantee that people have been honest in their responses to you. You can either ask closed or open questions. To set closed questions, you will need to create some possible answers – people then select one of these responses. Closed questions are quick and easy for people to complete and the results are easy to represent later. However, respondents are limited to the options that you have already chosen on their behalf. Open questions allow people to answer freely in any way they choose. They can be very useful, in that they allow people to respond freely about the subject, but it can be harder to analyse the answers.

		Impact score on a scale of 0–5 (where 0 is low and 5 is high)				
Site	Distance from car park (m)	Footpath erosion	Noise	Litter	Space (busy-ness)	Total score
1	0	5	4	3	4	16
2	100	5	2	4	3	14
3	200	4	1	2	2	9
4	300	3	1	2	2	8
5	400	2	1	1	1	5
6	500	1	0	1	1	3
7	600	1	2	2	3	8
8	700	0	1	0	1	2
9	800	0	0	1	0	1

▲ **Figure 23** Environmental impact assessment scores; sample points were taken every 100 m from the car park below the Stiperstones ridge.

Activities

1 Use Figure 9 (page 4) to describe the location of the Shropshire Hills AONB.
2 Suggest two separate reasons why it so popular with day trippers from Birmingham.
3 Study the following enquiry question about the Stiperstones:
What impact do visitors have on the environment at the Stiperstones?
   a) Suggest one way the enquiry question could be amended so that you can investigate whether visitors have an impact on businesses in Shropshire Hills AONB.
   b) How would you amend the sub-questions?

Processing and presenting evidence

You can process data in various ways. For example by calculating:

- totals and percentages
- averages (mean, mode or median)
- maximum, minimum and interquartile ranges (see page 17).

Your data needs to be represented using appropriate techniques to help you identify any patterns or trends. Here are some questions to consider when choosing a technique.

- Is the data expressed as actual values or percentages? You could use block charts for actual values and pie charts to show percentages.
- Is the data discrete or continuous? A block graph should be used to represent discrete data whereas a line graph should be used for continuous data.
- Does the data tell you something about different places? If so, could you locate your blocks or pie charts on a base map?
- Have you got two sets of data that are related to one another? If so, would a scattergraph (page 267) show a relationship?

Analysing the evidence

Analysis means studying the data and identifying any patterns or trends. For example:

- Is the data increasing, decreasing or consistent throughout your surveys?
- Is there any data that does not follow the pattern? These are **anomalies**. Can you explain why this might this be?
- How does the data relate to sites on the ground? Does the data show a **spatial** pattern? In other words, can you see patterns that vary over space?

Anomalies

Data that are different from or do not fit with patterns or trends shown in other data.

Spatial

An adjective that refers to the location or distribution of data. A spatial pattern is something that can be mapped.

Evaluating your enquiry

Evaluation means assessing the strengths and weaknesses of each step of the enquiry process. Your evaluation could involve taking the following steps:

- **Step 1**: Did you ask the correct question at the beginning? Were you able to collect the right type of data to allow you to draw conclusions?
- **Step 2**: Was your data representative and reliable? If you used questionnaires, did you get enough responses to properly represent the opinions of the public? If you used data from the internet or another source, can the source be trusted, or could the information be biased?
- **Step 3**: Was the enquiry well planned? Be aware that not collecting enough data, running out of time and not putting in appropriate effort is not human error but poor planning. In what ways could you improve your enquiry if you repeated it?
- **Step 4**: Are there any further questions that need to be answered? How might the enquiry be extended?

Activities

4 Study Figure 23.
 a) What type of sampling was used to collect this data?
 b) Suggest a hypothesis linking distance from the car park to environmental impacts.
 c) Represent the data using two different techniques.
 d) What are the strengths and limitations of each technique?

Enquiry

For a honeypot site that you are familiar with:

- Develop a question that you might study.
- What data might you collect?
- What sampling strategies will you use? Why?
- What analysis and presentation techniques might you use?

Landscapes and physical processes

Chapter 2 Landform processes and change

River processes

► **Figure 1** A number of different river processes are evident in this river.

From the moment water begins to flow over the surface of the land, gravity gives it the power to erode the landscape. **Erosional processes** occur where the river has plenty of energy so, for example, where the river is flowing quickly or when the river is full of water after heavy rain. Rivers that are flowing across gentle slopes (such as the river in Figure 1) tend to flow with greatest force on the outer bend of each curve (or **meander**). Water is thrown sideways into the river bank, which is eroded by both **hydraulic action** and **abrasion**. The bank gradually becomes undercut. The overhanging soil slumps into the river channel where this new load of material can be picked up and transported downstream by the flowing water. Figure 2 describes the main processes of transportation.

Erosional processes

These result from the power of a flowing river and wear away the bed and banks of the river channel. They include: hydraulic action, abrasion, attrition and solution. See Figure 3.

Transportation process	Sediment size or type	Typical flow conditions	Description of the process
Solution	Soluble minerals such as calcium carbonate	Any	Minerals are dissolved from soil or rock and carried along in the flow
Suspension	Small particles e.g. clay and silt	Suspension occurs in all but the slowest flowing rivers	Tiny particles are carried long distances in the flowing water
Saltation	Sand and small gravels	More energetic rivers with higher velocities	The sediment bounces and skips along
Traction	Larger gravels, cobbles and boulders	Only common in high energy river channels or during flood events	The bed load rolls along in contact with the river bed

▲ **Figure 2** The transportation of sediment.

Erosional processes

- **Hydraulic action** – water crashes into gaps in the soil and rock, compressing the air and forcing particles apart
- **Abrasion** – the flowing water picks up rocks from the bed that smash against the river banks
- **Attrition** – rocks carried by the river smash against one another, so they wear down into smaller and more rounded particles
- **Solution** – minerals such as calcium carbonate (the main part of chalk and limestone rocks) are dissolved in the river water. This process is also known as corrosion

▲ **Figure 3** Four processes of river channel erosion.

The process of deposition occurs where the river loses its energy. For example, where a river enters a lake and its flow is slowed by the body of still water. Deposition also occurs in very shallow sections of a river channel where friction between the river bed and the water causes the river to lose its energy and deposit its **load**. The process of deposition creates layers of sand and gravel that are often sorted by sediment size because the coarsest sediment is deposited first.

Load

The material that is transported down a river. Load can be silt, sand, pebbles or boulders.

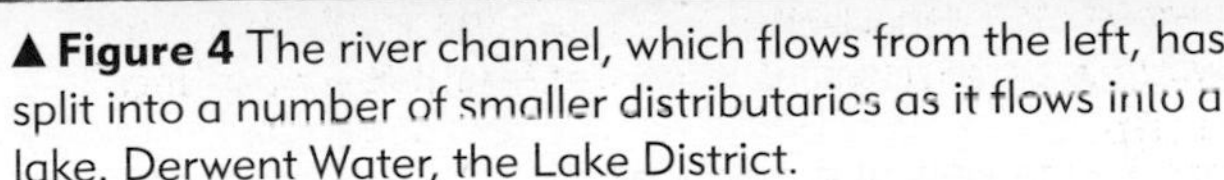

▲ **Figure 4** The river channel, which flows from the left, has split into a number of smaller distributaries as it flows into a lake. Derwent Water, the Lake District.

Activities

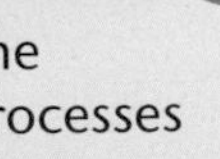

1. Study Figure 1. Use evidence from the photograph to suggest what river processes are occurring at A, B and C.
2. Explain why deposition occurs in very shallow water.
3. Draw four diagrams or cartoons to illustrate the ways in which a river transports material.
4. Study Figure 4. Suggest how erosion, transportation and deposition have each played a role in the formation of this landform.

River landforms

The processes of erosion, transportation and deposition result in the development of distinctive **river landforms**. In Figure 5, the river is flowing down a steep gradient and has cut a V-shaped valley. Much of the force of the water is directed downwards. **Vertical erosion** cuts into the river bed. The river cuts a narrow valley with steep V-shaped sides. The flow of water within the river channel also swings from side to side, creating some sideways erosion. Over time, this process means that the V-shaped valley is cut, or incised, into the hillside to form **interlocking spurs** rather like the teeth of a zip.

Stones lying in the river channel can be large and quite angular because there hasn't been enough time for the process of attrition to make them smooth. As a river flows downstream the process of attrition gradually reduces the overall size of the load.

▲ **Figure 5** Ashes Hollow, Shropshire, is a typical V-shaped valley.

River landforms

Natural features of the Earth's surface associated with rivers. V-shaped valleys and waterfalls are made by erosional processes. Floodplains are made by deposition. River landforms vary in scale. Meanders, for example, are often large landforms. They contain smaller features, such as slip-off slopes.

Activities

1 Study Figure 5. Describe how each of the following features was formed:
   a) V-shaped valley sides
   b) large angular boulders in the stream bed
   c) interlocking spurs.

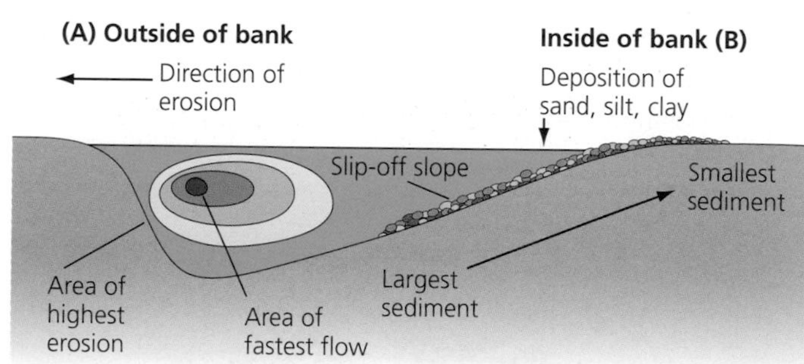

▲ **Figure 6** Processes at work in a meander.

How are river meanders formed?

Rivers flowing over gentle gradients tend to swing from side to side. The water flows fastest on the outside of the bend of each meander. This causes erosion of the banks rather than the bed, a process known as **lateral erosion**. The slower flowing water on the inside of each bend loses energy and deposits its load. The material is sorted, with the larger gravel being deposited first, then the sand and finally the silt. This process creates a **slip-off slope** or **point bar**, which is a pebble beach that slopes down into the river. Meandering rivers such as the Elan, shown in Figure 8, flow across a wide **floodplain**. This flat landform has been created over many thousands of years by the processes of lateral erosion and deposition.

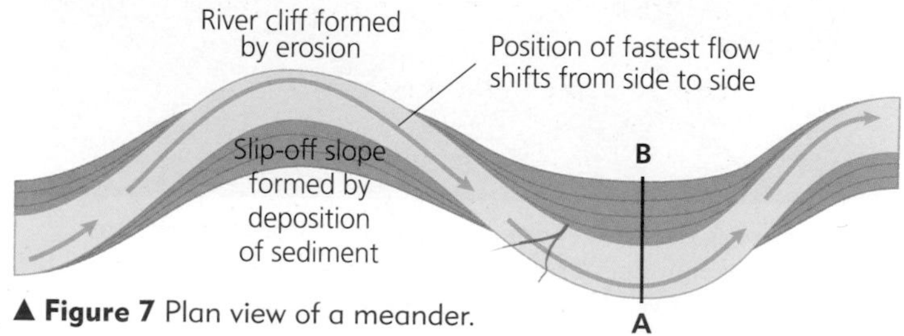

▲ **Figure 7** Plan view of a meander.

▲ **Figure 8** Meanders on the Afon Elan; the blue arrows show the direction of river flow.

▲ **Figure 9** Detail of the slip-off slope on the Afon Elan at point 3 in Figure 8.

Activities

2 Using Figure 8:
   a) Describe the slopes at points 1 and 2.
   b) Describe the processes at points 3 and 4.
3 Draw a sketch of Figure 9. Use Figures 6 and 7 to add suitable annotations to your diagram that show how this feature has formed.
4 Figure 10 shows the location of the meanders on the Afon Elan.
   a) Use Figures 8 and 10 to estimate the direction the camera was pointing.
   b) In which direction is the river flowing at Pont ar Elan?
5 a) Draw a cross section along the line X–Y on Figure 10.
   b) Describe the differences and similarities between the rivers shown in Figures 5 and 8.

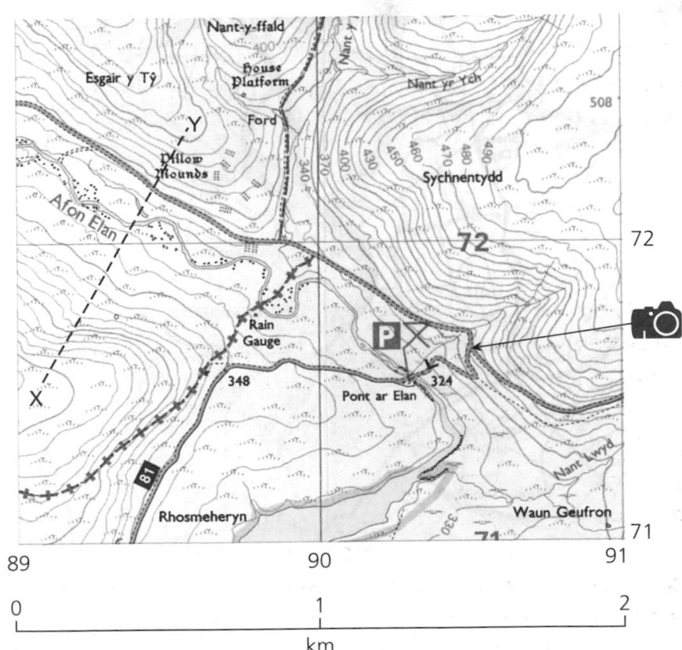

▲ **Figure 10** An OS map at a scale of 1 : 25,000 showing the meanders on the Afon Elan.

Investigating downstream changes

Rivers can change in shape and character as they flow downstream. It is possible to pose geographical questions about these changes. For example, when investigating change at the small sale:

How does river velocity change across a point bar and how does this affect sediment size?

Or, when investigating change at a larger scale:

How do cross-sectional area and river velocity change as you move downstream?

Choosing sample sites

The UK's rivers vary from just a few hundred metres in length to rivers like the River Severn which, at 369 km, is the UK's longest river. In an enquiry about how a river changes as it flows downstream it is important to make sure that your sample points are far enough apart to show change. For a river that is 10 km long you could collect data at 1 km, 5 km and 10 km, or at 3 km, 4 km and 5 km. The first sampling strategy would provide a representative sample of the whole of the river, whereas the second would only provide evidence of small-scale changes in just one short section of the river.

Activity

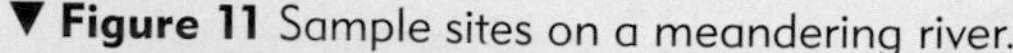

1 Explain why sampling at 3 km, 4 km and 5 km would not be representative of change along a whole river that is 10 km in length.

Calculating cross-sectional area

To measure the cross-section of a river channel you will need to set a horizontal line across the river and carefully measure down from this line to the ground. This is shown in Figure 11 where the horizontal yellow line represents the survey line from X to Y and the four vertical lines represent the first four measurements. Collect data when the river is low but take measurements for the dry land on either side of the river. Then, when your results are plotted on a graph, you can estimate the amount of water in the river when the river channel is full and about to flood.

Step 1 Stretch a tape measure at right angles across the river from one bank to the other, keeping it parallel to the surface of the water.

Step 2 Divide the width of the river by 10. This will create 11 equally spaced survey points. For example, in a river that is 4 m wide you will record the depth every 40 cm (1/10th) of the way across. This is an example of systematic sampling (see page 8).

Step 3 At each survey point measure the depth of the river. Make sure that your depth and width readings are both recorded in the same units (for example, both in metres).

Cross-sectional area (CSA) in square metres = width (m) multiplied by mean depth (m).

▼ **Figure 11** Sample sites on a meandering river.

Survey site	1	2	3	4	5	6	7	8	9	10	11
Distance from bank (m)	0	0.4	0.8	1.2	1.6	2.0	2.4	2.8	3.2	3.6	4.0
Distance from survey line to ground (m)	0.25	0.68	0.71	0.65	0.67	0.64	0.58	0.52	0.48	0.33	0.22
Depth of water (m)	0	0.28	0.31	0.25	0.27	0.24	0.18	0.12	0.08	0	0

▲ **Figure 12** Depth measurements taken in the river in Figure 11 at 0.4 m intervals.

Activities

2 Use Figure 12.
   a) Plot the profile of this river on graph paper. Remember to work downwards from your horizontal axis.
   b) What is the mean water depth?
   c) Calculate the cross-sectional area.
   d) If water levels rose by 20 cm, what would be the new cross-sectional area?

How do I calculate range, median and interquartile range?

The flow of water in the river channel has enough energy to transport sediment. As the water speed slows, for example, in shallow water on the point bar, energy is lost and sediment is deposited.

To test whether this process is happening in your river, you will need to sample some pebbles and record their size. Students collected a sample of 11 pebbles from sites A, B and C in Figure 11. Their aim was to discover how the size and range of pebbles changed across the point bar.

To calculate the range, median and interquartile range, you need to put your data into rank order. The data for site A is shown in Figure 14 in rank order.

Site	Pebble sizes (mm)										
A	45	52	12	67	34	75	42	81	65	40	24
B	44	37	28	56	61	43	38	28	35	42	36
C	37	34	26	40	24	35	29	42	38	18	20

▲ **Figure 13** Pebble sizes (mm) collected at random at sites A, B and C in Figure 11.

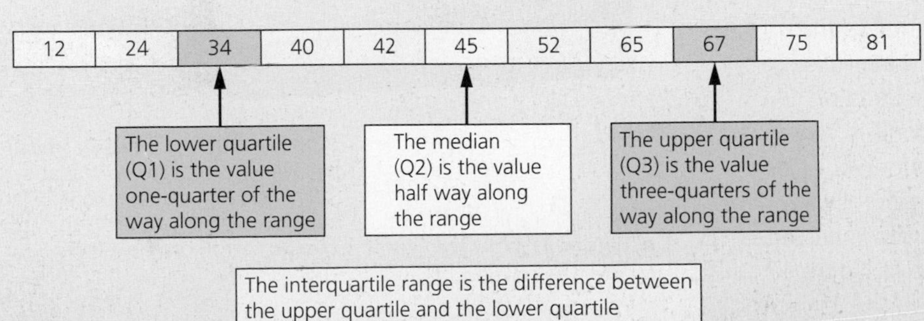

▲ **Figure 14** Pebble sizes for site A arranged in rank order.

Activities

3 Use Figure 13.
   a) For each sample site, calculate the:
      i) range
      ii) median
      iii) interquartile range.
   b) What conclusions can you draw from these results?

How are waterfalls formed?

Waterfalls form along the course of a river where there is a steep change in gradient of the river's channel. Many waterfalls in upland areas of the UK are due to landform processes that occurred at the end of the ice age, around 10,000 years ago. During the ice age, ice sheets expanded over large parts of northern and western parts of the UK. The extent of the ice can be seen in Figure 1, page 1. Glaciers flowed from these ice sheets towards the sea – similar glaciers can be seen in southern Iceland today. The glaciers carved deep, steep-sided or **U-shaped valleys** into the landscape. Figure 16 shows how this glacial landscape created the tall plunging waterfalls we see in parts of North and Mid Wales today.

▲ **Figure 15** Rhiwargor, in Mid Wales, is an example of a waterfall plunging into a U-shaped valley.

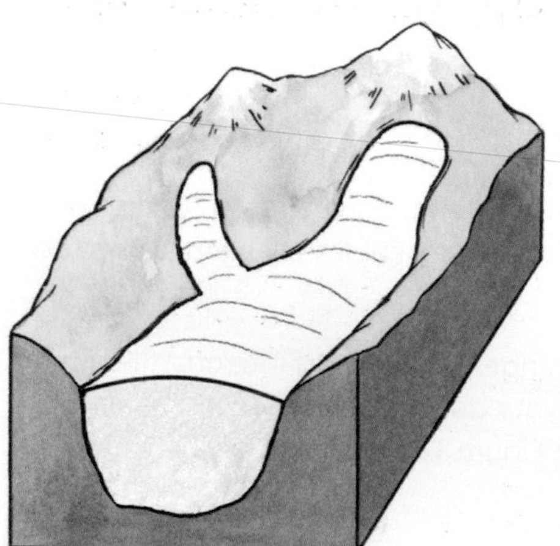

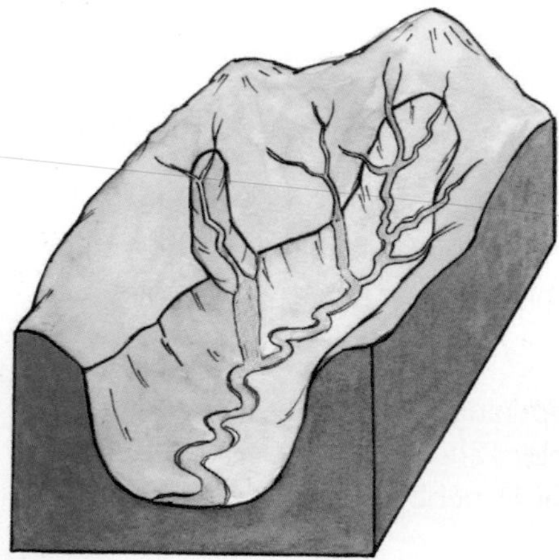

▲ **Figure 16** Glacial landscape and today's landscape.

Waterfall

A river landform that occurs where the river channel suddenly drops in level. At the base of a waterfall is a small-scale feature known as a plunge pool. The retreat of a waterfall creates a gorge.

Waterfalls formed by differential erosion

Waterfalls can occur where the river channel crosses from one rock type to another. If the rate of erosion of each rock type is different, then a waterfall is formed. Waterfalls have been formed in this way on the River Neath and its tributaries in an area of South Wales known as Waterfall Country.

It is the geology of Waterfall Country that is the main factor in the formation of these waterfalls. Carboniferous limestone is overlain by beds of sandstone and mudstone. The sandstone is very resistant to erosion whereas the mudstone is eroded more easily. A series of faults, running across the river channels, has brought the sandstone and mudstones alongside one another, as you can see in Figure 17.

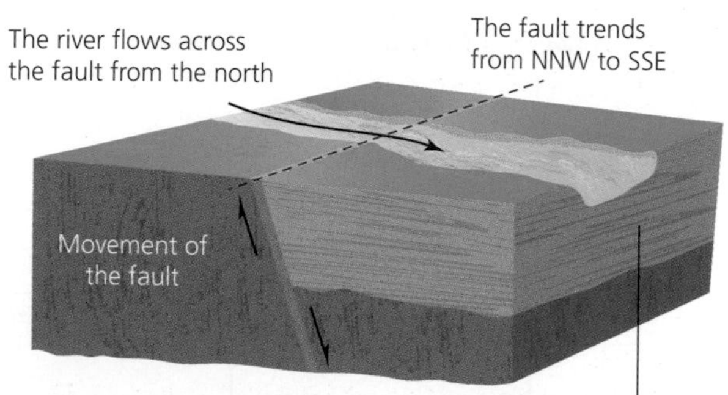

▲ **Figure 17** How the Sgwd yr Eira waterfall on the Afon Mellte was formed.

As the river plunges over the sandstone it pours on to the softer mudstone below. A combination of hydraulic action and abrasion erodes this rock relatively easily, creating a **plunge pool**. Abrasion at the back of the plunge pool undercuts the layers of sandstone. Eventually this overhang will fracture and the rocks will fall into the plunge pool where they are broken up by attrition. So each waterfall is gradually eroded backwards towards the river's source in a process known as **retreat**. Below each waterfall is a narrow valley with almost vertical sides. This feature is known as a **gorge**. The process of retreat has cut the gorge over many hundreds of years.

▲ **Figure 18** The Sgwd yr Eira waterfall on the Afon Mellte.

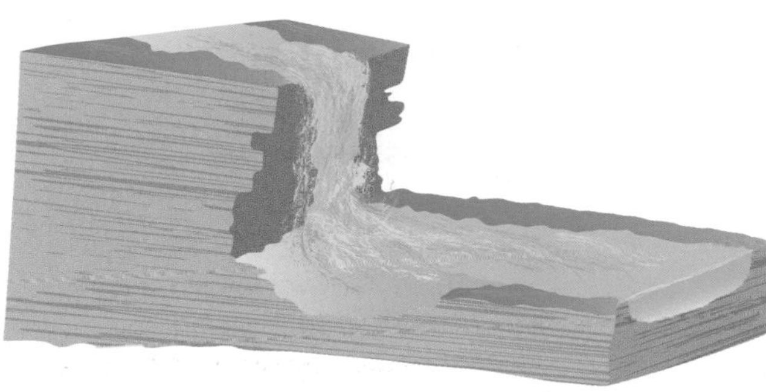

▲ **Figure 19** River processes at the Sgwd yr Eira waterfall

1 A 'cap rock' of sandstone is resistant – it erodes relatively slowly

2 The overhang is susceptible to collapse and retreat.

3 Abrasion deepens the plunge pool.

4 Attrition breaks down the eroded rock fragments as they are transported downstream.

5 Pebbles swirl around in hollows to create potholes by abrasion.

6 Beds of resistant rock create an irregular river bed of rapids and smaller waterfalls below the waterfall.

Activities

1. Describe how the faulting of the rocks in Waterfall Country has led to the formation of waterfalls.
2. a) Make a copy of Figure 19.
   b) Add the labels to suitable places on your diagram.
   c) Explain why the retreat of the plunge pool has, over thousands of years, created a gorge.
   d) Draw a series of diagrams to show the formation of the gorge.
3. Explain why the retreat of a waterfall leads to the formation of a gorge.
4. Explain why we need to understand how natural processes have changed since the last ice age in order to understand the landscape today.

Enquiry

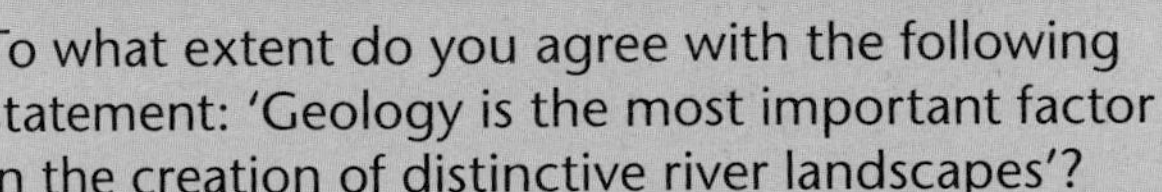

To what extent do you agree with the following statement: 'Geology is the most important factor in the creation of distinctive river landscapes'?

Make use of information on pages 16 to 19 to help you justify your answer.

Weblinks

www.world-of-waterfalls.com/europe.html – an interactive map showing the location of waterfalls across the UK.

How do waves erode our coastal landscapes?

Waves provide the force that shapes our coastline. Waves are created by friction between wind and the surface of the sea. Stronger winds make bigger waves. Large waves also need time and space in which to develop. So, large waves need the wind to blow for a long time over a large surface area of water. The distance over which a wave has developed is known as **fetch**, so the largest waves need strong winds and a long fetch.

The water in a wave moves in a circular motion. A lot of energy is spent moving the water up and down. So waves in deep water have little energy to erode a coastline. However, as a wave enters shallow water, it is slowed by friction with the sea bed. The water at the surface, however, surges forward freely. It is this forward motion of the breaking wave that causes the **erosional processes** that create coastal landforms such as the wave-cut platform in Figure 21.

Activity 

1 Make a copy of Figure 20 and add the following labels in appropriate places.
- Waves in deeper water
- Circular motion
- Breaking wave
- Water thrown forward
- Friction with the sea bed.

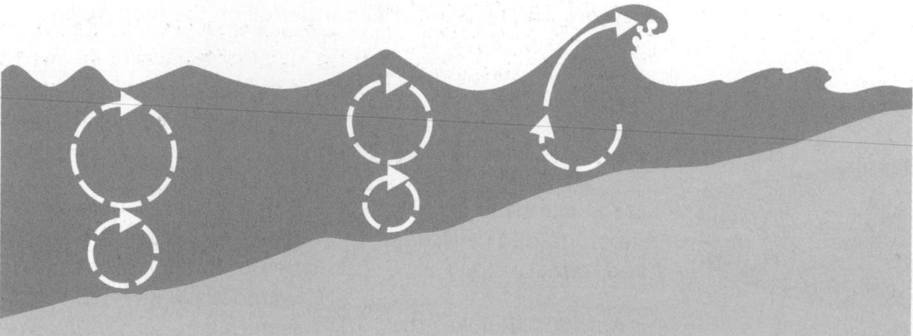

▲ **Figure 20** The motion of water in a wave.

Erosional processes

These result from the power of the waves, and are able to wear away the shoreline. They include hydraulic action, abrasion, attrition and solution. See Figure 22.

The repeated pounding of large waves at the foot of a cliff can cause enormous damage through the process of hydraulic action and abrasion. The repeated hammering effect of the waves on this narrow zone creates a **wave-cut notch**. Cliffs that are already weakened by joints or cracks can suddenly collapse in a rock fall which is a type of **mass movement**. The collapse causes the line of the cliffs to retreat inland. The **wave-cut platform** in Figure 21 has been formed by the gradual retreat of the cliffs.

▲▶ **Figure 21** The rocky wave-cut platform of the Glamorgan Heritage Coast.

Erosional processes

Hydraulic action – waves crash against the cliff, compressing the water and air into cracks and forcing the rocks apart.

Abrasion – waves pick up rocks from the sea bed or beach and smash them against the cliffs.

Solution – minerals such as calcium carbonate (in chalk and limestone) are slowly dissolved in sea water. This is also known as corrosion.

Attrition – sand and pebbles are picked up by the sea and smash against one another, wearing them down into smaller and more rounded particles.

▲ **Figure 22** Four processes of coastal erosion.

▲ **Figure 23** Evidence of erosion in limestone cliffs on the Glamorgan Heritage Coast.

Activities

2 Study Figures 21 and 22.

a) Use the correct erosion terms to complete the annotations below.
- Joints in the rock are widened in the process of .... which is when ....
- Boulders on the beach are rounded because ....
- This rock pool has been scoured into the rock by ....

b) Make a simple sketch of Figure 21 and add your annotations (above) to the sketch.

3 Consider Figure 23 and its annotations.

a) Write a list (or draw a timeline) that puts the events acting on this cliff in the correct sequence.

b) Make another list (or timeline) suggesting what will happen to this cliff in the next few years.

c) Over the next 100 years this coastline will retreat by about 20–40 m. Draw a story-board to show how this process of retreat creates the rocky wave-cut platform in front of the cliff.

Slope processes

Erosion by waves only occurs at the bottom of the cliff. The rest of the cliff face can be affected by other slope processes such as **weathering**, rock falls, landslides or slumping. Rocks, loosened perhaps by winter frosts or heavy rainfall, can fall suddenly to the beach below in a **rock fall**. Large sections of cliff, undermined by erosion at the base, can slide rapidly downwards on to the beach. A **landslide** such as this leaves a concave scar in the upper cliff, like a giant bite mark, and a fan-shaped pile of debris on the beach. A similar process, in loosely compacted rocks such as those found in the Holderness Coast, is called **slumping**. The effects of slumping are much the same as a landslide – bite-shaped chunks are removed from the top of the cliff.

Landslides usually occur in sedimentary rocks and are often triggered by extreme weather events – stormy seas that batter the cliff or heavy rainfall that adds extra weight to the cliff face. But landslides can happen at any time and are dangerous, of course. In June 2015 a young woman was killed by a rock fall at Llantwit Major – the cliffs shown in Figure 23. There have been many landslides and rock falls on this coast, the most recent are shown in Figure 25.

Debris from a landslide forms a fan at the base of the cliff	Beach is too narrow to absorb much wave energy
A narrow headland or arête between two landslides	Concave scarp at the top of the cliff

▲ **Figure 24** Cliffs on the Isle of Wight with numerous landslides.

Weathering

Processes that weaken joints in rocks. They include frost action, chemical reactions and the growth of plant roots.

Activities

1 a) Match the labels to features on Figure 24.
b) Suggest why these cliffs have no defences.
c) Suggest why a landslide or rock fall will prevent further wave erosion for a few months. Use Figure 22 to help you.

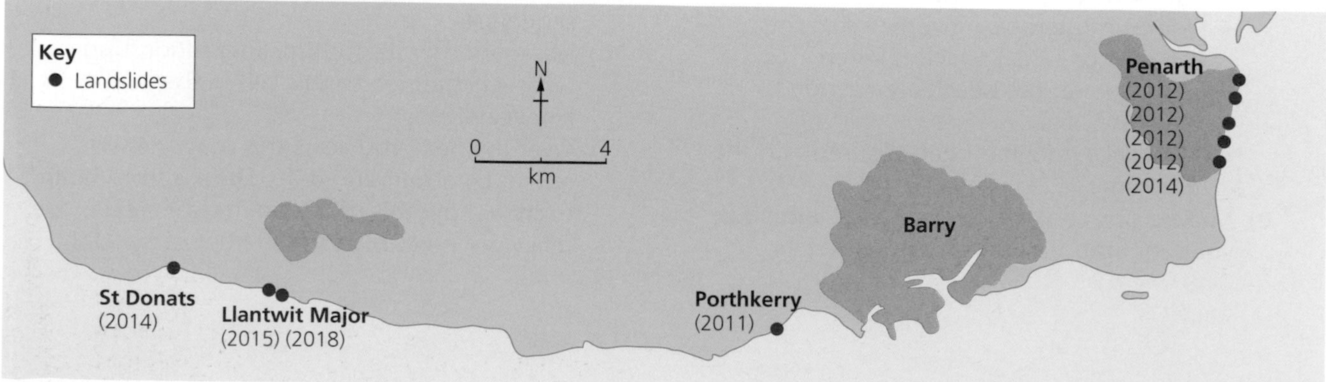

▲ **Figure 25** Recent landslides on the Glamorgan Heritage Coast, South Wales.

GEOGRAPHICAL SKILLS

Calculating and representing frequency

Coastal hazards such as floods, rock falls or landslides are examples of events that happen at random intervals. We tend to record the date and location of such events. Figure 25 shows the location and date of recent landslides on the Glamorgan Heritage Coast in South Wales. Recording such events allows us to see how common or frequent they are. Knowing this may help us to understand the level of risk that is created by the hazard.

Year	Landslides
2009	0
2010	0
2011	2
2012	3
2013	1
2014	4
2015	1
2016	2
2017	2
2018	1
Total	16

Recurrence interval (T) is simple to calculate. It can be found by dividing the number of years in a record (N) by the number of events (n).

$$T = \frac{N}{n}$$

So, for the data in Figure 26, the recurrence interval (to two decimal places) is:

$$T = \frac{10}{16} = 0.625 \text{ years}$$

Frequency describes the recurrence interval of an event. This may be defined as the average time lag between two events. The best way to represent frequency is with a frequency bar graph – with time (usually years) on the *x*-axis. An example, showing the frequency of landslides on the coast between Lyme Regis and Charmouth in Dorset, is shown in Figure 26.

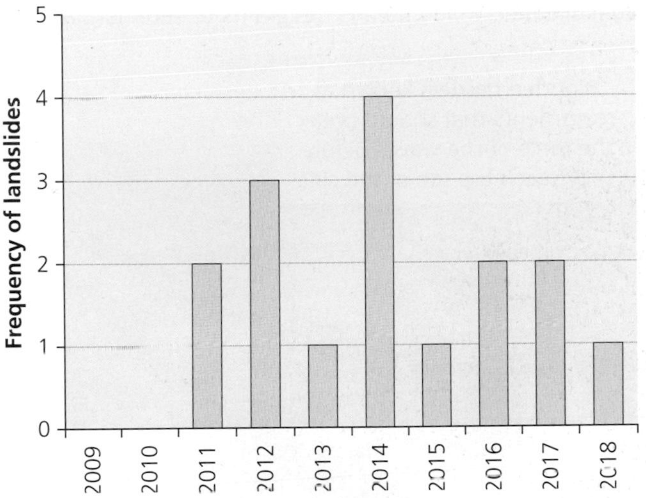

▲ **Figure 26** Landslides on the Lyme Regis–Charmouth coastline, Dorset.

Of course, 0.625 of a year isn't very helpful: we usually express parts of a year in months. So, if we multiply 0.625 by the number of months in a year:

$$T = 0.625 \times 12 = 7.5 \text{ months}$$

You must remember that what you have calculated is a **mean** time period. The landslides are random events and can occur at any time. Look again at Figure 26. There were no landslides in this location at all for two years, but there were four landslides in 2014.

Activities

2 Using Figure 25:
   a) Represent the frequency of the landslides on the Glamorgan Coast.
   b) Calculate the frequency.
   c) Which is the more hazardous – the coastline in Glamorgan or Dorset?

Enquiry

Which part of the UK coastline has most landslides?

Use the weblink to the BGS site to research the frequency of landslides on the UK coastline.

Weblinks

http://mapapps2.bgs.ac.uk/geoindex/home.html?theme=hazards – an interactive UK map showing the location of coastal landslides.

Cliffs in unconsolidated rocks

Geology is an important factor that affects the rate of change of the coastline. Rock type affects the strength of the cliff. This means that rock type affects the likelihood and rate of slope processes such as slumping. Many kilometres of the UK's coastline are made of layers of sand, silt and clay, deposited as ice melted at the end of the ice age. The extent of the ice is shown in Figure 1 on page 1. These young sedimentary rocks have not been compacted as much as older rocks and they are **unconsolidated**, which means the grains of sediment are not 'glued' together very well. This makes them much less resistant to erosion than older sedimentary rocks such as the Carboniferous limestone cliffs seen in Figure 29.

Figure 28 helps to explain why the cliffs at Happisburgh in Norfolk are not very resistant to erosion by waves. Once the toe of the slope has been eroded by the sea, the whole slope becomes unstable. It is then at risk of slumping. The chance of slumping is increased by periods of heavy rain, which adds mass to the cliff.

▲ **Figure 27** The coastline at Happisburgh, North Norfolk in 2011.

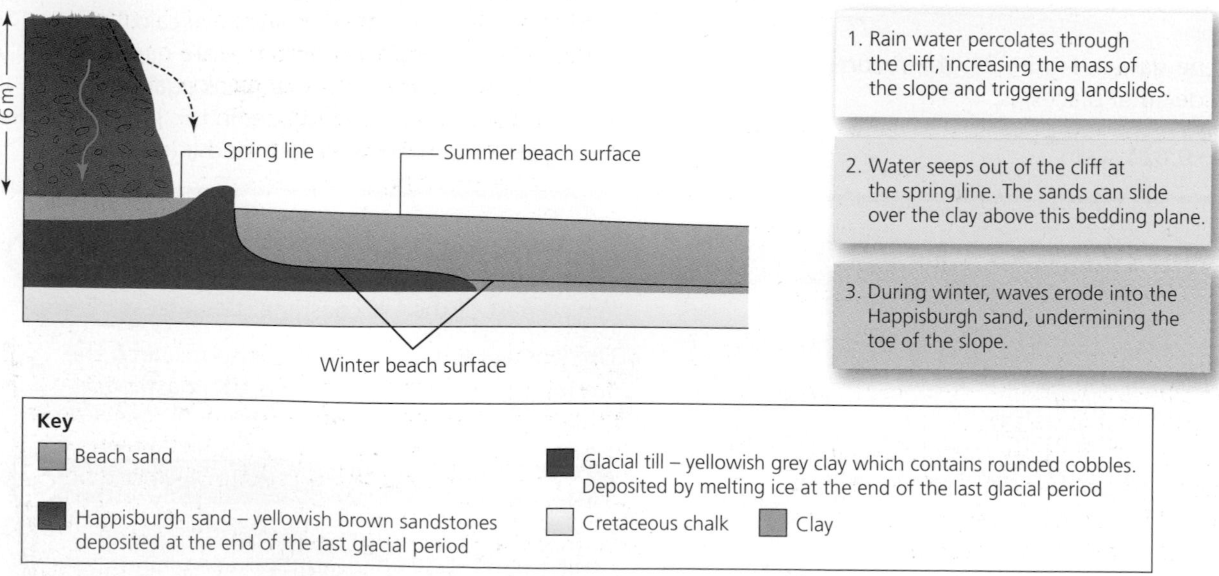

▲ **Figure 28** Cross section through the cliffs at Happisburgh, North Norfolk; these cliffs are made of loosely compacted layers of sand, silt and clay deposited at the end of the ice age.

Cliff landforms in resistant rock types

We have seen that young sedimentary rocks, such as those at Happisburgh, have a weak structure that makes them vulnerable to erosion. By contrast, older sedimentary rocks are compacted and consolidated, meaning they are more resistant to erosion. Limestone coastlines, like those in Figure 29, tend to form almost vertical cliffs. However, bedding planes and joints in the rock are lines of weakness in these cliffs. These lines are more easily eroded than the massive blocks of stone in-between them. Erosion along these lines can lead to the formation of caves, **sea arches** and **stacks**.

▲ **Figure 29** The Green Bridge of Wales, Pembrokeshire; a natural sea arch formed in a Carboniferous limestone cliff.

Activities

1 Read the following annotations and decide where they fit best on Figure 27.
- Waves have eroded the toe of the cliff here.
- The vegetation on this slope proves that it hasn't slumped for several months.
- Concrete blocks on the beach may protect the cliff from wave erosion.
- Evidence of gulley erosion by rain water on these slopes.

2 a) Make a copy of Figure 28.
b) Add the annotations 1 to 3 to suitable places on your diagram.
c) Use the diagram to explain why the rock type and structure of the cliffs at Happisburgh make them vulnerable to erosion and mass movement.

3 Historical records show that the cliffs here retreated by 250 m between 1600 and 1850. What is the average rate of erosion per year?

4 a) Make a sketch of Figure 29.
b) Label the following features on your sketch: cave, bedding planes, sea arch, stack.

5 Explain why, even though all of the rocks in Figure 29 are the same, the landscape is so varied.

Enquiry

How might the landscape in Figure 29 evolve over time? Discuss how the processes of wave action and mass movement might affect this coastline. Draw a story board to show how you expect it might change in the future.

Beach and sand dune processes

Beaches are constantly changing coastal landforms. The energy of the wind and waves is always moving sediment around and changing the shape of the beach. Each wave transports sediment up the beach in the **swash** and back down again in the **backwash**. All of this movement uses a lot of the wave's energy, so a wide, thick beach is a good natural defence against coastal erosion. Where the waves approach the beach at an angle, some of the sediment is transported along the coastline in a process known as **longshore drift**. This is shown in Figure 30. Where the coastline changes direction, for example at the mouth of an estuary, this process forms a landform known as a **spit**.

▲ **Figure 30** Transport of sediment by the process of longshore drift.

▲ **Figure 31** The beach at Borth seen from the cliffs to the south of the pebble ridge.

The sand and pebbles on a beach usually come from the local environment. Neighbouring cliffs may supply some sediment if they are being actively eroded by wave action. A lot of finer silts and sands are transported to the coast by rivers. This sediment is then deposited in the estuary or on an **offshore bar** at the mouth of the river. It will be washed onshore by the swash of the waves and deposited on the beach.

At Borth, on the Ceredigion coast, there is a pebble ridge making a spit on the southern side of the estuary. These pebbles came from cliffs to the south. Figure 33 shows the processes that are supplying and transporting material on this coastline.

▲ **Figure 32** The sand dunes at Ynyslas seen from Aberdyfi on the north side of the Dyfi estuary.

Activities

1 Describe the landforms seen at A, B and C in Figures 31 and 32.
2 Study Figures 30, 31 and 33. Use an annotated diagram to explain the formation of the pebble ridge on which the village of Borth is built.

Beach

A coastal feature that provides a natural defence against erosion and coastal flooding. A wide, thick beach absorbs wave energy. Building groynes makes a beach wider and thicker.

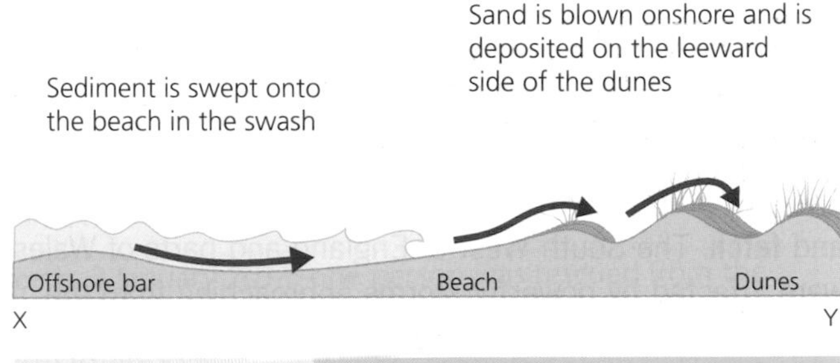

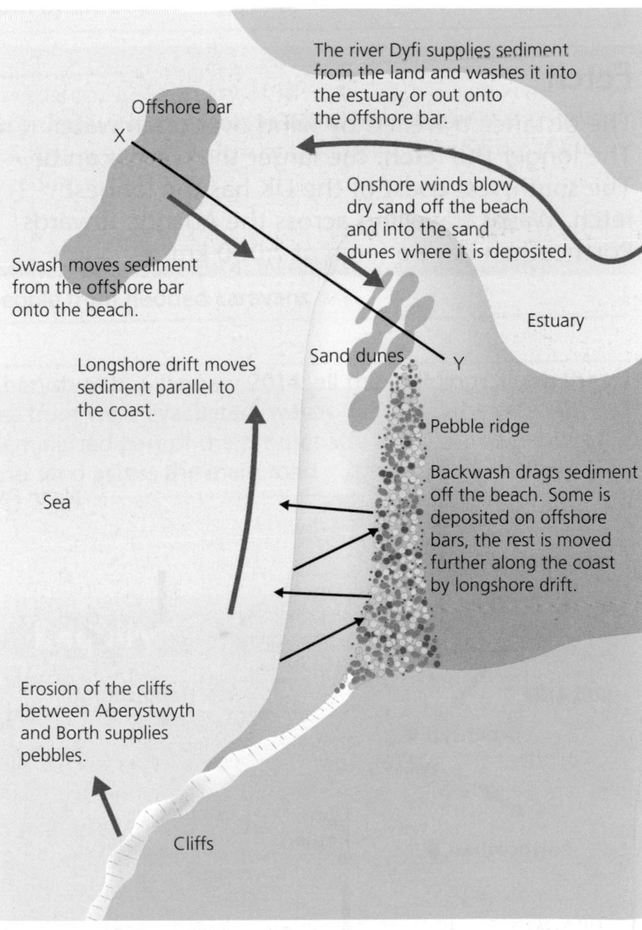

▲ **Figure 33** The transport of beach sediment at Borth and Ynyslas on the Ceredigion coast.

How geology affects rates of coastal change

Pen Llŷn (the Llŷn Peninsula) extends about 48 km into the Irish Sea. To the west of Pwllheli, the south coast of the peninsula is made up of a series of **headlands** and **bays** created by differential rates of erosion. The headlands consist of outcrops of very resistant (harder) rock. The headland at Llanbedrog (near Abersoch), for example, is composed of the igneous rock granite. Many of the bays, on the other hand, are made up of mudstones and shales. Some of these are covered by layers of loosely consolidated sand, silt and clay deposited at the end of the ice age – similar to the cliffs at Happisburgh in Norfolk (see page 24). Waves are able to erode these areas of less-resistant (softer) rock more rapidly.

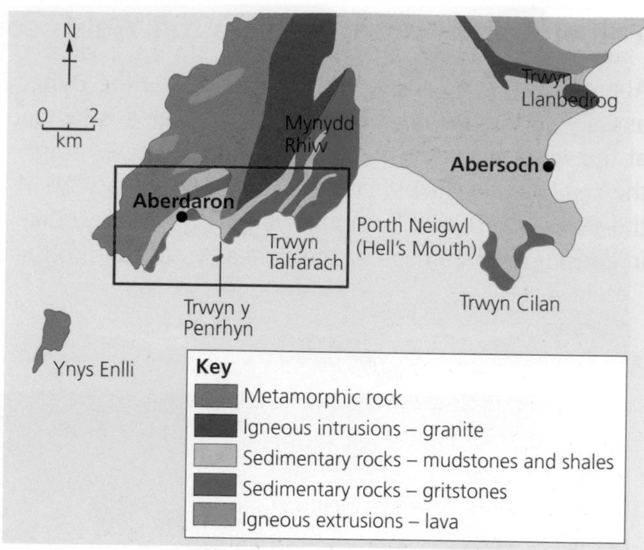

▲ **Figure 37** A simplified geological map of the western end of the Llŷn Peninsula; the red line shows the area covered by Figure 38.

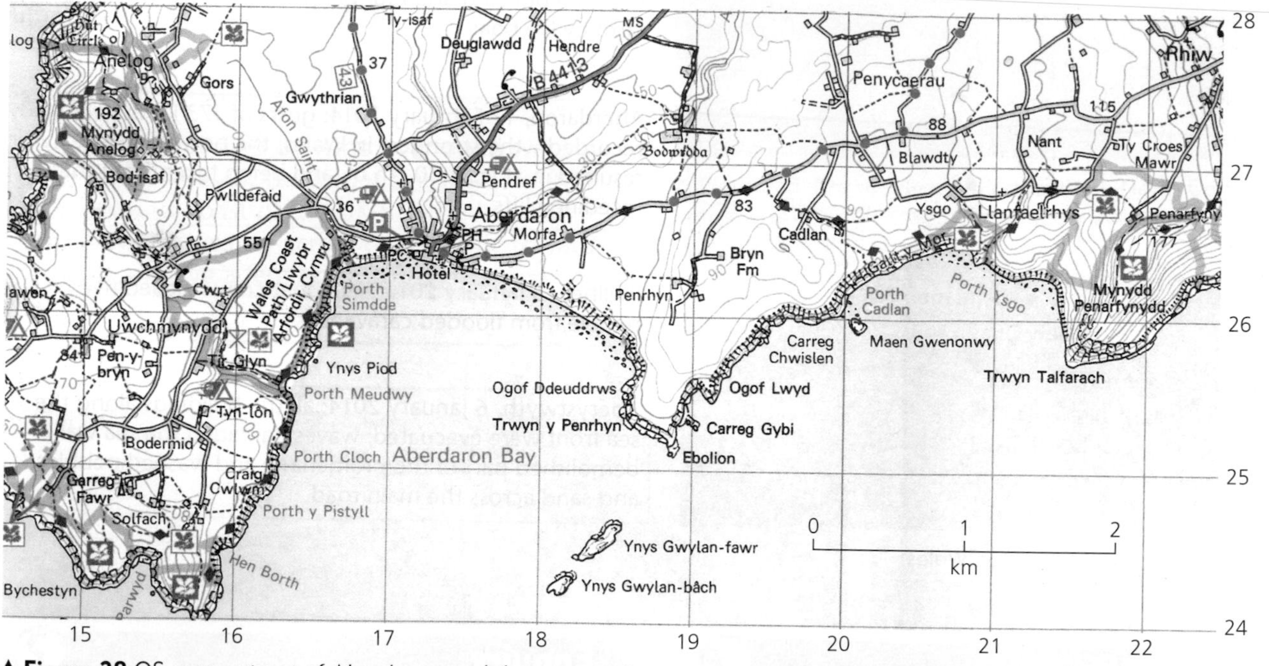

▲ **Figure 38** OS map extract of Aberdaron and the surrounding area at a scale of 1 : 50 000.

Headland

A rocky outcrop that sticks out into the sea. Headlands are made of rocks that are resistant to erosion.

Bay

A body of shallow water that forms where softer rocks have been eroded by the sea. Headlands and bays exist because rocks erode at different rates.

Weblinks

http://mapapps.bgs.ac.uk/geologyofbritain/home.html – use this interactive map to discover the geology of a coastline near to you.

Figure 37 is a geological map of the extreme western tip of the Llŷn Peninsula. The large headland immediately to the west of Aberdaron is formed of extremely tough metamorphic rocks. These rocks create high, almost vertical cliffs on all sides of the headland and on Ynys Enlli (Bardsey Island). The rate of erosion of these headlands is very slow. However, faults, vertical joints and cracks within resistant rocks are lines of weakness that can be eroded more rapidly than the rest of the cliff by processes such as hydraulic action. Eventually, the joints and faults are widened to form caves and blowholes. This means that even tough rocks can be eroded if their structure (joints, cracks and faults) makes them weak.

The largest of the bays is Porth Neigwl (Hell's Mouth). The bay is made of rocks that were easily eroded. The low cliffs along Porth Neigwl are made of loosely consolidated sands. Erosion by waves at the base of these cliffs causes slumping and their collapse onto the beach.

Porth Neigwl lies between two large headlands. In normal weather conditions, the headlands cause waves to slow down and lose energy as they enter the bay, so beach material has accumulated over the years.

Porth Neigwl was exposed to the full force of winter storms from the Atlantic in 2014. Storm waves eroded the cliffs but they also transported large pebbles and rocks up the beach slope. These were deposited in a steep pile at the top of the eastern end of the beach. This feature is known as a **storm beach**.

▲ **Figure 39** The beach and the cliffs at Porth Neigwl; the headland of Trwyn Cilan can be seen in the background.

Activities

1 Describe the relief of the coastal area shown in Figure 38.
2 a) Draw an outline map of this coastline.
  b) Label the physical features of this coastline on your map.
  c) Using Figure 37 to help you, explain why bays have formed at:
    i) Aberdaron
    ii) Abersoch.
3 Make a sketch of Figure 39.
  a) Label Trwyn Cilan and explain why it is here in an annotation.
  b) Annotate the slumped cliff to explain the processes happening here. Make use of Figure 28 (page 24).
4 Summarise two different ways that geology can affect coastal landforms using examples from this page.

How does human activity affect the coastline?

People have managed the coastline for centuries – building flood embankments and draining coastal marshes, for example. The intention is to protect the coastline from flooding and make harbours safer. But engineering strategies such as these can have **unintended consequences**. We saw on page 26 that sediment moves along the coastline by a process of longshore drift. Building a harbour wall or breakwater can trap sediment and prevent its movement along the coast. Beaches further along the coast become starved of new sand and erosion can accelerate.

▲ **Figure 40** The castle headland at Criccieth.

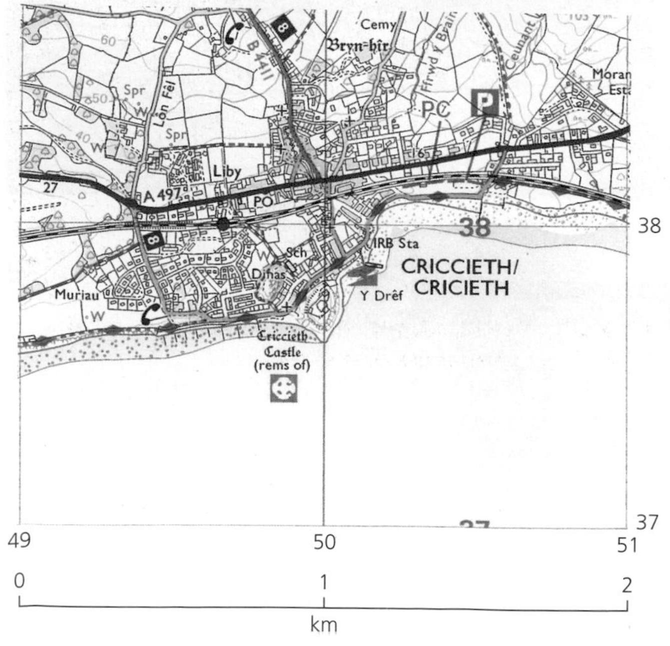

▲ **Figure 41** OS map extract of Criccieth at a scale of 1 : 25 000.

Human intervention at Criccieth (Cricieth)

The town of Criccieth lies on the south coast of Gwynedd overlooking Cardigan Bay. Criccieth Bay is an area of loosely consolidated glacial deposits (or till) and is easily eroded by storm waves. The castle headland, seen in Figure 40, is made up of volcanic rock and is very resistant to erosion.

A number of coastal management strategies have been put in place along Criccieth Bay. These include rock armour at the western end, a recurved sea wall along the promenade and groynes on the beach.

The beach on the western side of the castle is a mixture of sand, shingle and pebbles. The wooden groyne structures positioned along the beach ensure that erosion along the western promenade is minimised, and that sand is trapped between the groynes. As a seaside resort, Criccieth needs to retain the beach to encourage tourists. Beaches also play a vital role in coastal protection as they absorb wave energy, which in turn protects the land behind.

Unintended consequence

A side-effect of human activity that can, for example, affect coastal processes. Building groynes provides a good example. Groynes trap sand, making the beach thicker and wider, so protecting the coastline where the groynes have been built. However, groynes prevent the movement of sediment along the coastline by longshore drift. Therefore, an unintended consequence of building groynes is that beaches further along the coast are starved of sediment. These beaches become thinner and the rate of erosion can actually increase.

There are also two large groynes within Criccieth Bay, at the eastern end of the beach. One of these is very large, and Figure 43 shows how successful the groynes are at trapping sediment.

To the east of the two large groynes is a section of glacial till cliff. This can be seen in Figure 44. Beach material in front of this cliff is limited. As a result the cliff face is easily eroded by large storm waves. Are the groynes responsible for this? There are some places along the cliff where obvious collapses of material have occurred. This, however, adds material at the foot of the cliff and acts as a natural defence by increasing the size of the beach again.

▲ **Figure 42** Wooden groynes at the western side of Criccieth Beach.

Enquiry

Should the coastline around Criccieth be protected in the future?

- Use map evidence to argue the case for coastal defences.
- Research current defences here.
- Should more be done to manage the coastline?

▲ **Figure 43** One of the two large groynes on the eastern side of Criccieth Bay (grid reference 505381 in Figure 41).

Activities

1 Explain why people manage the coast.
2 Use Figure 41 to:
   a) give a six-figure grid reference for the castle headland
   b) estimate the length of the western beach that is protected by groynes.
3 a) Draw a sketch of Figure 42.
   b) Annotate your sketch to describe the management strategies that have been used along this stretch of coastline.
4 Study Figure 43.
   a) Explain why beach material is thicker to the right of the photograph.
   b) Is this an intended consequence? If so, why?
   c) Use Figures 41 and 43 to give the direction of longshore drift on this coastline.
5 Describe the features of Figure 44. Explain why erosion here may be an unintended consequence of human activity.

► **Figure 44** The glacial till cliff at the far eastern end of Cricceth beach after the 2014 storms.

Investigating landscape change

The coastline is a constantly changing landscape. Change can be very gradual and occur over many decades. Change over long periods can be measured by comparing primary fieldwork data with historical photographs or maps. However, the shape of a beach profile (its cross-sectional shape) can sometime change overnight. For example, rapid change occurs when beach sediment is eroded or deposited by an extreme storm event. Rapid change to a beach profile may be recorded by comparing measurements taken on two separate days.

Collecting primary evidence of a beach profile

The size and shape of beaches can be recorded by taking beach profile measurements, as shown in Figure 45.

- Person A stands at a safe distance from the edge of the sea holding a ranging pole.
- Person B stands holding a second ranging pole further up the beach. They must stand at the break in slope where there is a change in the angle of the beach.
- The distance between the two ranging poles is measured using a tape measure.
- The angle between markers at the same height on each ranging pole is measured using a clinometer.

Repeat this process at each break of slope until the top of the beach is reached.

When all the data has been collected you can plot the distances and angles for each break on a graph to show the profile of the beach. Data collected on different dates can then be compared.

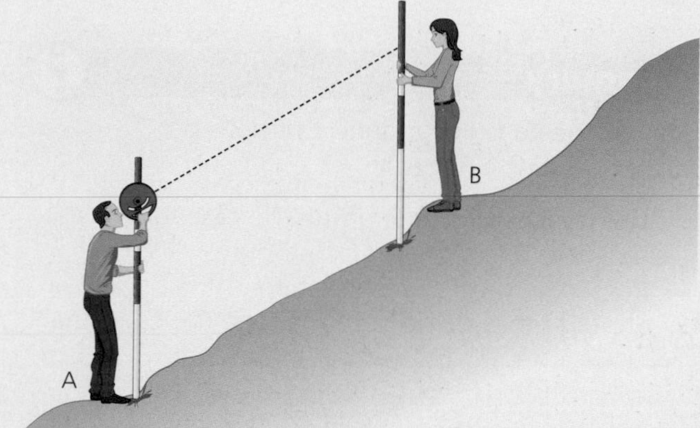

▲ **Figure 45** How to carry out a beach profile.

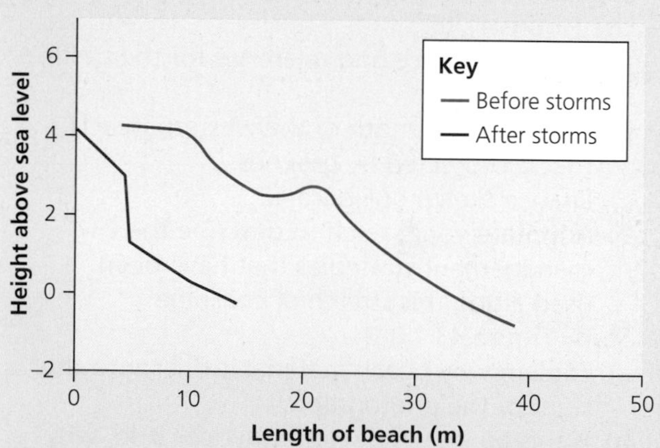

▲ **Figure 46** Beach profiles for different dates at Slapton Sands, Torcross.

Activities

1 Use Figure 46.

a) Describe how the beach profile changed after the storms. You will need to refer to the height and length of the beach.

b) How might you calculate the area of the beach by using several profiles?

Investigating the concept of place

Coastal locations provide an opportunity to investigate how people think about the environment – whether it be the physical environment of a natural coastal landscape, or the human environment of a seaside resort. This will involve collecting qualitative data (see page 8) by using, for example:

- bi-polar surveys (see page 104)
- and questionnaires (see page 10).

Posing questions for an enquiry

Study Figure 47, it shows the seaside resort of Rhyl on the North Wales coast. How do people of different ages think about this **place**: what features do they like and dislike? What words would they use to describe the unique or special features of this place? How many would be positive and how many would be negative? Do tourists to Rhyl have the same ideas about this place as local people? You could use a mixture of open and closed questions to investigate these ideas. If you record the approximate age of each person you could then sort your results to see whether younger people have a different view of the place than older people. Or whether tourists have a different view to local people.

Place

A geographical concept which is used to describe what makes somewhere special, unique or distinct. Each place includes many different features of the human and physical environment such as landscape features and landmarks, local styles of building, ecosystems and habitats, or local historical and cultural features. Each of these features may be relatively common across the UK. However, it is the unique combination of these geographical features that creates an identity for any one place.

▲ **Figure 47** Like many UK seaside resorts, Rhyl has an ageing population and suffers from the decline of tourism, seasonal employment and a shortage of high-paid jobs.

Activities

2 Study Figure 47.
   a) List the human and physical features of this environment.
   b) How do these features compare to other seaside towns?
   c) Create a bi-polar survey that could be used to assess people's views about this environment. Test it on colleagues in your class.

Enquiry

Design an enquiry for Rhyl.

a) Create an over-arching enquiry question about place.
b) Describe how you would collect the qualitative data you need and design data collection sheets.

THEME 1

Landscapes and physical processes

Chapter 3 Drainage basins

Drainage basin stores and flows

Climate, geology and human activity all play an important role in the river landscapes of the UK. The **porosity** and **permeability** of the rocks beneath our feet helps to determine how much water we see in the UK landscape. **Porous** rocks have tiny spaces known as pores between the grains of rock. Porosity is a measure of how much water can be stored in these pore spaces. Rocks such as sandstones can hold water in their pore spaces as a **groundwater store**. Permeability is a measure of how easily water can travel through a rock. **Permeable** rocks allow water to pass through them. Water travels easily through the vertical and horizontal joints and cracks that are common in permeable rocks such as sandstone and limestone. **Impermeable** rocks have few pores spaces or joints, so water tends to flow over them on the surface of the land. Most igneous rocks, such as granite, and metamorphic rocks, such as slate, are impermeable. Clay, which is a sedimentary rock, is also impermeable. In regions where the geology is impermeable, water can be stored at the surface. Lakes and rivers are natural **surface stores** of water. Rivers can be dammed to control flooding and create reservoirs for water supply. There are 168 large dams (defined as being more than 15 m high and holding at least 3,000,000 cubic metres of water) in the UK.

▲ **Figure 1** The Afon Elan flows through an upland area of Wales, where rocks are impermeable and soils are thin; the closest weather station to this drainage basin is at Cwmystwyth.

▲ **Figure 2** The Penygarreg reservoir and the Craig Goch dam in the Elan Valley.

	Jan	Feb	Mar	Apr	May	Jun	Jul	Aug	Sep	Oct	Nov	Dec
Aberystwyth	97	72	60	56	65	76	99	93	108	118	111	96
Cwmystwyth	192	139	158	108	97	116	116	135	151	187	206	213
Birmingham	74	54	50	53	64	50	69	69	61	69	84	67
Norwich	55	43	48	41	42	58	42	54	47	68	70	53

▲ **Figure 3** Monthly precipitation totals for UK weather stations along a west–east transect through Wales and England.

Activities

1 Using Figures 1 and 2, suggest **two** different ways that human activity has affected the landscape of the River Elan.

2 Describe the landforms visible in Figure 1 or 2 and the processes that formed them (see page 13).

3 Study Figure 3.
   a) Draw four precipitation graphs.
   b) Describe how precipitation changes from west to east across Wales and England.
   c) Give two reasons why the Elan Valley may have been chosen to build the Craig Goch dam.

Where are the water stores and flows in a drainage basin?

Very little precipitation falls directly into rivers. Most falls elsewhere within the **drainage basin** and its **tributaries**. Figure 4 shows flows of water through a typical drainage basin. Water either flows over the surface as **overland flow** or flows into the soil – a process known as **infiltration**. Once in the soil, water moves slowly downhill as **throughflow**. Some water percolates deeper into the ground and enters the bedrock where it continues to travel as **groundwater flow**. Rates of infiltration, throughflow and groundwater flow depend on a number of factors, including the:

- size and shape of the drainage basin and the steepness of its slopes
- amount of rainfall throughout the year and the intensity of rain storms
- amount and type of vegetation cover
- permeability and porosity of the soil and underlying rocks.

Drainage basin

The area of land from which a river collects its water. Water is stored in the drainage basin in rivers, vegetation, soil and rocks. Water flows overland, through soil or through bedrock.

▼ **Figure 4** Stores and flows of water in a natural drainage basin.

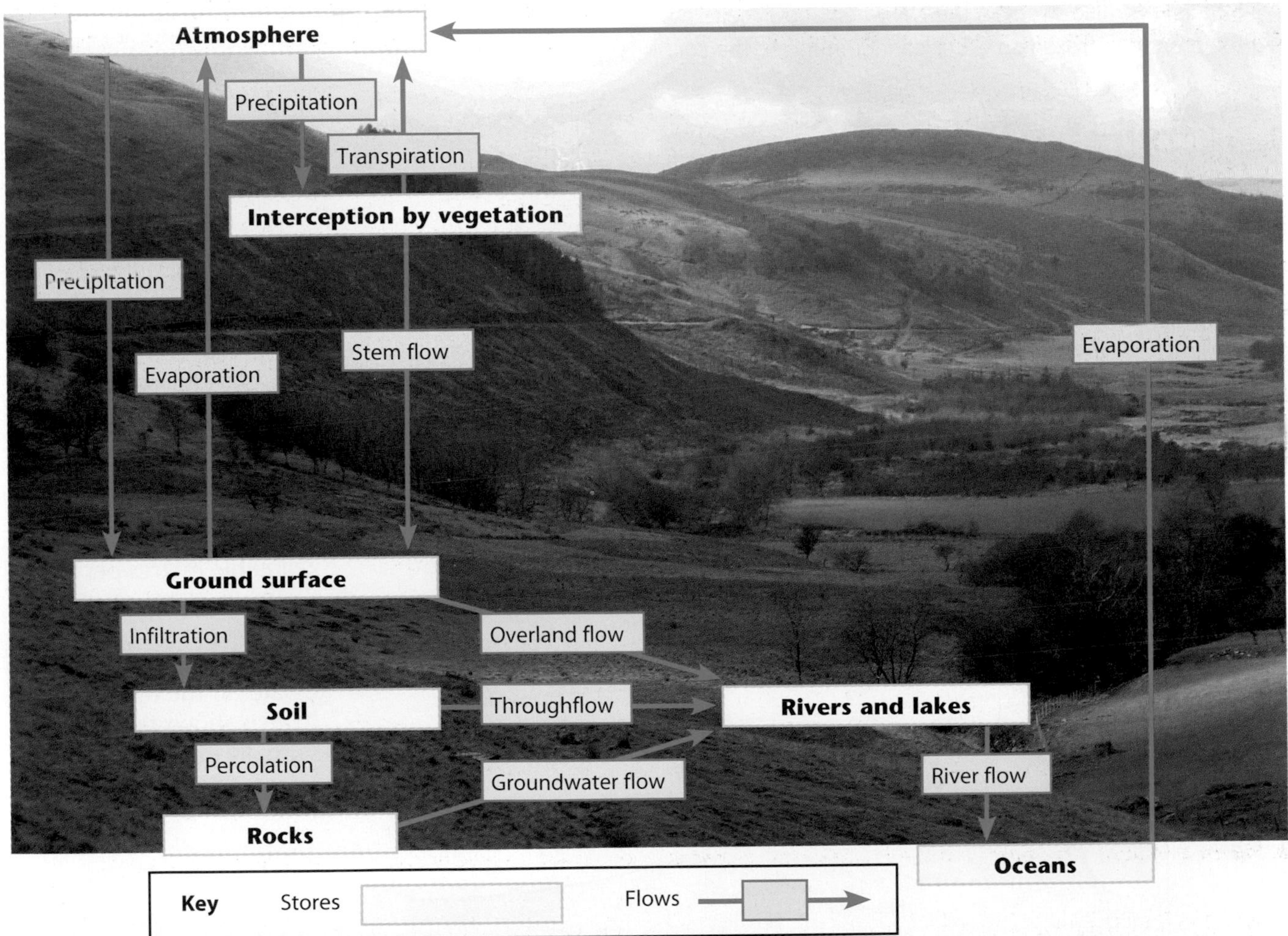

Activities

4 Use Figure 4 to name:
   a) three surface stores of water
   b) two places water is stored below the surface.

5 Suggest why precipitation falling into a drainage basin of impermeable rock is likely to reach the river much more quickly than rainwater falling in an area of porous rock.

Why do rivers flood?

In the UK we experience two types of flood:

- **Flash floods**, like the Boscastle flood event of 2004, when high volumes of rain fall in a very short period of time, causing a sudden rise in river levels. Flash floods sometimes occur in the summer in the UK when the ground is hard and baked dry. With these soil conditions, the rainfall is so intense that it cannot soak into the ground quickly enough so runs overland instead.
- **Seasonal floods**, like the Somerset Levels flood event of 2014, when river levels rise due to seasonal variations in rainfall. These types of flood usually occur after a long period of rain, when the ground is already saturated and cannot absorb any more water. Floods may also occur in the UK when snow melts but the ground is frozen so water cannot infiltrate the soil.

Do human actions increase flood risk?

Flooding occurs when water cannot infiltrate the soil. Paving over the soil creates an **impermeable** surface and reduces infiltration, so the growth of urban areas increases the risk of flooding. It is thought that paving over front gardens to create parking spaces may increase the risk of flash floods in urban areas. Vegetation helps to remove water from the soil before it reaches a river so cutting down trees or leaving fields bare in winter can increase the risk of seasonal floods. On the other hand, replanting upland areas with trees, a process known as **afforestation**, may help to reduce the risk of floods further downstream. It is thought that afforestation in mid Wales could help to reduce floods at Shrewsbury or Bewdley (see pages 50–1).

GEOGRAPHICAL SKILLS

How do you analyse a hydrograph?

Figure 5 is a flood **hydrograph**. It shows how a small river might respond to a flood event. The blue bar represents a sudden downpour of rain. In this example it takes two hours for overland flow from the drainage basin to reach the river channel. The amount of water in the channel begins to rise. This is shown by the **rising limb** of the graph. The time between the peak rainfall and the **peak discharge** is known as **lag time**. In Figure 5, the lag time is about 5 hours. The lag time and height of the peak discharge depend on the features of the drainage basin, such as porosity of the rocks, steepness of the slopes or the amount of vegetation. Some of these factors are illustrated in Figures 6 and 7.

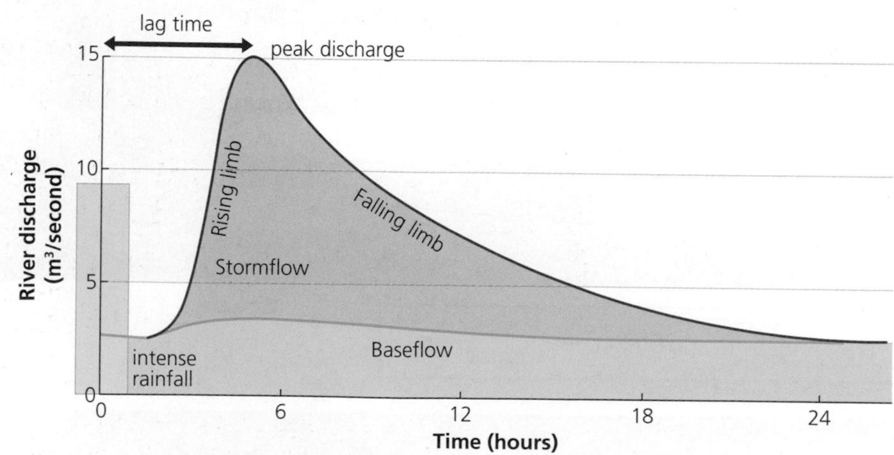

▲ **Figure 5** A simple flood hydrograph.

Hydrograph

A graph that shows how river discharge changes over time. The discharge (amount of water) is shown on the *y*-axis and is usually recorded in units of cubic metres per second, although sometimes the *y*-axis shows the depth of water in the river. The *x*-axis shows time – usually in hours or days.

Activity

1 Use Figure 6 to explain how cutting down a large forest could affect lag time and peak discharge in a nearby river.

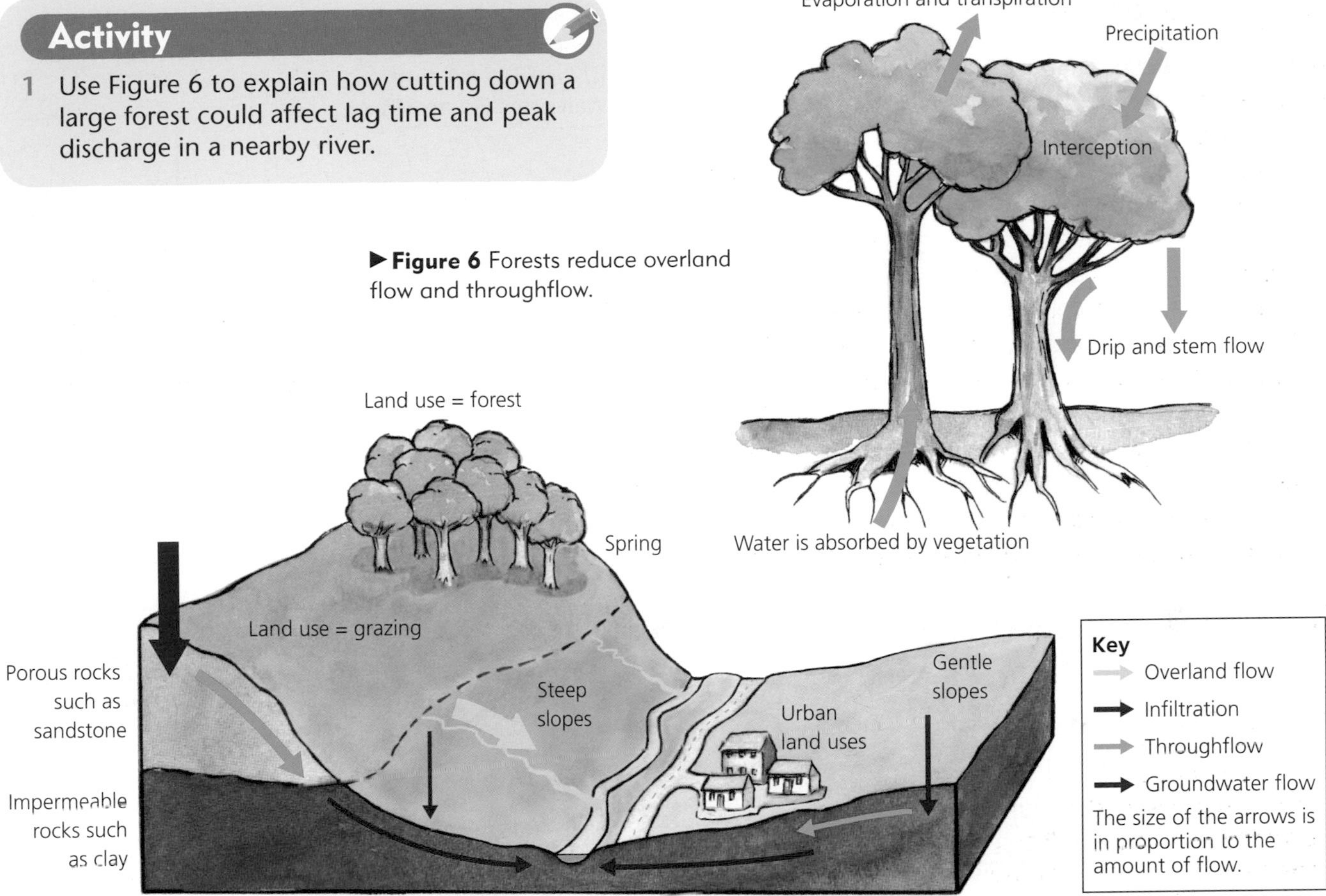

► **Figure 6** Forests reduce overland flow and throughflow.

▲ **Figure 7** Factors that influence infiltration and overland flow in the drainage basin.

Activities

2 Study Figure 5.
   a) Use times and discharge figures from the hydrograph to describe:
      i) the shape of the rising limb and the lag time
      ii) the shape of the falling limb and baseflow.

3 Use Figures 4, 5, 6 and 7 to help you copy and complete the following table.

Drainage basin factor	Impact on infiltration	Impact on overland flow and throughflow	Impact on lag time
Steep slopes			
Gentle slopes			
Porous rocks			
Impermeable rocks			
Urban land uses			
Planting more trees			

Enquiry

Can you predict the response of different drainage basins to a sudden downpour of rain?

- Sketch a pair of flood hydrographs to show the difference between similar sized drainage basins – one of which has urban land uses and one of which has lots of forests.
- Draw a second pair of hydrographs to compare the response of rivers in a drainage basin that has porous rocks compared with one that has impermeable rocks.
- Discuss your hydrographs with a colleague. Justify the shapes on your hydrographs. Make sure you can explain why you have predicted the shape of the rising and falling limbs and the possible length of the lag time.

How does geology affect flows and stores?

Geology is a major influence on how quickly water flows through a drainage basin and the amount of water that the drainage basin can store. The amount of water in a river is its **discharge** and this is measured in cubic metres per second, or **cumecs**. Study Figures 8 and 9. They show the **annual regime** in 2004 for two rivers that have similar sized catchment areas. However, the geology and annual rainfall of the two drainage basins are quite different and this affects the flow of water through each basin.

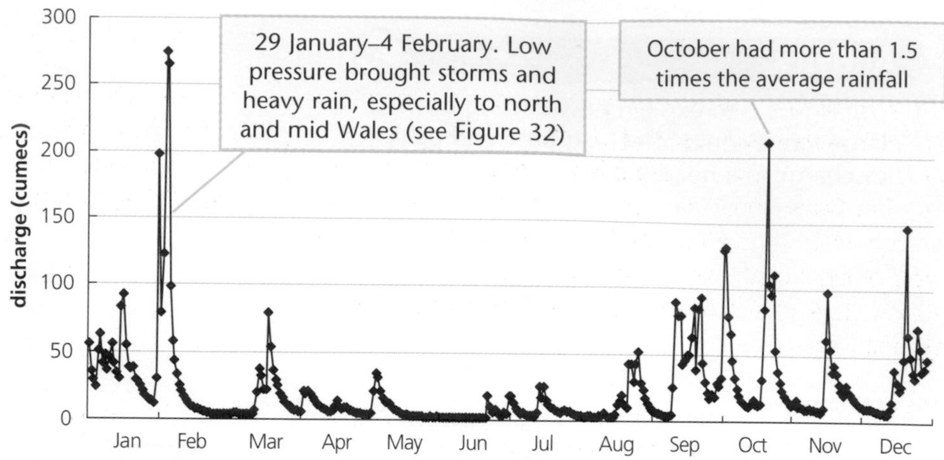

▲ **Figure 8** Hydrograph for the River Dyfi, Wales (2004).

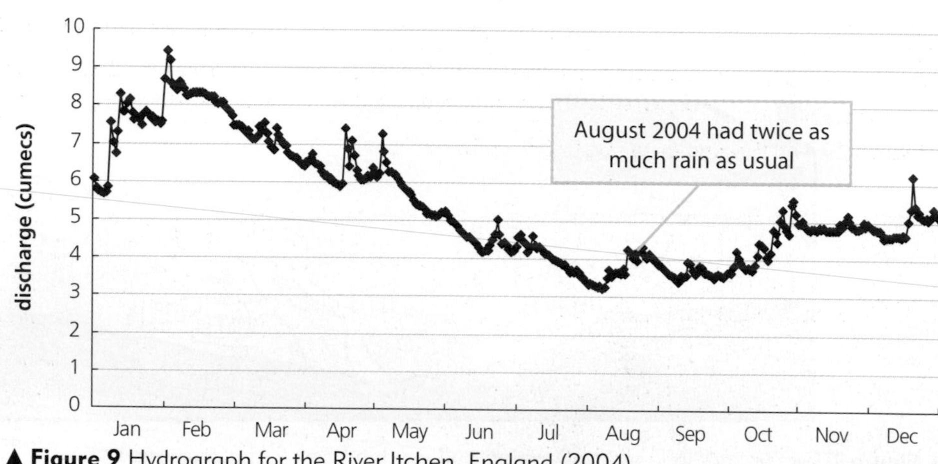

▲ **Figure 9** Hydrograph for the River Itchen, England (2004).

Factfile: The River Dyfi and River Itchen

River	River Dyfi	River Itchen
Location	West Wales	South East England
Total average rainfall	1,834 mm	838 mm
Geology	100% impermeable rocks	90% chalk
Size of drainage basin (above the gauging station)	471 km^2	360 km^2
Landscape	Steeply sloping hills and mountains reaching a maximum of 907 m above sea level	Rolling hills. Maximum height 208 m above sea level
Land use	60% grassland (sheep pasture); 30% forest; 10% moorland	Mainly arable (cereal) farmland with some grassland
Human factors affecting run-off	There are virtually no human influences on run-off	Run-off is reduced by some abstraction for water supply. Some water is used to recharge groundwater in the chalk aquifer

Activities

1. Compare Figures 8 and 9. Describe:
   a) one similarity
   b) three differences.
2. Use the Factfile to suggest how each of the following factors may have affected the flow:
   - rainfall
   - geology
   - landscape
   - land use.
3. Imagine you work for a water company. Suggest how each river could be used for water supply.

How does extreme weather affect flows and stores of water?

Parts of the UK experienced record-breaking levels of rainfall during November and December 2015. A number of **low-pressure systems** brought wet weather across the UK. The warm, moist air masses came from the Azores – a warm region of the Atlantic to the west of Portugal. They were brought by the jet stream (see page 173), which stayed stuck in a southerly loop to the west of the UK for several weeks.

After weeks of rain the ground was saturated – meaning that further rainfall could not soak into it. River levels rose and a series of floods affected communities across North Wales, Northern England and Scotland.

4–5 December	Storm Desmond arrives. More than a month's rain falls in parts of Cumbria. Flooding in Carlisle, Appleby and other parts of Cumbria.
12 December	Environment Agency issues more than 70 flood warnings, mainly in Cumbria and Lancashire.
22 December	Storm Eva arrives. Cumbria receives more rain and some communities flood again.
26–27 December	People are evacuated from their homes in Lancashire and Yorkshire, including in Hebden Bridge, Leeds, Greater Manchester and York. There is also flooding in Llanrwst in North Wales.

▲ **Figure 10** The timeline of events in December 2015.

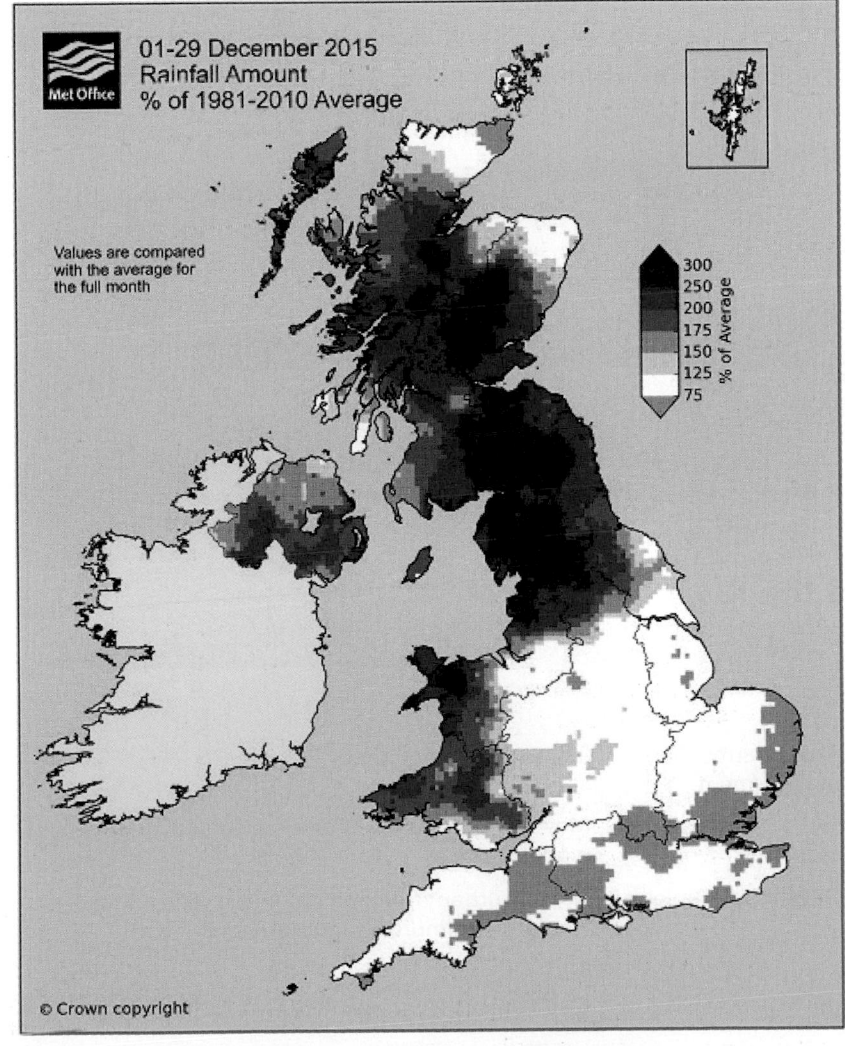

▲ **Figure 11** Record-breaking rain fell during December 2015.

Low-pressure system

A condition of the atmosphere that brings wet and windy weather to the UK. These systems begin when warm, moist air rises from the ground. This creates air pressure that is lower than normal at the centre of the system and air is drawn in to fill the space, which can generate strong wind. Low-pressure systems are also called cyclones.

Activities

4 Explain why the UK suffered flooding in late 2015.

5 Use Figures 10 and 11, and an atlas, to label an outline map of the UK showing the main areas affected by flooding.

6 Describe the pattern of rainfall shown on Figure 11. Use an atlas to help you:

a) name three upland regions that had more than 250 per cent of usual rainfall

b) name three regions that had less than 75 per cent of usual rainfall.

The impacts of the 2015 floods

During December 2015 rivers were in flood across parts of Wales, northern England and Scotland. Rivers behave very differently when they are in flood. Water levels are not just higher – the discharge is many times greater than during normal flow. During such extreme events rivers can make sudden and dramatic changes to the shape of the river channel. With so much extra energy the river is able to erode and transport huge quantities of material. Banks are undermined and even more rock falls into the torrent. Trees too can be swept into the swollen river. The Raise Beck, a mountain stream in Cumbria, widened its channel by several metres, eroding part of the A591. A similar stream flooded at Glenridding, Cumbria. The force of its flow washed hundreds of tonnes of gravel and large pebbles across the streets.

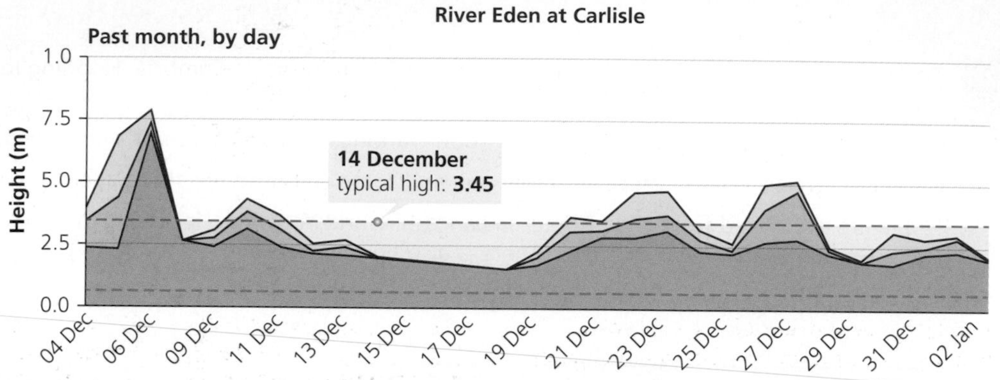

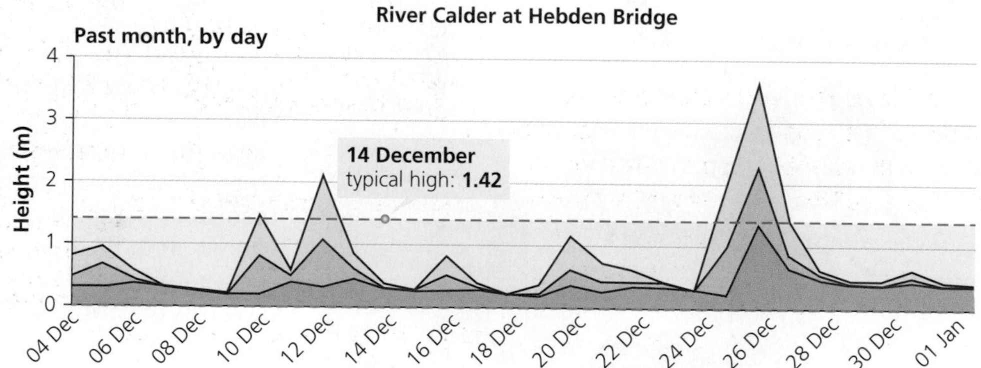

▲ **Figure 12** Hydrographs for the River Eden at Carlisle (Cumbria) and River Calder at Hebden Bridge (Yorkshire); the blue lines indicate levels recorded on each day; the green line is the typical high level for the month of December.

Factfile: The River Eden and the River Calder

	River Eden at Carlisle	River Calder at Hebden Bridge
Size of catchment (km²)	2,286	171
Geology	Limestone to east and impermeable gritstones to west of catchment	Impermeable gritstones
Land uses	Mainly moorland used for grazing Towns of Carlisle, Penrith and Appleby in valleys	Mainly moorland used for grazing and steep, wooded slopes A lot of urban development in the valley floors; towns of Todmorden and Hebden Bridge
Mean flow (cumecs)	53.3	4.1
Max discharge before 2015	988 cumecs when the river was 5.52 m deep	59.6 cumecs when the river was 3.09 m deep

▲ **Figure 13** The Raise Beck eroded part of the A591 between Ambleside and Keswick; this stream runs off the steep hills of the Lake District in a V-shaped valley and then runs alongside the road.

How were people affected?

Roads were closed either due to flood water or because bridges were in a dangerous condition. Transporting rocks and trees downstream, the river battered the piers of bridges – several bridges collapsed during the December floods. Trees and branches got stuck under bridges, preventing the flow of water. Bridges can begin to act like dams and the extra weight of water behind them can cause structural damage. Road and bridge closures caused delays for drivers and many got stuck in flood water, damaging engines beyond repair. In early January 2016, one month after the floods began, Cumbria had:

- fifteen sections of road closed because the road was washed away or covered by a landslide
- eight footbridges closed and waiting for safety inspections
- twenty bridges closed due to damage and a further three awaiting inspection.

An electricity substation on the River Lune near to Lancaster was flooded, leaving the whole of Lancaster without power for almost three days. Emergency generators were used to keep the power on in Lancaster Hospital and for 19,000 people in the area. The University of Lancaster was closed and students sent home one week before the end of term.

40 bridges were closed due to flooding or damage

341 mm of rain fell at Honister, Cumbria, in 24 hours (between the 4th and 5th of December)

500 homes were flooded in York by storm Eva

5,200 homes were flooded in Cumbria and Lancashire by storm Desmond

20,000 homes were without power over Christmas – mainly in Yorkshire

55,000 homes were without power in the Lancaster area during storm Desmond

▲ **Figure 14** The floods in numbers.

Activities

1 a) Compare the flood hydrographs for the two rivers shown in Figure 12. Use page 38 to help you.
  b) Suggest reasons for the main differences you have identified.

2 Describe the changes that occurred in the channel of the Raise Beck. Use pages 12 and 13 to add detail to your description.

3 Describe the impacts of the flooding on:
  a) infrastructure
  b) the environment.

4 Suggest how each of the following groups of people may have been affected:
  a) elderly people living alone
  b) people living in remote rural areas.

How did flood defences cope in 2015?

The town of Carlisle in Cumbria is built at the confluence of three rivers – the Eden, Petteril and Caldew. In January 2005 it was badly flooded with almost 2,000 homes damaged, foul water from the sewers mixed with the flood water, and the fire service had to borrow wooden rowing boats to reach people trapped by flood water. Following these floods:

- The Environment Agency spent £38 million on improving flood defences for Carlisle. Flood embankments 6.2 m high and 10 km long were built alongside the river and 30 flood gates installed.
- United Utilities replaced 4 km of old sewers at a cost of £13 million.
- The fire service was equipped with inflatable boats and dry suits, and trained to rescue people from fast-flowing water.

During the 2005 flood, some water became trapped by the existing flood defences and couldn't flow back into the river. Ten larger drains were installed to stop that happening again.

▲ **Figure 15** The fire service rescuing residents of flooded homes in Carlisle, December 2015.

Do flood defences work?

Despite all of the investment, why did the town flood again in 2015? The Environment Agency argues that the flood defences in Carlisle were good value for money, but local people were angry that their homes were flooded, and not everyone agrees that this is the best way to defend our towns from flooding.

Floods threatened Carlisle in 2009 and again in 2012. On each occasion, Carlisle's flood defences worked well. It is estimated that £180 million of flood damage was avoided. The flood defences cost £38 million, so they have more than paid for themselves. I think that's good value for money.

Mike Harper – Environment Agency

This has been an exceptional period of rainfall but we must acknowledge that, with our changing climate, such weather events are happening more frequently. This is why flood and coastal risk management remains a key priority for this [Welsh] government.

Carwyn Jones – First Minister for Wales

▲ **Figure 16** Two points of view on the 2015 floods.

Year	Flooding event
1963	Flooding
1968	Flooding
1979	Flooding
1980	Flooding
1984	Flooding
2005	Flooding
2009	Floods prevented
2012	Floods prevented
2015	Flooding

▲ **Figure 17** Recent flood events in Carlisle.

GEOGRAPHICAL SKILLS

Calculating percentage of average

Average rainfall figures for each month can be calculated by recording actual rainfall figures over long periods of time (usually at least twenty years) and then finding the mean value. December 2015 was a record breaker – the wettest month ever on record for the UK. To calculate how much wetter than average, you need to divide the actual amount for any month by the average for that month. So, for November 2015 at Hazelrigg:

$249.8 \div 114.7 = 218\%$

	2015	Previous wettest	Average (mean)	2015 % of average
November	249.8	244.8 (2009)	114.7	218
December	306.4	234.5 (1986)	112.8	
November and December	556.2	357.5 (1986)	227.4	

▲ **Figure 18** Rainfall (mm) for Hazelrigg in Cumbria, 2015, compared to previous wettest figures and averages for November and December.

Activities

1 Using Figure 15, describe two risks created by flooding.
2 Describe three ways that flood defences in Carlisle were improved after 2005.
3 Calculate the frequency of flooding (see page 23) in Carlisle:
   a) with flood defences
   b) if flood defences had not been built after 2005.
4 Use Figure 18 to calculate the 2015 rainfall for Hazelrigg as a percentage of average for:
   a) December
   b) November and December
5 Use Figures 16, 17 and 18 to suggest two reasons why the floods occurred, despite the flood defences.

Future flood prevention

In 2019 work began on raising and extending the flood walls in Carlisle. The work will cost £25 million and should protect more than 1,600 homes and businesses. But are we spending money in the correct places and on the right kind of scheme?

George Monbiot, an environmental author, argues that flood defences are being built in the wrong places. Instead of defending our towns in the valley floors, we should be spending money managing rivers in the upper catchments. He argues that dredging and straightening rivers to protect farmland makes the flood risk worse because water moves down river faster and rushes into the nearest town. He gives the River Liza in Ennerdale as an example. This river used to be dredged – but not anymore. Instead it has been allowed to meander and deposit stones and logs in the river channel, creating a **braided** shape – like plaited hair. The river now looks much more natural. During a flood event, like the one in 2009, the wide braided channel of the River Liza stores flood water and releases it slowly – protecting places downstream from the sudden rush.

▲ **Figure 19** The flood storage basin at Thacka Beck, Penrith.

Penrith is another Cumbrian town that is vulnerable to river flooding. A small stream, the Thacka Beck, runs under the town in a culvert. If the culvert became blocked, the town flooded. A flood defence scheme was completed in the town in 2010 at a cost of £5.6 million. The scheme protects 263 homes and 115 shops, and other businesses. It has two parts:

1 A total of 675 m of culverts were replaced so that water will not back up into the streets.
2 A flood storage reservoir was constructed before the Beck enters the culvert. The reservoir is simply a hollow in a field. In normal weather conditions the pond is just 12 cm deep, but it is capable of storing 76,000 cubic metres of water until flood levels in the river have gone down.

Activities

6 a) Explain why some argue that dredging and straightening has the unintended consequence of increasing the risk of floods.
   b) Describe how flood storage schemes could help protect towns. Give two different examples.
7 The construction of the Thacka Beck flood storage scheme was supported by Cumbria Wildlife Trust. Suggest the benefits of this scheme for local residents and wildlife.

Enquiry

Should higher embankments be built alongside the River Eden in Carlisle?

What should be done in the future? Should alternative schemes be tried? Give reasons for your answer.

The distinctive landscape of the Somerset Levels

The Somerset Levels is a distinctive flat landscape covering 250 square miles. The Levels are only 8 m above sea level and much of this landscape would be flooded twice a month by high spring tides if it wasn't for flood defences at the coast. The Romans built flood defences and they dug ditches to improve the drainage. Over the years, rivers have been dredged and water pumped out, changing the wetland into productive farmland used for livestock and arable crops. Some wetland remains and is conserved as nature reserves. The flat land of the Somerset Levels is vulnerable to:

- coastal flooding by high tides and storm surges, for example in 1919
- river flooding after prolonged periods of rain, for example in 2014.

What caused the 2014 floods?

High rainfall totals over the winter of 2013–14 saturated the soils with water. More rainfall during January and February 2014 ran overland into the already flooded rivers. At each high tide, the rivers backed up because flood water couldn't escape quickly into the Bristol Channel. Local people claimed that the rivers and drainage ditches of the Levels hadn't been cleared of silt and mud since the 1990s. This reduced the capacity of the river channels to hold water. Overhanging trees slowed down the rivers' discharge.

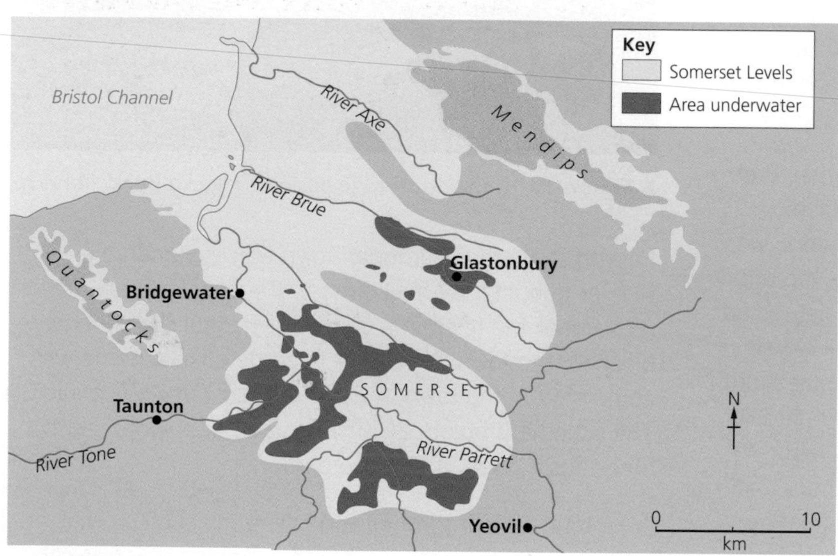

▲ **Figure 20** Map showing the location of the Somerset Levels and the flooding of 2014.

Weblinks

www.metoffice.gov.uk/public/weather/climate-historic/#?tab=climateHistoric

This website gives access to historic weather data from 40 weather stations across the UK including Yeovilton in Somerset.

http://nrfa.ceh.ac.uk

This is the National River Flow Archive. You can view and download discharge data for many UK rivers. Use this data to create hydrographs like Figure 21.

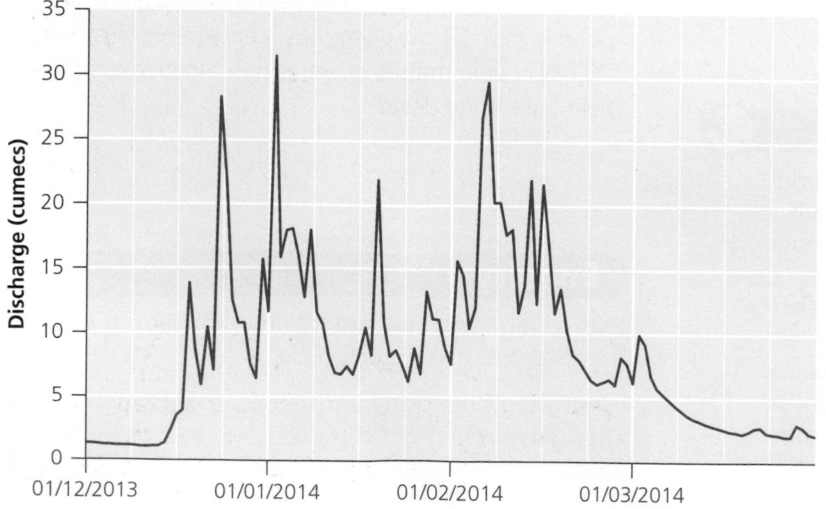

◀ **Figure 21** Discharge in the River Tone at Bishops Hull (December 2013 to end March 2014).

Activities

1 Use Figure 20 to:
   a) Describe the drainage patterns in the Somerset Levels
   b) Estimate the proportion of the Levels that were flooded in 2014.

Year	Tidal/ coastal	River
1885	✓	
1919	✓	
1968		✓
1981	✓	
1990	✓	
2005		✓
2007		✓
2008		✓
2012		✓
2014		✓

▲ **Figure 22** Historic flood events in the Somerset Levels.

▲ **Figure 24** Large areas of the Levels were under water for weeks in 2014.

Year	Month	Precipitation (mm)
2013	Jun	25.7
	Jul	40.4
	Aug	15.2
	Sep	67.6
	Oct	115.8
	Nov	42.9
	Dec	121.4
2014	Jan	166.4
	Feb	131.2
	Mar	38.4
	Apr	62.2
	May	59.6
	Jun	79.0
	Jul	56.8
	Aug	74.6
	Sep	3.2
	Oct	99.8
	Nov	108.0
	Dec	34.8
2015	Jan	75.4
	Feb	41.8
	Mar	22.0
	Apr	25.8
	May	55.2

▲ **Figure 23** Monthly precipitation totals at Yeovilton (2013–15).

Activities

2 a) Explain why the 2014 floods of the Somerset Levels happened. You should be able to identify physical and human factors.
   b) Study Figure 22. What is happening to the frequency of floods in the Levels?
   c) Suggest two different reasons that might explain this frequency pattern.
3 Use Figure 21 to describe the discharge pattern of the River Tone.
4 a) Draw a graph to represent the monthly precipitation patterns at Yeovilton shown in Figure 23.
   b) How unusual was the rainfall pattern during the period November 2013 to February 2014?
   c) Compare your completed graph to Figure 21. Use these two graphs and your understanding of tidal movements to suggest reasons for the peaks in discharge.

Enquiry

Analyse winter discharge patterns of rivers in the Somerset Levels. Use the National River Flow Archive website for your investigation. How do patterns in the last winter compare to Figure 23? Suggest reasons for any similarities or differences.

Hard engineering

The immediate responses to the Somerset floods were:

- Thirteen large pumps from Holland were used to remove 7.3 million tonnes of water per day from the flood plain.
- Residents of flood-damaged homes were given emergency help and insurance advice.
- Teams were brought in to start emergency repairs to infrastructure such as rail and road networks, electricity supplies, telephone systems and sewers.
- Livestock was rescued and moved to other areas.

In mid-2014 the UK Government announced a £100 million recovery programme for the Somerset Levels. This plan included a range of **hard engineering** options to improve flood defences. The aims of the Somerset Levels and Moors Flood Action Plan are shown in Figure 26.

▲ **Figure 25** Pumps from Holland were used to pump water back into Somerset's rivers from the floodplain.

Medium term aims – up to 2020

- A dredging operation on the Rivers Parrett and Tone cost £6 million. A total length of 8 km was dredged.
- Embankments were built around vulnerable villages. £180,000 were spent protecting the ten houses in Thorney.
- The height of the A372 road was raised and 44 km of damaged roads were repaired.
- Better preparation and planning has been carried out to protect residents and business owners.

Longer-term aims – up to 2035

- To build a flood barrier at Bridgwater by 2024. The estimated cost is at least £60 million. This would reduce the impact of high tides that prevent flood water from escaping from the Levels.
- To investigate the costs and benefits of building a £50 million flood storage scheme upstream from Taunton. This would capture and hold back rainwater from draining into the River Tone. A large-scale tree-planting programme in the upper catchment would be part of this scheme.

▲ **Figure 26** The Somerset Levels and Moors Flood Action Plan.

Hard engineering

Schemes that use artificial structures to prevent flooding. These include dredging sediment from the river to increase the capacity of the river to hold water. Another example of hard engineering is a flood embankment alongside the river, built to keep water from flooding onto the floodplain.

◀ **Figure 27** Dredging and bank stabilisation on the River Parrett.

Factfile: Impacts of the floods

- Half of all businesses in Somerset were affected either directly or indirectly by flooding.
- Damage to residential property cost up to £20 million.
- Costs to local government, the police and rescue services totalled £19 million.
- The rail line between Taunton and Bridgwater was closed for four weeks, costing the local economy £21 million.
- Over 80 roads were submerged for weeks with a cost of £15 million to the local economy.

Enquiry

Who should pay for the Somerset Levels and Moors Flood Action Plan? Discuss the responsibility of the following groups before making your recommendation:

- Those directly affected by flooding
- All the residents of Somerset
- The tax payers of the UK.

Activities

1 Draw and label a sequence of simple cross-sectional diagrams to show how dredging of the rivers would make them more efficient and reduce the risk of flooding.
2 Evaluate the arguments for and against continued dredging of rivers in the Levels.
3 Use the information on these pages to analyse the Somerset Levels and Moors Flood Action Plan. Do this by completing a table like the one below:

	Advantages	Disadvantages
Economic		
Social		
Environmental		

Our computer models suggest that the dredging programme will make the rivers 90% effective in flushing out flood water. Currently the silt build up since 2009 makes them only 60% efficient.

Ashley Gibson, Somerset Water Management Partnership

Part of the blame must be down to the Environment Agency. Why on earth did they cut back on the dredging programme in the first place?

Ian Liddell-Grainger, MP

Dredging has an impact on the habitat in and around the river. The fish are severely disturbed during dredging and the water is murky for days after. Reed beds alongside the river are ripped up so birds and small mammals lose habitat.

Royal Society for the Protection of Birds

It would have been cheaper to maintain the rivers properly instead of cutting back the dredging process in 2009. The clean-up and all the repairs will cost the tax payer millions.

National Farmers Union

I would urge the authorities not to take dramatic action until they have considered the impact of climate change and sea level rise. A 12 cm rise in sea level could easily make the proposed Bridgwater barrage a waste of money.

Lord Krebs, Climate expert

I object to paying an extra £25 a year in local taxes to fund flood prevention. The council should not allow building on flood plains. I have heard that over 900 homes have been built on vulnerable land in Somerset since 2001.

Resident of Glastonbury not affected by the flood

The Environment Agency are taking a lot of the blame for reducing the dredging programme in recent years but government cutbacks in public spending required them to lose 1,700 jobs. We can't have it all ways.

Western Morning News

We must allow the rivers to flood naturally. Some of the wetland habitats are unique. Sometimes nature needs to come before economics and people.

The Somerset Wildlife Trust

▲ **Figure 28** Different views from stakeholders on what should be done.

Land use zoning

A serious flood in 2000 prompted flood defences to be built at Frankwell, Shrewsbury on the River Severn. These included earth embankments, concrete flood walls and demountable flood barriers. These barriers (seen in Figure 29) are made of aluminium panels. They can be slotted together before the flood arrives. The demountable barriers successfully held back 1.9 m of flood water during a flood in February 2004 and 74 properties were protected. **Land use zoning** is also used in Shrewsbury. In Figures 29 and 30, you can see that the car park has been allowed to flood. This reduces the level of flood water in the river channel, so that the threat of flooding elsewhere is reduced.

Land use zoning

Where land uses that have a low value, such as car parks or playing fields, are not protected by flood walls. These zones provide safe areas for water to be stored during a flood event so that water is kept away from more valuable land uses such as homes.

▲ **Figure 29** A total length of 700 m of flood embankments and walls has been built where the river enters the town to prevent floods in Shrewsbury. A further 155 m of river bank is protected using demountable defences.

▲ **Figure 30** Guildhall and Frankwell car park, SY3 8HQ, the same location as shown in Figure 29.

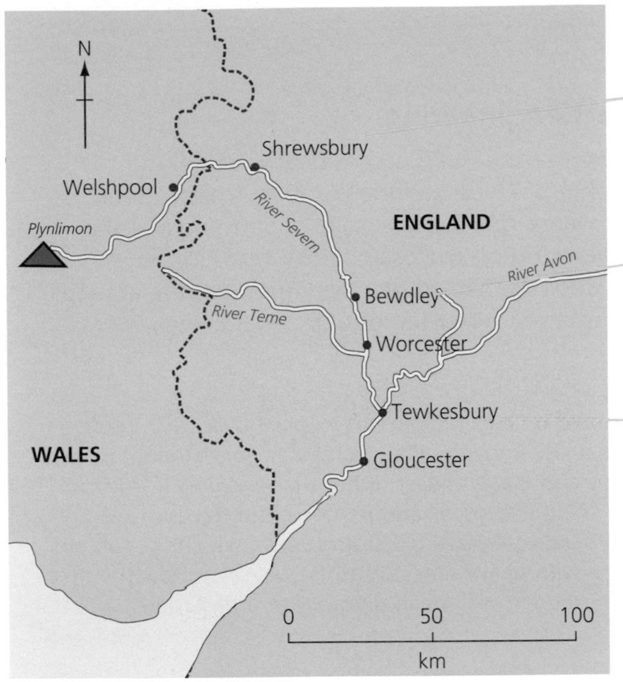

Shrewsbury: Flood damage in 2000 (severe), 2002, 2004 and 2007 (severe) and in 2014. Major flood defences were constructed in the spring of 2004. A further £20 million for extended flood defences for Shrewsbury was announced in February 2015, the work will be completed by 2020.

Bewdley: Flood damage in 2000 and 2002. Severe flooding was avoided in 2004 and again in 2014 by the use of demountable flood barriers.

Tewkesbury: Both Tewkesbury and **Gloucester** were badly affected by floods in 2000 and 2007. A five year multi-million pound flood prevention scheme protected both towns when the River Severn rose to dangerous levels in 2012. Only isolated villages downstream from Tewkesbury suffered any damage in 2012 and 2014.

◀ **Figure 31** Historic floods (2000–15) on the River Severn.

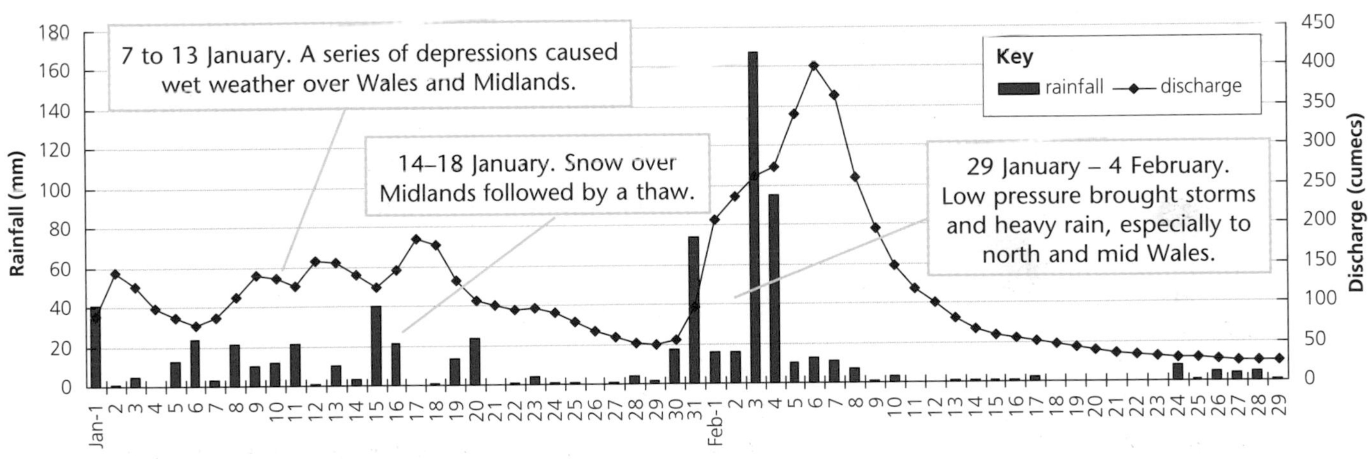

▲ **Figure 32** Flood hydrograph for the River Severn at Bewdley (January–February 2004). Rainfall data is for Capel Curig, North Wales.

Activities

1. Describe the course of the River Severn.
2. Suggest why planners might prevent the building of new homes on the flood plain of the River Severn.
3. Suggest why local residents might prefer demountable barriers to walls and embankments.
4. Explain why land use zoning is used in Shrewsbury. Consider the economic and environmental benefits.

Enquiry

How do weather events affect seasonal flood patterns?

- Carefully describe the shape of the flood hydrograph between 29 January and 29 February. Use Figure 5 on page 38 to help your description.
- Describe how each of the three weather events described in the labels affected the flow of the river. Focus on the time lag as well as the gradient of each rising limb.
- Suggest how the Environment Agency uses rainfall data from Wales to warn householders in Bewdley about future flood events.

Soft engineering

Scientists are debating what can be done to protect UK towns and cities from future floods. Many believe that **soft engineering** schemes should play a much greater part in our future flood defences. This means rivers would behave more naturally. Moorlands, forests and floodplains could be allowed to store more water and release it slowly downstream.

Soft engineering

Schemes that help nature deal with a flood problem. These include planting trees to soak up water. Another example of soft engineering is blocking off the drainage systems in the UK's uplands so that these moorland areas are able to store more water for longer.

▶ **Figure 33** Alternative points of view on solving the flood problem.

Planner

Householders should be encouraged not to pave over their gardens. Paving and tarmac are impermeable. Rainwater goes straight down into storm drains and into the river rather than soaking slowly into the soil. People could use gravel and permeable surfaces instead of tarmac. We also need to replace old storm drains which are too old and small to cope with heavy rain storms. However, motorists won't like that because it will mean digging up urban roads!

Spokesperson for RSPB

In the past, farmers in upland areas of England and Wales added drains to their fields to improve the amount of grass that could be grown. However, these field drains increased the flow of rivers further downstream. We are involved in a scheme to restore the old peat bogs in upland Wales. We are using bales made from heather to block the drains. This will slow down the overland flow and force water to soak back into the soil. Not only will this help reduce the risk of floods, but it will also improve the moorland ecosystem and will help to protect rare birds of prey like the merlin and hen harrier.

River scientist

Hard engineering schemes, like flood walls and embankments, speed up the flow of water. These schemes may funnel water along to the next community living further downstream and actually increase their risk of flooding. What we need to do is to return river valleys to a more natural state. We should use floodplains as temporary water stores so that flooding can occur away from built-up areas.

House builder

Homes can be made more flood proof with measures such as putting plug sockets higher up the walls and replacing wooden floors and carpets with tiles.

Resident in Shrewsbury

I'm really pleased with the new flood defences. My property has flooded in the past but was protected during 2007. The Shrewsbury flood defence scheme cost £4.6 million but I think it was worth it.

Government housing minister

We need to build an extra 3 million homes in the UK by 2020. Some of these houses will have to be built on greenfield sites. However, we should restrict building on floodplains in the future.

▼ **Figure 34** The movement of water through upland drainage basins in mid Wales was altered when field drains were added. The size of the arrows is in proportion to the amount of each flow.

a) The natural system

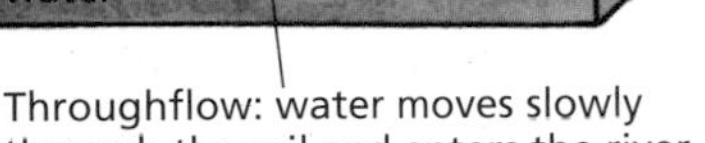

b) Field drains were added to improve drainage

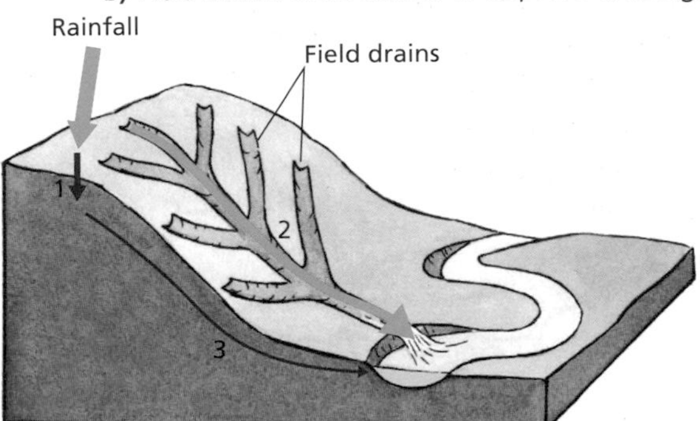

Activities

1 Study Figure 34.
  a) Make a copy of the second diagram.
  b) Add labels that explain water flows at 1, 2 and 3.
  c) Explain how the differences in the two diagrams would affect the flow of water in the river downstream.

Enquiry

You have been asked to advise Tewkesbury Council on flood prevention. What do you think should be done to prevent future floods in the town?

- Use what you have learned in this chapter, and the points of view in Figure 33 to complete a copy of the table.

Possible solution	Short-term benefits and problems	Long-term benefits and problems	Who might agree and disagree with this solution
Building flood defences like those in Shrewsbury			
Restoring bogs and moorland in mid Wales by blocking drains			
Tighter controls on building on floodplains and paving over gardens			
Allowing rivers to flow naturally and spill over onto the floodplain			

- Now you need to recommend your plan. What do you think should be done and why do you think your plan will work? Use the following table to plan your answer.

Key questions to ask yourself	My answers
Is my plan realistic and achievable?	
Which groups of people will benefit from my plan?	
How will the environment be affected?	
Why is this plan better than the alternatives?	

THEME 2 Rural–urban links

Chapter 1 Rural–urban continuum

Where do people live in the UK?

Urban places are busy built environments. People live close together at a high population density. In contrast, rural places have more open space. Rural population densities are lower or sparse. Within rural areas there is a **hierarchy** of settlements – hamlets, villages and small towns – but these settlements tend to be smaller than in urban regions.

Study Figure 2. Cities such as Liverpool or Chester provide services and employment opportunities for people living in smaller towns nearby, such as Wrexham and Llangollen, and the rural areas that surround them. In this way, the towns and rural communities of northeast Wales and Cheshire are linked to the **urban spheres of influence** of Liverpool and Chester.

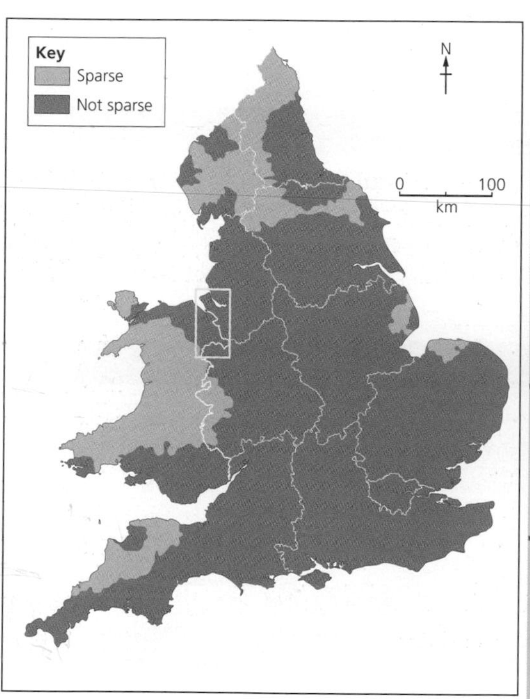

▲ **Figure 1** Sparse populations in England and Wales; the yellow box outlines the shape of Figure 2.

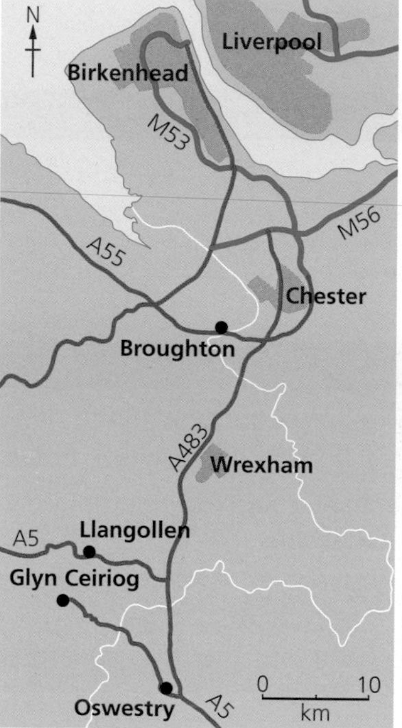

Chester – population 329,608. Shops in Chester attract customers from a huge area across Cheshire, North Shropshire and North Wales. Chester University has 15,000 students.

Broughton – population 6,000. A small rural town in Wales within easy commuting distance of Chester. Airbus has a factory here that makes aircraft wings and employs 6,000 people. People commute to the factory from Chester, Birkenhead and North Wales.

Wrexham – population 42,000. Wrexham is the largest town in North Wales and is the administrative centre of Wrexham County Borough. Wrexham has four secondary schools.

Llangollen – population 4,000. A small rural town popular with tourists visiting the Clwydian Range AONB.

Glyn Ceiriog – population 800. The Ceiriog Valley is a sheep farming community with a population density of 0.2 people per hectare. The village has one shop, which includes the post office.

▲ **Figure 2** Urban and rural places on the northern border between England and Wales.

Urban sphere of influence

The influence over the surrounding area that every urban area has through its close links with nearby towns and villages. These links may be physical (such as roads and railways), social (such as families or friendships) or economic (such as links between a bank or solicitor in the city and a small business in a nearby town). People living within the urban sphere of influence may commute to the city regularly for work. Others visit the city to use its services. For example, cities provide specialist educational services such as universities, and specialist health services, such as hospitals and clinics, which attract people from a wide surrounding area. Urban spheres of influence exist because cities have jobs and services that do not exist in smaller towns and villages. For example, if you live in a rural area you may have a GPs' surgery nearby but, if you need to go to Accident and Emergency, you probably do not have one near where you live, so you have to go to one in your nearest large town or city.

My wife and I live in a small hamlet close to Llangollen. We both used to work in Wrexham but are now retired from full-time work. We usually do our shopping in Oswestry or Wrexham. The nearest theatre is in Mold and our closest airports are in Liverpool and Manchester. The closest premier football clubs are Liverpool and Everton.

▲ **Figure 3** The rural town of Llangollen.

	Rural places (%)			Urban places (%)		England and Wales
Age	Glyn Ceiriog	Llangollen	Broughton North East	Garden Quarter, Chester	Upton, Chester	
75+	9.6	12.5	8.3	2.8	11.3	7.8
65–74	11.3	12.6	9.9	3.2	9.5	8.6
45–64	31.9	31.1	24.8	11.8	24.0	25.4
30–44	18.0	16.2	20.9	14.4	20.4	20.6
20–29	7.0	9.2	12.7	45.1	10.2	13.7
16–19	5.3	4.3	5.4	17.6	4.3	5.1
10–15	7.1	5.4	7.0	1.5	7.4	7.0
0–9	9.8	8.7	11.0	3.6	12.9	11.9

▲ **Figure 4** Population data for rural and urban places in the region of Llangollen and Chester.

Activities

1 Describe the distribution of sparse populations in:
   a) Wales
   b) England.
2 Use evidence from Figures 1 and 2 to identify the one place labelled on Figure 2 that has a sparse population.
3 Study Figures 2 and 3 to help you suggest the different spheres of influence of:
   a) Glyn Ceiriog village shop
   b) Chester University
   c) Airbus at Broughton
   d) Liverpool Football Club.
4 Use evidence from Figures 2 and 3 to suggest why people might want to retire to Llangollen.
5 Explain why people living in rural places such as Llangollen have to travel longer distances and more frequently than people who live in urban places such as Liverpool.

Enquiry

How does the population structure vary between urban and rural places?

- Analyse the data in Figure 4 and suggest a possible hypothesis.
- Represent the data in Figure 4 using a suitable style of graph.
- What conclusions do you reach?
- Investigate Garden Quarter, Chester. Why does it have an unusual population structure?

The urban–rural continuum

Rural and urban may seem to be opposites but, in reality, there is an **urban–rural continuum** – a sliding scale between urban places and the most remote rural regions. Some rural places are easily accessible to large cities. The rural communities of Kent and South Oxfordshire in England, for example, are close to major transport routes into London. This makes **commuting** and access to shops and entertainment easy. The population of accessible rural areas such as these is growing – a process known as **counter-urbanisation**. At the other end of the continuum are some remote and very sparsely populated rural areas. It is not possible to commute easily from these remote places. The population here is ageing as younger people leave the countryside to seek work in the city.

Different types of rural places

Rural places have common features, such as open spaces and village communities, but they also have other features that can be used to classify them into different types. Figure 6 shows examples of four very different types of rural place:

- *Deep green*: These are remote and isolated rural places with poor road networks. They have lots of open space and very sparse populations.
- *Rapid change*: These rural areas are less densely populated and include some larger towns. Many people living here are commuters who work in urban areas rather than the countryside.
- *Leisure and amenity*: Some of the UK's most beautiful scenery and National Parks are contained in these rural places. They are in remote parts of the UK.
- *Coastal retirement*: The population of these seaside towns includes a significant proportion of people who moved here when they retired.

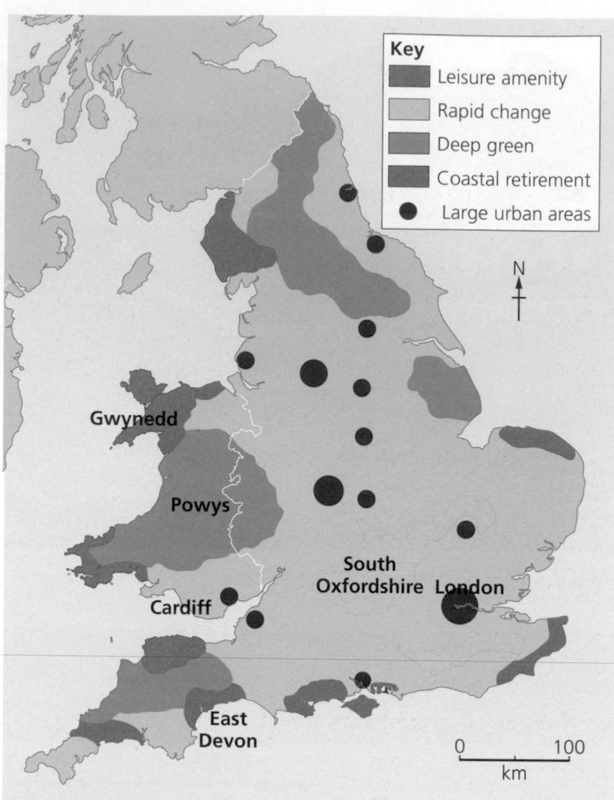

▲ **Figure 6** The location and distribution of different types of rural place in England and Wales.

> **Counter-urbanisation**
>
> A process that occurs when people move out of cities and into villages and towns in rural areas. They may feel pushed out by congestion, air pollution or crime. They may be attracted by a better quality of life in the countryside. Smaller towns in the yellow areas of Figure 6 have grown rapidly since the 1960s as a result of counter-urbanisation.

▲ **Figure 5** The 'deep green' sparsely populated region of Powys, Mid Wales.

What are the strengths and challenges of rural life?

Each of the types of rural place shown in Figure 6 has its own strengths and challenges. For example, the very sparse population of deep green places makes it difficult for small rural businesses and services to stay open. These rural places face the challenge of the closure of village shops, schools and post offices. Public transport is poor, so those who cannot drive are worst affected when such services close.

Area	Population density (persons per hectare)	Retired population (%)
East Devon	1.63	22.36
South Oxfordshire	1.98	13.9
Powys	0.26	17.9
Gwynedd	0.48	16.4
Riverside, Cardiff	53.3	7.5

Figure 7 Population features of different rural areas of England and Wales; the figures for Riverside, an inner-urban area of Cardiff, are provided for comparison.

Area	Financial occupations (%)	Professional occupations (%)	Farming, forestry and fishing (%)	Accommodation and food (%)
East Devon	2.35	6.19	3.06	7.25
South Oxfordshire	2.80	11.29	1.03	4.42
Powys	1.43	4.44	8.67	6.42
Gwynedd	1.13	3.66	3.46	9.82

Figure 8 Selected occupations for different rural areas of England and Wales (per cent).

Sparse local services

Ageing populations

Poor mobile coverage

Poor road links

Young adult population increasing

Commuters can use rail services to London

Beautiful scenery

Strong, safe communities

▲ **Figure 9** Strengths and challenges of rural life.

Weblinks

www.ukcensusdata.com – this site provides easy access to data from the Office for National Statistics. You can search by postcode or place name.

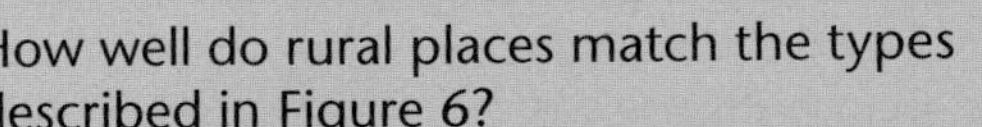

Enquiry

How well do rural places match the types described in Figure 6?

Use the weblink, or any other site that uses census data, to research the population and work characteristics of each of the following rural places. Which rural type does each best match?

- East Lindsey in Lincolnshire
- South Lakeland in Cumbria
- North Norfolk
- Pembrokeshire

Activities

1 Study Figure 6.
  a) Describe the distribution of each type of rural area.
  b) Suggest reasons for the distribution of deep green and rapid change areas.

2 Study Figures 7 and 8. They show data for specific places within each of the four types of rural areas shown on Figure 6.
  a) Represent the data to summarise the main similarities and differences.
  b) Use the data to draw conclusions about the people and occupations that typify each of these rural places.

3 a) Sort the statements in Figure 9 into strengths and challenges.
  b) For four of these statements:
    i) decide which of the rural types it best describes
    ii) describe the consequences for rural residents.

Consequences of counter-urbanisation

One consequence of counter-urbanisation has been an increase in the number of people commuting daily to and from work. Many commuters live in the region surrounding the city and travel into it each day to work. A smaller number of commuters travel in the opposite direction, leaving their home in the city to work in a nearby town. Almost 11.3 million people in the UK travel between their home in one local authority and their workplace in a different local authority each day of the working week.

Large cities, such as London, Liverpool or Cardiff, have many more jobs available than their surrounding region, and a particularly large sphere of influence. It's often cheaper to live in a smaller town and commute into the city – even if the commute is quite long. Other people only commute occasionally, preferring to **telework** from their rural home. Teleworkers use mobile technology and the internet to use their home as an office.

Commuting factfile

- Average commuting times in the UK are going up: in 2003 it was 45 minutes; by 2017 it had risen to 64 minutes.
- 616,000 people commute into the city of London each weekday.
- 1.8 million people in the UK (one in ten UK commuters) travel for over 3 hours a day. These are so-called 'extreme commuters'.

Area of UK	Average time spent commuting (to and from work) in minutes
Central London	110
Outer London	74
South East	62
Wales	52

Figure 10 Average UK commute times.

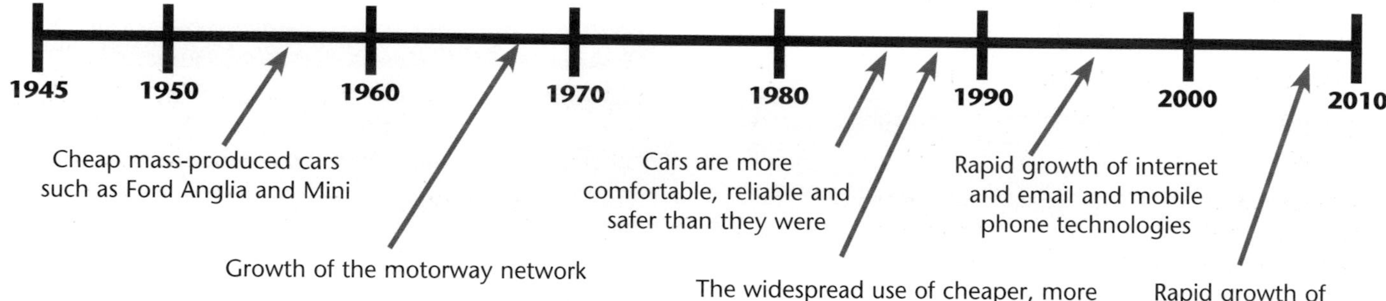

▲ **Figure 11** Technological changes have increased the connections between rural and urban places.

Factors that affect commuting
The difference in house prices between a city and its surrounding region
Fast rail links
The rising cost of fuel and the rising cost of rail fares
The availability of good 3G and 4G signals and free wi-fi on train services
The affordability of fuel-efficient cars
Flexible working hours that allow workers to begin the working day any time between 7 a.m. and 10 a.m.

▲ **Figure 12** Factors that affect commuting.

Activities

1 Study Figure 11. Use it to explain why:
   a) the number of commuters has increased since the 1960s
   b) more people are able to work from home now than before 1980.
2 For each factor in Figure 12, suggest how it may affect patterns of commuting.
3 a) Suggest why London has the longest commute times in the UK.
   b) Suggest two different reasons why the length of the average commute in the UK is increasing.

Patterns of commuting in southeast Wales

Cardiff has a very wide sphere of influence – people from all over Wales occasionally travel to Cardiff for its shops, to attend concerts and sporting events, and to visit its museums. This sphere of influence is felt most strongly across southeast Wales. People in this region regularly commute into and out of Cardiff, as shown in Figure 13.

Local council	Total workplace population	Daily commuters to Cardiff	Daily commuters from Cardiff
Vale of Glamorgan	37,300	20,600	4,700
Rhondda Cynon Taf	71,500	20,400	3,800
Caerphilly	56,500	15,100	3,200
Newport	70,200	6,900	6,700
Bridgend	62,900	5,400	ND
Merthyr Tydfil	21,900	2,300	ND
Torfaen	35,400	1,900	1,900
Monmouthshire	43,200	1,700	ND
Swansea	111,400	1,600	ND
Neath Port Talbot	45,200	1,500	ND
Blaenau Gwent	19,200	1,000	ND
Cardiff	217,000	0	0
Other		4,500	13,700

▲ **Figure 13** Commuting patterns in and out of Cardiff; ND means there is no data for these commuter flows (they are likely to be small).

▲ **Figure 14** Map of Cardiff and the surrounding local councils.

Cardiff city region factfile

- Cardiff has an estimated population of 357,000 (in 2019).
- Half the population of Wales, 1.49 million people, live within 32 km of Cardiff city centre.
- The city has a workforce of 189,000.
- Almost 78,000 people commute into Cardiff, and 33,900 people commute out of Cardiff, each weekday.

Enquiry

How are commuter patterns affected by transport infrastructure?

Use the internet to find the main road and rail links into Cardiff. How do these transport routes help explain any patterns shown in Figure 13?

Activities

4 Use the factfile to suggest one benefit and one problem created by Cardiff's strong sphere of influence within southeast Wales.

5 a) Make a simple copy of Figure 14.

b) Use the data in Figure 13 to make a choropleth map of daily commuters to Cardiff.

c) Describe the patterns shown by your map.

d) Describe another technique that could be used to represent this data.

Issues created by commuting

Almost 63,000 people commute into Cardiff each weekday by car. Queues on the A470, from the north, and on the A48, from the east, can cause traffic delays of up to an hour to get into Cardiff. Most other commuters arrive by train. Once in the city, they rely on bus, cycle or pedestrian routes to get them safely to their place of work. But cycling, and even walking, can be dangerous when streets are packed with commuter traffic.

▲ **Figure 15** The Taff Trail; a green corridor of parks provides a safe environment for commuters to walk or cycle through Cardiff from north to south.

Cardiff has three park-and-ride schemes
The Taff Trail
Cycle routes
Bus lanes
Car share schemes
Integrated systems where bus routes connect with train stations and cycle routes
Flexi hours so that businesses start and finish at different times
Locating major employers, such as the BBC, close to train stations
Water taxis take passengers from Cardiff Bay, up the River Taff to the city centre

▲ **Figure 16** Nine solutions to Cardiff's traffic problems.

Cardiff Central Square to be redeveloped as integrated transport hub

An area of central Cardiff is being regenerated at a cost of £400 million. The site outside the Central Train Station will be transformed as buildings are demolished and new offices are built. As many as 10,000 jobs could be created by the project. The plan includes a new building for the BBC, which opened in 2019. The regeneration also includes other new office buildings, a 150-bedroom hotel, shops, 200 residential homes and a new central bus station. All will be within minutes' walking distance of the city centre.

Construction of a new bus station began in 2019. It will link to the existing train station on wide, flat pavements with no steps to allow easy pedestrian movement. The design of the new station will include: live information screens; facilities for travellers to include assistance points, toilets and restaurants; safe storage for bicycles; pleasant waiting areas; and protection from the rain by overhangs from the new office buildings.

▲ **Figure 17** The regeneration of Central Square will include an integrated transport hub. The image shows the new BBC headquarters.

Activities

1 a) Discuss the nine solutions to Cardiff's traffic problems in Figure 16.
b) Make a diamond nine diagram (like the one on page 83) and place the solutions in the diagram, placing those that you think are essential at the top of the diagram.
c) Explain why you have chosen your top three solutions.

Enquiry

'The regeneration of Central Square will solve all of Cardiff's transport problems.'

To what extent do you agree with this statement? Use evidence from the factfile on page 59 and Figure 17 to support your view.

The issue of second homes

Counter-urbanisation is one reason why the number of second homes in the UK has increased in recent years. Around 1.6 million people in England and Wales own a second home in the countryside that they use at the weekends or for holidays. Many of these are in the most picturesque parts of the countryside, such as the Lake District in England or Pen Llŷn (the Llŷn Peninsula) in Wales. The lack of affordable housing in Gwynedd, Wales, is linked by some to the sale of rural houses as second homes. Holiday homes are empty for long periods of time. As regular demand for services falls, village pubs close and are converted to homes, bus services are axed, and local shops and banks may also close. The rise of internet banking has also led to the closure of high street banks in rural places.

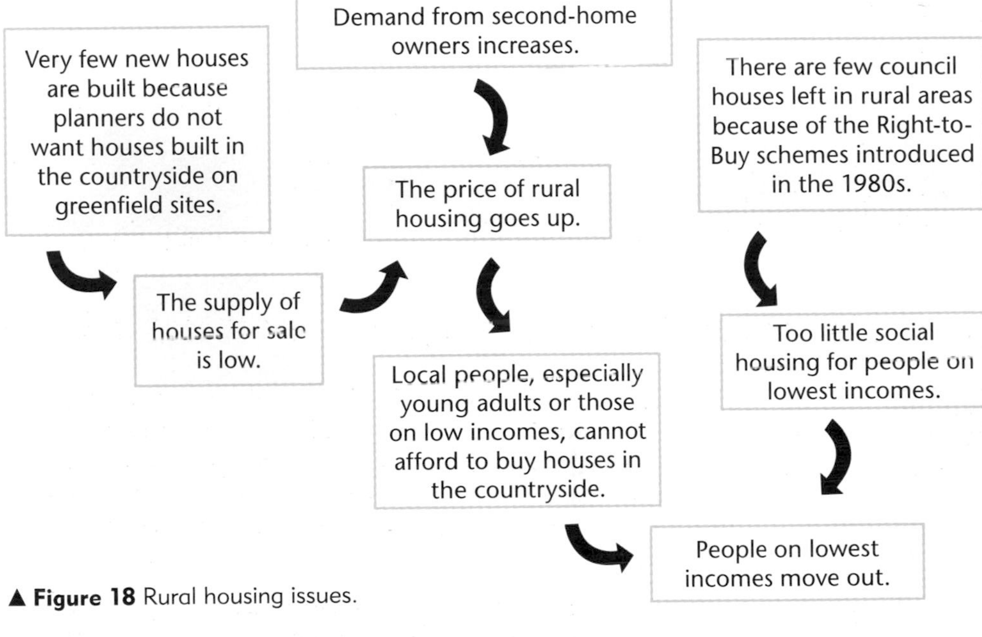

▲ **Figure 18** Rural housing issues.

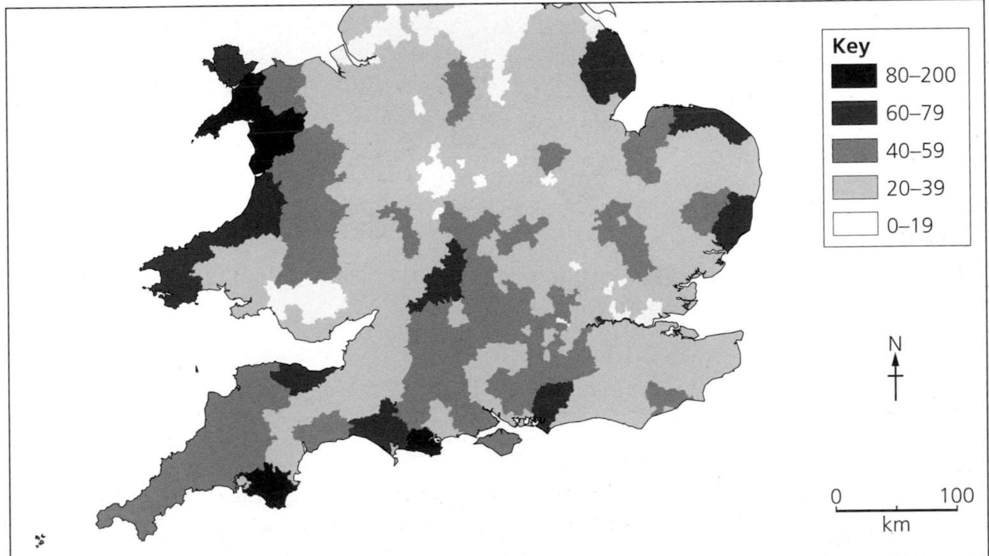

▲ **Figure 19** The number of second homes (per 1,000 usual residents) in Wales and southern/central England.

Activities

2 Study Figure 18.
   a) Identify three main reasons for the lack of affordable housing in rural places.
   b) Explain why the sale of houses as holiday homes may lead to the closure of village shops and other services.

3 Compare Figure 19 to the map of National Parks, Figure 10 on page 4, and Figure 6 on page 56. What conclusions can you reach?

Issues facing remote rural places

Remote rural places can be affected by a number of problems:

- low incomes and part-time work
- lack of services, especially health services, schools and adult education
- infrequent public transport and high transport costs.

Poor roads, isolation and sparse populations are all factors that can lead to a decline in rural jobs and rural services. These problems are examples of rural **deprivation**. The lack of jobs, leisure facilities and shops in the countryside are all push factors for younger adults. In addition, the high cost of buying a house may prevent young families on lower incomes from staying in a rural area. This leads to a falling birth rate. As a result, the population of some rural areas is both ageing and decreasing – a process known as rural **depopulation**.

Deprivation

Communities are described as being deprived if they lack features that are usually regarded as necessary for a reasonable standard of living. Low wages are a cause of deprivation in some urban places. Lack of public transport, health care and education services are reasons for deprivation in some rural places.

Too few rural jobs and opportunities.

Rural to urban migration is greater than migration into the rural area.

Declining rural populations.

Reduced demand for schools, shops and other services.

Collapse of rural services.

Further rural depopulation.

▲ **Figure 20** How rural depopulation creates unsustainable communities.

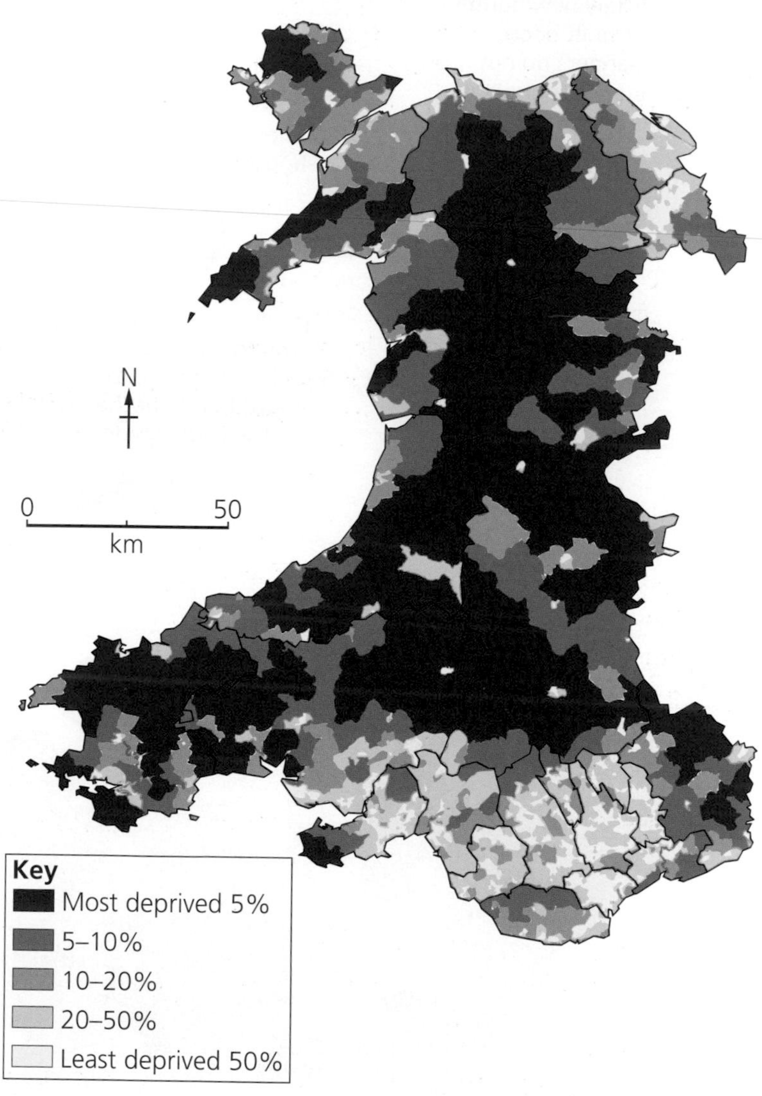

▲ **Figure 21** Rural deprivation: local authorities in Wales have been ranked by their isolation from important services (including education, health and post offices); those coloured red have the worst access to these services.

Isolation from health services

Cities provide an important range of health and education services to the surrounding rural region – this is an important function of their sphere of influence (see page 54). Hospitals, located in the city, provide specialist care that is available to people across very large rural regions. We saw on pages 56–7 that many rural places have an **ageing population**. This creates issues:

- Elderly people are more likely to have long-term (or chronic) health issues. These include diabetes, reduced mobility due to arthritis, and dementia.
- The sparse population usually means that specialist treatment or care is not locally available – patients have to travel long distances. For example, there are currently only three health centres serving the whole of Pen Llŷn (the Llŷn Peninsula) in northwest Wales. Two of these are in the larger settlements of Pwllheli and Nefyn. The main hospital, however, is some 40 km away in Bangor.

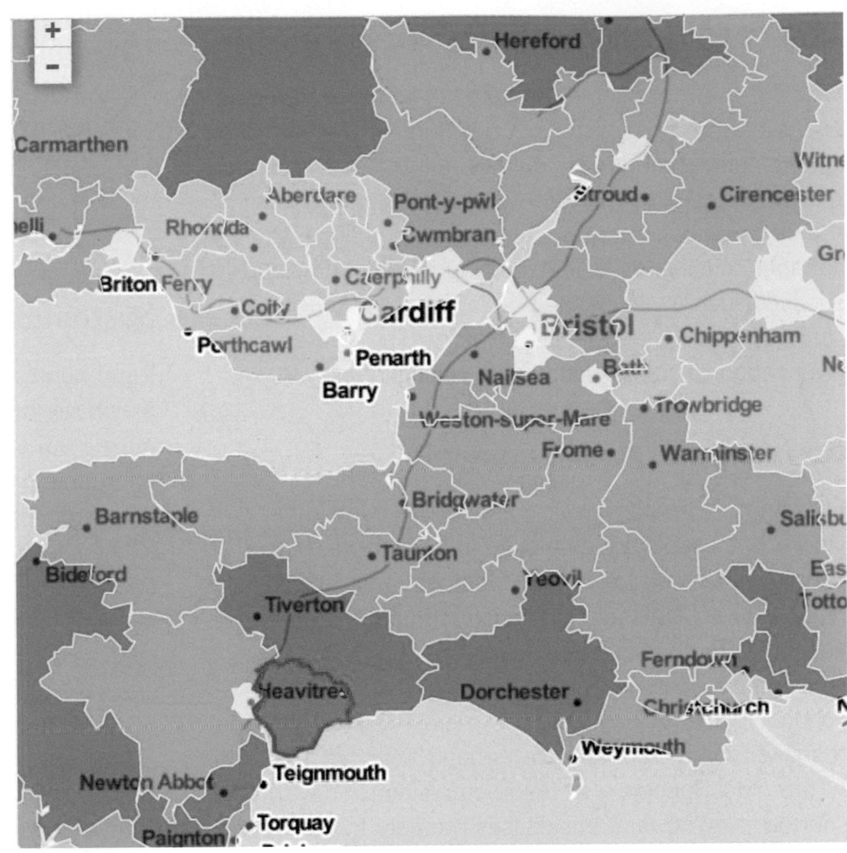

▲ **Figure 22** The percentage of the population aged over 65 in southeast Wales and southwest England.

Key
Per cent
- <12.5
- 12.5–16.6
- 16.7–20.2
- 20.3–24.3
- 24.4+
- No data

	Powys		Cardiff	
Population	132,976	–	346,090	–
Population aged over 65	a	22.7%	c	13.16%
Population aged over 75	b	10.5%	d	6.6%
Population density (per hectare)	0.2567	–	24.6534	–
Area (hectares)	518,037	–	14,038	–

▲ **Figure 23** Comparing the population over 65 in rural and urban areas of Wales.

Activities

1 Give two main reasons for rural depopulation. Use Figure 20 to help you.
2 Use Figure 21 to describe the distribution of the:
   a) most deprived places
   b) least deprived places.
3 Compare Figure 21 to Figure 1 on page 54 and Figure 6 on page 56. Suggest reasons for the patterns shown on Figure 21.
4 a) Make a copy of Figure 23.
   b) Calculate the missing population figures and add them to cells a, b, c and d.
5 a) Describe the patterns shown on Figure 22.
   b) Use Figure 6 on page 56 to help explain these patterns.

Enquiry

'The issue created by the ageing population is greater in rural areas than in urban ones.'

To what extent do you agree with this statement? Use evidence from Figures 22 and 23 to help your answer.

Are rural communities sustainable?

We have seen that some rural areas face issues created by the closure of services, second-home ownership or an ageing population. Local councils need to use strategies that will boost local businesses and create new jobs in the countryside. They also need to provide access to suitable housing, as well as helping to maintain rural services such as access to post offices, banks, bus services, schools and health services. These strategies are needed to create **sustainable communities**.

Boosting the rural economy

Full-time, well-paid jobs are rare in the countryside. Only a few people are employed in agriculture. Many people who live in the countryside commute to work in nearby towns. One solution is **diversification** of the rural economy. Tourism helps to sustain existing jobs in the countryside, such as work in catering, hotels and shops. Technology has allowed home owners in the countryside to make money from tourism through websites like Airbnb. Most people visit the countryside for day trips or short breaks, so rural businesses promote food festivals, music festivals or walking festivals to attract visitors to return on more occasions throughout the year.

Diversification

The creation of new businesses and more varied jobs in the local economy.

▲ **Figure 24** Crowds enjoying an annual weekend festival, held in September, in Bishop's Castle, Shropshire.

Sustainable community

A community designed to give everyone access to the services they need. In a sustainable community, the local economy is doing well and residents feel safe and secure. For more details, see Egan's Wheel, on pages 72–73.

Sustainable housing

Rural homes are expensive for local young adults who may have low incomes or part-time work. One way to make rural areas more sustainable is to build more affordable housing. Local councils can insist that a proportion of all new housing that is built is available for people on lower incomes. Shared ownership is one solution, where the property is part owned by the resident, and part owned by a social housing provider such as a Housing Association. The population of rural areas is, on average, older than urban areas (see pages 66–67). As the rural population ages, and some elderly people lose mobility within their home, traditional homes can become unsuitable because of their stairs and narrow doorways. To be sustainable, rural communities need more suitable homes for elderly residents, including houses that are accessible to people in wheelchairs and homes where residents have emergency care from a warden.

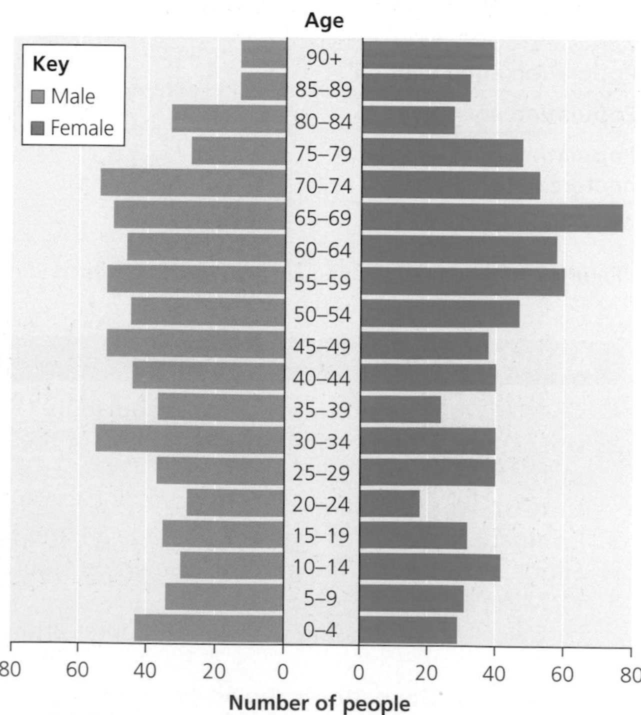

▲ **Figure 25** A population pyramid for Bishop's Castle, Shropshire.

Maintaining rural services

Rising costs, increasing use of the internet and population change have all caused the decline of services in many rural communities. Many banks, post offices and village shops have closed and been converted to homes. These closures threaten the sustainability of rural communities. For example, since 2010, a third of bank branches have closed and rural areas have been hit hard. People can, of course, use one of the 70,000 cash machines (ATMs) across the UK to withdraw money, but 300 of these are being closed every month. One solution in rural areas is for services to move under one roof to save costs. For example, post offices move inside village shops or inside village pubs. Where banks have closed, customers are able to use the Post Office to withdraw cash. However, in order to save money, Barclays decided to end this service (see Figure 27). Barclays received so many complaints about this decision that they say they have reversed it for now.

▲ **Figure 26** This phone box, in a rural location in Cornwall, has been repurposed as a library.

Ban on withdrawing cash at post offices

With bank branches and cash machines disappearing at an alarming rate, customers have been repeatedly reassured they can visit one of 11,500 post offices to deposit and withdraw cash, pay in cheques and check their balances.

However, Barclays, which made £3.5 billion in 2018 has moved to stop customers withdrawing cash at post offices. Barclays estimate they will save £7 million with the move, but now face a mass exodus of customers who depend on post office banking.

Barclays has already closed 481 branches since 2015 — leaving many communities dependent on post offices. It is estimated that this move will hit the most isolated and vulnerable communities.

▲ **Figure 27** A news article in the *Daily Mail* (October 2019).

Activities

1 How might rural festivals, like the one shown in Figure 24, contribute to a sustainable rural community?
2 Explain why rural areas need to diversify their economies.
3 a) Use Figure 25 to describe the population structure of Bishop's Castle.
  b) Suggest how housing in rural areas needs to change to make them sustainable for all people who live in the countryside. Support your answer with evidence from Figure 25.
4 Explain why it is important to maintain services such as libraries and post offices in rural areas.
5 Study Figure 27. What are the advantages and disadvantages for banks like Barclays of closing rural branches and ending the agreement with the Post Office to provide banking services?

Enquiry

'Investments in leisure and tourism are the best way to solve the problems of remote rural areas.'

How far do you agree with this statement? Make use of examples from this chapter and Theme 1, Chapter 1 to help you answer.

THEME 2

Rural–urban links

Chapter 2 Population and urban change

How and why is the UK population changing?

Since 1964 the population of the UK has grown by over 10 million. The population is now around 65 million and is slowly increasing. Changes to the size of the UK's population are due to a number of factors:

- An increase in the age that people live to, known as **average life expectancy**. Improving health care has led to fewer deaths and longer average lifespan. Better health care also means lower infant mortality rates. In 1980 there was an average of eleven deaths per thousand live births. This had fallen to four deaths in 2014.
- Migration both in and out of the UK, especially from Commonwealth countries and members of the EU. European migration laws allow movement of workers between members of the EU. For example, 50,000 EU migrants came from Bulgaria and Romania in the year ending June 2015. Of these, 84 per cent came to the UK for work-related reasons.
- The changing birth rate. The **birth rate** in the UK has been relatively low since the 1980s. Women choose to have small families so that they can concentrate on a career. However, the birth rate has increased slightly since 2004 as young adults have moved into the UK and had families of their own.

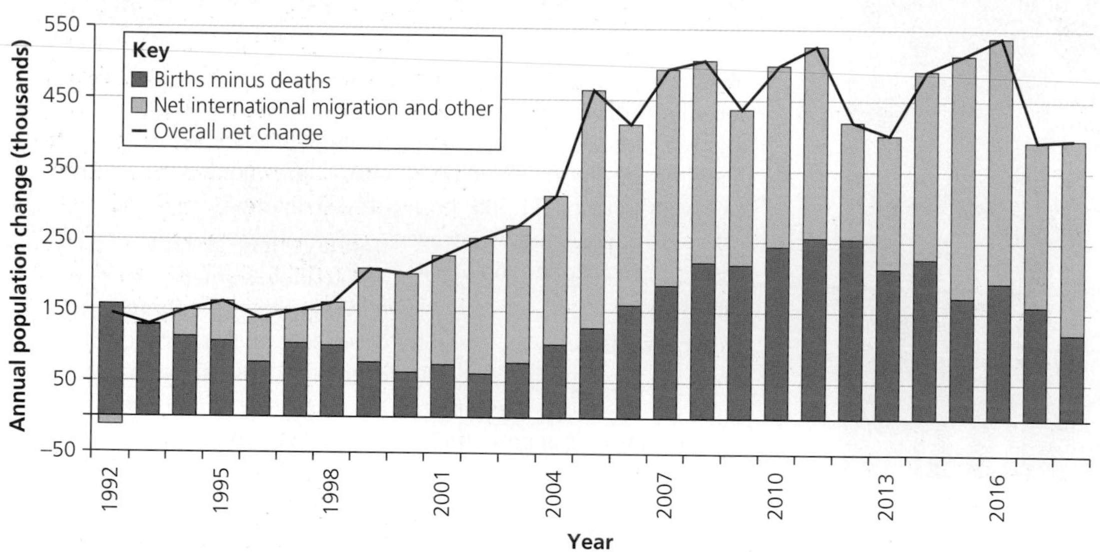

▲ **Figure 1** Annual UK population change; the line graph shows the number (in thousands) of extra people in the UK population each year; the bars show whether they are due to natural change (births minus deaths) or migration.

Activities

1 a) Identify three main reasons why the UK population is growing.
  b) Discuss these reasons. Can you classify them as economic, political or social?

2 Study Figure 1. In which period of years was growth of the population:
  a) slowest
  b) fastest
  c) due mainly to natural change
  d) due mainly to migration?

3 a) Choose a suitable technique to represent the data in Figure 2.
  b) What conclusions can you reach?

4 Study Figure 3.
  a) Copy the statements in the green boxes.
  b) Match the statements in the blue boxes and connect them with a suitable phrase.

5 Discuss each statement in Figure 4. Explain why each of these ideas might help to respond to the issues created by an ageing population.

How should the UK prepare for an ageing population?

People in the UK are living longer. In 1995 there were less than 9 million people aged over 65 in the UK. Forecasts vary but it seems likely that by 2030 there will be about 13 million. Not only are more people living into retirement, many more people are living into old age. In 1951 there were only about 300 people over 100 years old. By 2030 there could be as many as 36,000 centenarians. The **ageing population** will have consequences for UK society, shown in Figure 3. Government, businesses and individuals all need to prepare for these consequences, and possible responses are shown in Figure 4.

Year	Aged 0–15 (%)	Aged 16–64 (%)	Aged 65 and over (%)
1984	21	64	15
1994	21	63	16
2004	20	65	16
2014	19	64	18
2024	19	61	20
2034	18	58	24
2044	17	58	25

▲ **Figure 2** The UK's ageing population – actual and predicted figures.

Many older people have valuable work skills and experience	Valuable contribution to business, education and training
There are a growing number of wealthy older people	Commitment to support local communities through voluntary work
The proportion of people of working age is shrinking while the proportion of people who are retired is growing	Feelings of isolation and lack of value
The number of older people who live alone is increasing	Create and maintain jobs when they spend money on holidays or leisure activities
The number of older people who have complex or long-term health issues is growing	Require expensive health care or support from family and carers
People who are retired have flexibility about when and how often they work	The government receives less money but pays more in state pensions

▲ **Figure 3** Some consequences of the UK's ageing population.

Ageing population

Life expectancy in the UK is increasing, which means more people are living longer. At the same time, the fertility rate (which is the average number of children that each woman will have in her lifetime) remains level at about 1.7. This means that the population structure of the UK is changing, as the proportion of people over the age of 65 is gradually increasing. This is what we call an **ageing population**. It is most noticeable in rural and coastal areas of the UK (see Figure 5).

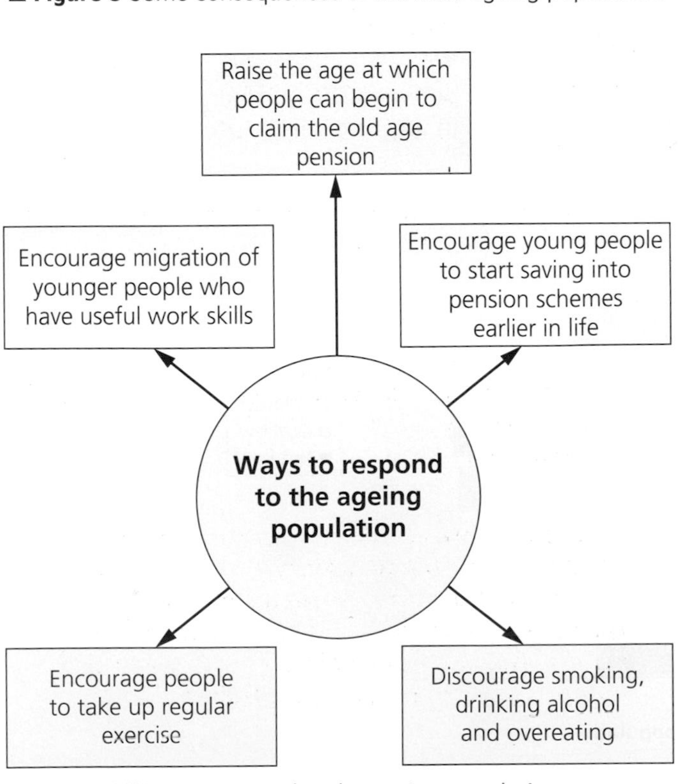

▲ **Figure 4** Ways to respond to the ageing population.

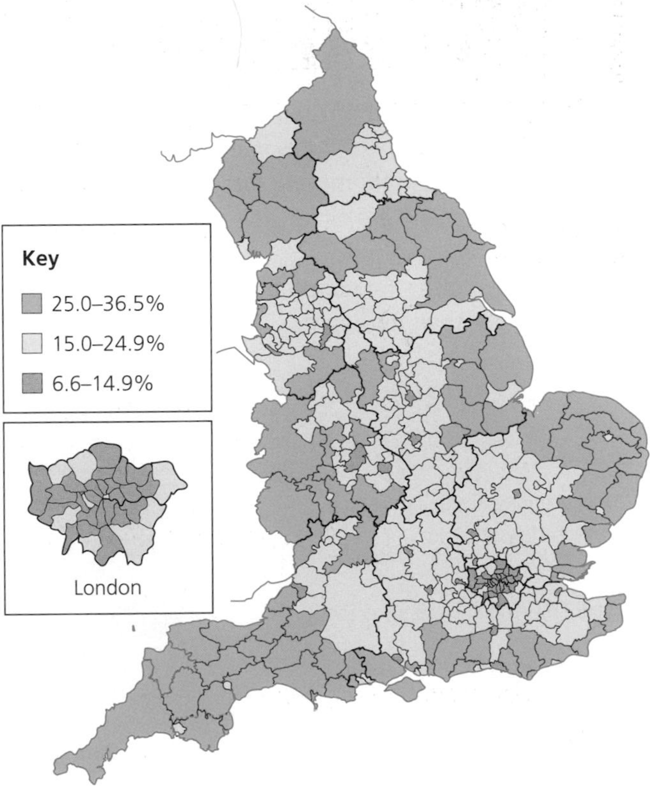

▲ **Figure 5** Percentage of people aged over 65 in 2024.

Migration into and within the UK

Migration is the movement of people from one place to another in order to make a new home or start a new job. There are two main types of migration that affect the population of the UK:

- Migration within the UK, for example, the movement of young adults from their parents' homes to attend university or get a job.
- Migration into and out of the UK. For example, people moving into the UK from India to take a highly skilled job in the health service or people moving out of the UK to retire in a warm climate such as Spain.

Migration within the UK

A survey by Open Access Government in 2019 suggests that, on average, people move home every 23 years in the UK. The most common reasons people have for moving within the UK are:

- to complete their education
- to start a new job
- to move somewhere new when they retire from work.

Studies of migration within the UK in 2018 show that:

- Most rural places in the UK have more people moving in than moving out. This is the process of counter-urbanisation that was discussed on pages 56–61.
- Most towns and cities have roughly the same number of people moving in and out. People tend to move relatively short distances. For example, migration in and out of Bristol is mainly from and to Gloucestershire, Bath and Somerset.
- Most of the migrants moving into towns and cities are young adults who are either attending university or seeking work.
- The largest cities (or conurbations) have more people moving out than moving in.

Migration into and out of the UK

People move into and out of the UK from abroad. People moving into the UK are immigrants and those leaving the UK are emigrants. In March 2019, it is estimated that the number of immigrants who had entered the UK in the previous 12 months was 612,000. Over the same period, it is estimated that 366,000 people emigrated. This leaves a **net migration** figure of +246,000.

The UK Government believes that immigration benefits the UK. Like many HICs, the UK has an ageing population, with a smaller proportion of people of working age. Ageing populations are expensive because of pension and health costs. Governments need increased taxation from working people to pay for these.

UK	1,021,257
India	21,207
Philippines	18,584
Ireland	13,320
Poland	9,272
Portugal	7,178
Nigeria	6,770
Italy	6,396

Spain	5,899
Romania	4,451
Zimbabwe	4,049
Pakistan	3,975
Greece	3,194
Ghana	2,570
Germany	2,427
Malaysia	2,298

▲ **Figure 7** Most common countries of origin of National Health Service (NHS) staff (2019).

Net migration

The difference between the number of immigrants and emigrants in any one year.

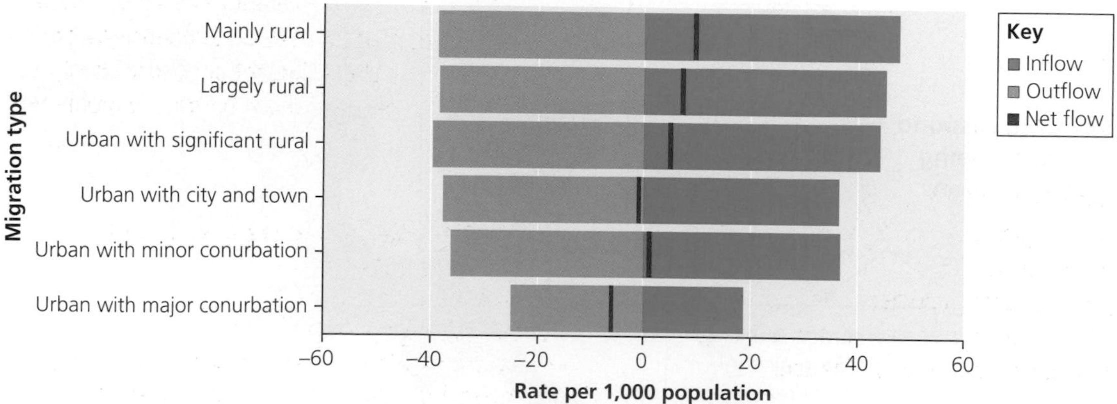

▲ **Figure 6** Migration within the UK (number of migrants per 1,000 population).

Activities

1 Study Figure 6. Use it to describe the number of people who move in and out of:
   a) areas that are mainly rural
   b) areas that are mainly urban with a major conurbation.
2 Use what you understand about migration within the UK to suggest what kind of new housing needs to be built in:
   a) rural and coastal locations
   b) towns and cities.
   Use evidence from Figure 6 to support your answer.
3 Study Figure 7. What are the advantages and disadvantages of using:
   a) a pie chart
   b) a flow line map
   to present these data?

The impacts on the UK economy

The globalised UK economy has led to higher immigration. Its manufacturing economy declined in the 1970s and 1980s, and was replaced by a growing service economy. Now, two groups of workers migrate to the UK for jobs here. Half of these are **highly skilled**.

- Many come to take up well-paid jobs in the UK's **knowledge economy**. These are jobs requiring expertise in, for example, finance and banking, law, and IT. Between them, people in these jobs produce 35 per cent of all UK exports.
- The majority of these jobs are in the city of London, where banks, law firms, shipping and biotechnology companies need employees recruited from the world's most highly qualified people.
- London now 'imports' experts from overseas, as there are not enough in the UK.

The other half are **unskilled workers**. There is plenty of work:

- Many dirty, difficult and dangerous jobs are rejected by UK workers.
- The lifestyle of many British families is sustained by unsocial hours jobs ranging from childcare to house cleaning and pizza delivery.
- Jobs are filled by immigrants in sectors such as farming, construction, hotels, restaurant and tourism, which suffer from seasonal shortages of labour.

▲ **Figure 8** The City of London, where the knowledge economy is most concentrated.

The social and cultural impacts of immigration

With 37 per cent of its population born overseas, London has the world's second largest urban immigrant population, after New York. Immigration is much debated by politicians and newspapers.

- Some journalists and politicians claim there are too many migrants: that migrants take jobs from local people and strain services such as schools and housing. In fact, the impact on services is more than made up by taxes paid by migrants.
- Others emphasise benefits of immigration. Many believe that the UK is improved by migrants' skills and cultural contributions. They claim that diverse friendships, varied restaurants and the cultural impact on British sport, music and media more than make up for any problems.

Activities

4 Draw a spider diagram to show economic benefits to the UK of:
   a) skilled migrants
   b) unskilled migrants.
5 Draw and complete a table to show possible benefits and problems for the UK economy if the UK reduced immigration.

Does the UK need to build more new homes?

In 2018 there were 165,000 new homes built in the UK. That number has risen from 135,000 in 2013, but it is not enough. As the population of the UK increases we need more homes. If too few are built, demand is greater than supply and the price of buying a new home rises faster than people's wages. The government wants to see an extra 240,000 homes being built every year. But where should these new homes be built? The greatest demand is in the South East of England. The economy is strongest in the South East, so this is an area where many younger people are moving to from other parts of the UK and from other EU countries.

Activities

1 Study Figure 9.
   a) Draw proportional bars on graph paper to represent the percentage change in population in each region.
   b) Cut out the bars and stick them onto an outline map of England and Wales.
   c) Explain why this representation of Figure 9 is more useful than a simple bar graph.
2 On average, 2.4 people live in each house in the UK. Use Figure 9 to calculate how many new houses are needed in your region by 2026.
3 Describe the location of the Thames Gateway development.

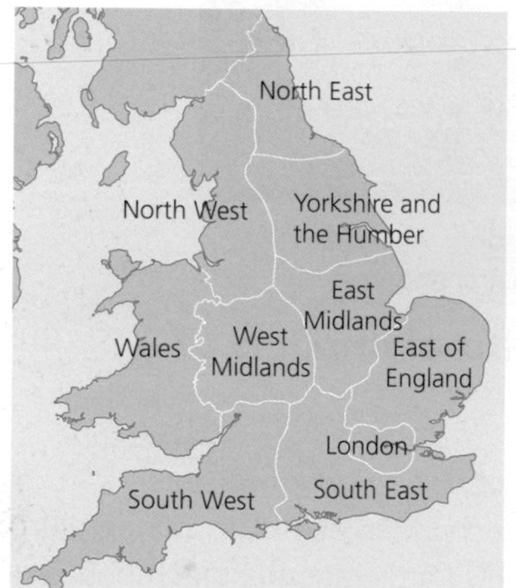

Region	Population change over 10 years	Percentage change
London	774,000	8.8
East	444,000	7.3
South West	364,000	6.6
South East	574,000	6.4
East Midlands	283,000	6.0
West Midlands	314,000	5.4
Yorkshire and the Humber	190,000	3.5
North West	243,000	3.4
North East	50,000	1.9
England	3,238,000	5.9
Wales	97,000	3.2

▲ **Figure 9** Projected population increase in the regions of England and Wales (2016–26).

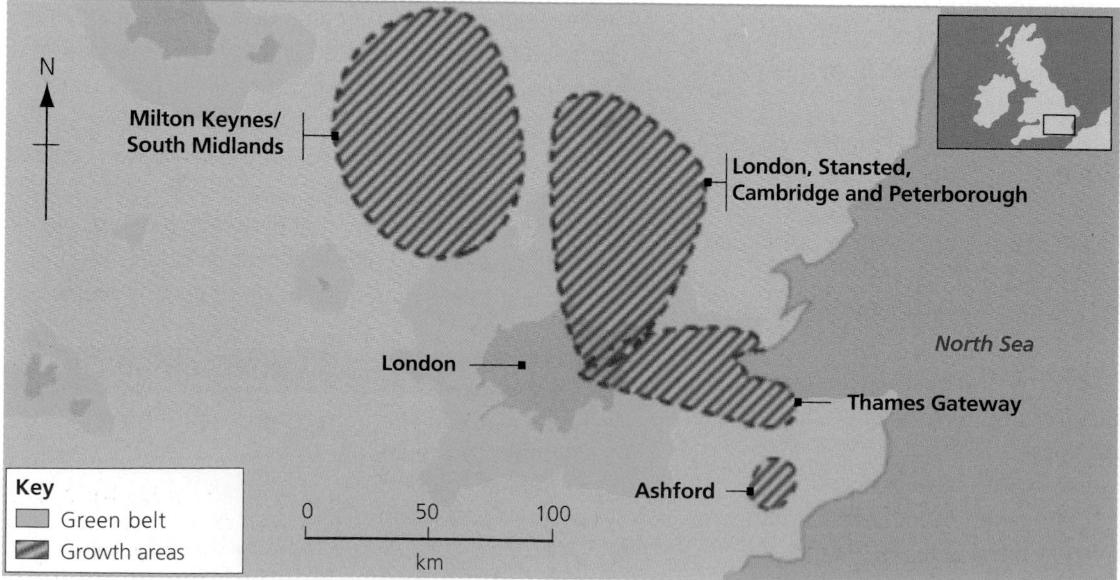

▲ **Figure 10** The key areas for new homes in England.

Are new garden cities the answer?

In 2014 the UK Government announced that three new **garden cities** would be built in England. Each would have about 15,000 new homes. The first garden cities were built in the 1930s at a time when there was a lot of new housebuilding. These towns had a lot of open space – hence the name. The new garden cities will feature high quality homes, lots of green space and access to local jobs and services. They will be built at:

- Bicester, near Oxford
- Northstowe, in Cambridgeshire
- Ebbsfleet, in the Thames Gateway (Kent).

Ebbsfleet, in Kent, is part of the Thames Gateway region. This region is well connected to London – it is only 20 minutes by rail from Ebbsfleet. The UK Government is keen to see new housebuilding in the Thames Gateway – perhaps as many as 90,000 new homes by 2030. It has committed £200 million to help fund the Ebbsfleet Development Corporation that will oversee the project and pay for new infrastructure in Ebbsfleet, such as roads.

The proposed garden city of Ebbsfleet will be designed as though it were a cluster of smaller traditional villages. Each 'village' will have its own primary school that is within walking distance of all homes in its catchment. Each 'village' will also have green open space, allotments, sports pitches and a community building for the use of groups of local residents. Building at Ebbsfleet has begun. By early 2019, a total of 1,669 new homes had been built in Ebbsfleet. When finished, this new garden city will house about 30,000 people in 15,000 new homes.

The Thames Gateway is low lying and at risk of coastal flooding (see pages 144–5). The cost of new sea defences could be £500 million.

Much of the development land in the Thames Gateway is brownfield. Part of Ebbsfleet will be built on a disused quarry.

Neighbouring small villages will become part of the new, larger town. They may lose their distinctive character.

A new bus service, called Fastrack, will run every 5–10 minutes during the day between Ebbsfleet and other towns in the Thames Gateway.

The government wants the Thames Gateway to be a low-carbon region. It hopes that people will work locally.

▲ **Figure 11** Aspects of development in the Thames Gateway.

▲ **Figure 12** Existing homes on the edge of an area of scrubland that has been identified as the possible site of the new Ebbsfleet garden city. Part of this site is a disused quarry.

Activity

4 Use information about the Thames Gateway and Ebbsfleet to complete a table like the one below. You will write more in some boxes than others.

	Arguments for building new homes in Ebbsfleet	Arguments against building new homes in Ebbsfleet
Economic		
Environmental		
Social		

Enquiry

Do the new garden cities sound as though they will be sustainable communities?

Use Figure 15 (page 73) to help you to justify your ideas.

How can we create sustainable urban and rural communities?

Figure 13 shows some features of modern eco-housing. Houses like this are environmentally sustainable. However, to make a **sustainable community**, planners need to consider other factors too. They need to design places for people to live in that have good transport networks, local services (like schools and shops), green spaces and jobs where people have less need to commute.

▼ **Figure 13** Eco-homes at the Beddington Zero Energy Development (BedZED) in Surrey. BedZED was the first and largest carbo-neutral eco-community in the UK.

A sustainable community has …

- some affordable housing for people on lower incomes
- been built on a brownfield site rather than a greenfield site
- jobs available locally
- public transport available to everyone
- schemes to reduce car ownership such as increased parking costs
- green technologies to reduce heating costs and carbon emissions
- local facilities for people of all ages, e.g. crèche, youth group, community centre
- some buildings designed for elderly or disabled people with wide doorways for wheelchair users and ground-floor bedrooms and bathrooms

▲ **Figure 14** Possible features of a sustainable community.

Egan's Wheel

Egan's Wheel describes eight features of a sustainable community. These include things you would expect, such as local jobs, well-designed homes (eco-friendly, perhaps) and good transport links. It suggests that residents feel safe and included in decision making by the local council.

Sustainable community

A community that manages its resources so that the needs of current residents are met while ensuring that enough resources are saved for future residents. The community's resources include its economy (jobs and businesses), its environment (buildings, clean air and water) and its people. A sustainable community has the features described in Egan's Wheel.

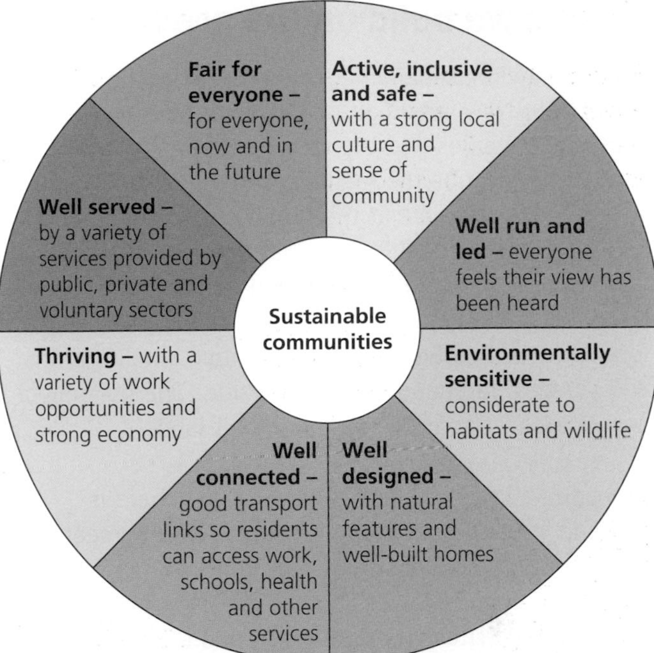

▲ **Figure 15** Egan's Wheel.

▲ **Figure 16** Evidence of sustainable urban communities in Cardiff.

Activities

1. Study the features of the eco-housing in Figure 13. Explain how each feature contributes to either environmental or economic sustainability.
2. Explain the main differences between an eco-home and a sustainable community.
3. Discuss Figure 14.
   a) For each feature in the diagram, suggest how it might be sustainable.
   b) Suggest at least two of the features that might be controversial. Which groups of people might come into conflict over these suggestions?
   c) Suggest at least two more features that you think are necessary in a new sustainable community.
4. Study the photographs in Figure 16.
   a) For each photograph, use Egan's Wheel to identify the element of sustainability that is being met.
   b) Choose one of these photographs. Write a 50-word caption that explains how the feature in the photo is contributing to a sustainable community.

Enquiry

How sustainable is Cardiff?

- Design at least five pairs of bi-polar statements (see pages 3–4) that you could use to survey the sustainability of a community. You should use Figure 15 for some ideas.
- Use your bi-polar survey to assess each of the images of Cardiff in Figure 16.

Should we build on the green belt?

A lot of new housebuilding took place after the Second World War. New suburban homes were built on the edges of UK cities and the term 'suburban sprawl' is used to describe the resulting rapid growth of the suburbs. UK planners at the time were so concerned about the loss of countryside that they prevented further loss by creating wide **green belts** around many UK cities. Green belts currently occupy 13 per cent of total land area in England. They contain smaller towns and villages, farmland and countryside. The building of new homes is restricted on green belt land.

The population of the UK is growing and there is demand for new homes. Should these new homes be built on **greenfield sites** – land that has not been used for building before? Or should homes be built on **brownfield sites** – land that has had a former use such as a factory, warehouse, dock or quarry and which is now unused or derelict? New housebuilding on greenfield sites may be opposed by local people who are concerned that new housing will spoil the rural character of their local community. Protests about local planning issues are commonly called **NIMBYism** – which stands for Not in My Back Yard.

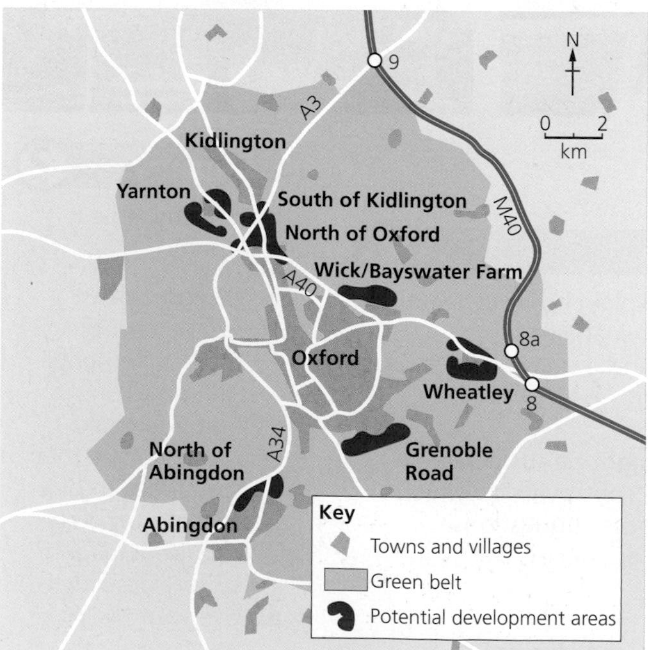

▲ **Figure 17** Development areas suggested by Oxford City Council that are within Oxford's green belt.

Region of England	Gross disposable household income per head (£)	Average house price (£)
North East	15,809	131,000
North West	16,861	162,000
Yorkshire and the Humber	16,119	161,000
East Midlands	16,932	193,000
West Midlands	16,885	195,000
East of England	20,081	189,000
London	27,825	472,000
South East	22,568	319,000
South West	18,984	253,000

▲ **Figure 18** The cost of housing compared with gross disposable household income per person (latest figures).

Oxford's employers, who include BMW Mini, schools, hospitals and the university, are finding it difficult to recruit workers because of the high cost of housing.

Over 50 per cent of the workforce commute into the city. In Oxford, new building on greenfield sites leap-frogs the green belt. This means that families who occupy the new housing in places like Bicester have a particularly long commute to work into Oxford. The pressure on our roads and public transport is not sustainable.

Oxford's universities have an international reputation for excellence in both research and teaching. We need cheaper housing so we can recruit the best research workers. These researchers are the reason for so many high-tech firms locating here.

▲ **Figure 19** Members of Oxford City Council explain why building on the green belt may be necessary.

New homes in Oxford?

Where demand for houses is greater than supply, the price of homes has risen rapidly. Study Figure 18. House prices in the North East are about 8 times gross disposable houehold income. In London, houses are about 17 times gross disposable houehold income. Houses are least affordable in London and in cities in the South East such as Oxford.

Oxford City Council believes that Oxford needs between 24,000 and 32,000 homes to be built by 2031. Controversially, it would like to build many of these within the green belt. South Oxford District Council has opposed the plan. It recognises that new homes are needed but doesn't want them built on green belt land.

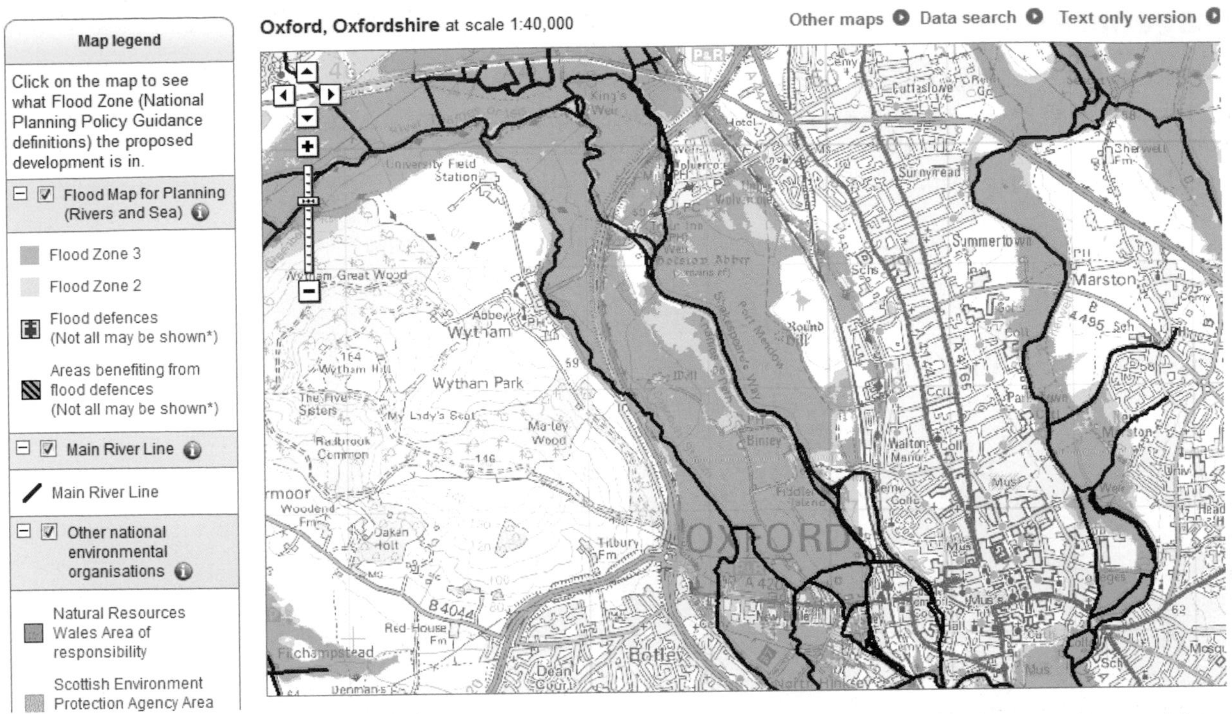

▲ **Figure 20** A screenshot from the Environment Agency website. The city centre is in the bottom right corner of the shot. Notice the open space to the north-west of the city which forms a green corridor along the River Cherwell.

The development will extend the village of Wheatley. The village will lose its distinctive character.

The site is alongside the A40 so access into Oxford is easy.

The land is open countryside and has never been built on before.

The land is sloping and poor quality farmland used for grazing.

Steep slopes on some of the site will make the building of homes expensive.

▲ **Figure 21** Features of the proposed development site at Wheatley.

Activities

1 Write a 200-word press release which explains why Oxford City council thinks it is necessary to build in the green belt. Use Figure 19.
2 Study Figure 18.
   a) Choose a suitable technique to represent this data.
   b) Use an atlas to analyse the pattern shown by this data.
3 a) Describe the location of the development site at Wheatley. Use Figure 17.
   b) Use Figure 21 to list two reasons why you think the development should go ahead.
   c) Give two reasons why you think the development should not go ahead.
   d) Suggest two different groups of people who might object to this development.
4 a) Use Figure 20 to describe the areas of Oxford that are vulnerable to flooding.
   b) Explain why it is important to maintain green corridors of land that have not been developed within cities like Oxford.

Enquiry

How important is the green belt?

Design a survey to investigate people's views of house building on green belt land. How could you design the sampling strategy (see page 8) to ensure that the views of different groups of people are included?

Urban regeneration

During the 1970s and 80s, it became unfashionable to live in the inner urban areas of many UK towns and cities. People didn't really want to live in areas of older inner-city housing, especially in the docks and waterfront areas of the UK's ports like Butetown in Cardiff or Salford Quays in Manchester.

However, these once run-down or derelict **brownfield sites** have been redeveloped with new homes and businesses. People are moving back into the inner-city – a process known as re-urbanisation.

Urban regeneration in Ipswich

▲ **Figure 22** A view of Ipswich waterfront from one of the blocks of flats.

Ipswich is one of the fastest growing towns in the UK. Ipswich is partly growing through natural population increase but also through significant inward migration of people.

Ipswich Waterfront development is the biggest urban regeneration project in the East of England. The site was formerly an industrial dock area with warehouses and factories, but it had become increasingly derelict since the 1970s. Ipswich Borough Council worked in partnership with a number of developers to renew this brownfield site. Old warehouses have been refurbished and turned into shops, restaurants and flats. A range of new buildings has been constructed for homes, leisure use and education.

Brownfield site

A plot of land that has disused buildings (such as factories or warehouses). These sites are often near the middle of towns and cities so, if new homes are built here, the residents will be close to shops, services and jobs. This means that it is possible to create sustainable communities on brownfield sites. They already have services such as electricity, sewage and water. However, previous land uses may have left behind ground pollution. This can be expensive to clean up.

Activities

1 a) Represent the data shown in Figure 23.
b) What factors create the need for so many new dwellings in Ipswich?

	Population			Estimated number of new dwellings required
	2001 Census	2021 (predicted)	Change	
Suffolk as a whole	670,200	733,600	+ 63,400	61,700
Ipswich	117,400	138,700	+ 21,300	15,400

▲ **Figure 23** Population growth of Ipswich and housing need.

Some derelict sites were contaminated by waste from their former industrial use. For example, the Orwell Gasworks Quay was the site of a town gasworks where coal had been converted to gas. The land was polluted and it cost up to £270,000 per hectare to remove the waste and make it safe.

Anglo-Saxons lived here in the seventh century so archaeological surveys had to be carried out. Remains of historical value had to be conserved. This cost £1.2 million per hectare.

The ground is mainly soft sands and gravels deposited at the end of the ice age. This means that the foundations for new buildings had to be driven deep into the ground, which was expensive.

The site is next to a tidal estuary. Flood defences had to be built to protect central Ipswich from tidal floods. These cost £53 million. They protect 10 hectares of brownfield land which could not have been redeveloped without these flood defences.

Buildings of historical interest, such as old warehouses, had to be conserved and modernised. This was more expensive than building new homes and offices.

Building on greenfield sites on the edge of Ipswich may have been cheaper but it leads to urban sprawl. As our towns grow larger, this creates more transport problems as people have to commute further to work.

If we fill in the empty spaces in our towns and cities with pockets of new housing then the city doesn't grow any larger and people can live close to their work and leisure facilities in the city centre.

▲ **Figure 24** Advantages and disadvantages of developing the waterfront sites in Ipswich.

The location is superb. I can walk to my office in the town centre in minutes and I no longer need to use my car during the week.

I have all the entertainment I need on my doorstep. The bars are lively at night and on a sunny day you can sit and watch the activity in the marina while enjoying a coffee with friends.

▲ **Figure 25** The views of residents.

▲ **Figure 26** The medieval street pattern, narrow roads and old sewers are under pressure as the waterfront development has created more traffic in this area of the inner city.

Activities

2 Study Figures 24, 25 and 26. Use information from these resources to complete a table like the one below. You will write more in some boxes than others.

	Advantages of using brownfield sites	Disadvantages of using brownfield sites
Economic		
Environmental		
Social		

3 Use Figure 15, page 73, to explain why developing a brownfield site is often more sustainable than developing a greenfield site.

Enquiry

For a new housing or other urban development close to your school, design a table which has two columns featuring positive and negative aspects of the scheme. Try to consider social, economic and environmental factors when you review the successes and failures of the scheme.

How and why is retailing changing?

The UK's retail industry employed about 2.9 million people in 2018. The majority of retail sales are still made in shops located in the high street, in shopping centres or in out-of-town retail parks. The retail industry is changing due to a mixture of economic, cultural and technological factors. These factors are changing how and where we shop.

Economic factors

Rates (which are a kind of local tax) have to be paid on all shops. Rates can be very high in city centre locations. The costs are so great that they have forced some retailers out of business. This is one reason why one in ten shops in 2019 is vacant. Charity shops, on the other hand, pay 80 per cent less in rates than other shops. This means that, when a retailer goes out of business, the vacant shop on the high street may be filled by a charity shop.

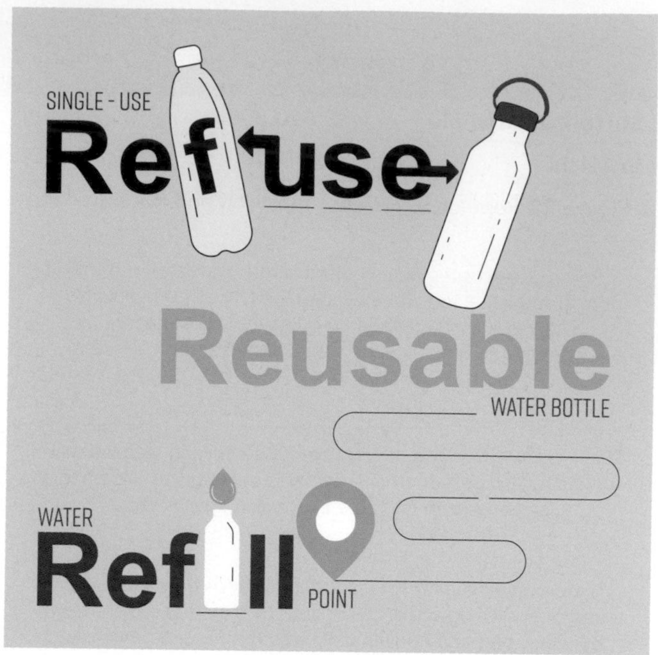

▲ **Figure 27** Tackling the issue of single-use plastic is an example of a cultural issue that affects retailing.

Cultural factors

The way we shop and what we buy is influenced by trends in the media. For example, shoppers have become increasingly concerned about the impact of plastics on the environment. This has led to pressure on retailers to reduce the amount of single-use plastic bags and plastic packaging used in their shops. Retailers are also influenced by other cultural trends among consumers such as vegetarianism, buying local products, and avoiding the purchase of products made in sweatshops.

Technological factors

Around 18 per cent of all retail sales in the UK are now made online. Online sales of clothes, furniture, groceries, music and books are all increasing. Consumers can also pay for flights and holidays online, as well as doing their banking. The sale of goods and services online is another reason for the decline of sales in high street shops and the closure of high street banks and travel agents.

The rise of online shopping 24/7 has created a new set of logistical problems for retailers, especially supermarkets. How do you get perishable goods to the customer quickly and efficiently? The answer is to have effective distribution services that can take advantage of the UK's motorway network.

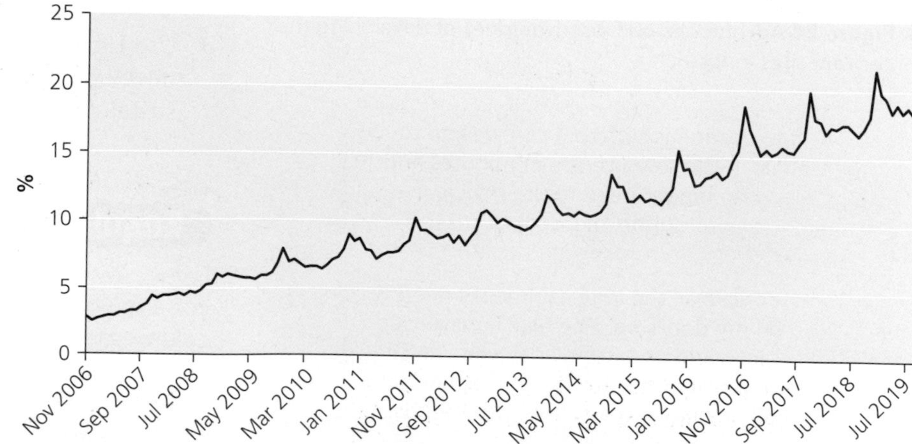

▲ **Figure 28** Online sales as a percentage of total retail sales (%).

Most of Morrisons' stores were in the north of England until 2004, when the company bought Safeway and expanded its stores into locations in the south of England. It needed to improve its distribution system so that it could supply its new stores. Two new distribution centres were built in Sittingbourne (in the South East) and Bridgwater (in the South West).

> I run a shop and website selling antiques. Sales are declining, especially in the shop, and other dealers I know have closed their shops. I think the problem is the rise of online auctions. It's really easy for anyone to sell things online using one of these sites. You don't have bills like rent, or rates like you would if you had a shop.

> I live in a rural area of Wales. I do my banking online. It's a convenient way to manage my money and I can check my account whenever I want. However, I was annoyed when my local branch closed. It's now a 25 mile round trip to visit my nearest high street bank which is something I need to do if I need financial advice.

> I work as a courier – driving a van delivering parcels. I spend most of my working day driving up and down the UK's motorways. I do a lot of miles and my old diesel van will need replacing soon. I'm not sure what I will buy next. I can't really afford a new hybrid vehicle.

▲ **Figure 29** Points of view on the rise of online services.

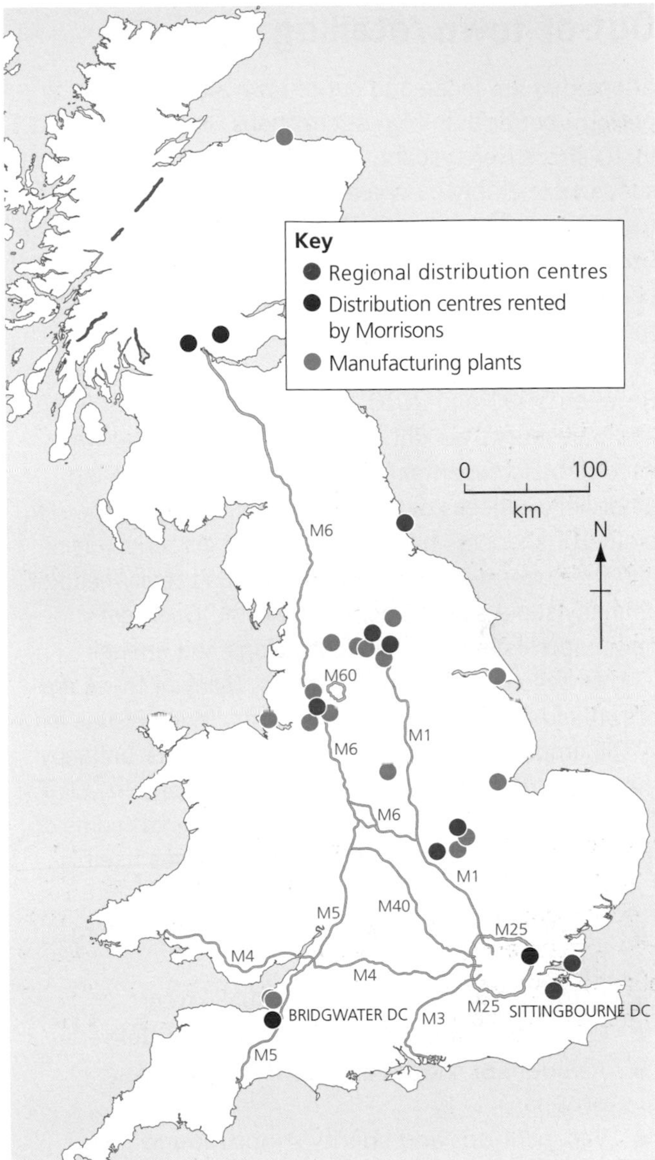

▲ **Figure 30** Morrisons regional distribution centres and manufacturing plants.

Activities

1. Discuss the use of plastics by the retail industry.
   a) List the ways that plastics are used in retailing.
   b) Suggest ways in which the use of plastics could be reduced by retailers.
2. Study Figure 28. Describe the trend of this graph.
3. List some positive and negative impacts of online shopping. Consider the economic, social and environmental consequences. Use evidence from this page.
4. Using Figure 30:
   a) Describe the location of the Bridgwater Distribution Centre.
   b) Explain why Morrisons chose this location for a new distribution centre after 2004.

Enquiry

'The positive impacts of online retailing outweigh the negative impacts.'

To what extent do you agree with this statement? Consider the impacts on people (consumers and employees), the economy and the environment. Use evidence from this page to support your answer.

Out-of-town retailing

Shops that are located in out-of-town sites continue to perform better than high street shops. In 2019, around 1 in 10 shops were vacant: 11.2 per cent of all high street shops were empty compared with 4.9 per cent of out-of-town shops. This may be due partly to their accessibility. Out-of-town shops are usually easy to get to by car and parking is usually free. This means that customers find them convenient for their big, weekly grocery shop.

Between 1990 and 2010, large retailers such as Tesco and Sainsbury's bought a lot of land. Some land was used to build supermarkets, but other sites were left empty in a process described as land banking. However, by 2010, shoppers had begun to order groceries online and the supermarkets realised they didn't really need as many super-size stores. In 2014, the *Guardian* newspaper estimated that Tesco alone had enough land to accomodate 15,000 homes. Many of these are brownfield sites in suburban locations. Much of this land is still empty. In many cases, local people are unhappy that large, derelict sites have been left empty. In 2019, Tesco hadn't decided whether it should sell off some of its spare land or use the land to build houses itself.

- Parking close to the shops may be difficult to find.
- Large surface car parks are usually free.
- Good road links make access easy for deliveries.
- A variety of shops cater for different ethnic groups.
- Vacant shops and charity shops reduce consumer choice.
- Pedestrianised streets enable people to interact with one another.
- Narrow streets make deliveries difficult.
- Older buildings are difficult to convert to modern stores.

▲ **Figure 31** Comparing out-of-town and high street locations.

Activities

1 Study Figure 31. Use it to make a list of the advantages and disadvantages of out-of-town and high street locations.
2 Study Figure 32.
   a) Describe the site of this supermarket.
   b) Explain why the retailer chose this site.

▲ **Figure 32** An out-of-town supermarket.

Changing shopping patterns

The way people shop for their groceries is affected by a combination of cultural, technological and economic factors. The financial crash of 2007 and benefit changes (such as the introduction of Universal Credit) have had impacts on how we have shopped over the last decade. For example, people on lower incomes may have to save money to pay rent or wait to receive benefits. As a consequence, poorer households tend to shop little and often.

The large supermarkets, such as Tesco and Sainsbury's have responded to this by opening smaller stores in town and city centres (like Tesco Express or Sainsbury's Local) rather than more large, out-of-town supermarkets. At the same time, there has been a sharp rise in the number of food banks. These are charities that give emergency food parcels to people who cannot afford to buy enough food to feed themselves and their families. In a 2019 survey of people who use Trussell Trust Foodbanks, 33 per cent said that their income was too low to cover essentials such as food and 38 per cent said that they relied on food banks because they were experiencing benefit delays or benefit changes.

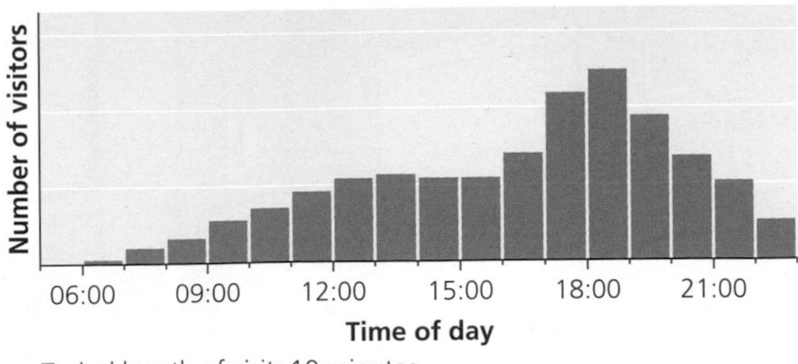

▲ **Figure 33** Footfall at Tesco Express, Cardiff city centre store, CF10 1AB.

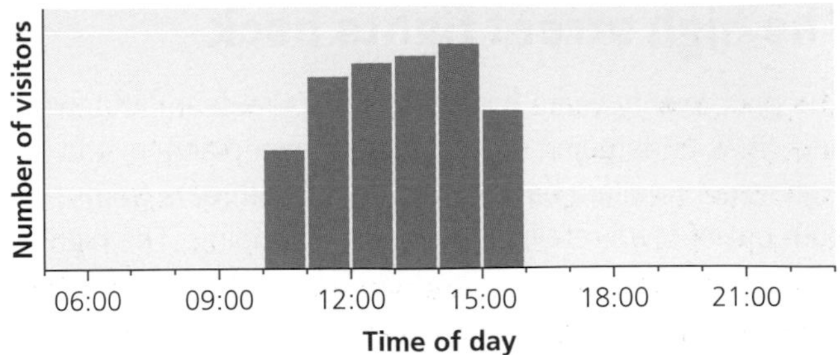

▲ **Figure 34** Footfall at Tesco Extra, Cardiff out-of-town store, CF14 3AT.

Financial year ending March	Number of three-day emergency food parcels
2014	913,138
2015	1,084,604
2016	1,109,309
2017	1,182,954
2018	1,332,952
2019	1,583,668

▲ **Figure 35** Number of three-day emergency food parcels given by Trussell Trust foodbanks.

Activities

3 Study Figures 33 and 34.
   a) Compare the patterns shown on these two graphs. Think about differences and similarities.
   b) Explain why retailers like Tesco have opened new, small supermarkets in town and city centres. Use evidence from Figure 33.

4 Study Figure 35.
   a) Describe the pattern shown in these data.
   b) Suggest one way in which these data could be presented. Justify your choice.

Enquiry

'Economic factors are more important than technological or cultural factors in changing the way we shop.'

To what extent do you agree? Use evidence on this page to support your answer.

The high street fights back

The past twenty years have seen high streets in UK towns and cities undergoing rapid change. Town planners and high street retailers have had to adapt to threats from both out-of-town retailing and online shopping. The high street has fought back to improve the quality of the urban environment for shoppers. A variety of strategies have been used to win back high street shoppers. Some schemes have provided safer streets for pedestrians by restricting access for cars or adding more CCTV cameras. Local councils have worked with developers to create covered shopping malls where it doesn't matter if it's raining outside!

The city centre of Lancaster has recently been improved with new paving in some pedestrianised streets and better street furniture using high quality, strong materials. Changes to road layouts meant that old signposts were confusing. New wayfinding (or signage) information has been installed for shoppers and visitors to the town. A camera was also installed to measure footfall – the number of pedestrians in key locations. It proves that 185,000 people use the city centre each week and that Wednesday is the most popular day – which is market day. This would seem to prove that people still like to shop from independent traders as well as the big chain stores.

▲ **Figure 36** New wayfinding information was installed in Lancaster in 2015.

Activities

1 Study Figure 37 carefully. Write an annotation for each of the features that has a number. Use your annotation to explain how this improves the environment for shoppers.
2 Explain how the features shown in Figures 37 and 38 can improve the quality of the retail environment.

▲ **Figure 37** The high street environment in Lancaster.

▲ **Figure 38** Town centres use Christmas markets to attract more shoppers. This one is in Winchester.

GEOGRAPHICAL SKILLS

Diamond ranking

Diamond ranking, or diamond nine, is a useful technique to use when you have been asked to make a decision. Sometimes it's not easy to try to rank or prioritise ideas when there is no obvious answer. Use this technique to group your ideas, putting your favourite ideas near the top, and the ones you think are less convincing at the bottom of the diamond.

Enquiry

What are the best strategies for improving the retail environment?

Study the strategies in Figure 39.

a) Make a diamond nine diagram and place the strategies in the diagram, putting those that you think are essential at the top of the diagram.
b) Justify your choice of the top three strategies by explaining how they will benefit the experience of shoppers and improve the economy of the high street.

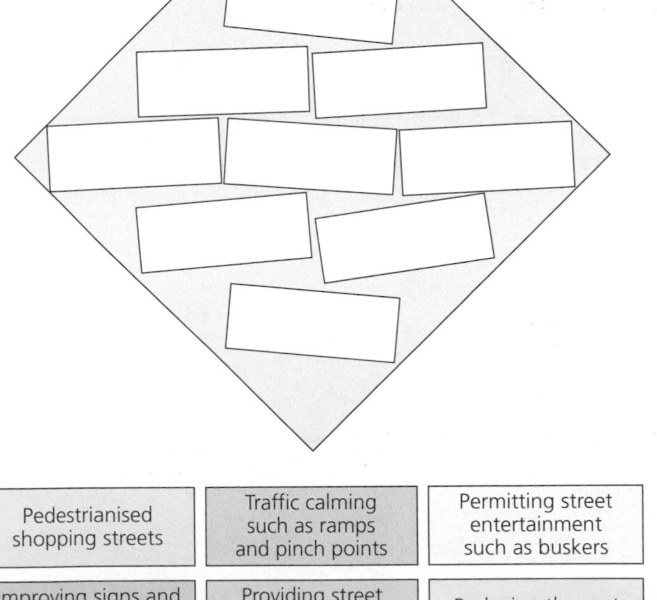

Pedestrianised shopping streets	Traffic calming such as ramps and pinch points	Permitting street entertainment such as buskers
Improving signs and wayfinding information for pedestrians	Providing street furniture such as flower beds and benches	Reducing the cost of short-stay parking
Creating park-and-ride schemes	Special high street events such as French or Christmas markets	Allowing pop-up shops to sell from vacant shops

▲ **Figure 39** Strategies to improve the high street.

Does each town centre need its own distinct identity?

Each town's **identity** is created by the mixture of features that give it its own character. A town's identity is what gives it its unique sense of place. Many are concerned that a town loses its identity when locally owned independent shops are replaced by shops that are part of a national chain.

Bailey Street (left side)	Bailey Street (right side)
Timpson – dry cleaning and key cutting	Bailey's – café and bakery
Radio – café	Age UK – charity shop
VACANT	Cascade – amusement arcade
Shades – curtains/fabrics	Pound Stretcher – discount shop
The Flower Gallery – florists	Card Factory – greeting cards
The Oak Furniture Shop – beds and furniture	Specsavers – opticians
Polka Dot – travel agent	
RJ Christian – jewellers	
The Works – book shop	

▲ **Figure 40** A simple way to record shops. Note that the photographs in Figure 42 were taken in the same street.

Clone town

The clone town survey is part of a national survey to identify the number of independent shops that are left in the UK's high streets. It is a measure of the range of different types of shop (diversity) and of the proportion of independent shops compared to chain stores (identity). Follow these steps to calculate a 'clone town' score.

Step 1 Use a recording sheet like Figure 43 to survey between 40 and 60 shops.

Step 2 Find the total number of **types** of shop in your survey and the total number of **independently** owned shops in your table.

Step 3 Multiply the total number of independently owned shops by 75, then divide this by the total number of shops in your survey.

Step 4 Add the total number of types of shop to the figure you found in Step 3. This gives your clone town score. Scores below 50 are clone towns. Scores above 65 have a good diversity of shops and enough independent shops to give the town identity.

▲ **Figure 41** Clone town score.

▲ **Figure 42** Shops on Bailey Street, Oswestry, Shropshire. Bailey Street is pedestrianised. It has not been photographed on Google Street View.

Type of shop	Independently owned ✓	Part of a chain ✓
Food (butcher, baker, supermarket)		
Newsagent/tobacconist		
Books/stationery		
Department/catalogue store		
Pub/bar		
Off-licence		
Professional e.g. insurance/ accountancy		
Estate agent		
Healthcare/pharmacy/optician		
Household goods (furniture, kitchen)		
Clothes/shoes		
Cinema/theatre		
Electronic/IT, e.g. phones, computers		
Pets/pet supplies/vet		
Hair salon/beauty treatment		
Toys/sports/cycling/outdoor leisure		
Car accessories/petrol		
Builders merchant/DIY store		
Florist/garden centre		
Dry cleaning/launderette		
Travel agent		
Camera/photo processing		
Post Office		
Other, e.g. betting, amusements, antiques, charity shops, jewellers		
TOTAL		

▲ **Figure 43** Clone town survey recording sheet.

Activities

1 Study Figure 42.
   a) Identify one feature that gives this street an identity.
   b) Use Figures 40 and 43 to help you create a tally chart for the shops in Figure 42. In your tally chart, you should identify:
      i) the number of independently owned shops
      ii) the number of different types of shop.
2 Use Google Street View to complete a virtual clone town survey of Oswestry. You can add your results to Figure 43. This means that you will need to classify between 25 and 45 shops.
3 Discuss the features that give your local town centre an identity of its own.
   a) List three features of the built environment and three features of the human environment that create identity.
   b) To what extent do you agree that it is important for town centres to have their own identity? Why does this issue seem to be important?

Enquiry

Create an enquiry that could be used to investigate perceptions of identity in your own town centre. Your plan should include:

- an overarching enquiry question
- your sampling strategy
- ideas for how you would gather both quantitative and qualitative data (see page 8).

THEME 2 Rural–urban links

Chapter 3 Global cities

What are the global patterns of urbanisation?

Over half (54 per cent) of the world's population lives in cities. But it's not always been this way. At the start of the twentieth century most of the world's population lived in the countryside or smaller towns. During the second half of the twentieth century the population of the world's cities, and the physical size of their built-up areas, grew – a process known as **urbanisation**. This process was particularly rapid in the cities of the **newly industrialised countries** (NICs) of Asia and Latin America.

Urbanisation has been due to a combination of:

- migration from rural to urban areas
- the natural increase of the population when there are more births than deaths each year.

According to the United Nations, there are 548 cities in the world with a population of at least 1 million people. Of these, 33 are **mega-cities**. Asia has 20 megacities, six of which are in China, with five in India. Latin America has six megacities, and Europe only two.

◀ **Figure 1** The centre of Mumbai, India. The neighbourhood of Bhendi Bazaar is in the foreground. The low-rise buildings in this neighbourhood are traditional dormitories known as chawls.

Newly industrialised countries (NICs)

NICs are characterised by: rapid urban growth; growth of the manufacturing sector of the economy; strong trading relationships with other countries; and the operation of foreign-owned multi-national companies (MNCs) within the country.

Mega-city

A city that has a population of at least 10 million people.

1990		2015		2030	
Tokyo, Japan	32.53	Tokyo, Japan	38.00	Tokyo, Japan	37.19
Osaka, Japan	18.39	Delhi, India	25.70	Delhi, India	36.06
New York–Newark, USA	16.09	Shanghai, China	23.74	Shanghai, China	30.75
Ciudad de México (Mexico City)	15.64	São Paulo, Brazil	21.07	Mumbai, India	27.80
São Paulo, Brazil	14.78	Mumbai, India	21.04	Beijing, China	27.71
Mumbai, India	12.44	Ciudad de México (Mexico City)	21.00	Dhaka, Bangladesh	27.37
Kolkata, India	10.89	Beijing, China	20.38	Karachi, Pakistan	24.84
Los Angeles–Long Beach-Santa Ana, USA	10.88	Osaka, Japan	20.24	Cairo, Egypt	24.50
Seoul, South Korea	10.52	Cairo, Egypt	18.77	Lagos, Nigeria	24.24
Buenos Aires, Argentina	10.51	New York–Newark, USA	18.59	Ciudad de México (Mexico City)	23.86

▲ **Figure 2** The world's ten largest mega-cities, population in millions.

What will happen next?

The world's urban population is expected to rise from 3.9 billion (in 2014) to 6.0 billion by 2045. The largest urban growth is expected to be in India, China and Nigeria where an extra 404 million, 292 million and 212 million urban dwellers will be added respectively by 2050. The fastest growing cities are not mega-cities but those with a population under 500,000. Many of the world's fastest growing cities are expected to be in sub-Saharan Africa and Asia.

Year	Mumbai, India	Kinshasa, DRC
1950	2.86	0.20
1960	4.06	0.44
1970	5.81	1.07
1980	8.66	2.05
1990	12.44	3.68
2000	16.37	6.14
2010	19.42	9.38
2020	22.84	14.12
2030	27.80	20.00

▲ **Figure 3** Population (millions) of Mumbai, India and Kinshasa, Democratic Republic of Congo (DRC), which is one of Africa's fastest growing cities.

Enquiry

What is the geographical future for urban areas around the world?

- Represent the data in Figure 4. A good way to do this would be to draw proportional bars on an outline map of the world.
- Use your map to write a 200-word report. Make sure that you contrast the rates of urban growth in different regions of the world.

Region	1990–95 (%)	2015–20 (%)	2045–50 (%)
Sub-Saharan Africa	4.09	3.83	2.78
East Asia (including China)	3.28	1.91	-0.06
South Asia (including India)	2.95	2.40	1.37
Western Europe	0.78	0.47	0.12
South America	2.43	1.16	0.36

▲ **Figure 4** Average annual rate of change of the urban population (per cent).

Activities

1. Write definitions for the following terms: urbanisation mega-cities NICs
2. Study Figure 2.
   a) Choose a suitable graph to present this data.
   b) Describe what has happened to the distribution of the world's largest cities.
3. Study Figure 3.
   a) Draw a pair of line graphs to represent the growth of these two cities.
   b) Describe the similarities and differences of these two graphs.
4. Is rapid urbanisation a good thing? Predict the possible consequences of rapid population growth in a city such as Kinshasa, which is in a Low Income Country (LIC). You should consider the advantages and disadvantages.

So what are global cities?

All cities have a regional influence. They interact with the area that surrounds them, for example:

- attracting daily commuters to work
- providing specialist services such as hospitals and universities
- acting as transport hubs for rail networks or regional airports.

However, some cities have greater influence than others. **Global cities** are those cities which interact with other places at the global scale. Figure 5 illustrates some of these global connections. The Globalisation and World Cities Research Network has identified over 300 cities that have inter-connections with other parts of the globe. The UK has 14 global cities. London is ranked first but the next highest in the UK is Manchester at 78th. Cardiff is in 248th place. Figure 6 shows those Global Cities that have the most important global connections.

Globalisation

The process of globalisation connects places economically, socially, politically or culturally. Global cities play an important role in this process. Cities have always been connected by trade and migration. However, as communications and transport improve, the process is accelerating. It is as though the world is shrinking as it becomes better connected and places become less isolated.

Migration and culture:
Global cities attract economic migrants from all over the world. Migration leads to cultural diversity. More than 100 languages are spoken in 30 of London's 33 boroughs. The 2011 Census showed that 22 per cent of London's residents do not speak English as their main language. That's just over 1.7 million people.

Governance and decision making:
Business managers in one city can make decisions that affect people worldwide. For example, Tata is an Indian TNC with its headquarters in Mumbai and businesses in over 100 countries. Politicians and civil servants can also have worldwide decision-making roles. The UN employs 41,000 people. 6,389 are based in their headquarters in New York.

The HSBC headquarters in Hong Kong

Finance and trade:
The world's most important global cities are financial centres. Banks have their head offices here. Dealers working at financial markets like FTSE buy and sell commodities on the world markets.

Transport hubs:
The top global cities are all well-connected to the rest of the world by major airports or ports. These allow the flow of people, tourists and trade. About 1,400 flights take off or land at London's Heathrow airport every day.

Ideas and information:
Many of the world's global cities are the home for major broadcasting companies. Newspapers, TV stations and filmmakers are based in global cities. BBC World News is an international TV channel broadcasting 24 hours a day to over 300 million households in 200 countries.

Global cities

▲ **Figure 5** How global cities connect to the rest of the world.

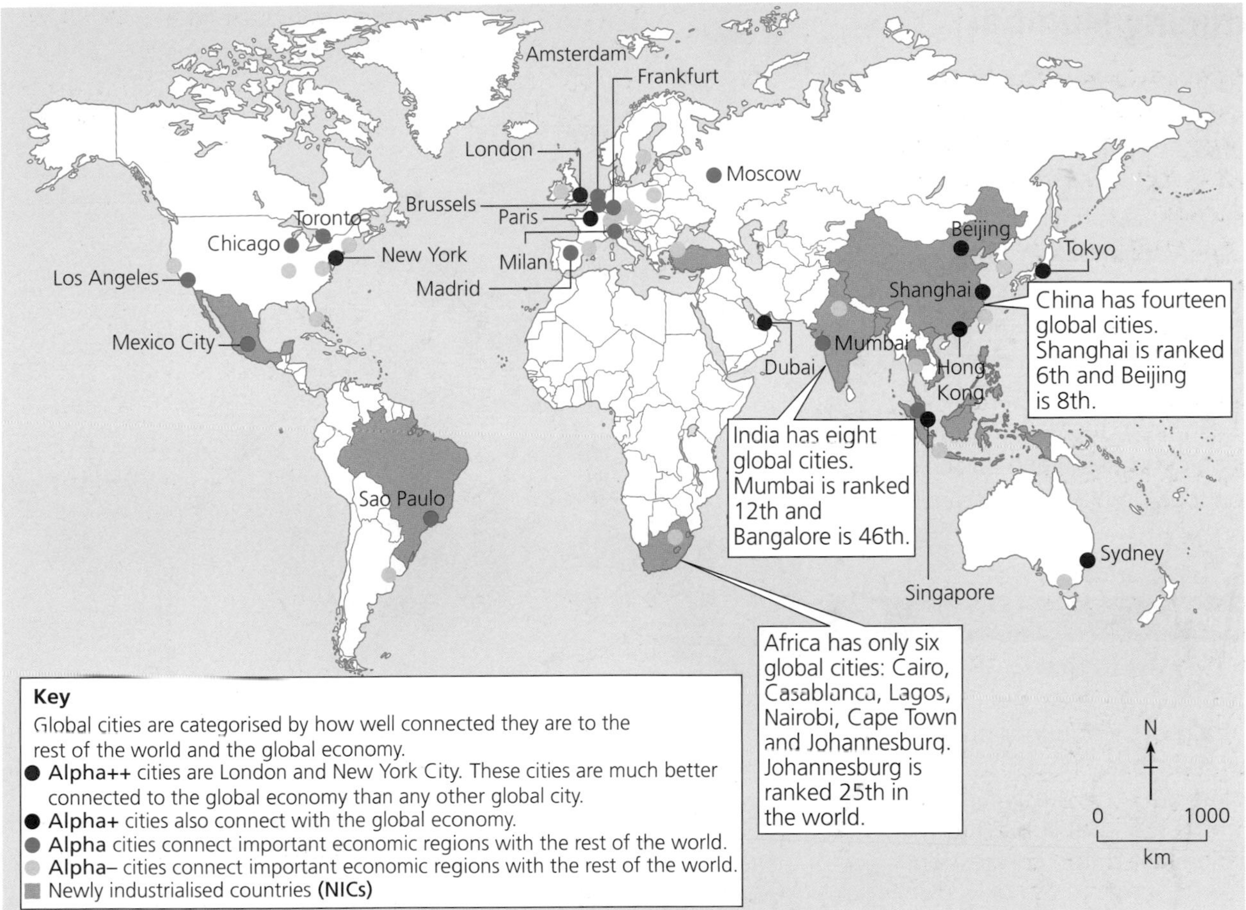

▲ **Figure 6** The location and distribution of the world's alpha (top-ranked) global cities.

Year	Population
1950	8.36
1960	8.19
1970	7.51
1980	7.66
1990	8.05
2000	8.61
2010	9.70
2020	10.85
2030	11.47

▲ **Figure 7** Population (millions) of London, the top-ranked global city.

Activities

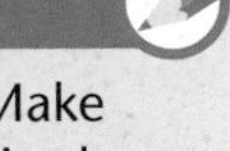

1 Explain the difference between mega-cities and global cities. Make sure that you give examples of cities that fall into both categories by using Figures 2 and 6.

2 Study Figure 6.
   a) Describe the distribution of the world's alpha++ and alpha+ cities.
   b) What percentage of global cities shown on Figure 6 are located in NICs?
   c) Suggest why there are so many global cities in India and China.

3 a) Use the data in Figure 7 to draw a line graph of London's population growth.
   b) Use your graph to estimate the year in which London became a mega-city.
   c) Compare and contrast London's population growth to that of Mumbai and Kinshasa (shown in Figure 3 on page 87).

Global city

A city that plays an important role in the world's economy.

Enquiry

Why do you think London is the world's most important global city? Make a list of ten ways that London is:

- connected to other parts of the UK
- interconnected with other parts of the world.

Introducing Mumbai

Mumbai is India's largest city and the country's best connected global city. Mumbai is also a mega-city, with an estimated population of 24 million in 2018. The city of Greater Mumbai is built on a low-lying island in the Arabian Sea. As the city grew, it sprawled northwards and eastwards across Thane Creek to form a large metropolitan region. 465 km of suburban railway links Central Mumbai to its suburbs on the mainland. However, there are only four rail crossings onto the island and this creates congestion for Mumbai's daily 7.5 million commuters.

Activities

1. Use Figure 8 to draw a sketch map of Mumbai. Include the city centre, CST, airport, container port and Navi Mumbai.
2. Explain why the site of Greater Mumbai has contributed to the problems of traffic congestion.

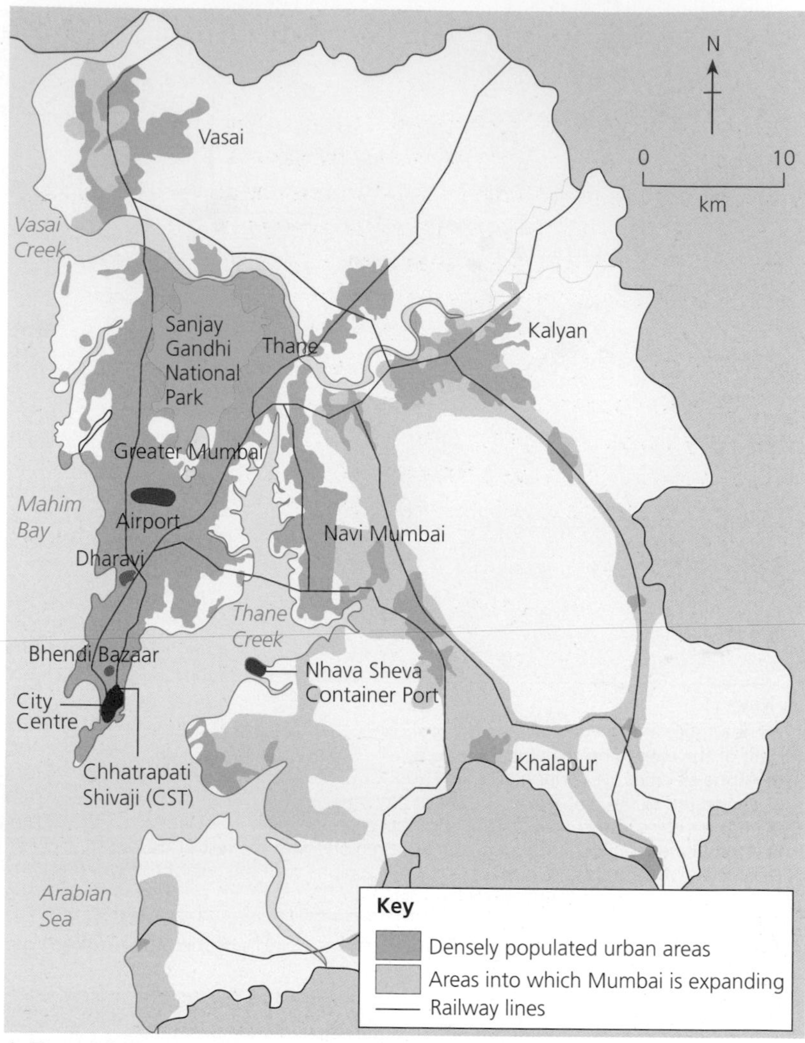

▲ **Figure 8** The Metropolitan Region of Mumbai.

◀ **Figure 9** A severely crowded train in Mumbai.

Transport issues – an inevitable consequence of rapid growth?

Large cities like Mumbai need efficient **mass transit** or (rapid transit) systems so that commuters can get into work quickly and safely. Mumbai's rail system is one of the busiest in the world. At peak times trains carry three times the number of passengers they were designed for. People hang from doorways and even ride on the train's roof. Overcrowding is more than just uncomfortable – it is dangerous. People die when they fall from trains or are hit when they are crossing the tracks. At least nine people are killed every day on Mumbai's railways. How can the system be improved? Trains run every three minutes at peak times. It may not be possible to add any more trains without risking accidents.

Possible ways to improve Mumbai's rail system
Encourage businesses to offer flexi-hours instead of regular 9 a.m.–5 p.m. hours.
Reserve more seats on trains for elderly passengers.
Improve toilet facilities.
Prevent people from travelling on the roofs of trains.
Remove fast food kiosks and other vendors from all platforms.
Increase platform length.
Improve ventilation in carriages.
Demolish informal housing next to railway tracks.
Fit proper doors to all trains so passengers cannot travel half-in and half-out.

▲ **Figure 10** Possible ways to improve Mumbai's rail system.

Mass transit

Transport systems, such as commuter trains or underground trains, designed to carry very large numbers of people.

◀ **Figure 11** Many people live dangerously close to Mumbai's railways in informal housing.

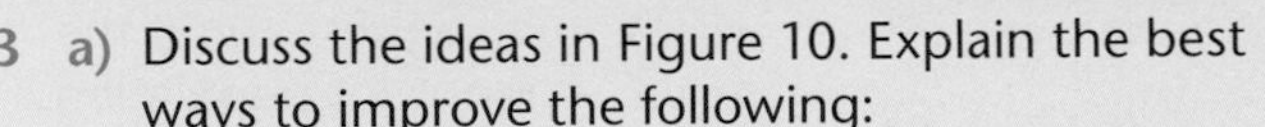

Activities

3 a) Discuss the ideas in Figure 10. Explain the best ways to improve the following:
   i) rail safety
   ii) passenger comfort
   iii) train times.

b) Use a diamond nine technique (see page 83) to rank all nine ideas. Which three do you think should be given the highest priority to improve services? Justify your choice.

4 Explain why demolishing informal housing next to railway lines could:
   a) improve rail safety
   b) improve train times.

5 Study Figure 11.
   a) Describe the buildings in this neighbourhood.
   b) Suggest three ways that people who live here are affected by the quality of their environment.

Why has Mumbai grown?

We saw on page 87 how quickly Mumbai's population has grown. Like other Indian cities, Mumbai has grown due to a combination of natural increase and **rural-to-urban migration**.

Natural increase of Mumbai's population

Natural increase is the growth of population that occurs when there are more births than deaths. Natural increase has been the main reason for Mumbai's growth during the twentieth century. One simple way to investigate this is to examine average family size, a statistic which is known as **fertility rate**. If women, on average, have more than two children, then the population will grow. Fertility rates in Mumbai tend to be slightly lower than in rural areas of Maharashtra. In 2007, the fertility rate was:

- 2.2 in rural areas of Maharashtra
- 1.8 in urban areas of the state.

Year (or period)	Fertility rate
Average 1974–82	4.03
Average 1984–90	3.45
Average 1994–2000	2.60
2004	2.20
2010	2.00
2018	1.80

▲ **Figure 12** Changing fertility rates in Maharashtra.

Rural-to-urban migration in Maharashtra

India's rail system has some of the lowest fares in the world. It costs only 250 rupees (about £2.50) to travel from Kolkata to Mumbai. Cheap rail travel is one **pull factor** that encourages rural-to-urban migration to Mumbai. People living in rural areas of India are attracted by the jobs and better training opportunities available in cities such as Mumbai. Poverty, the poor standard of housing, health care and sanitation are all **push factors** that can force people to move away from rural areas.

Fertility rate

The average number of children a woman will have in her lifetime.

Rural-to-urban migration

The movement of people from the countryside to towns and cities. The people who move are usually young adults. Migrants say that they move because they want to join a loved one who lives in the city or because they are looking for work.

Activities 

1 a) Select a technique to represent the data in Figure 12.
b) Describe the trend in fertility. What does this suggest about the reasons for population growth in Mumbai?
c) Suggest why fertility is lower in urban areas of Maharashtra.

◄ **Figure 13** Migrant flows into Mumbai from other Indian states.

Factfile: Why do people move to Mumbai?

Out of 1,000 people who were interviewed about why they moved to Mumbai:

- 538 moved because of marriage
- 187 moved because a family member/parent moved
- 173 moved to find work
- 35 moved to improve their education/attend university
- 62% of males gave work or business-related reasons
- 80% of females moved to marry or accompany a family member.

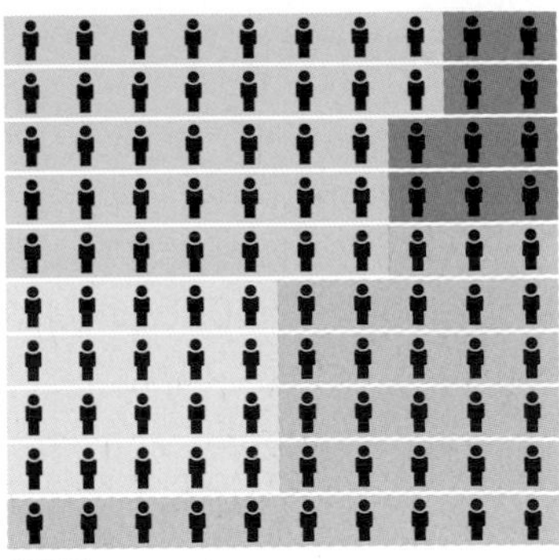

◀ **Figure 14** Where do migrants to Mumbai come from?

Name of Indian state	Number of households	Number of households with no bathroom	Number of households with a computer
Maharashtra	23,830,580	3,478,681	3,174,031
Uttar Pradesh	32,924,266	14,761,001	2,664,447
Gujarat	12,181,718	3,967,358	1,077,510
Karnataka	13,179,911	1,807,662	1,692,253
Rajasthan	12,581,303	5,595,753	869,923
Bihar	18,940,629	11,849,779	1,334,565
Tamil Nadu	18,493,003	6,625,321	1,956,630
Andhra Pradesh	21,024534	6,910,308	1,763,555
Kerala	7,716,370	1,097,456	1,214,644
West Bengal	20,067,299	12,869,502	1,668,757

▲ **Figure 15** Selected data for the Indian states which are the source of Mumbai's migrant population.

GEOGRAPHICAL SKILLS

Calculating percentages from raw data

The census data in Figure 15 is **raw data**, meaning data that has not been processed. It would be interesting to see whether poverty in Uttar Pradesh is a push factor. For example, are there more households in Uttar Pradesh than Maharashtra that do not have a bathroom? To make a comparison we need to process the raw data to find the percentage of homes that have no bathroom. You do this by dividing the number of households with no bathroom by the total number of households in that state and then multiplying by 100. For Uttar Pradesh this would be:

14,761,001 ÷ 32,924,266 = 0.4483

0.4483 × 100 = 44.83%

Activities

2 a) List five facts about the origin of Mumbai's migrants.
  b) In African countries, most migrants only move short distances. To what extent is this true in India?

3 Describe the main push and pull factors for migration to Mumbai.

Enquiry

Is poverty a push factor in rural to urban migration?

- Process the data in Figure 15 into percentage figures.
- Draw a scattergraph to show the percentage of migrants (shown in Figure 13) on the vertical axis and the percentage of households with no bathroom on the horizontal axis.
- Comment on the trend shown by your scattergraph. What conclusions can you draw?

Why is Mumbai a global city?

Mumbai's economy is well connected to other locations both within India and abroad:

- The Hindi film industry, known as Bollywood, is based in Mumbai and is thought to employ 175,000 people.
- Tata Steel, which employs people in over 100 countries, has its headquarters in Mumbai.
- Nhava Sheva is India's largest container port. Goods made in India can be shipped to Felixstowe, via the Suez Canal, in nineteen days. Container ships from Kolkata, on India's east coast, take 28 days. This gives Mumbai a clear advantage in its global rankings.
- Mumbai International Airport is well placed to fly businessmen and women between Europe, the Middle East and Asia. London is nine hours, Dubai three hours and Hong Kong six hours by air.

Year	India	USA
2011	1,255	819
2012	1,602	738
2013	1,724	738
2014	1,966	707
2015	1,845	791
2016	1,903	789
2017	1,986	821
2018	1,813	576

▲ **Figure 16** Number of films made per year in India and the USA.

▲ **Figure 17** Jaguar Land Rover is the UK's largest car manufacturer. It is a subsidiary company of Tata, which has its HQ in Mumbai.

Activities 

1. a) Use a suitable technique to represent the data in Figure 16.
   b) Compare the trends shown by film making in India and the USA.
2. Explain why the location of Mumbai has given the city an economic advantage over Kolkata and Chennai in its global rankings.

Informal sector

That part of the economy that is not fully controlled by government. People working in this sector are often self-employed or in small businesses. They do not pay tax but they do not receive any state benefits either. Perhaps 80 per cent of workers in India work in this sector. They provide essential goods and services that are used by the formal sector.

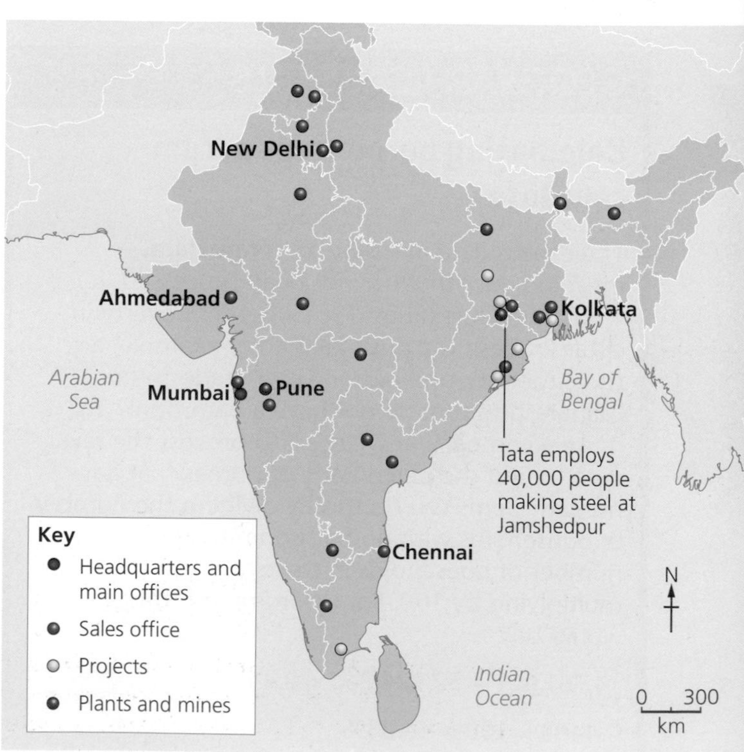

▲ **Figure 18** The location and distribution of Tata Steel in India.

Mumbai: formal or informal occupations?

An emerging middle class of young, professional, well-paid men and women is creating urban and economic change in India. Many are young graduates who left rural lives behind when they went to university in a city like Mumbai. They take jobs in central and state government, banking and financial industries, the IT industry, textile manufacturing, jobs related to the sea port, and the Hindi movie industry. These are all formal occupations that receive a regular wage. However, Mumbai's economy also has a very large **informal sector**. Street vending, rickshaw driving and recycling waste are all examples of informal jobs. These jobs are not regulated by the state. You don't necessarily need a qualification to do them and you probably don't pay tax. However, informal jobs have no paid holidays, pensions or sickness benefits, and there are no rules to protect your health and safety at work.

Mumbai is a city of huge contrasts between rich and poor. It is said to have the third most expensive office space in the world. It is also home to millions of people living in poverty. Informal housing occupies seven per cent of Mumbai but houses 60 per cent of its population, so it is extremely overcrowded. These slums develop on unwanted land: marshland that is vulnerable to flooding during the monsoon or alongside busy railway tracks. Dharavi is on such an informal settlement – it recycles Mumbai's rubbish.

▲ **Figure 19** A young ragpicker in Dharavi, one of Mumbai's slums. Sorting recycled plastics is an example of an informal occupation; ragpickers recycle 80 per cent of Mumbai's waste and together contribute around £700 million to Mumbai's economy each year.

Activities

3 Suggest how each of the following is creating urban and economic growth in India:
   a) low wages in rural occupations
   b) university education
   c) demand for consumer goods
   d) the emerging middle class.

4 Study Figure 20. Explain why poverty in the informal sector may prevent India from:
   a) developing greater wealth in the economy
   b) improving education, training and health care facilities

5 Using Figure 19 to help you, suggest how the informal sector helps to support Mumbai's growing economy.

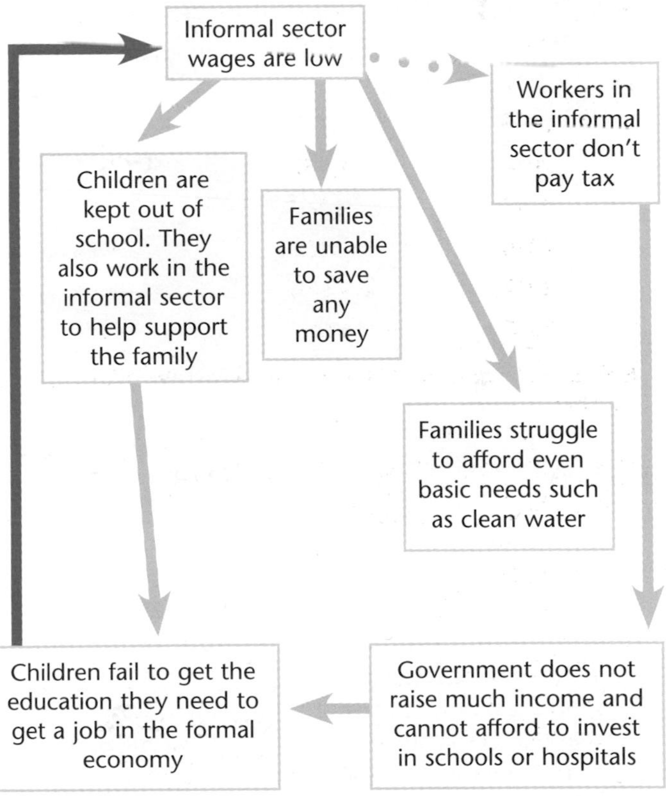

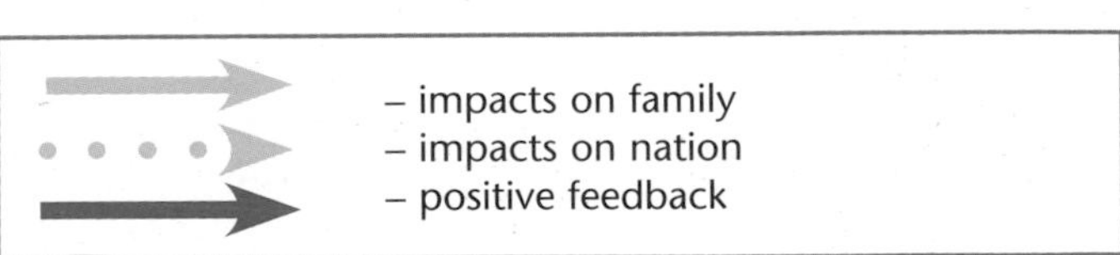

▲ **Figure 20** The relationship between the informal sector and the wider economy.

What are Mumbai's housing problems?

Overcrowding, poor sanitation and homes that are in danger of collapse, flooding or fire – these are some of the issues facing the urban poor of Mumbai. There are three types of housing where occupants are at risk because of the poor conditions:

- **Chawls** are a type of four- or five-storey tenement building. Families live in single rooms arranged along a corridor. Basic toilet facilities are shared by the tenants on each corridor. Many chawls were built between 1920 and 1956. They are overcrowded and poorly ventilated. But they are affordable.
- **Squatter homes**, also known as **slums** in India, are simple single or two-storey buildings built without planning control. Sanitation is very poor: 73 per cent of residents share communal latrines.
- **Pavement dwellers**, many of whom are children, live in huts which narrow the pavement. Pavement dwellers pay rent to criminals who control the pavements. The structures are illegal and may be demolished by the authorities.

Opinions vary about how these issues can be solved. Some people believe that **self-help** projects can improve housing or sanitation. Others think that **wholesale clearance** and redevelopment are the best solutions. Which is the most sustainable solution?

◀ **Figure 21** Traditional chawl tenements in Bhendi Bazaar that will be demolished.

Activities

1. Use Figure 8 on page 90 to describe the location of Bhendi Bazaar.
2. Suggest how each of the following might affect the safety and health of local residents:
   a) overcrowding in the chawls
   b) having to share toilet facilities
   c) insecurity of tenure for pavement dwellers.
3. Study Figure 15 on page 73. It shows Egan's Wheel. Explain how each of the following elements of sustainability might be met by the redevelopment of Bhendi Bazaar:
   a) housing and built environment
   b) social and cultural
   c) economy.

▲ **Figure 22** Market traders in Bhendi Bazaar.

Is wholesale clearance and redevelopment the answer?

Bhendi Bazaar is a mixed area of chawls and 1,250 shops and stalls. It is estimated that 20,000 people live here. The chawls are old and overcrowded. There is no proper waste disposal system and water is only supplied for a few hours each day. An ambitious plan will demolish 250 buildings and replace them with 17 high-rise tower blocks. Construction work started in 2016. The ambitious project should be complete by 2026.

The new development is planned to be sustainable:

- There will be a mixture of houses and shops so people can continue to work locally.
- Wide roads and tree-lined pavements will replace narrow alleyways.
- There will be open spaces for parks, green spaces and play areas.
- Mosques will be retained and enhanced.
- Car parking and connections to public transport are planned.

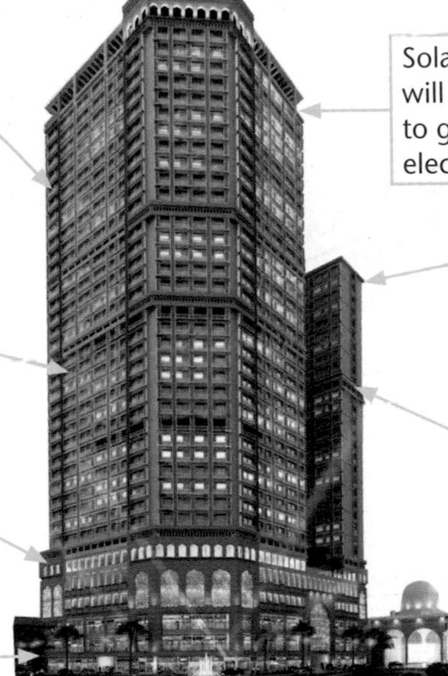

◄ **Figure 23** Sustainable features of the new multi-storey buildings at Bhendi Bazaar.

Activities

4 a) Discuss the strengths and limitations of this redevelopment scheme.

b) Can you suggest three different ways it could be improved?

My home in Bhendi Bazaar has already been demolished and I now live in the 'transit' housing while I wait to be permanently rehomed. Life is much better here. We have moved into a furnished room. It has a carpet, curtains and a cupboard. There is a small kitchen with hot water and a washing machine. It is so much cleaner and quieter here! My children can play cricket in the open space in front of the tower block. We even have internet connection.

Mariya, tenant of the temporary 'transit' block

I am a street vendor and I sleep in front of one of the shops. Tenants will get a new home. But I am fearful that I will get nothing. There are thousands of pavement dwellers like me who make a living in the street market. If I can no longer live and work here I may have to go back to my village in Uttar Pradesh.

Taha, a second-hand wristwatch seller

Large fire engines cannot get into districts with chawls like Bhendi Bazaar or squatter settlements like Dharavi. We need to buy mini-fire tenders to get into the narrow, congested alleyways. We also need to build new fire stations across the city. Mumbai has only 33 fire stations. It needs at least 100 fire stations to meet global standards.

Aziz, city planner

In 2015 a fire quickly spread through Kalbadevi which is another congested neighbourhood of chawls. Widening of roads is crucial to make the area safer. I understand that the clearance of Bhendi Bazaar will allow them to widen the roads from the current width of 6 to 8 metres to up to 16 metres.

Mohammed, a firefighter

▲ **Figure 24** Opinions on the redevelopment of Bhendi Bazaar.

Introducing Cardiff

Cardiff is one of fourteen global cities in the UK and the largest city in Wales. We saw on pages 59 and 60 that Cardiff has a strong local sphere of influence. A population of 1.49 million people live within 32 km of Cardiff city centre. Many use the city regularly for work, shopping or entertainment. However, Cardiff's sphere of influence is larger – it also has national and global links. Cardiff is home to the Welsh Government, which has responsibilities for education, health and the economy for the whole of Wales. Cardiff is well connected by rail to London, and flights from Cardiff's airport fly to several European destinations. The Principality Stadium is a world-class sporting and music venue that has hosted international events including the Rugby World Cup in 2015. One large multinational – Admiral, the insurance company – has its main office here. Several BBC TV programmes, including Sherlock and Doctor Who, are filmed on location here and in film studios based in Cardiff. These TV programmes are sold all around the world. These media and cultural links help to place Cardiff on the list of global cities.

GEOGRAPHICAL SKILLS

Describing locations

To describe a location means to be able to pinpoint something on a map. Describing a location on an Ordnance Survey (OS) map is easily done by giving a grid reference. However, describing a location on a map that has no grid lines requires a different technique.

First, you need to give a broad indication of the location by describing in which part of the map the viewer should be looking. Always use geographical terms such as 'in south Wales' rather than 'at the bottom of the map' or 'near to the Bristol Channel'.

Then, to describe the exact location, you should use another significant place on the map and give:

- the distance from that other place in kilometres
- the direction using points of the compass.

For example, on Figure 25, Bridgend is 28 km to the west of Cardiff.

Cardiff's growth

Like many cities in the UK, Cardiff went through a period of rapid growth between around 1850 and 1920 as people moved into the city to find work in industries related to the sale of coal from Cardiff Docks. Terraced houses were built to provide homes for these dock workers and many of these homes can still be seen today in the inner-urban areas close to the city centre.

During the period from 1930 to the mid-1980s Cardiff went through a phase of suburban growth. Better public transport and more widespread car ownership meant that people could live further from their place of work. New housing was built, filling in the spaces between the edge of the city and existing small villages such as Radyr and Whitchurch. These villages are now part of the urban area of Cardiff.

Between the mid-1980s and today, Cardiff has gone through a phase of **re-urbanisation**. This time new housing was built on the brownfield sites once occupied by the dock-related industries in Butetown next to Cardiff Bay.

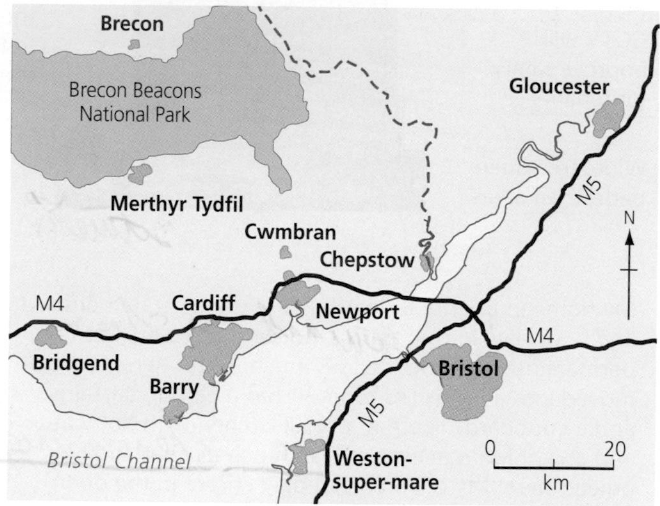

▲ **Figure 25** The location of Cardiff.

Re-urbanisation

People now see city centres as a good place to live. New apartments have been built on sites that used to have offices, factories or warehouses. The population of inner urban areas is increasing. This process is called re-urbanisation.

Year	Population
1801	1,870
1851	18,352
1861	48,965
1871	57,363
1881	96,637
1891	128,915
1901	164,333
1911	182,259
1921	222,827
1931	226,937
1941	No data
1951	243,632
1961	283,998
1971	293,220
1981	285,740
1991	296,900
2001	305,353
2011	346,000
2017	362,800

▲ **Figure 26** Population of Cardiff.

In 1841, 87,000 tonnes of coal was being shipped from Cardiff Docks. By 1862, the docks were exporting 2 million tonnes of coal each year and, by 1883, this figure had risen to 6 million. The maximum figure was 10.7 million in 1913. By 1946 it had fallen to 1 million tonnes. In 1970, Bute East Dock was closed. In 1980, the M4 was completed to the north of Cardiff.

▲ **Figure 27** Some significant dates in Cardiff's development

Activities

1 a) Identify five links that Cardiff has with the rest of the UK and/or the rest of the world.
   b) Classify these links as economic, political or social.
2 a) Select a suitable method to represent the information in Figure 26.
   b) Describe the changes in Cardiff's population carefully. In which decades did Cardiff's population rise most rapidly and in which did it decrease?
   c) Use the information in Figure 27 to create four labels for your graph. Your labels should help explain why Cardiff's population changed.
3 Study Figure 28. Describe how Cardiff changed between 1920 and the present day. Use the following specialist terms in your answer: suburban sprawl, inner urban and re-urbanisation.

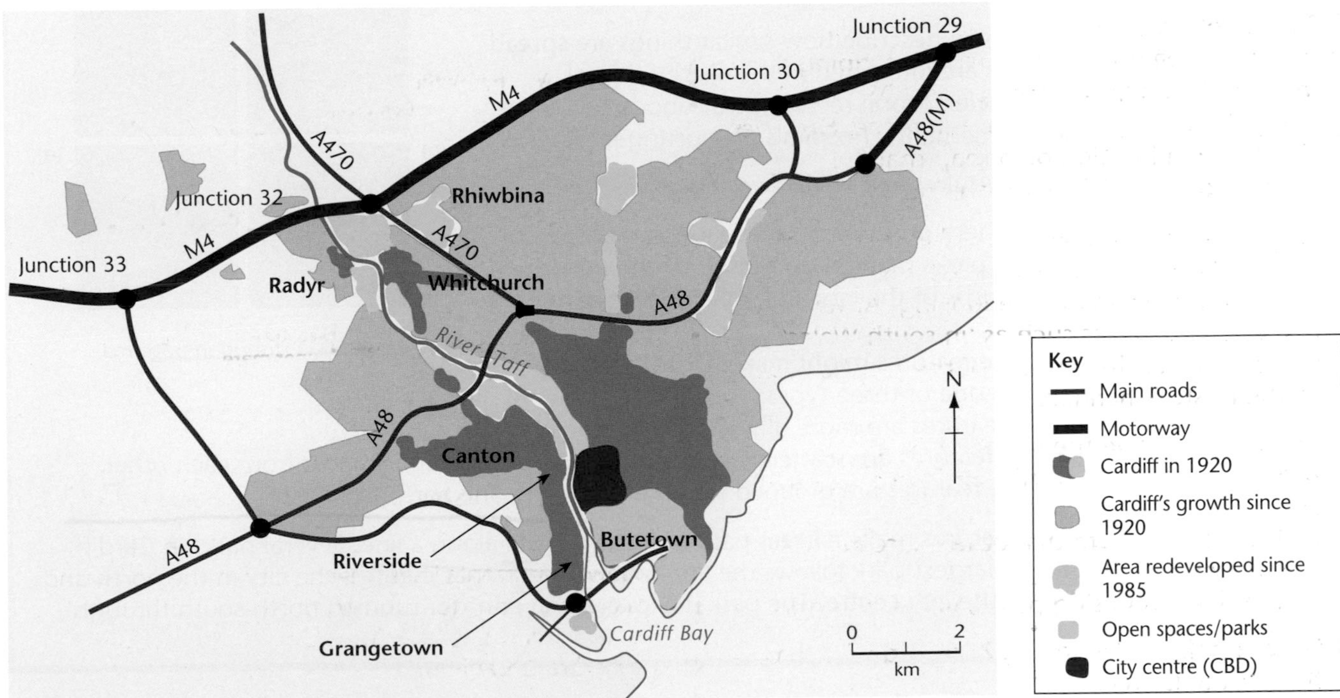

▲ **Figure 28** Cardiff's built-up area; the named neighbourhoods are described in later pages.

Cultural patterns

Cardiff is a multicultural city. Between 1800 and the 1930s **economic migrants** moved from other European countries and the countries of the British Empire to Cardiff. Many were sailors who worked on the ships that exported coal from South Wales. Most settled in Butetown, close to the docks, in an area that was then called Tiger Bay. Today, a total of 8 per cent of Cardiff's population are members of ethnic minorities. People have settled in the city from over 50 different countries. The two largest communities are descended from South Asian (India, Pakistan and Bangladesh) and Somali migrants.

The Somali population in Cardiff has a population estimated to be a little under 10,000. Most of them live in a relatively small neighbourhood within the inner urban area in the wards of Grangetown and Riverside. Somalis choose to live in this district to be close to other family members. The area has many shops that cater for the Muslim population such as halal butchers and fast food shops that sell food prepared using halal meat. The area has a number of mosques and Muslim cultural centres. This district has a wide variety of different sized houses and flats for rent and sale at a variety of prices.

▲ **Figure 29** A shop in Riverside, Cardiff, selling halal meat.

GEOGRAPHICAL SKILLS

Describing distributions

To describe a distribution is to describe how similar things are spread across a map. Geographers are interested in the distribution of natural features such as glaciers, coral reefs and volcanoes; as well as human features such as settlements, hospitals or sporting facilities.

Describing a distribution requires you to do two things.

1 You need to describe where on the map the features are located. For example, of the ten parks marked on Figure 31, the majority are in the northern suburbs of the city and only four are within the inner urban area.
2 Describe any pattern the features might make. Distribution patterns usually fall into one of three types:
   - Regular, where the features are more or less equally spaced.
   - Random, where the features are scattered across the map at irregular distances from each other.
   - Clustered, where the features are grouped together into only one part of the map.

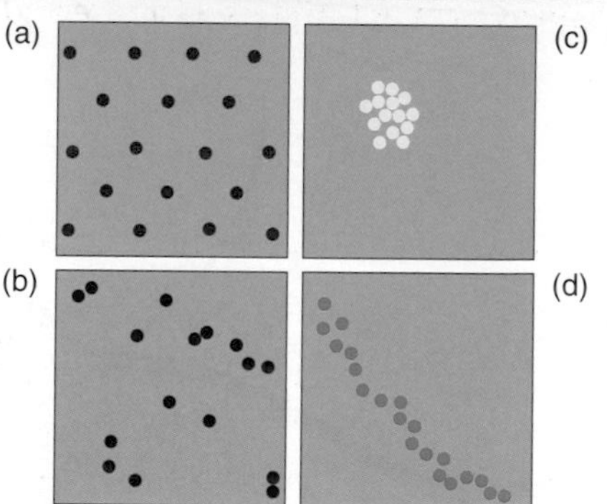

▲ **Figure 30** Distribution patterns.

In addition to this, some features make a **linear** pattern if they all fall along a line. Several parks in Cardiff make a linear pattern. The largest park follows the line of the River Taff as it enters the city in the north and flows southwards towards the city centre. The park makes a green corridor running north-south through the city.

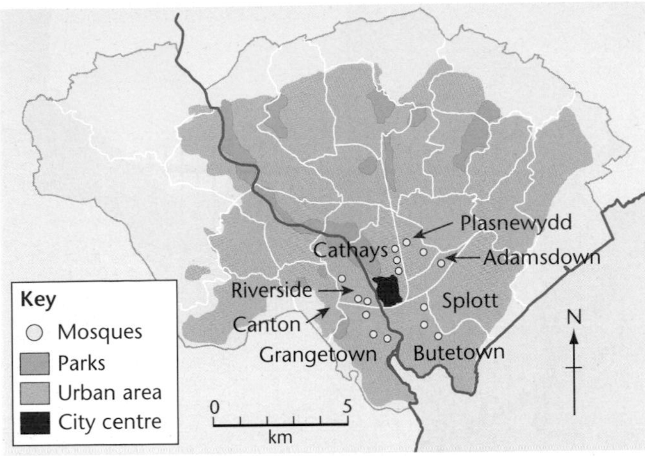

▲ **Figure 31** Map showing the distribution of mosques in Cardiff. The white lines show the ward boundaries.

The history of Somali migration to Cardiff

Somali migrants first arrived in Cardiff in the period 1880 to 1900. Many of them worked in the ships that were exporting coal from the docks in Butetown (which was then known as Tiger Bay). At this time Somali seamen settled in other UK ports, as well as Cardiff, such as the London docklands, Bristol, Hull and Liverpool. By 1945 there were around 2,000 Somali sailors and their families living in Cardiff. In 1991 a long civil war began in Somalia. Refugees from this conflict moved first to refugee camps in other African countries such as Kenya and Ethiopia. Eventually some of these people migrated to the UK where they joined the existing Somali community.

Ward	Black	Asian	Mixed ethnicity
Adamsdown	3.4	5.8	3.5
Butetown	13.4	8.1	8.3
Canton	0.8	4.7	1.8
Grangetown	4.2	13.2	3.8
Plasnewydd	1.5	9.5	1.7
Riverside	2.8	15.6	2.4
Splott	1.8	3.3	2.9

▲ **Figure 32** Wards with significant ethnic populations.

Economic migrants

Those who move home to find work. They are often young adults. A lot of people move from one region of the UK to another in search of work. Other economic migrants move into the UK from other countries. Most of these are from Commonwealth countries, such as India, or other European countries, such as Poland. Economic migrants contribute positively to the UK economy by doing a range of jobs in areas such as health, social care and agriculture.

Activities

1 Match the four distribution patterns shown in Figure 30 to the following terms:
- random
- regular
- linear
- clustered

2 a) Match the following geographical features to one of the four distribution patterns shown in Figure 30:
- motorway service stations
- high street banks in a large town
- primary schools in a city.

b) Suggest why these features are distributed in this way.

3 Use Figure 31 to describe the distribution of mosques in Cardiff. Suggest what this map tells you about the distribution of the Muslim population of Cardiff.

4 Use the text on these pages to draw a timeline of Somali migration to Cardiff.

5 Explain why the Somali population of Cardiff is found in a relatively small area of the city. Make sure you give one historical reason and one social reason.

Enquiry

How does Cardiff's growth compare to that of Mumbai?

Consider the importance of natural increase and migration patterns in both cities.

Urban challenges and solutions

Like other HIC global cities, Cardiff faces urban challenges that need solutions. For example, one challenge is how to improve the urban environment in the more deprived parts of the city. Another is to provide enough housing, especially housing that is affordable to people on lower incomes.

Tackling deprivation

Some neighbourhoods of Cardiff lack conditions that help give people a comfortable quality of life. They may lack safe places for children to play. Residents may be affected by poor housing conditions or poor health. In addition, some of these neighbourhoods have environmental issues such as poor air quality due to busy traffic.

Cardiff Council uses Neighbourhood Renewal Schemes to improve the physical environment of areas suffering **deprivation**. Local residents are involved in the planning stage to discover what is needed. The schemes are designed to make the local environment safer, more attractive and better used by local people. For example, there have been schemes to improve pavements, plant trees, install better street lighting and provide skate parks. Another project, designed to improve safety and reduce crime in many locations across central and southern Cardiff is the Alley Gating Programme (2019–2021). This scheme installs metal gates across the alleys that run behind the rows of Victorian terraced houses.

▲ **Figure 33** Metal gates are installed as part of the Alley Gating Programme.

▲ **Figure 34** The Millennium Walkway, in Cardiff city centre, is a location where graffiti artists are allowed to paint.

Deprivation

Lacking the social, economic or environmental features that give people a comfortable quality of life. Deprivation is a measure that takes into account a wide range of factors such as income and job type, health, safety and the physical condition of the local environment.

Activities

1 Discuss the schemes shown in Figure 35.
  a) Identify who would benefit from each scheme.
  b) Use the diamond ranking technique (see page 83) to rank all nine ideas. Which three do you think would be most effective at tackling deprivation? Think about the social, economic and environmental impacts of the scheme.

The Alley Gating Programme.
Improving street lighting.
Allowing graffiti in some areas.
Installing an all-weather football pitch in Splott.
Traffic calming by widening pavements and installing speed bumps.
Working with local residents to create a community garden.
Installing lowered kerbs for pedestrians and cyclists.
Planting trees.
Installing a children's playground in Lisvane.

▲ **Figure 35** Examples of schemes used in Cardiff to regenerate deprived neighbourhoods.

Housing issues

Cardiff council is spending £280 million to provide more housing for people on lower incomes. It plans to provide an extra 2,000 homes in 20 sites across Cardiff. Some of these houses will be brand new. Other houses will be purchased by the council and then rented out to tenants. This is the largest **social housing** scheme in the city since 1945.

Cardiff council will provide 21 temporary homes to homeless families by converting old shipping containers. This is a quick solution to the urgent demand for housing but it is controversial. Where containers have been used in other parts of the UK, tenants complain that the containers are cramped, cold in the winter and too hot in the summer. Other UK cities are converting unused office blocks into social housing.

Social housing

Homes that are provided at an affordable price for rent. Social housing is either provided by the local council or by a housing association.

Norwich council houses win Stirling architecture prize

The 105 creamy-brick homes are designed to strict environmental standards, meaning energy costs are around 70% cheaper than average. The walls are highly insulated and the roofs are cleverly angled at 15 degrees, to ensure each terrace doesn't block sunlight from the homes behind, while letterboxes are built into external porches, rather than the front doors, to reduce any possibility of draughts. The back gardens look on to a planted alley, dotted with communal tables and benches, while parking has been pushed to the edge of the site, freeing up the streets for people, not cars.

▲ **Figure 36** Newspaper article from the *Guardian* about an award-winning development of council houses in Norwich.

Activities

2 Suggest why UK cities need to provide extra social housing.
3 Study Figure 36 and the information on page 73 about Egan's Wheel. Identify the features of this development that may have made it an award-winning project.

Enquiry

Do you think it is right to convert old shipping containers or unused offices into homes for homeless families? To what extent do you think this is a sustainable solution to the challenge of providing enough social housing?

Bi-polar surveys

Some features of the urban environment are easy to count and quantify. An example would be the number of cars passing along a street in ten minutes. However, fieldwork enquiries sometimes involve features that cannot be quantified so simply. People's perceptions of the urban environment are one example of something we might want to measure. Because perceptions vary a lot from one person to another we say they are subjective. So how do we assess perceptions?

▲ **Figure 39** Flats and neighbourhood shops in RIV07 (grid reference 169765) on Figure 34.

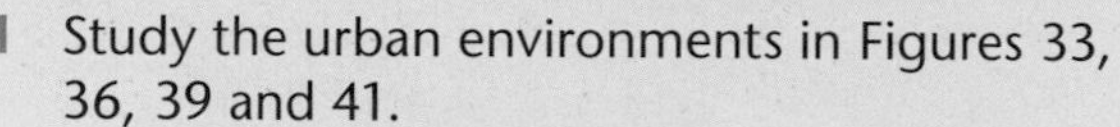

Activities

1 Study the urban environments in Figures 33, 36, 39 and 41.
   a) Use the bi-polar assessment in Figure 40 to calculate a score for each urban environment in these four photographs.
   b) Share your scores with at least four members of your class. Calculate a mean score for each photograph.
   c) Create a sketch map of Riverside using information from Figure 34. Plot the location of Figures 33, 36, 39 and 41 on your map. Choose a suitable method to represent your mean scores on this map.

Creating a simple bi-polar survey

Step 1 Choose categories for the bi-polar statements. These will depend upon the specific focus of your study, but may include factors such as presence/absence of natural vegetation, street lighting, upkeep of building and condition of pavements. In some cases more personal attitudes may be investigated such as fear of crime or feelings of safety. Pairs of opposing statements are put at either end of a scale as can be seen in Figure 40. A bi-polar scale has a range of values, for example, from -5 to +5. The positive and negative values indicate a person's perception of an environment. The zero is usually removed from the middle of the scale to discourage people from choosing the safe middle option.

Step 2 Choose the location for your survey. Your choice of location may be the result of a pilot study, having previously visited the area, or as a result of some previous knowledge of your study area. If you are unfamiliar with your site then you might use aerial photography (Bing Maps, Google Earth, for example) or an application such as Google Street View to help you choose your sites before going on the field trip. You could use secondary data (like Figures 37 or 38) from http://www.neighbourhood.statistics.gov.uk to help select your sites. If so, you would want to make sure that you visited at least eight sites in Riverside, Cardiff. This is at least one in each LSOA.

Step 3 Collecting the bi-polar scores. The simplest method is to complete the survey yourself. However, in order to investigate differing perceptions you will need to compare your score with other people. These could be members of your class or members of the public. You can combine scores together to create a 'class score' or by calculating the mean score for each location. If so, these mean scores could then be represented on a base map of the area that you visited.

	+5	+4	+3	+2	+1	−1	-2	−3	−4	−5	
Attractive urban environment											Unattractive urban environment
Safe for pedestrians											Unsafe for pedestrians
Natural features nearby											No natural features nearby
Thriving communities with job opportunities											Declining communities with few job opportunities

▲ **Figure 40** An example of bi-polar statements.

Creating a more sophisticated bi-polar assessment

Here are three suggestions to try. In each case, think about the strengths and weaknesses of these assessments compared to a standard bi-polar survey.

1 **Different perceptions** Conduct a standard bi-polar survey as yourself – a teenager. Then imagine what it might be like to perceive the urban environment as though you were someone else. Try doing the bi-polar survey again but imagining yourself as someone with limited mobility, for example, or as a single parent with a very young child. How would your perception change?

2 **Photo surveys** Decide on five categories, or features, for your bi-polar survey. For example, open spaces, building design, pedestrian safety, leisure features and road traffic. Don't write opposing statements. Inspect your area carefully, taking as many photographs as you can. Make sure you record where each photo was taken. Then choose the pair of photos that represent the best and worst examples of each category or feature.

3 **Weightings** Not all of the categories or features recorded by a bi-polar survey have equal importance to people when they think about the quality of the environment. For example, you may decide that open space is a more important feature than road traffic or litter. Try ranking or weighting your bi-polar statements. For example, if you think the amount of open space is twice as important as litter then you should multiply your bi-polar scores for open space by a two. Discuss these weightings to get an agreement as a class or group. How might your weightings be different if you were an OAP rather than a teenager?

▲ **Figure 41** Flats and neighbourhood shops in RIV02 (grid reference 179759) on Figure 34.

Enquiry

Write four new pairs of bi-polar statements. Use them to survey the neighbourhood around your school.

- Conduct the survey as yourself.
- Now imagine how another person might perceive this neighbourhood. Imagine yourself as:
  i) a single parent with a young child
  ii) an elderly resident.
- What are the strengths and limitations of this more sophisticated style of bi-polar survey in trying to collect different perceptions?

Chapter 1 Tectonic processes and landforms

What are the largest scale tectonic processes?

The Earth's crust may appear to be a stable and unchanging place but, in fact, it is constantly changing and in motion. In 1835 Charles Darwin said: 'Nothing, not even the wind that blows, is so unstable as the crust of this earth.' His visit to the volcanic Galapagos Islands inspired him to think about the forces that shape the Earth. We now understand that the Earth's crust is made up of several very large **plates** that move relative to one another. Plates move towards each other on **destructive plate margins** and away from each other on **constructive plate margins**.

▲ **Figure 1** The crater of Villarica and Lanín (a stratovolcano) in Chile; this tectonic landscape is typical of the fold mountains in the Andes, South America.

Why do the plates move?

Heat from deep within the Earth causes molten rock, or **magma**, to rise up through the **mantle**, which is the thick layer beneath Earth's thin, solid crust. It is this rising magma that causes volcanic eruptions in places like Iceland in the Mid-Atlantic. Magma rising through the mantle may also explain the movement of the plates that make up our crust. There are two theories: slab pull and convection.

Slab pull is one explanation for plate movement. In this theory, plates move because gravity pulls them apart. As magma rises beneath a constructive plate margin it heats the rocks of the crust and forces them to bend upwards. The crust is raised, forming the **mid-ocean ridge**, which is about 2,500 m higher than the ocean floor. Gravity pulls the crust and it slides down the slope away from the centre of the mid-ocean ridge, allowing further eruptions. An even greater gravitational force pulls at the crust at the other end of the plate. This occurs where the oldest and densest ocean crust bends downwards and slides back into the mantle. This process is known as **subduction**. The huge mass of the subducting slab pulls the rest of the ocean plate away from the mid-ocean ridge.

Subduction

The process where one plate is destroyed as it slowly sinks underneath another plate.

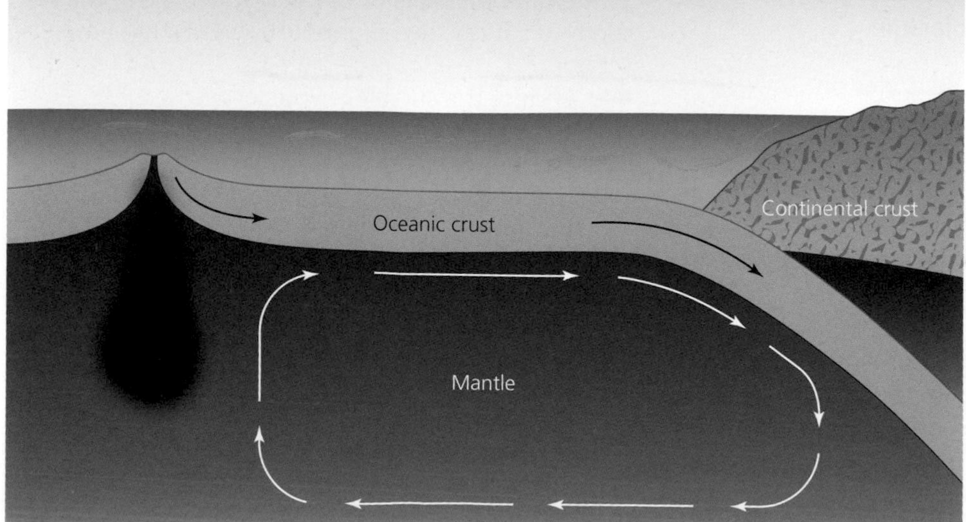

A plume of hot magma rises through the mantle

The oceanic crust is warmed and forced upwards by the magma, creating a mid-ocean ridge

The ocean crust cools, becomes denser and slides away from the ridge under gravity

A deep ocean trench is formed where the oceanic crust flexes downwards under the continental crust

The immense weight of the oceanic crust pulls the plate as it subducts into the mantle

▲ **Figure 2** The processes that drive plate movement.

Convection is an alternative explanation for plate movement. It is a process that transfers heat energy. You can see it in a boiling saucepan: the contents of the pan are heated by the hob beneath; the heat rises through the contents of the pan and you see bubbles burst at the surface, which is a little cooler; the cooler fluids sink back into the pan towards the heat source, and the whole process starts again. This creates a circular convection current. It is thought that a similar process may be happening in the mantle. The Earth's core is the heat source and magma rising through the mantle transfers this heat energy away from the source. If convection currents exist within the mantle then they may help to explain the movement of the plates seen in Figure 2.

Activities

1 Make a copy of Figure 2.
   a) Add the labels to suitable places on your diagram to describe the process of slab pull.
   b) Add your own labels in a different colour to explain the process of convection.

Destructive plate margin

A zone in the Earth's crust where plates are converging (moving towards each other).

Large-scale features of destructive plate boundaries

Ocean trenches are formed where the ocean crust bends downwards and slides back into the mantle on destructive plate margins. Ocean trenches are long, narrow, deep features in the seabed. Most are deeper than 6 km. The Mariana Trench, in the western Pacific Ocean, is 11 km deep, 2,500 km long and has an average width of just 70 km. Subduction is not a smooth process. Friction locks the plates together. As pressure increases, friction is finally overcome and the plates move with a violent jerk. These large earthquakes can have huge impacts. It was this type of subduction zone earthquake that caused the 2011 tsunami in Japan (see page 128).

Fold mountain ranges, such as the Andes of South America, are another large-scale feature of destructive plate margins. The Andes are 7,000 km long and several peaks rise to over 6,000 m. They are formed as continental crust is squashed together and folded upwards. Heat from the mantle and friction caused by subduction melt the oceanic plate beneath the Andes. The resulting magma rises through the crust above and forms a chain of large volcanoes, like those in Figure 3.

Activities

2 a) Describe the landscape in Figure 1.
   b) Use Figure 3 to describe the processes responsible for this landscape.
3 Explain why scientists need to understand the process of subduction.

Enquiry

Are all sections of the mid-ocean ridge system similar?

- Research these two sections:
  a) Mid-Atlantic Ridge
  b) East Pacific Rise.
- Describe the similarities and differences between these two sections of the mid-ocean ridge system.

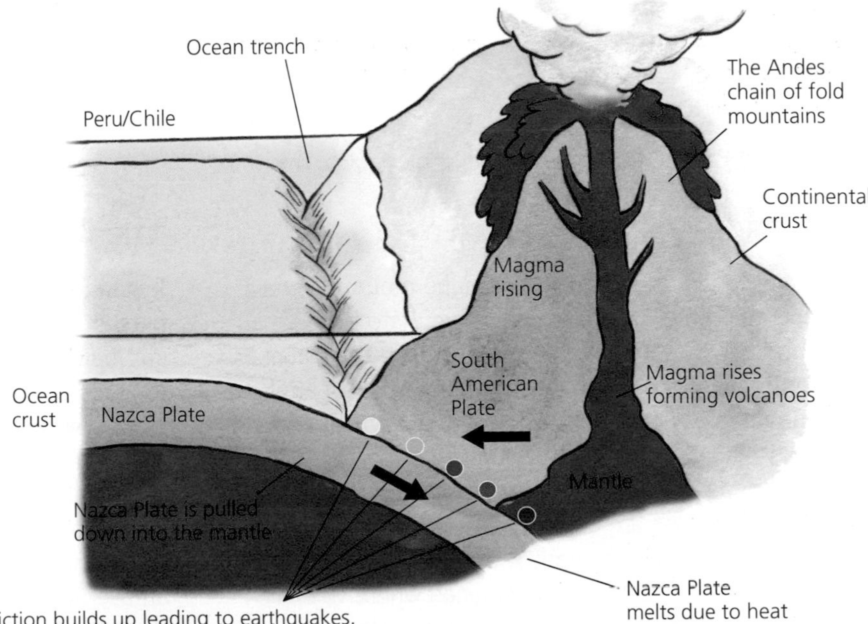

▲ **Figure 3** How ocean trenches and fold mountains are formed in South America.

What large tectonic features are found at constructive plate margins?

Iceland, in the mid-Atlantic, is formed by volcanic eruptions on a **constructive plate margin**. Earthquakes occur along these plate boundaries caused by the wrenching apart of plates and the movement of magma forcing its way through **fractures** in the crust. These earthquakes are usually of a small to medium magnitude but they are felt strongly as they occur close to the surface. However, as most constructive plate boundaries are in mid-ocean, these earthquakes usually pose little threat to people.

Constructive plate margin

A region in the Earth's crust where plates are pulling apart (also known as a divergence zone).

Activities

1 Use Figure 4 to name pairs of plates that form:
   a) constructive plate margins
   b) destructive plate margins.
2 Use Figure 4 and an atlas to name:
   a) four countries that are on destructive plate boundaries
   b) four countries that are on constructive plate boundaries.
3 a) Make a sketch of Figure 5.
   b) Match the labels to the five numbered features in the photograph.

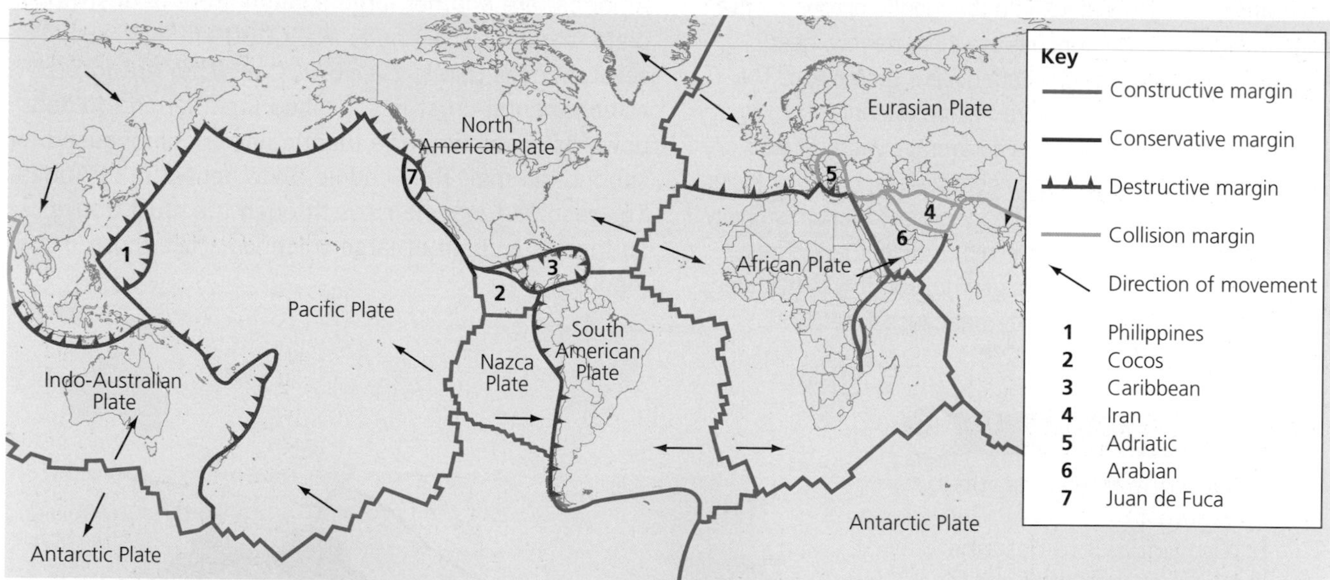

▲ **Figure 4** Plate margins and the direction of plate movement.

The mid-ocean ridge system

The mid-ocean ridge system is a huge mountain chain formed along the Earth's constructive plate boundaries. The ridge is an average of about 2,500 m high (about twice the height of Snowdon) and over 65,000 km long. It wraps itself around the world a little like the seam on a cricket ball. The central region of the ridge is deeply cracked and fissured where the crust is pulling apart. Volcanic eruptions are common along the ridge. Most of the mid-ocean ridge is about 2,500 m beneath sea level, but the ridge is visible in a few places where volcanic activity has created islands, such as in Iceland.

The formation of rift valleys

Iceland sits astride the Mid-Atlantic Ridge on a **volcanic hotspot**. Iceland is being gradually torn in two and this can be clearly seen in Thingvellir National Park. Over time a giant fissure, 7.7 km long, has opened in the Earth's surface. As the fissure has widened, the land at the centre has gradually collapsed downwards. This process has formed a **rift valley** (or **graben**) with a flat valley floor and steep **escarpments** along either side. The rift continues to grow. The valley walls are pulling apart at an average of about 7 mm per year and the valley floor is subsiding at about 1 mm per year.

The flat rift valley floor is formed by subsidence

Steep escarpments on the western edge of the rift

The gentle slopes of a shield volcano

Deep fissures in the valley floor indicate further divergence

Tilted slabs of crust subside along the edge of the rift

◀ **Figure 5** A view northwards from the western escarpment of the rift valley at Thingvellir, Iceland.

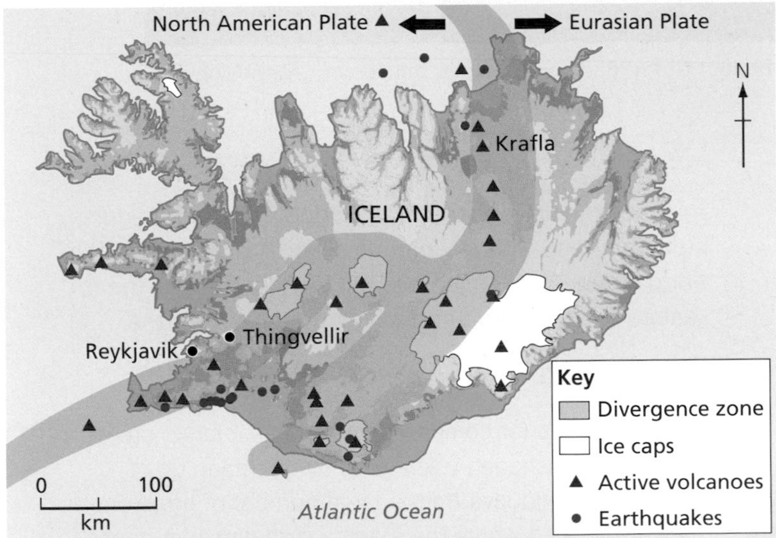

▲ **Figure 6** Map of Iceland showing where the divergence zone crosses the country.

▲ **Figure 7** Formation of the rift valley at Thingvellir, Iceland.

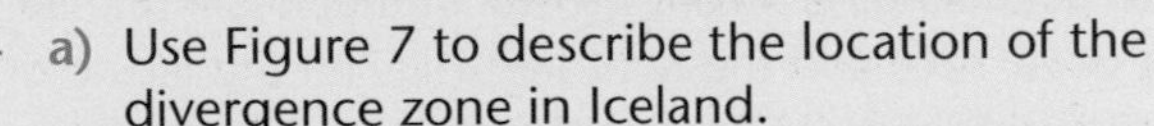

4 a) Use Figure 7 to describe the location of the divergence zone in Iceland.
b) Suggest a hypothesis that explains the pattern made by volcanoes and earthquakes in Iceland.
c) Where would you expect to find the oldest rocks in Iceland and why?

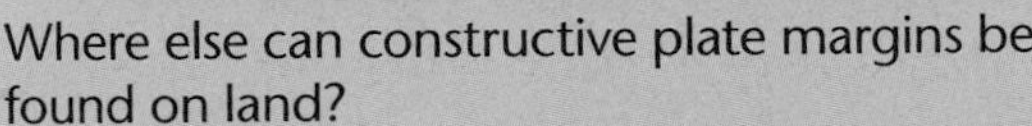

Where else can constructive plate margins be found on land?

Research the features of the African Rift Valley. Create a poster describing the tectonic features found here. Include:

- a sketch map showing plate boundaries and the direction of plate movement
- a description of the rift valley and one active volcano in this region.

Features of volcanic landscapes

Volcanoes are created on both constructive and destructive plate boundaries when magma is forced to the surface. The shape of the volcano depends on the type of magma and its gas content.

▲ **Figure 8** The Erta Ale shield volcano in Ethiopia has a summit at 613 m and gentle slopes; it is famous for having a lava lake in its crater.

Shield volcanoes

Magma from constructive plate boundaries tends to have a low **viscosity** and low gas content. The lava spreads out across wide, almost flat, **lava fields** rather than building steep-sided cones. Volcanoes formed in this way are called **shield volcanoes** because of their gentle slopes and circular shape. Examples of shield volcanoes include Erta Ale in Ethiopia, Skjaldbreidur in Iceland and Mauna Loa in Hawaii. Mauna Loa is the world's largest volcano. It is 9,170 m high from the seabed to its summit and 125 km across its base.

▲ **Figure 9** Mount Merapi in Indonesia is a stratovolcano; notice the deeply eroded river valleys that have been cut through layers of soft ash from numerous eruptions.

Stratovolcanoes

A **stratovolcano** (or composite volcano) is a large, steep-sided, conical-shaped volcano. They are made up of layers of ash and lava from a large number of eruptions. They are formed where the magma contains a lot of gas. This gives the lava a high viscosity, making it less able to flow freely. During an eruption the sticky lava may make a steep-sided dome.

Stratovolcanoes are commonly found on destructive plate boundaries – Mount Merapi in Indonesia (Figure 9, and page 120) and the Soufrière Hills on the Island of Montserrat (page 118) are examples. However, they can also occur on mid-ocean ridges, such as Mount Pico in the Azores (page 112). Some stratovolcanoes occur close to large populations and their explosive style of eruption makes them dangerous. Vesuvius in Italy and Merapi in Indonesia are examples.

The shield volcanoes of Hawaii

The Hawaiian Islands are a chain of very large mountains formed by volcanic activity, but they are not close to a plate boundary. The Hawaiian Islands have formed over a **volcanic hotspot** in the middle of the Pacific Plate. The nearest plate boundary is 3,200 km away. Figure 10 shows that, as the Pacific Plate moves slowly across the hotspot, magma rises through the crust forming a volcanic island. As time passes, the plate moves but the hotspot stays in the same place, so that new shield volcanoes are formed behind a line of increasingly older islands that contain extinct volcanoes.

Volcanic hotspots

Locations on the Earth's crust that are particularly active. It is thought that hotspots occur where super-hot plumes of magma rise through the mantle. The plume finds a weakness in the crust and breaks through to create volcanic activity. Some hotpots are on plate boundaries. One example is Iceland, which is a hotspot on the Mid-Atlantic Ridge. Hotspots can also occur in the middle of plates.

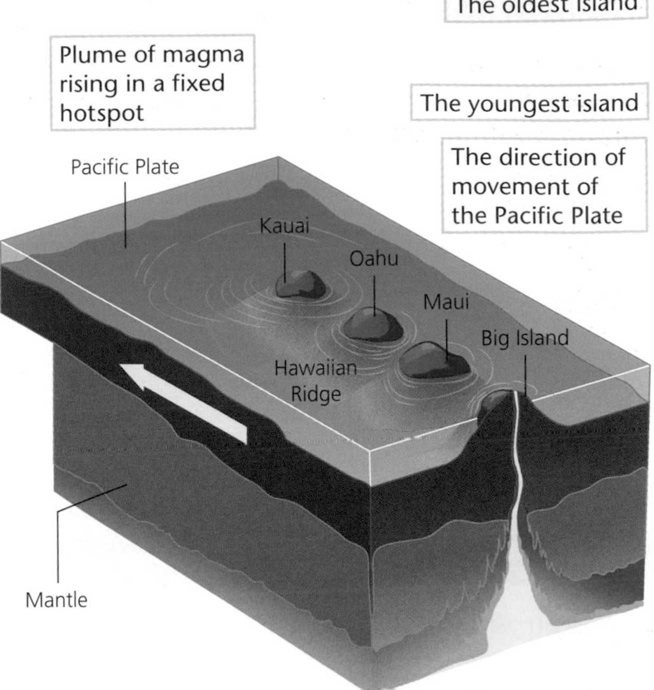

▲ **Figure 10** Hotspot formation of volcanic island chains, such as the Hawaiian Islands.

Geysers

In many volcanic regions, water interacts with heat in the ground to create thermal features such as **geysers**, hot springs, fumaroles and mud pots. These small-scale features are formed when rainwater or snowmelt percolates downwards and meets hot rocks. The water is heated and expands. This increases the pressure and forces a mixture of boiling water and steam back up through fissures in the ground. It sometimes emerges in spectacular hydrothermal explosions that send columns of steam high into the air. Repeated explosions of a geyser can create a small crater, a few metres in diameter.

Activities

1. Describe the main similarities and differences between shield and stratovolcanoes.
2. Why are stratovolcanoes considered more dangerous than shield volcanoes? What damage could shield volcanoes cause?
3. a) Make a copy of Figure 10.
   b) Add the labels to suitable places on your diagram.
4. Draw a simple labelled diagram or flow chart to show how a geyser works.

Enquiry

Create a hierarchy of volcano types according to their size and danger. Describe any problems you had deciding on places.

◀ **Figure 11** Steam, in the centre of the image, forcing a bulge of water out of a small hydrothermal crater; this geyser, at Strokkur in Iceland, erupts every 8–10 minutes.

Understanding volcanic landscapes

Shield and stratovolcanoes are large volcanic features, but volcanic landscapes contain many separate landforms of different **scales**. It is the combination of these different landforms that creates a distinctive and unique landscape.

◀ **Figure 12** A view of Pico, in the Azores, from Faial; both are volcanic islands located on the Mid-Atlantic Ridge.

A small rift valley (graben)

A 2,351 m high stratovolcano with a typical steep-sided conical shape

A group of at least three cinder cones, each about 100–250 m high

Gently sloping lava fields on the lower slopes of the stratovolcano; lava tubes have been discovered here

A small lava cone that grew within the crater of the stratovolcano during its last eruption

Cinder cones

The smallest, most common type of volcano is a **cinder cone**. These cone-shaped hills can be found on their own (such as Parícutin, Mexico) or on the slopes of other, larger volcanoes. They are made from cinders (or **scoria**) created when lava is ejected into the air from a single vent. The lava spray cools very quickly as it is thrown through the air, falling to the ground as hot cinders. These build up around the vent to form a steep-sided, circular cone. Cinder cones usually form during a single eruption. This explains why they don't reach the size of composite volcanoes, which evolve over many eruptions.

Lava tubes

Figure 5, page 116, shows a lava flow in Hawaii. The surface of the lava has been chilled by the air and is solidifying. This layer insulates the lava below, which remains liquid – you can see red-hot lava at over 1,000 °C in the centre of the photo. Underground rivers of lava can flow just below the solidified crust as far as 50 km from the volcano's crater. When the eruption stops, the liquid lava drains away and an empty **lava tube** is left. These lava tubes are typically oval in cross section, 5–10 m wide, 1–5 m high and hundreds of metres in length. Some tubes have branching patterns where tributary rivers of lava have joined one another.

▲ **Figure 13** A lava tube on the lower slopes of Mount Pico, the Azores; notice the formations on the roof where hot lava has solidified as it dripped from the ceiling of the tube.

Calderas

In some violent eruptions, so much lava and ash is ejected that the **magma chamber** that supplies the volcano empties. The upper part of the volcano collapses into the empty space below creating a circular depression which is larger than a typical volcanic vent. These features are called **calderas** (from the Spanish word for caldron). Calderas vary greatly in size. The caldera on Faial, in the Azores, shown in Figure 14, is 2 km across and 400 m deep. The Yellowstone caldera, in North America, was formed by a much more explosive eruption and, consequently, is much larger. This caldera is 55 km by 72 km across.

▲ **Figure 14** A view into the caldera on Faial, the Azores.

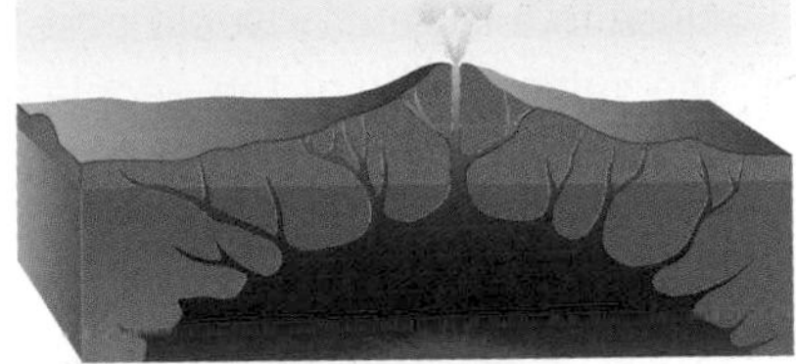

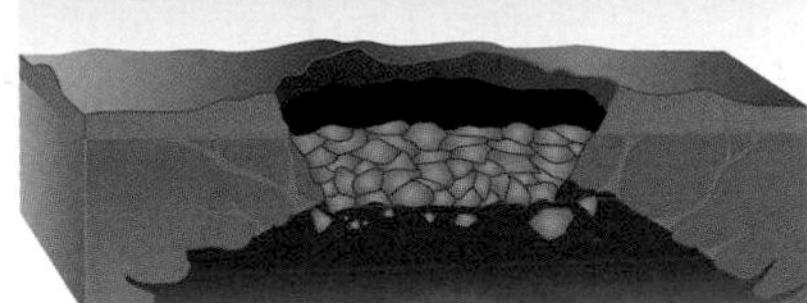

▲ **Figure 15** Formation of a caldera.

The steeply sloping wall of the caldera is 400 m high
A small cinder cone with a central vent is at the bottom of the caldera
A recent landslide on the steep slope has removed all vegetation
Piles of scree at the foot of the caldera wall from old landslides

Scale

Scale is an important geographical concept. Landscapes are often a complex mix of different features and landforms. Scale helps geographers to break the landscape down into its different parts and describe the relative size of these geographical features and the processes that create them.

Activities

1. a) Make a sketch of Figure 12.
   b) Add the correct label to each of the numbered points.
2. a) Make a copy of Figure 15.
   b) Add your own labels to the diagrams to describe the formation of a caldera.
3. a) Match the correct label to each of the numbered points in Figure 14.
   b) Which of the features in Figure 14 is the most recent and why?
   c) Why must the formation of the cinder cone be more recent than the last large eruption of this volcano?
4. Explain why an understanding of scale is important when trying to interpret the landscape in Figures 12 and 14.
5. Draw a series of three diagrams to show the formation of a lava tube.

Chapter 2 Vulnerability and hazard reduction

Why are some communities vulnerable to tectonic events?

Tectonic processes create a variety of potentially dangerous events, which include volcanic eruptions, earthquakes and tsunami. The threat to people created by these events depends on a number of factors:

- the strength or **magnitude** of the event
- population size and density, and its closeness to the event – an underwater volcanic eruption on a mid-ocean ridge is not a hazard, but a volcanic eruption close to a large city would pose a huge threat
- the **vulnerability** of the people affected by the event; vulnerability is linked to factors such as poverty.

Vulnerability

A measure of someone's inability to cope with, or recover from, a disaster such as an earthquake or other tectonic hazard. Some groups of people are more vulnerable to hazards than others. Poverty, age, gender and disability are all factors that can affect vulnerability. For example, someone's poverty may mean that their house is badly built and unable to withstand ground shaking during an earthquake.

Capacity

The opposite of vulnerability. It describes someone's ability to survive a hazard or recover from it quickly. Capacity is increased where individuals or communities have the resources they need to cope with the hazard. Resources can be financial or material, such as strong building materials. Capacity can also be improved by factors such as education, technology and disaster preparedness and planning.

Activity

1 Use Figure 1 to describe the location of the world's most active volcanoes.

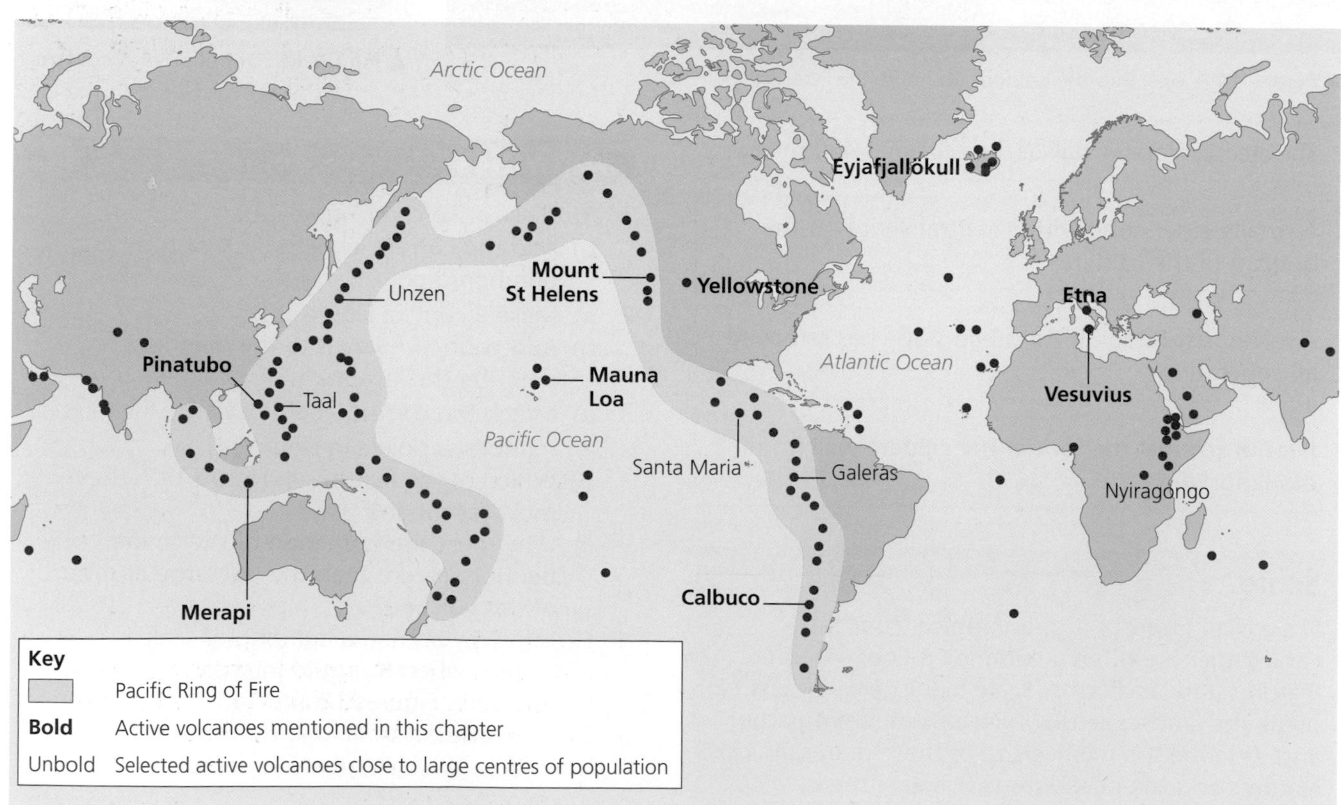

▲ **Figure 1** The location of the world's most active volcanoes.

Country	Number of active volcanoes	Cities with a population greater than 300,000	Gross national income (GNI) per person (US$)	Percentage of people living on less than US$1.90 per day
Chile	108	6	14,670	0.7
DR Congo	6	16	490	76.6
Iceland	35	0	60,740	0
Indonesia	147	32	3,840	5.7
Japan	118	32	41,340	0
Papua New Guinea	67	1	2,530	38.0
Philippines	53	30	3,850	7.8

▲ **Figure 2** Selected countries that have a significant number of active volcanoes.

How can we reduce vulnerability to tectonic events?

Across the world there are 67 large cities (with a population greater than 100,000) that are close to active volcanoes. These include Tokyo in Japan, Manila in the Philippines and Mexico City, which all have populations greater than 10 million. The number of large cities at risk from earthquakes is even greater. There are 38 major cities at risk from earthquakes in India alone. To reduce vulnerability to a tectonic event, organisations need to:

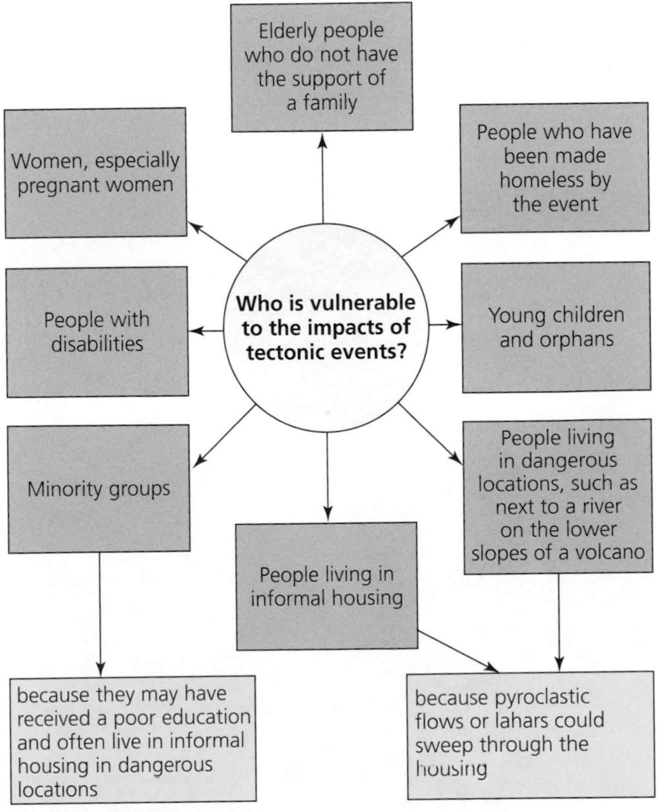

▲ **Figure 3** Why are some groups of people more vulnerable than others to natural disasters?

- *Reduce the impact of the hazard*: it is possible to reduce the impact of a volcanic hazard by monitoring volcanic activity, making predictions about possible eruptions and evacuating people. In the longer term, risk can be managed by mapping the hazards and restricting the movement of people in zones that are at risk. This has been attempted on the volcanic island of Montserrat in the Caribbean.
- *Build* **capacity** *to cope with the tectonic hazard*: for example, educating people about what to do during an earthquake so that they are more likely to survive, or creating a disaster plan so that emergency services, hospitals and local authorities all know what to do during and after a major earthquake. In Japan and California, USA, a lot of money is spent on the organisation of disaster plans in the event of an earthquake.
- *Tackle the root causes of vulnerability*: this means that governments need to reduce poverty and inequality in society so that everyone has the same opportunities and the same level of protection during a disaster.

Activities

2 Study Figure 2.
   a) Explain why volcanoes pose a greater threat in Japan than in Iceland.
   b) Suggest which two countries may have the greatest vulnerability to volcanic activity. Justify your choice.

3 Choose four different groups of people in Figure 3.
   a) For each group explain why they may be more vulnerable to natural disasters.
   b) For one of these groups, suggest how their capacity could be improved to cope with an earthquake.

Volcanic hazards

The magnitude of volcanic eruptions is measured using the **Volcanic Explosivity Index (VEI)**. It measures the height and volume of the plume of material ejected from a volcano. Each point on the scale is ten times the magnitude of the one before it.

VEI	Amount of material ejected	Height of plume	Example
0	<10,000 m^3	<100 m	
1	>10,000 m^3	100 m–1 km	Nyiragongo, DR of the Congo (2002)
2	>1,000,000 m^3	1-5 km	
3	>10,000,000 m^3	3-15 km	Soufrière Hills, Montserrat (1995)
4	>0.1 km^3	>10 km	Calbuco, Chile (2015)
5	>1 km^3	>20 km	
6	>10 km^3	>20 km	Pinatubo, Philippines (1991)
7	>100 km^3	>20 km	
8	>1,000 km^3	>20 km	

▲ **Figure 4** The Volcano Explosivity Index (VEI).

Lava flows

Lava flows rarely cause death or injury. They usually move slowly and people can simply move out of their way. A lava flow will, however, destroy anything in its path, including buildings, roads and crops. Blocked roads can hamper aid efforts or prevent evacuation.

Kilauea, one of Hawaii's five volcanoes, has been erupting since 1983. Lava from this eruption has destroyed 214 buildings, covered over 100 km^2 of land and buried 14 km of roads, but has not directly caused a single fatality. It is possible to divert lava flows but this isn't generally done in Hawaii for spiritual reasons. The Mauna Loa volcano observatory in Hawaii is protected by barriers as it contains very expensive and important equipment. It is the only lava barrier in Hawaii.

Activities

1 Describe the landscape in Figure 5.
2 Give three reasons why a decision may be made not to divert a lava flow.

Enquiry

How do the magnitudes of recent eruptions compare?

- Research each of the following eruptions:
  a) Eyjafjallajökull, Iceland (2010)
  b) Mount St Helens, USA (1980)
  c) Tambora, Indonesia (1815)
  d) Stromboli, Italy (most recent)
- Record the amount of material ejected, and add each event to a copy of Figure 4.
- Describe the economic and social impacts of one of these events.

Lava flows often threaten towns and villages on the slopes of Mount Etna in Italy. Diverting these flows is costly, and deciding which land to save and which to sacrifice raises complex legal, political and social issues. There are a number of successful methods that can be used to divert lava flows, including:

- spraying lava with water to cool and solidify it, so it acts as its own barrier to the molten lava behind
- creating earth embankments to channel the direction of the lava flow
- dropping concrete blocks via helicopter to divert the flow
- blowing up lava tubes to widen them and spread out the flow so it solidifies quicker and doesn't travel as far.

▲ **Figure 5** A slow-moving lava flow on Big Island, Hawaii.

Ash clouds

Calbuco is a stratovolcano in Chile. It erupted violently in April 2015 creating a huge **ash cloud** that can be seen in Figure 7. Ash clouds form when an explosive eruption throws a mixture of gas, fractured rock and tiny lava droplets high into the air. The lava droplets cool rapidly and solidify into tiny fragments of glass-like ash. These ash particles are sharp and abrasive. If sucked into the engine of a passenger jet it can cause its engines to stall. In 2010 the eruption of Eyjafjallajökull in Iceland caused an ash cloud that rose up to 10 km high. The cloud spread over much of Northern Europe, forcing the cancellation of around 100,000 flights at a cost of around US$2.6 billion to the aviation industry.

Falling ash collects on power and communication lines as well as on buildings. When rain falls the ash becomes heavier, causing structural damage and power blackouts. Ash can have health effects too. It can pollute water supplies, and inhalation of volcanic ash can lead to lung damage. Crops can also be damaged by ash, leading to food shortages and higher food prices.

▲ **Figure 6** Map showing the extent of Calbuco's ash cloud on 25 April 2015.

▲ **Figure 7** The Calbuco volcano in southern Chile erupted in 2015 after lying dormant for 43 years.

1893–94	1895	1906	1907	1909	1911–12	1917	1929	1961	1972	2015

▲ **Figure 8** Years in which Calbuco, Chile, has erupted.

Ash cloud

A large-scale volcanic hazard caused by the eruption of tiny fragments of volcanic rock into the atmosphere.

Eruption of Calbuco, Chile (2015)

- The ash plume rose higher than 15 km.
- Pyroclastic flows travelled up to 7 km.
- Lahars (volcanic mud flows) travelled 15 km.
- International flights were delayed or cancelled.
- The town of Ensenada (population 1,600) was covered in 50 cm of ash; some houses collapsed from the weight of the ash on their roofs.
- About 5,000 people were evacuated.

Activities

3 Use the Factfile to describe:
   a) the magnitude of the Calbuco eruption in 2015
   b) the economic and social impacts of the event.

4 a) Plot a frequency graph using the data in Figure 8.
   b) Calculate the frequency of eruptions of Calbuco.

5 Use Figure 6 to describe:
   a) the location of Calbuco
   b) the approximate area and location of the ash cloud.

6 a) Explain why flights across Europe were cancelled during the Eyjafjallajökull eruption in 2010.
   b) Suggest how this may have affected business travellers and tourists.

Why are pyroclastic flows so hazardous?

Pyroclastic flows are fast-moving clouds of super-heated gas, ash and rock. They are caused by the collapse of a column of ash. Alternatively, they occur when a dome of cooling lava collapses. This triggers the release of energy from the magma chamber beneath, causing the sudden and violent eruption of ash and gas.

Pyroclastic flows are deadly. The British Geological Survey, which monitors Soufrière Hills in Montserrat, has recorded temperatures between 100 and 600 °C inside the flows.

Pyroclastic flows on Montserrat typically travel at around 110 kph down the volcano, carrying rock fragments ranging in size from fine ash to boulders the size of a small car. It can demolish buildings and the high temperatures start fires. Even on the edges of the flow, people are in danger from burns and the inhalation of hot, choking gases. Although pyroclastic flows are clouds of solid material they move like a liquid, hugging the ground and following contour patterns downwards much like an avalanche. However, they also have enough mass and energy to travel uphill on the lower slopes of a volcano. This makes their paths difficult to predict.

▲ **Figure 9** The collapse of an ash column can trigger pyroclastic flows.

Ash rises in a column up to 20 km into the atmosphere; as pressure is released from the magma chamber, the gas in the rising column expands; it's like shaking a bottle of pop before you take the lid off

The weight of the ash cannot be sustained by the eruption; it collapses due to the effect of gravity

The pyroclastic flow is a mixture of heavier rock fragments, ash and gas; It rolls down the slopes, hugging the ground

Lighter rock fragments and ash billow upwards above the main part of the flow

▲ **Figure 10** A pyroclastic flow on Soufrière Hills, Montserrat.

Activities

1 Describe how pyroclastic flows could affect:
   a) people
   b) the economy
   c) the environment.
2 a) Make a copy of Figure 9.
   b) Add the annotations to suitable points on your diagram.
3 Suggest suitable labels or annotations for each of the three points in Figure 10.

Hazard mapping

The level of threat of tectonic events, such as pyroclastic flows, increases not only with the magnitude of the event but also with the vulnerability of local communities. If communities can be removed from these areas, then the threat disappears. **Hazard mapping** involves creating a map that identifies threat levels. It allows local authorities to:

- limit access to hazardous areas, for example, allowing people to only have access during daylight hours or only allowing scientists to enter the most dangerous zones
- control the development of areas considered to be at risk from tectonic events, for example, preventing the building of new homes in hazard zones.

The Soufrière Hills volcano in Montserrat erupted violently in 1995. Since 2012, there have been no eruptions, although scientists continue to monitor the volcano carefully. The southern part of the island has been divided into six zones, shown in Figure 11. Zone V is out of bounds except to scientists, although daytime access to some areas of this zone is permitted when volcanic activity is low. There are also maritime exclusion zones because pyroclastic flows can reach the coast and travel out to sea. Many residents have left the island since the 1995 eruption began – many moved to the UK.

Weblink

www.mvo.ms – website of the Montserrat Volcano Observatory.

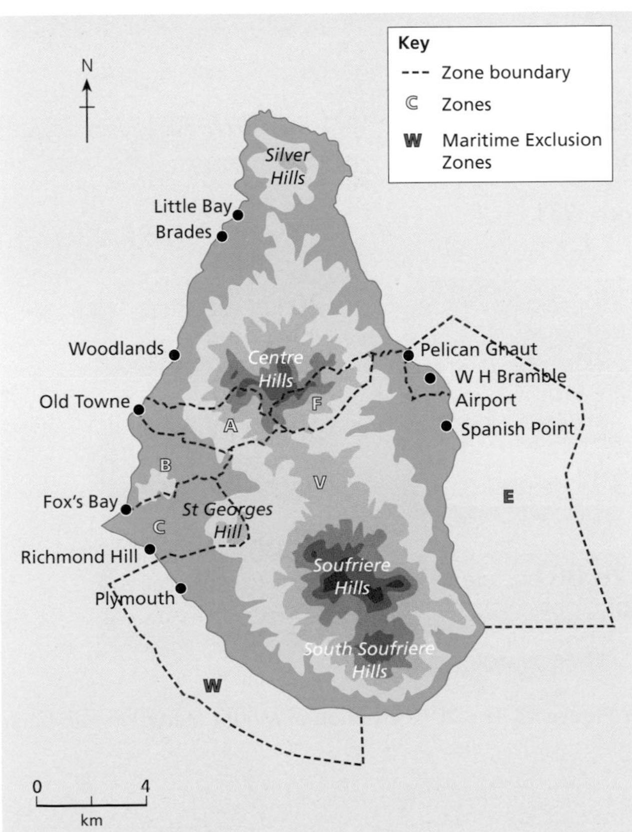

▲ **Figure 11** Hazard map for Montserrat; only the northern part of the island is the safe zone.

Year	Population
1951	14,100
1961	11,900
1971	11,600
1981	11,900
1991	10,800
2003	4,500
2019	5,200

▲ **Figure 12** The population of Montserrat.

Activities

4 Study Figure 11.
   a) Estimate the percentage of the island which is in the safe zone.
   b) How might the maritime exclusion zones affect local people?
   c) The old capital city Plymouth and WH Bramble airport had to be abandoned. What problems would this have caused for the people of Montserrat?

5 a) Represent the data in Figure 12 in a suitable graph.
   b) Describe what has happened to the population since the eruption began in 1995.

Enquiry

What is the hazard level in Montserrat today?

- Use the weblink to the Montserrat Volcano Observatory.
- Make a sketch map of Figure 11. Colour code your map to show the present level of hazard.
- What level of activity is permitted in each zone under this level?

Volcanic hazards of Mount Merapi

Mount Merapi is a stratovolcano located in Central Java. This is one of the most densely populated areas of Indonesia. Mount Merapi is Indonesia's most active volcano, producing pyroclastic flows and ash clouds. It is formed by the subduction of the Indo-Australian Plate beneath the Eurasian Plate at a destructive plate boundary.

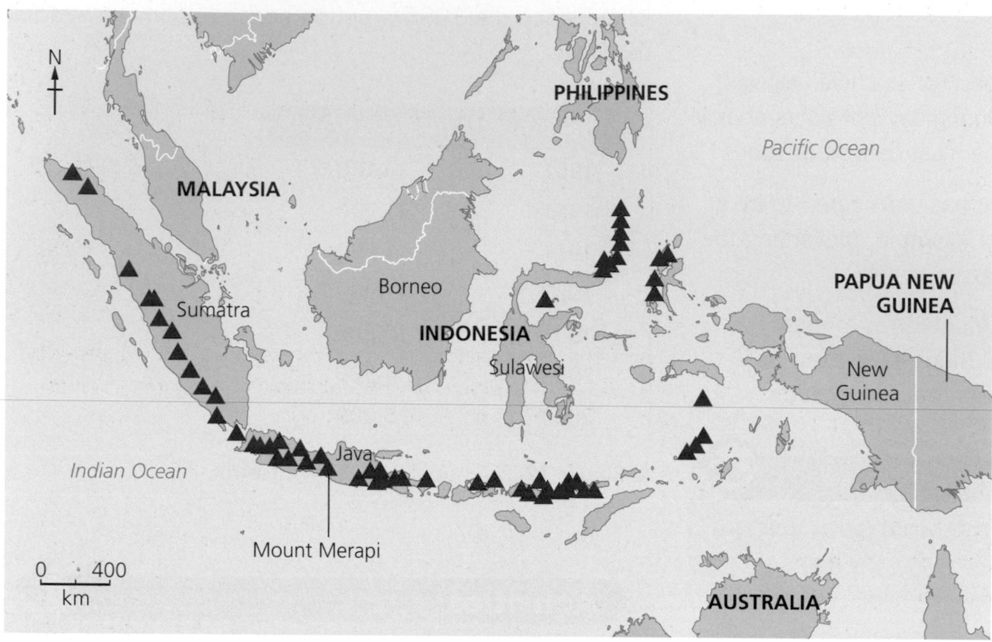

▲ **Figure 13** The location of Mount Merapi, Indonesia.

Activities

1 Use Figure 13 to describe:
   a) the location of Mount Merapi
   b) the distribution of volcanoes in this region.

2 a) Plot a frequency graph using the data in Figure 14.
   b) Calculate the frequency of eruptions of Mount Merapi.

3 Describe the effects of the 2010 eruption. Consider the effects on:
   a) people
   b) the economy.

1933–35	1939–40	1944–45	1948–49	1953–58	1961	1967–69	1971	1972–90
1992–98	2001–2	2006	2007	2010–11	2012	2013	2018	2019

▲ **Figure 14** Frequency of eruptions of Mount Merapi, Indonesia, since 1933.

Impacts of the October 2010 eruption

Mount Merapi's 2010 eruption was magnitude VEI 4. The blast destroyed homes and crops, with pyroclastic flows travelling up to 10 km from the summit. Thousands were left homeless. Ash falls of 30 cm were recorded in villages up to 20 km away and caused breathing difficulties. Planes were grounded in Western Australia because of the risk of damage to aircraft and hundreds of international flights had to be cancelled to and from Indonesia. Financial losses have been estimated at around US$700 million.

In the weeks after the eruption, the price of food rose and thousands had difficulty affording the increase in food prices. The 700 emergency shelters provided were cramped and there wasn't enough clean drinking water or toilets. This led to poor sanitation and the risk of outbreaks of disease such as cholera.

27 million cubic metres of ash and rock deposited in the River Gendol

20 km exclusion zone

320,000 people evacuated

353 deaths

26,00 hectares of crops covered in ash

1,900 livestock killed (mainly dairy cattle)

▲ **Figure 15** The 2010 eruption of Mount Merapi in numbers.

Hazard monitoring

Hazard monitoring involves scientists trying to predict when the next eruption will take place. Data from monitoring stations on Mount Merapi is sent to the Merapi Volcano Observatory in Yogyakarta City. The data is processed and the alert level of the volcano is adjusted on a daily basis so that local people can be warned if an eruption becomes more likely.

Six observation posts around Mount Merapi take regular photographs to show the evolution of the lava dome and record rock fall activity. Scientists can attempt to predict the direction of future pyroclastic flows based on the changing shape of the dome. Satellites provide high-resolution images of visible changes while thermal infrared images show increases in temperature due to rising magma. **Tiltmeters** analyse ground movements and five observation stations send a laser beam to mirrors placed around the volcano to measure the smallest change in ground shape. The ground may bulge when magma is filling up the magma chamber beneath the volcano. Mount Merapi has eight **seismometers** – instruments that record earthquakes. **Geomagnetic** monitoring is carried out by four stations: they take measurements of geomagnetic intensity once every minute. There are also several fixed-point **geochemical** monitoring sites, which sample gas emissions.

▲ **Figure 16** Search and rescue teams look for survivors in a riverside community that was home to 200 villagers, about 8 km from Mount Merapi, Indonesia.

Hazard monitoring

Volcanoes are observed using a variety of different scientific instruments. The data that is collected may give the scientists clues about whether or not a volcano is about to erupt. Scientists use this data to give local people enough time to evacuate.

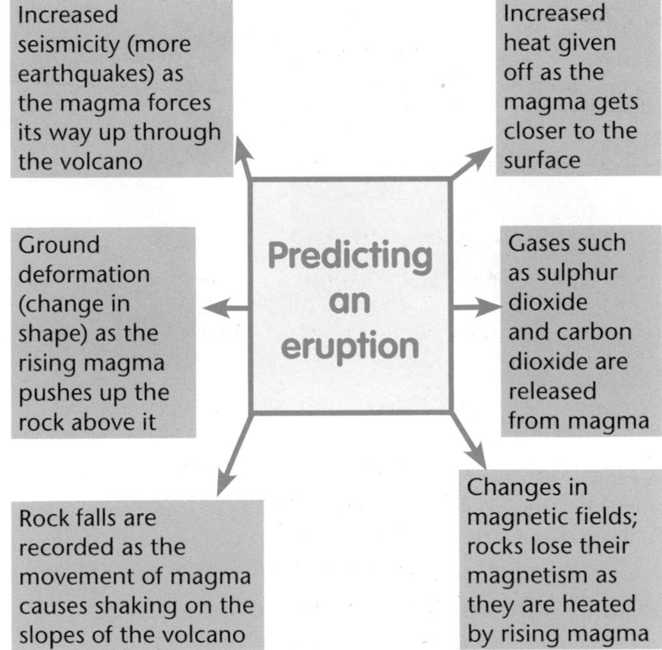

▲ **Figure 17** Scientists monitor volcanoes to try to predict when an eruption will take place.

Activities

4 Describe the landscape in Figure 16. How has the eruption affected the environment?

5 What is the cause of the observable changes in a volcano before an eruption?

6 Describe three different ways that volcanoes are monitored. For each one, describe what is monitored and what this tells scientists about the volcano.

Enquiry

What processes formed Mount Merapi?

Draw an annotated sketch to show how Mount Merapi formed. Use Figure 3 on page 107 to help you.

Coping with the threat from lahars

Pyroclastic flows deposit layers of loose rock and ash. After a large eruption these deposits can be hundreds of metres thick. When mixed with water they create a destructive type of mudflow known as a **lahar**. Lahars can be formed in two different ways:

- Snow and ice on the upper slopes of the volcano can be melted by the eruption. The melt water mixes with the ash deposits creating the lahar.
- Rainfall, during or after the eruption, mixes with pyroclastic deposits.

Lahars can travel at 65 kph, which is faster than a river in flood. The mixture of ash, rock and water has the consistency of wet concrete. The density of a lahar causes far more damage than water alone. People, buildings, roads and bridges are simply swept away or buried. Farmland is ruined, putting a strain on local food supply. Lahars can travel long distances from the volcano. Lahars can continue to be created long after the volcano has stopped erupting, every time heavy rain mixes with rock and ash deposited in river valleys.

	Jan	Feb	Mar	Apr	May	Jun	Jul	Aug	Sep	Oct	Nov	Dec
Precipitation (mm)	353	335	310	211	127	89	41	5	30	94	229	340
Temperature (°C)	26	26.5	26.5	27	27	26	25	25.5	26.5	27	27	26.5

▲ **Figure 18** Average precipitation and temperature for Yogyakarta, Indonesia.

▲ **Figure 19** Local people remove sand from a river during the eruption of Mount Merapi, Indonesia, to make sand bags to try to prevent flooding.

Lahar

A fast-flowing mudflow caused when volcanic ash is mobilised by water. Lahars are an example of a large-scale volcanic hazard. They can cause deaths more than 20 km from a volcano.

Activities

1 a) Describe how a lahar can be formed during an eruption.
  b) Explain why the risk of lahars continues for many months after an eruption.
2 Describe the similarities and differences between a lahar and a river flood.
3 a) Use Figure 18 to plot a climate graph for Yogyakarta.
  b) Label the months in which lahars would have been a hazard following the October 2010 eruption.

Why do people continue to live here?

Mount Merapi erupts frequently but most eruptions are relatively small. The average eruption is of magnitude VEI 3, so effects are usually localised. Volcanic ash is rich in minerals. Once it has weathered it creates fertile soil for agriculture. The area between the volcano and the city of Yogyakarta is a National Park that is popular with tourists. Some local people work in the tourist industry as guides, taking tourists into the new conservation areas where it is considered too dangerous to live.

Sector of the economy	(%)
Agriculture	16
Manufacturing	14
Tourism	21
Other service industries	49

▲ **Figure 20** Economic sectors and their value to Yogyakarta's economy.

Reducing the vulnerability of riverside communities

Indonesia is a newly industrialised country (NIC) with a rapidly growing urban population. Planners have struggled to keep up with the demand for affordable new housing. It is estimated that as much as 80 per cent of all new housing is in informal settlements. Much of this housing is built in locations that are not really suitable or safe, for example, riverside communities that are at risk of flooding in Indonesia's wet season. After volcanic eruptions, these communities are at deadly risk from lahars.

One non-governmental organisation (NGO) that is trying to help the situation is Arkomiogia – a group of community architects working in the riverside community of Kali Jawi in Yogyakarta. This project was begun by a group of women who started a micro-credit scheme so they could make regular savings. The women's group met with architects, planners and local government officials regularly to plan a new community centre. This centre can be used for village meetings and as a shelter during a flood or lahar. The new community centre has several features to make it practical and sustainable, which can be seen in Figure 21.

1 Concrete stilts raise the building above flood height

2 The bamboo construction uses three species of bamboo that are all commonly found in Java

3 Steel bolts firmly hold the flexible structure together

4 Community members were able to be involved in the simple construction

5 The traditional style of building fits in with local architecture

6 The large roof provides shade and shelter from rain; open spaces provide ventilation

▲ **Figure 21** The new hazard-resistant community centre in Kali Jawi.

Activities 

4 Suggest why people living in informal housing in Yogyakarta are most vulnerable to ash falls and lahars.

5 Study Figure 21. Describe how one or more features of the building are designed to:
   a) be sustainable
   b) suit the climate of the region
   c) improve the capacity of Kali Jawi to cope with flooding and lahars.

Year	Yogyakarta	Semarang	Surakarta
1970	343,000	627,000	408,000
1980	397,000	1,009,000	468,000
1990	412,000	1,243,000	503,000
2000	397,000	1,427,000	491,000
2010	388,000	1,558,000	499,000
2020	402,000	1,761,000	533,000
2030	503,000	2,188,000	668,000

▲ **Figure 22** Population of cities close to Mount Merapi.

Enquiry 

Which of the cities in Figure 22 is most at risk of lahars?

- Use an interactive world map, such as Google Maps, to find the distance of each city from Mount Merapi.
- Describe the drainage pattern of rivers in this region.
- Use this information to put the three cities in rank order of risk. Justify your choice.

The earthquake hazard

Earthquakes are measured on the **moment magnitude scale** (M_W). It measures the product of the distance moved by a fault and the force needed to move it. Each number on the scale is ten times more powerful than the one before it.

The strength of ground shaking during an earthquake is reduced with distance from its origin – a point underground known as the **focus**. This means that earthquakes close to the Earth's surface will be felt more strongly than an equally strong earthquake with a deeper focus. Similarly, ground shaking is reduced with distance from the **epicentre**, the point on the surface that is immediately above the focus.

The amount and strength of ground shaking during an earthquake is affected by the hardness of the rock the seismic waves travel through. Earthquake waves travel quickly through hard rock. When they reach softer ground they slow down and the amount of shaking becomes amplified.

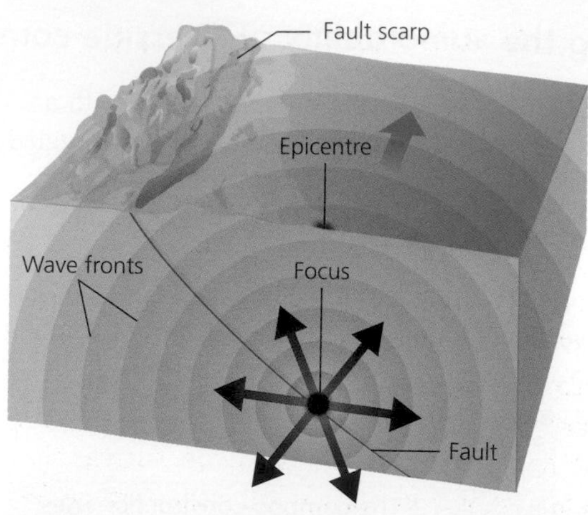

▲ **Figure 23** Cross section through a fault showing the features of an earthquake.

Weblink

https://earthquake.usgs.gov/earthquakes/browse/ – the US Geological Survey records details of earthquakes around the world.

Date	Country	Magnitude (M_W)	Impacts	Depth (km)	GNI per capita (US$)*
26/05/19	Peru	8.0	2 deaths, 833 homes damaged	123	6,530
06/09/18	Fiji	7.9	No impacts	670	5,860
14/05/19	Papua New Guinea	7.6	0 deaths, 100 homeless	10	2,530
22/02/19	Ecuador	7.5	1 death, 22 houses damaged	145	6,120
28/09/18	Indonesia	7.5	2,077 deaths, 206,000 homeless	20	3,840
16/06/19	New Zealand	7.3	No impacts	46	40,820
20/12/18	Russia	7.3	No impacts	17	10,230
24/06/19	Indonesia	7.3	No impacts	212	3,840
14/07/19	Indonesia	7.2	14,500,000 homeless	19	3,840
06/07/19	USA	7.1	0 deaths, 54 homes damaged	8	62,850

▲ **Figure 24** Magnitude, location and depth of selected large-magnitude earthquakes (2018–19).

*World Bank latest figures

Activities

1 Study Figure 23. Describe the difference between the focus and epicentre of an earthquake.
2 Explain why people living in homes built on soft ground are more vulnerable to earthquakes.
3 Study Figure 24. Use evidence in this table to explain the link between the depth of an earthquake and its impacts.

Enquiry

Why are some earthquakes more deadly than others?

- Use the USGS website (link in the Weblink box above) to investigate the location of the earthquakes that had no impacts compared with one that caused deaths. What conclusions can you reach?

Can technology be used to reduce the impacts of earthquakes?

The capacity to survive an earthquake is increased where individuals or communities have the resources or technology they need to cope with the hazard. Japan is a high income country (HIC) that has financial resources it can invest in technology and hazard preparedness. Many **aseismic** buildings have been constructed, such as the Yokohama Landmark Tower, to withstand earthquakes. These structures are flexible and strong rather than being rigid and brittle like concrete. Infrastructure, such as roads, can also be built to flex and withstand earthquakes.

So, as countries become more economically developed, do they become safer? Not necessarily. In many NICs, such as India or China, economic development does not guarantee that everyone has a reduced risk from earthquakes. Vulnerability is not reduced for everyone in society.

Shillong is a city in northeast India. In 1897 an earthquake of magnitude 8.3 M_W killed about 1,500 of the 50,000 people who lived there. It has been estimated that an earthquake of similar size today could perhaps kill 90,000 people of the current 400,000 population. Why are so many more people at risk today? The reason is technology. In 1897 most people lived in traditional single-storey bamboo homes. These were flexible and lightweight, like modern steel-frame structures. Now, most people live in cheaply constructed multi-storey buildings constructed from concrete. Many are on steep slopes. During an earthquake, these buildings could collapse, crushing and killing people inside.

▲ **Figure 25** Housing in Shillong, India, today.

Disaster planning

After an earthquake people face a number of further hazards, for example:

- from unstable buildings, which have been weakened by the earthquake
- the need of medical attention for their injuries
- risks associated with exposure because they have lost their homes
- risks associated with the lack of basic resources, such as food, clean water and sanitation

A swift and targeted response is needed to minimise these effects. This is harder to achieve when:

- the area is very remote
- the country does not have enough money to invest in disaster planning and preparation.

Specialist emergency teams trained to use equipment specifically designed to locate survivors buried under rubble are found in most HICs at risk of tectonic disasters, such as Japan, Italy and the USA. In these countries, educational programmes in schools and workplaces teach people how to react during an earthquake, tsunami or volcanic eruption. Annual events like the 'Great ShakeOut' in California involve whole communities in raising awareness of what to do during and after a major earthquake. In Japan, coastal towns have tsunami escape routes clearly signposted.

Aseismic

A term applied to structures that are designed to withstand the shaking of the ground during an earthquake. Some traditional buildings are aseismic.

Activities

4 Give two reasons why a large-magnitude earthquake in Shillong could be deadly for so many people. Use evidence from Figure 25.

5 Suggest why some groups of people in India remain vulnerable to earthquakes despite the rapid economic growth of the country. Use Figure 3, page 115, to help you.

6 How might the capacity of vulnerable communities in Shillong be improved?

The Nepal earthquake of April 2015

On 25 April 2015, a 7.8 M_w earthquake occurred in the Gorkha district of Nepal. The epicentre was close to the Nepalese city of Kathmandu, but people in India, China and Bangladesh were also affected by the powerful quake. A number of powerful aftershocks caused further damage and injuries, including one of 7.3 M_w close to Mount Everest on 12 May.

▲ **Figure 26** The 2015 Nepal earthquake in numbers.

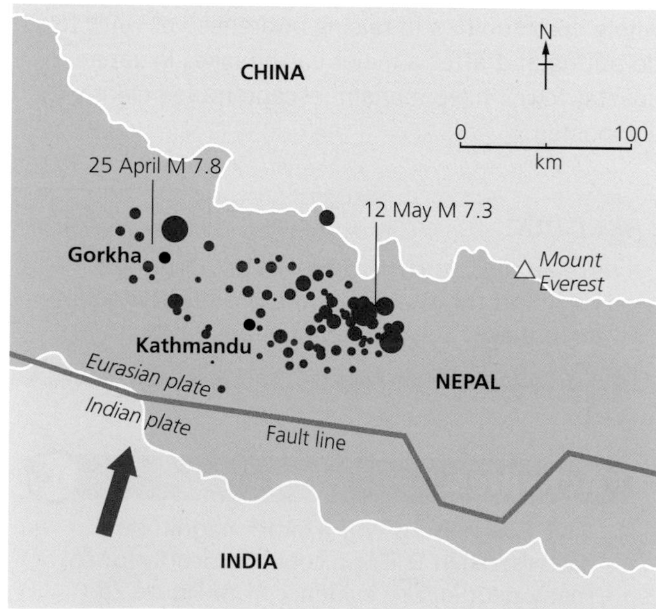

▲ **Figure 27** The location of the 2015 Nepal earthquakes. Larger earthquakes are represented by larger circles.

What were the effects of this earthquake?

The earthquake was caused by the Indian Plate moving under the Eurasian Plate in a sudden movement. As a result, Kathmandu is now 3 m further south than it was before the earthquake.

The violent shaking of the ground during the earthquake caused a lot of damage to buildings and **infrastructure**. Whole villages were destroyed as buildings collapsed. Many roads were fractured by the Earth's movement. Others were blocked by landslides that slid down the steep mountain slopes because of the earthquakes. Hundreds of people were killed by avalanches of snow and ice that were triggered by ground shaking. One of these avalanches was over 2 km wide.

Aid workers struggled to reach people who needed help because of the mountainous terrain and damaged roads. Much of the help had to come by helicopter. This meant that smaller amounts of supplies could be delivered and fewer people could be rescued in each trip. Many medical centres were destroyed and medical supplies hard to access. Fresh food and clean water were scarce. Reconstruction after the earthquake was slow. It is estimated that the earthquake caused US$10 billion of damage. Nepal is a low-income country (LIC), so there is little spare money in the economy to pay for these repairs.

Infrastructure

The essential services needed by all communities, such as clean water, electricity supplies, sewerage systems, roads and bridges.

Activities

1. Look at Figure 27.
   a) Describe the location of the epicentre of the earthquake.
   b) Describe the area affected by severe ground shaking. Use the scale line to estimate its area.
2. Describe how the Nepal earthquake affected:
   a) people
   b) the economy.
3. Explain why it was difficult to get aid to some communities after the earthquake.

How can technology help reduce risk?

Electricity and telephone lines were broken during the earthquake. This damaged communications, which could have slowed down the relief effort. However, a mixture of old and new technologies allowed the public to help. Amateur radio operators, using radios running on solar power and batteries, were able to give information about conditions in remote areas to the Nepalese police. This was made possible by the US chapter of the Computer Association of Nepal, who supplied the equipment needed to relay the radio messages. People also used social media to describe conditions in remote areas. Humanity Road, an NGO, kept a log of social media messages to relay important information to the rescue teams.

Other projects are in place to reduce the damage caused by future earthquakes. Traditional Nepalese houses use alternating bands of stone and wood; the wooden bands stitch the buildings together and allow some flexing during an earthquake. However, cut stone is expensive and many of the trees in Nepal have been cut down for firewood, so poorer people have built cheap concrete houses. These houses have no flexibility and many collapsed during the earthquake. Abari is a project that aims to revive traditional building techniques along with the benefits of modern technology. Specially treated bamboo is used instead of wood, and people are taught the traditional methods, thus making the buildings more affordable and sustainable.

▲ **Figure 29** A stone and bamboo school built by Abari and Learning Planet, which survived the earthquake in Gorkha.

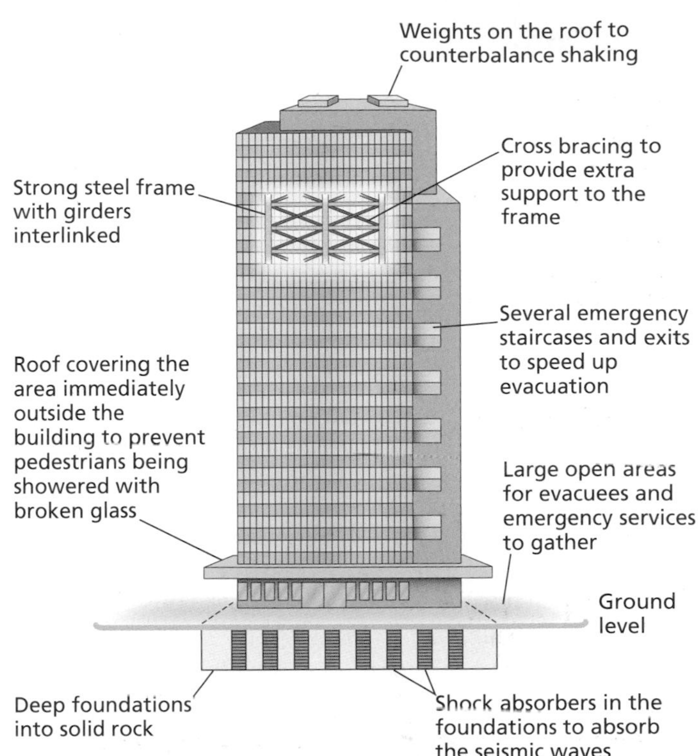

▲ **Figure 28** Features of an earthquake-resistant building.

Activities

4 Describe how technology was used to help the relief effort in Nepal.

5 Study Figure 28. Suggest how three of the features of this building improve its safety during an earthquake.

6 Compare the features of the Abari project (in Figure 29) with the Kali Jawi project (page 123). What are the similarities and differences?

Enquiry

'In a low income country, like Nepal, people in densely populated urban areas are more vulnerable than people in remote rural areas during a powerful earthquake.'

To what extent do you agree with this statement? Consider why the effects of an earthquake could be worse in a densely populated urban area. What advantages would an urban area have over a rural area during the hours and days after the earthquake?

Tsunami

Tsunami are caused by violent movements of the ocean floor, such as those during an earthquake. This movement displaces huge amounts of water upwards, which creates a series of waves. Tsunami waves travel at 25 m per second in deep water. As a tsunami reaches shallow water it slows down and increases in height in a process called shoaling. Debris carried in these large, powerful waves can be carried far inland across coastal plains and up river valleys, causing injury to people and severe damage to buildings. The danger can last for several hours after the first wave.

The 2011 Tōhoku earthquake and tsunami

The 11 March 2011 undersea earthquake off the coast of Tōhoku, Japan, had a magnitude of 9.0 M_w. Less than half an hour later the coast was hit by the subsequent tsunami. Waves reached heights of up to 10 m and travelled up to 10 km inland. The number of confirmed deaths was just under 20,000. Most were caused by drowning. The massive surge overtopped Japan's sea defences and destroyed three-storey buildings where people believed they were safe. Fukushima Daiichi, a nuclear power plant to the south of Sendai, had lost power to its cooling rods during the earthquake and was running on back-up power. When the tsunami struck it destroyed this back-up power, eventually causing a full meltdown of a number of reactors and releasing harmful radiation. Thirty-six nuclear reactors across the country were closed after the earthquake, resulting in electrical blackouts across much of the country.

▲ **Figure 30** Damage in Kesennuma, Miyagi Prefecture, after the March 2011 earthquake and tsunami.

Tsunami

A series of powerful waves created when the seabed is moved suddenly during an earthquake or landslide.

How are tsunami monitored?

The Pacific Tsunami Warning System (PTWS) was set up after the 1960 Chile earthquake. This earthquake caused a tsunami responsible for the deaths of hundreds of people across the Pacific region. There was no warning. The PTWS uses a network of seismographs and ocean buoys to detect earthquakes and changes in sea levels that may indicate the formation of a tsunami. Once a tsunami is detected, warnings can be given to local centres around the Pacific region. These centres can then warn local people by radio or text message, giving them time to evacuate the area. A similar system was set up in the Indian Ocean in 2006 following the Boxing Day tsunami of 2004, which killed over 230,000 people.

	Tōhoku, Japan	Solomon Islands, South Pacific
Date	11 March 2011	01 April 2007
Magnitude (M_w)	9.0	8.1
Depth (km)	30	10
Distance from shore (km)	129 km east of Sendai	45 km south-southeast of Gizo
Deaths	20,000	52

▲ **Figure 31** Comparing two tsunami events.

Activities

1. How much greater in magnitude was the 2011 Tōhoku earthquake than the Nepal earthquake of April 2015 (page 126)?
2. Describe the impacts of the Tōhoku earthquake. Use Figure 30 to help you.
3. Explain why the magnitude and scale of this event had such an impact.

The 2007 Solomon Islands tsunami

The Solomon Islands are a group of 492 islands in the Pacific Ocean to the northeast of Australia. The Solomon Islands are one of a group of countries known as the **Small Island Developing States** (SIDS). These countries are particularly vulnerable to natural disasters. On 1 April 2007, a powerful earthquake created a tsunami; it was detected by the PTWS but the wave struck the island of Ghizo before the warning could be issued. The small coastal town of Gizo was swamped by waves several metres high, which flooded 50–70 m inland, destroying a number of timber-framed buildings. The island of Ghizo is 370 km away from the capital, Honiara. The isolated location made it difficult for rescue workers and aid to reach the scene quickly. Australia sent US$54.1 million in aid. The nearest major airport in Australia is Darwin, which is 2,887 km away.

▲ **Figure 32** An informal settlement on Ghizo built to replace houses lost in the 2007 tsunami.

Small Island Developing States

A group of 58 countries with a combined population of 65 million people. Many SIDS are very small and some are located in remote and isolated parts of the world. Most SIDS are vulnerable to climate change and other natural disasters, such as earthquakes and tsunamis. Standards of living among small islands differ widely, with GNI per capita ranging from US$1,380 in Comoros to US$58,770 in Singapore.

Activities

4 Use Figure 31 to compare the two tsunami events.

5 Using Figure 33 and the information about SIDs on the left, suggest three reasons why many SIDS are vulnerable to natural hazards.

6 a) Explain why people in Ghizo were vulnerable to the tsunami of 2007.

b) Describe the buildings in Figure 32.

c) Suggest why people in Ghizo remained vulnerable to other natural hazards after the tsunami.

SIDS	Volcano?	Earthquake?	GNI per person (US$)	Poverty (% below US$1.90 a day)
Cape Verde	Yes	Yes	3,420	8.1
Comoros	Yes	Yes	1,380	17.9
Fiji	Yes	Yes	5,860	1.4
Haiti	No	Yes	800	25.0
Papua New Guinea	Yes	Yes	2,570	38.0
Saint Lucia	Yes	Yes	9,560	4.7
Solomon Islands	Yes	Yes	2,020	25.1

▲ **Figure 33** Hazards in selected small island developing states (SIDS).

Enquiry

Why are some coastal communities more vulnerable than others?

- Research two of the SIDS in Figure 33.
- Describe the tectonic hazards each country faces.
- Explain why these communities are vulnerable and suggest ways of building capacity.

Chapter 1
Managing coastal hazards

Erosion and coastal flooding during extreme weather events

The UK's coast is at risk of a variety of coastal hazards including:

- erosion, landslips and rock falls (see pages 20 to 23)
- storm surges during severe weather events
- flooding due to sea level change.

The coastline of the UK is densely populated. Cities such as Newport, Hull and London are all located on estuaries that are at risk of coastal flooding. England alone has 3.1 million people living in seaside towns, such as Blackpool or Bournemouth. Consequently, large parts of the UK population can be affected by coastal hazards. These hazards become more likely during severe weather events.

▲ **Figure 1** Rapid coastal erosion of the cliffs at Hemsby during the 2013 storm caused the collapse of seven homes.

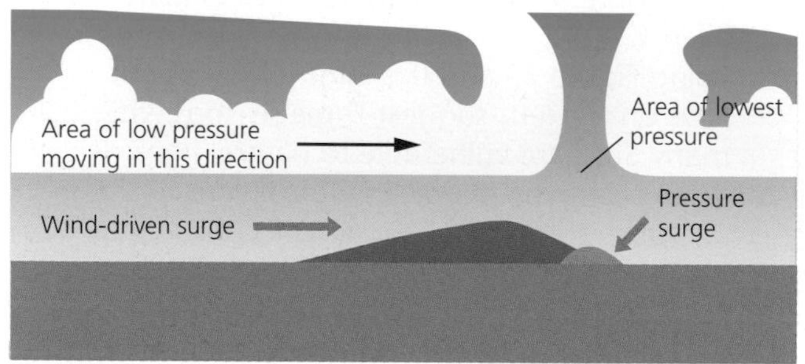

▲ **Figure 2** Storm surge due to low pressure.

Activities

1. Explain why the shape of the North Sea increases the risk of storm surges in Essex, Kent and the Thames Gateway areas.
2. a) Make a sketch map of Figure 3.
   b) Use an atlas to add the eight labels describing the effects of the storm surge to the correct locations on your sketch.
3. Use Figure 4 to describe the location of areas affected by the storm as it moved across the UK.
4. Analyse Figure 5.
   a) At what times was high tide expected on 5 December?
   b) At what time did the storm surge reach Lowestoft?
   c) How much higher was the storm surge than the expected height of high tide?

The storm surge of December 2013

Low pressure in the atmosphere has the effect of raising sea levels. When air pressure falls by 1 millibar (mb), sea levels rise by 1 cm. So, a deep depression of 960 mb will cause sea levels to rise by 50 cm. Strong winds create large waves that are pushed in front of an advancing area of low pressure creating even higher water levels. This effect is known as a **storm surge**, and is shown in Figure 2. If a storm approaches the coast at high tide, an event that happens twice a day, then the risk of flooding is increased. The UK's North Sea coastline is particularly vulnerable to storm surges. The southern part of this sea is shallow and shaped like a funnel. When low pressure travels southwards across the North Sea, the bulge of the storm surge can increase in height as water is forced through this shallow funnel.

During December 2013 coastal communities along the North Sea Coast faced the worst storm surge since 1953. Erosion is more rapid during extreme weather events, as you can see in Figure 1. Other effects of the storms are shown in Figure 3. Nevertheless, the Environment Agency said 800,000 homes had been protected by:

- accurate weather forecasts that gave people time to evacuate their homes
- coastal defences that held back some of the storm surge.

▲ **Figure 3** The locations affected by the 2013 storm surge.

1,000 sandbags were distributed to home-owners in Aldeburgh, Suffolk.	In the Humber region, 400 homes were affected by flood water.
Seven houses were destroyed in Hemsby, Norfolk when the cliff beneath them collapsed into the sea.	In Great Yarmouth, Norfolk, residents of 9,000 homes were advised to evacuate overnight.
The sea wall was breached at Jaywick, Essex. Firefighters and 10 rescue boats helped to evacuate 2,500 homes.	Boston, Lincolnshire was flooded; 223 people were evacuated from their homes.
In Kent, 200 homes were evacuated in Faversham and another 70 in Seasalter.	The sea wall at Scarborough, Yorkshire, was damaged.

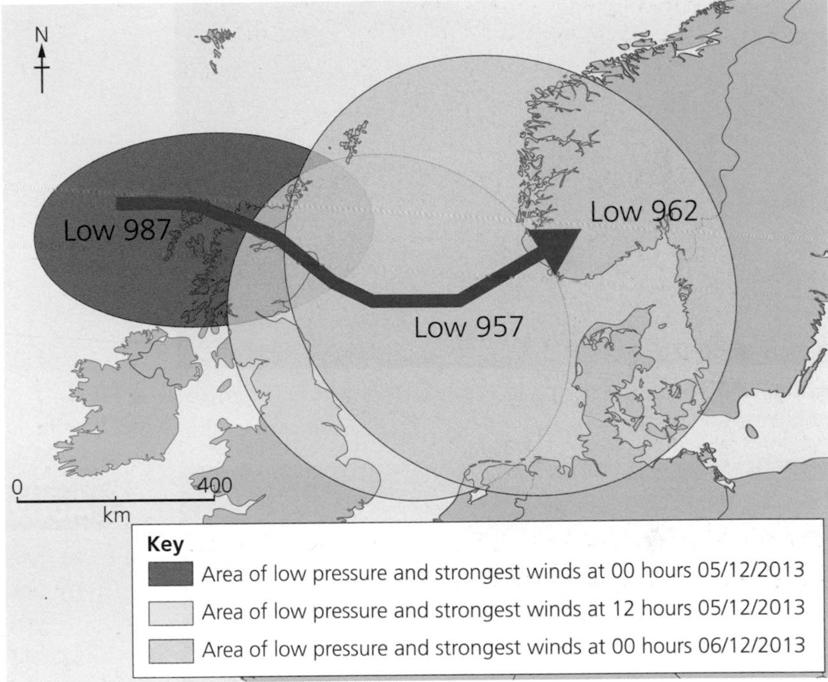

▲ **Figure 4** The course of the December 2013 North Sea storm.

Low pressure

Low pressure in the atmosphere is caused by air rising. Low pressure systems are also called cyclones. They bring strong winds, cloud and rain.

Storm surge

A temporary, localised rise in sea level caused by low pressure in the atmosphere.

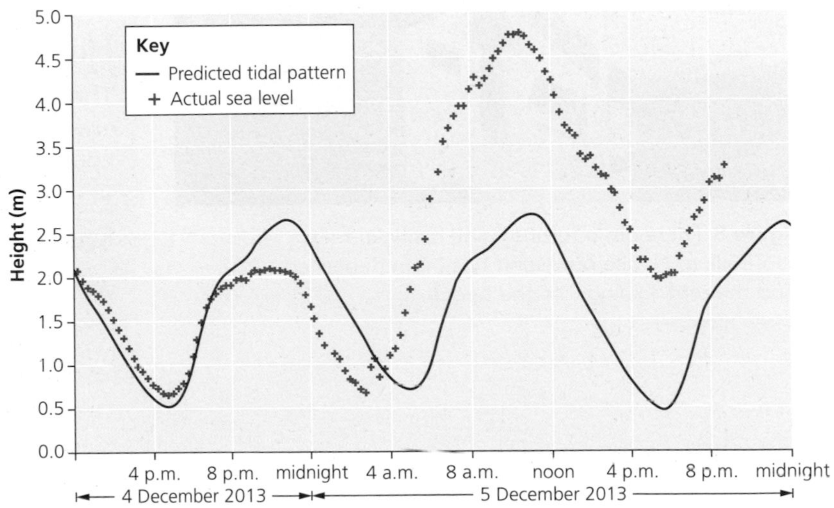

▲ **Figure 5** Sea levels at Lowestoft, Suffolk, during the storm surge.

How do we manage our coasts?

The usual way to manage coastlines has been through a combination of hard and soft engineering strategies. **Hard engineering** means building structures that prevent erosion and fix the coastline in place. The concrete sea wall and boulders in Figure 6 are a typical example. Wide beaches soak up a lot of wave energy and are a natural defence against coastal erosion. **Soft engineering** strategies mimic this by encouraging natural deposition to take place along the coastline. In Figure 8 you can see that an artificial rock reef has been built parallel to the coastline. This encourages deposition on the beach behind.

▲ **Figure 6** The sea wall at Sea Palling, Norfolk is an example of hard engineering.

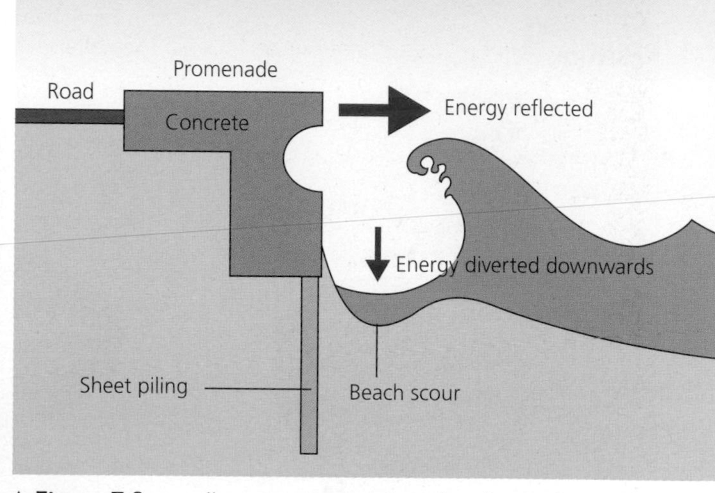

▲ **Figure 7** Sea walls can cause erosion of sediment from the beach.

▲ **Figure 8** There are a total of nine artificial reefs at Sea Palling. Notice how sand has been deposited behind the reef, joining it to the beach.

Activities

1 a) Make a sketch of Figure 6.
b) Label two hard engineering features on your sketch.
c) Explain why boulders have been positioned at the top of the beach.

Hard engineering

The use of artificial structures to reduce coastal erosion or the risk of coastal flooding. Hard engineering includes the use of strategies such as sea walls, embankments and artificial reefs.

Soft engineering

Working with nature to protect the coastline. Soft engineering includes strategies such as beach nourishment and maintaining healthy sand dunes.

The importance of the inter-tidal zone

Estuarine landscapes, such as the one in Figure 9, contain many tidal creeks, salt marshes and mud banks. These features are exposed at low tide but at high tide they can store huge quantities of water. This is the **inter-tidal zone** and it acts as a natural buffer during storms – soaking up wave energy during a storm surge before the waves can reach more valuable land further inland.

There is a much narrower inter-tidal zone in the UK than there used to be:

- Many salt marshes were reclaimed in the past to create new farm land. Old earth embankments have kept the sea off these low-lying fields for centuries.
- Some salt marshes are being eroded by the sea. This is a particular problem along the Essex and Thames gateway coastlines where the land is subsiding so sea levels are rising faster than elsewhere in the UK.

Inter-tidal zone

The strip of coastline that lies between high tide and low tide.

Managed retreat can be used to create new inter-tidal zones of salt marsh. The process begins by punching holes through the old earth embankment. The invading sea water moves slowly across the land at high tide. As it flows in, it deposits mud. As it flows out, it creates tidal creeks like those in Figure 9. This process recreates natural mudflats and salt marshes that will store water and act as a buffer to erosion in future flood events.

▲ **Figure 9** The inter-tidal zone of the Lune estuary, Lancashire. Tidal creeks and mud flats, here at low tide, can store huge quantities of water and help prevent flooding and erosion.

Before

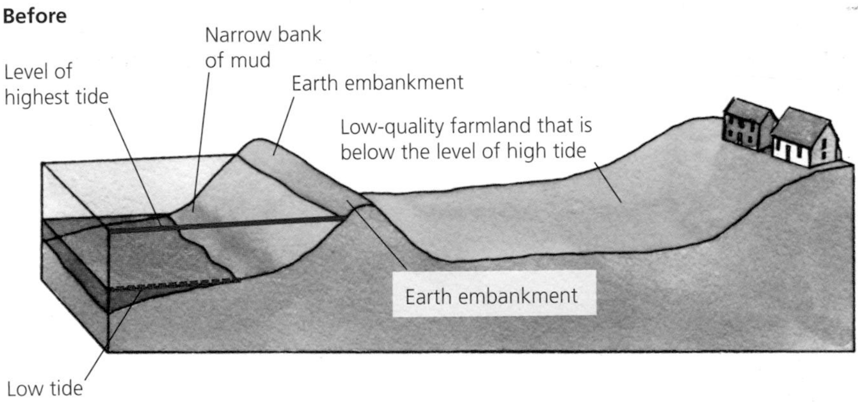

After

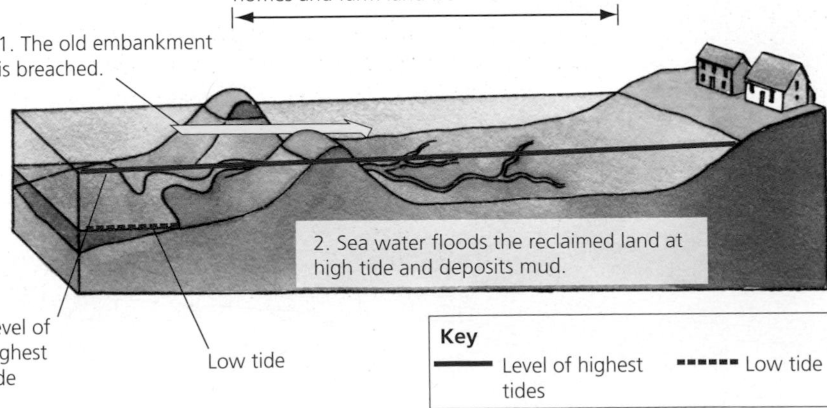

▲ **Figure 10** How managed realignment creates wide inter-tidal zones that act as natural stores for flood water at high tide.

Activities

2 a) Make a sketch of Figure 9.
b) Annotate your sketch to explain why the inter-tidal zone is an important natural defence against erosion and flooding.

Enquiry

Should we create more inter-tidal zones along the UK's coastline? Outline the arguments for and against managed retreat.

Managed retreat

Managed retreat (also known as managed realignment) is a controversial choice. In 2014 the UK's largest **managed retreat** sea defence scheme was completed at Medmerry, West Sussex. This scheme has effectively realigned the coastline 2 km further inland. The scheme began with the destruction of the existing sea wall, allowing some land to flood naturally at high tide. The newly flooded area will be able to absorb the energy of the waves and will reduce the risk of coastal flooding. A new 7 km sea wall has been built further inland. The Environment Agency says homes are much better protected as a result. However, at £28 million, the scheme was not popular with everyone, as can be seen in Figure 13.

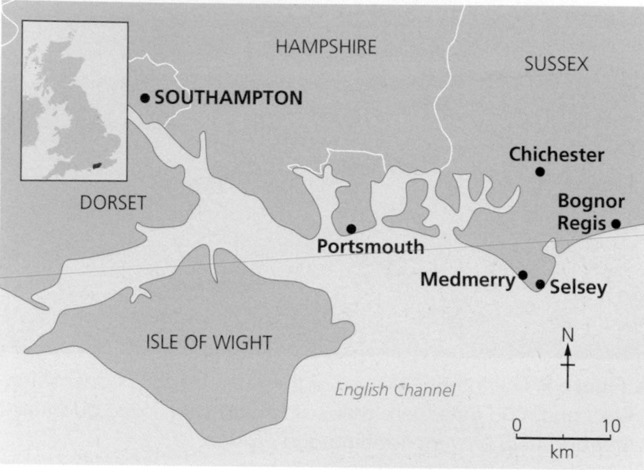

▲ **Figure 11** The location of the Medmerry realignment scheme.

Managed retreat

A form of coastal defence in which gaps are deliberately made in existing coastal flood embankments. This allows the high tide to flood onto the land behind the embankment. Managed retreat tends to be used where the value of the land that is flooded is low. This strategy widens the inter-tidal zone. This means that the energy of the waves will be absorbed by new salt marshes behind the old flood defences.

▲ **Figure 12** Medmerry seen from the air.

Has the Medmerry scheme been successful?

During the winter of 2014–15, weeks of rain and fierce storms lashed the south coast. Allan Chamberlain, estate director at the adjacent Medmerry Holiday Village, was shocked by how successful the scheme was. 'It's the first winter in years we haven't had to deal with surface flooding. The rainwater drains into the new marsh beautifully.' The bonus for the Holiday Village and the nearby Bunn Leisure Homes complex is that the new nature reserve is attracting even more tourists. Bookings are up and the tourist attractions can stay open for longer now that the area is largely flood free.

Once you give land back to the sea, there's no getting it back, so if this doesn't work, we will have given up that land for nothing. I would like to see the Environment Agency look at other alternatives such as constructing rock barriers out in the ocean in front of the coast to break wave energy.

Ben Cooper – a resident of nearby Selsey

Three productive farms producing oilseed rape and winter wheat will have to be sacrificed to the sea. The UK is not self-sufficient in food. The idea of letting perfectly good agricultural land disappear into the sea is wasteful and short-sighted.

Local farmers

We have had to endure terrible flooding this winter and we are quite upset about the whole thing. Why is the Environment Agency spending £28 million on creating a coastal nature reserve at Medmerry when they could use the money to dredge rivers and reduce the risk of flooding where we live instead?

Somerset residents badly affected by floods in 2013/14

▲ **Figure 13** Opposition to the Medmerry scheme.

Cost-benefit

Defending our coastline against erosion and flooding is costly and decisions about coastal defences are complex. Local councils need to decide whether the cost of protecting the coastline is worth it in the long run. They look at a range of factors:

- The cost of building defences and then maintaining them.
- The value of the land that is being protected.
- The cost of replacing infrastructure, such as roads, or buildings, such as homes and schools, if the land was flooded by the sea.

Geographers describe this kind of decision as cost-benefit analysis. Managed retreat involves the deliberate flooding of low-value farmland so that more valuable land can be protected in the long term.

Activities

1 Use Figure 11 to describe the location of Medmerry.
2 Summarise the benefits of realignment at Medmerry for:
   a) local homeowners and businesses
   b) the environment including wildlife.
3 Explain why some local people may have opposed realigning the coast here.

Enquiry

How sustainable is the decision to realign the coast at Medmerry?

Explain how the sustainability of this scheme could be measured over the next 50 years.

Shoreline Management Plans

It is the responsibility of the local councils of England and Wales to prepare a **Shoreline Management Plan (SMP)** for each section of coast. Figure 14 summarises the main options in a SMP. In deciding whether or not to build new coastal defences (or repair old ones), the local council needs to weigh up the benefits of building the defences against the costs. They may consider factors such as:

- How many people are threatened by flooding or erosion and what is their property worth?
- How much would it cost to replace infrastructure such as roads or railway lines if they were washed away?
- Are there historic or natural features that should be conserved? Do these features have an economic value, for example, by attracting tourists to the area?

The Environment Agency (in England) and Resources Wales are national bodies that provide advice on coastal management. This is essential because decisions made to prevent erosion in one location can have impacts on locations further along the coast. For example, groynes that protect a seaside town can starve beaches of sediment further along the coast, causing accelerated erosion.

Option	Description	Comment
Do nothing	Do nothing and allow gradual erosion.	This is an option if the land has a lower value than the cost of building sea defences.
Hold the line	Use hard engineering such as timber or rock groynes and concrete sea walls to protect the coastline, or add extra sand to a beach to make it more effective at absorbing wave energy.	Sea-level rise means that such defences need to be constantly maintained, and will eventually need to be replaced with larger structures. For this reason hard engineering is usually only used where the land that is being protected is particularly valuable.
Retreat the line	Punch a hole in an existing coastal defence to allow land to flood naturally between low and high tide (the inter-tidal zone).	Sand dunes and salt marshes provide a natural barrier to flooding and help to absorb wave energy. They adapt to changing sea levels through erosion at the seaward side and deposition further inland.
Advance the line	Build new coastal defences further out to sea.	This requires a huge engineering project and would be the most expensive option. The advantage would be that new, flat land would be available that could be used as a port or airport facility.

▲ **Figure 14** The options available to local councils when they prepare a Shoreline Management Plan.

◀ **Figure 15** Wooden groynes on Borth beach in 2009.

Activities

1 Use Figures 15 and 16.
   a) Describe these structures.
   b) Suggest how they helped to protect Borth from erosion and flooding.

Management at Borth, Ceredigion

The village of Borth is built on the southern end of a pebble ridge, or spit, that sticks out into the Dyfi estuary. Sand is trapped on the beach by wooden groynes. The sand absorbs wave energy and prevents waves from eroding the pebble ridge. However, the groynes are in poor condition and are at the end of their working lives. What should be done?

The Ceredigion SMP divides the coast up into small management units (MU). Figure 17 shows the extent of five of these MUs.

▲ **Figure 16** The wooden sea wall at the top of the pebble ridge (2009).

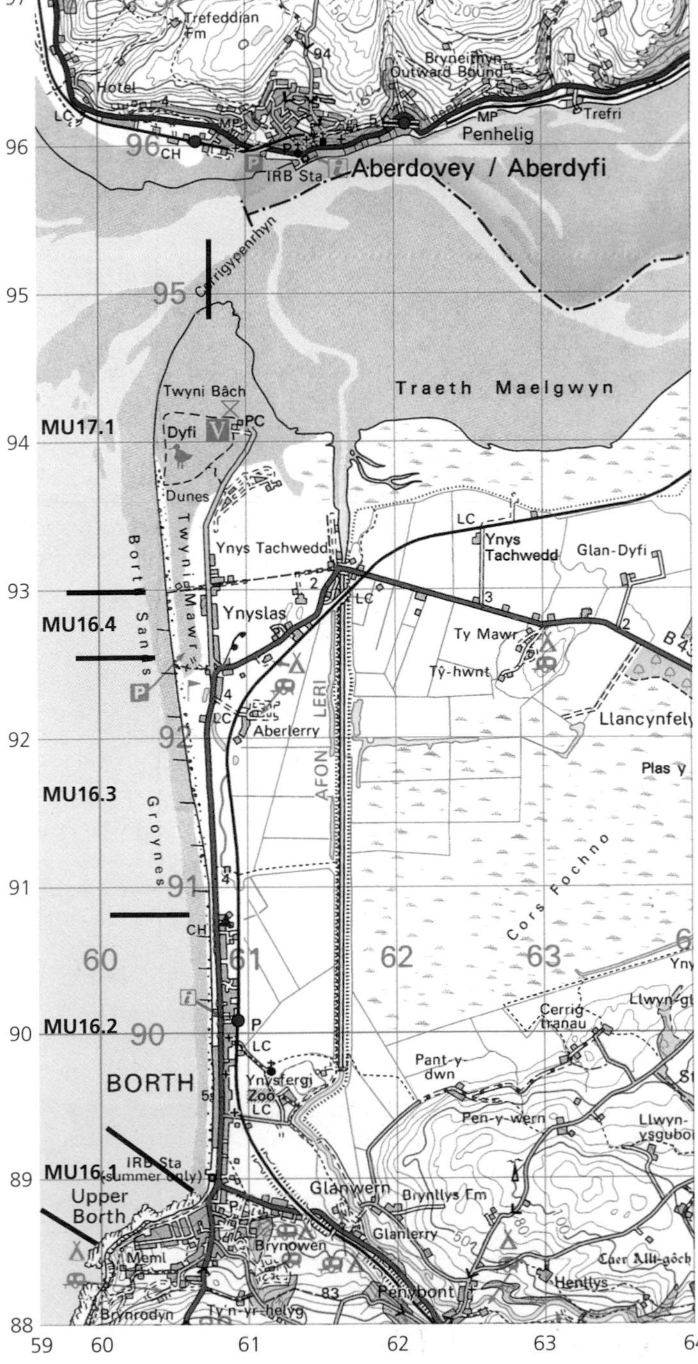

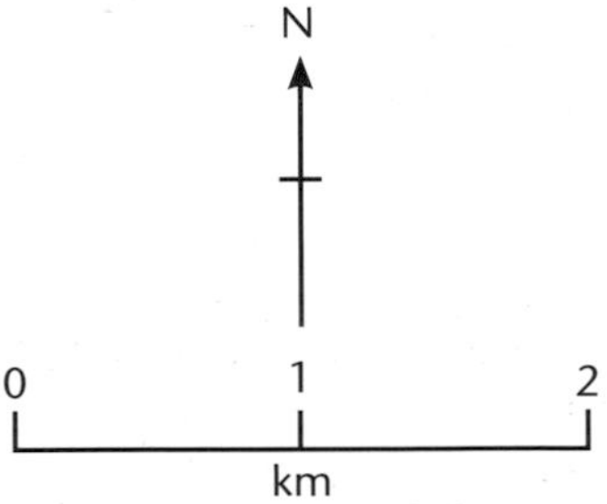

◀ **Figure 17** An Ordnance Survey extract of Borth. Scale 1:50,000.

Activities

2 Work in pairs.
Use Figure 17 to provide map evidence which suggests that this coast is worth protecting. Copy and complete the table below and add at least five more pieces of evidence.

Management Unit	Map evidence
16.2	Railway station at 609901 would be expensive to replace
16.3	
16.4	The campsite in 6192 provides local jobs
17.1	

3 Which is the best SMP option for Management Unit (MU) 16.2?
Use information from Figure 14 and evidence in Figure 17 to help you make your choice.

What coastal management is appropriate for Borth?

Ceredigion Council decided that there were two possible options for Management Unit (MU) 16.2 that needed further consideration. Read the points of view in Figure 19 before deciding what you would do.

Do nothing	Loss of property and economic loss in the short term. Change to Borth Bog.	Consider further
Hold the line	Current policy which protects property and businesses. Coastal processes disrupted with reduced longshore drift.	Consider further
Retreat	Retreat would affect homes that are immediately behind the existing line of defence.	Not considered further
Advance	No need to advance the line except to improve the tourist facilities.	Not considered further

▲ **Figure 18** The initial decision of the Ceredigion Council for MU16.2.

Scientist

Sand from the southern end of the beach is gradually being eroded by longshore drift, moving it northwards. This process is happening faster than new sand is being deposited. The beach is getting thinner and is less able to protect the pebble ridge (on which Borth is built) from erosion. If the council does nothing then the pebble ridge will be breached by storm waves and the town of Borth, and Borth Bog (Cors Fochno) will be flooded by the sea. This could happen in the next ten to fifteen years. The peat bog at Cors Fochno will be covered in sea water at high tide and its existing ecosystem lost. Over the next few years erosion will punch more holes through the pebble ridge. A new spit of pebbles will eventually form further to the east. The sand dunes at Ynyslas will probably be cut off and form a small island.

B&B owner

The beach and landscape of the spit, including the sand dunes at Ynyslas, are an important economic asset to the village. It's this natural environment that attracts thousands of holidaymakers each year. If the council does nothing then my home and many others will be flooded and local people will lose their livelihoods.

Scientist

The peat bog at Cors Fochno should be protected from flooding. It is a nationally and internationally important ecosystem. It has protection as a Special Area of Conservation and is also recognised by UNESCO. 'Do nothing' is an unacceptable option.

Local councillor

We calculate that property in Borth village is worth £10.75 million. On top of this there are many local businesses which would lose their income from tourism if we do nothing. The cost of holding the line is around £7 million. However, we are concerned that building new groynes will prevent longshore drift. We need to consider the impact of that. Currently the sediment moves to Ynyslas where it provides a natural defence to the whole estuary (including the larger village of Aberdyfi) from south-westerly storms.

▲ **Figure 19** Views on the future management of MU16.2.

Activities

1 Using a table, summarise the economic, social and environmental impacts of doing nothing or holding the line in Management Unit 16.2.
2 State which option you would recommend. Explain why you think your option is best for this stretch of coast.

Enquiry

Which is the best SMP option for Management Unit 17.1?

Analyse information from Figure 17 and evaluate the views in Figure 19 to help you justify your choice.

What happened next in Borth?

A decision was made to 'hold the line' and continue to protect the community at Borth. The old wooden groynes (seen in Figure 15) were constructed in the 1970s. These were badly worn so they were removed as four new rock groynes were built. These will trap sediment moved by longshore drift and maintain a wide beach in front of the town. A rock reef, parallel to the shore, was constructed at the southern end of the spit. The reef will help to break the force of the waves. As waves wash around each end of the reef they will lose energy, encouraging the deposition of sand and shingle behind the reef. The reef has been described as multi-purpose. As well as defending the coastline it should improve wave conditions for surfing. If the reef works, Borth's tourist economy could benefit.

Activities

3 a) What are the main similarities and differences between the beach at Borth now (seen in Figure 20), and as it was in 2009 when the photo in Figure 15 was taken?

b) Make a sketch of Figure 20. Annotate your drawing to show how the reef protects the town of Borth.

4 Describe the main benefits of the new coastal defence scheme for Borth and the local economy.

5 Do you think the new defences have been a success? Make use of the extract from a local resident's blog to help justify your answer.

The new coastal defences at Borth also include:

- replacing the wooden sea wall at the top of the beach
- adding new shingle to widen the beach.

The scheme was completed in 2015 at a cost of £18 million. Of this, £5 million came from the European Regional Development Fund – funding from the EU designed to promote growth in deprived regions. All of the rock was purchased from a local quarry.

▲ **Figure 20** The rock reef at the southern end of the spit at Borth, 2015.

Borth's sea defences a success

Some of the large boulders within the defences have moved slightly. The pebble banks that were extended and built up as part of the defences have also been reshaped by the waves, but the damage to houses and other properties was minimal. Some were flooded and there was plenty of debris in the road behind them, but along the stretch of beach where the defences are situated there was certainly less damage.

Flooding and storm damage along this stretch was quite a common occurrence too, and that was during storms that were much less ferocious than the one we've just experienced, so I'm almost certain that without the new sea defences the damage would have been much worse.

This recent storm was the first real test of the sea defences in Borth and it looks as though they have proved their worth. They may require a little reshaping and work to restore the banks following this storm, but surely that is better than rebuilding homes. They may not be the prettiest of things and they may not have worked as a surf reef, but they do seem to have fulfilled their primary purpose of acting as a sea defence in this instance.

▲ **Figure 21** A local resident comments on the reef in his blog on 11 January 2014 after a severe winter storm.

Coastal hazard mapping

Flood preparedness can help save lives and prevent injury. The ability of coastal communities to cope during a storm surge can be improved when:

- a warning of an extreme weather event is given
- people know which areas will be affected, what to do and how to protect themselves
- the emergency services and hospitals have a plan of action in the event of disaster.

The UK has sophisticated services that provide weather forecasts and warnings of extreme weather and flooding. Forecasts are provided by the Flood Forecasting Centre at Exeter, which provides a service 24 hours a day, seven days a week. Their forecasts are passed to the Environment Agency (England) and Natural Resources Wales, who inform the public. Each of these bodies provides two types of online hazard map:

- One showing areas that are prone to river and coastal flooding. This type of map can be used to plan an evacuation route long before the flood happens.
- A live map that uses warning symbols in places where floods are likely. The Met Office also has a live weather map that uses warning symbols. A red weather warning means there is danger to life.

Online hazard maps alert the public to the danger of coastal flooding. Hazard mapping also allows local authorities to control the development of areas considered to be at risk, for example by preventing the building of new homes in areas at risk of erosion or coastal flooding.

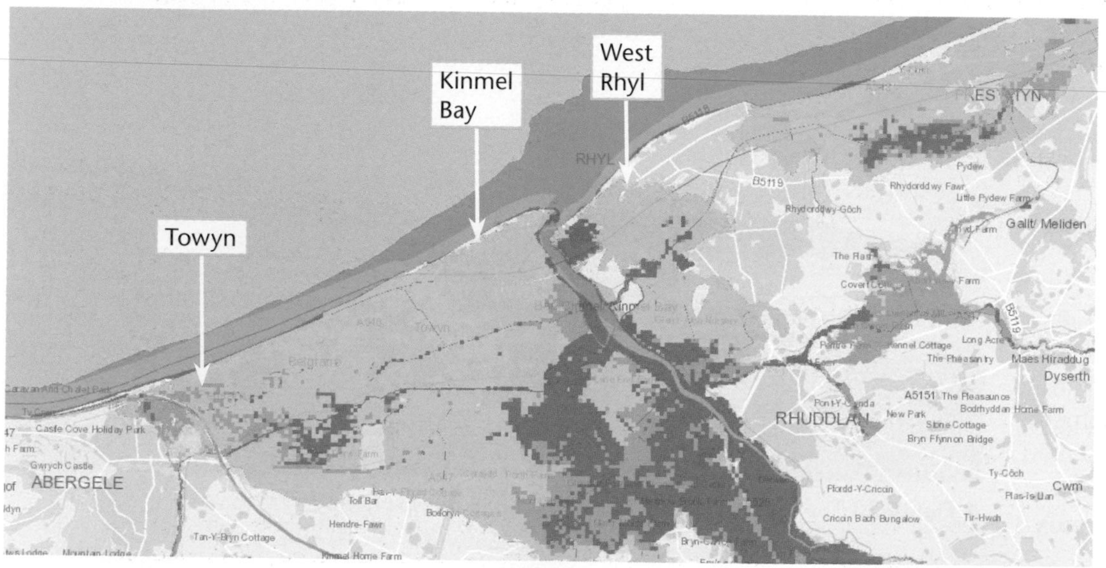

▲ **Figure 22** Screenshot of the Natural Resources Wales flood hazard map for Kinmel Bay and West Rhyl.

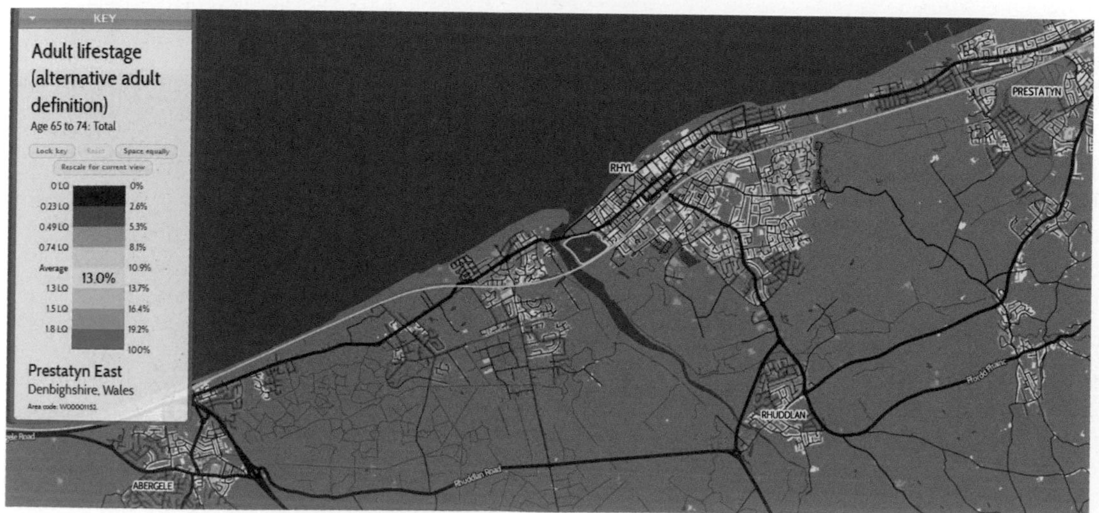

▲ **Figure 23** Screenshot from Data Shine. It shows the proportion of people aged 65–75 in Kinmel Bay and West Rhyl. Green colours are above the national average.

Kinmel Bay – a community at risk of coastal flooding

Natural Resources Wales calculate that 80,000 properties are potentially at risk of flooding around the coastline of Wales. Towyn, Kimnel Bay and Rhyl in North Wales are all examples. This area was badly affected by coastal flooding in 1990 and again in 2013. The storm surge of 5 December 2013 caused coastal flooding: 400 people were evacuated from their homes to Rhyl Leisure Centre. RNLI crews and the fire service rescued 25 residents from flooded bungalows in Rhyl, while caravan sites in Kimnel Bay were flooded and a number of caravans, which were empty for the winter season, were destroyed. At least 400 homes were without power. Two severe warnings – meaning there was a danger to life – were given. People were alerted by phone, text, email and Twitter. Volunteer flood wardens in the area helped with the evacuation.

Activities

1 a) Describe two different uses of hazard mapping in coastal flood zones.
  b) Explain why people living in flood zones should have a personal evacuation plan.
2 Describe the extent of the potential flood zone in Figure 22.
3 Compare the patterns shown on Figures 22 and 23.
4 a) Represent the data in Figure 24 using a suitable technique.
  b) Compare the health of people living in the flood zone in North Wales with those living in Cardiff.
  c) Suggest why residents in the flood zone in North Wales may be at greater risk from floods than other people.

	Towyn	Kinmel Bay	Rhyl West	Cardiff	Wales
Very good health	34.0	40.9	38.8	50.4	46.7
Good health	31.2	30.3	29.9	31.1	31.1
Fair health	21.1	17.6	18.4	12.1	14.6
Bad health	10.6	8.5	9.9	4.8	5.8
Very bad health	3.1	2.7	3.0	1.6	1.8
One person in household with long-term health problem or disability (no dependent children)	31.8	27.3	35.0	21.5	25.2

▲ **Figure 24** Selected health data for the flood zone in North Wales compared to Cardiff and Wales as a whole. The 2011 census asked people to describe their general health over the preceding twelve months as 'very good', 'good', 'fair', 'bad' or 'very bad'. Figure 24 shows the percentage who responded in each category.

Towyn	Kinmel Bay	Rhyl West
LL22 9HW	LL18 5BB	LL18 1LP
LL22 9LR	LL18 5AS	LL18 3AH
LL22 9LX	LL18 5EQ	LL18 3ET

▲ **Figure 25** Selected postcodes.

Weblink

https://naturalresources.wales/flooding – follow the link from this page to the river level flood map, which also shows coastal flooding.

Enquiry

How vulnerable is Kinmel Bay and the surrounding area to coastal flooding?

- Use an online UK atlas (such as Google Maps or Bing Maps) to find the businesses located at each postcode in Figure 25.
- Which of these businesses is in the flood zone shown in Figure 22? You could use the weblink to Natural Resources Wales to check the map.
- Summarise the economic and social impacts of flooding at the postcodes you have researched.

Chapter 2 Vulnerable coastlines

Why are some coastal communities more vulnerable than others?

Coastal environments pose a number of potential hazards, such as landslides, flooding and sea level rise. The level of risk depends on a number of factors:

- physical factors, such as the frequency and magnitude of extreme weather events, climate change and geology
- the number of people who may be affected
- the **vulnerability** of the people affected by the event; vulnerability is linked to factors such as poverty.

Vulnerability

A measure of someone's inability to cope with, or recover from, a disaster such as coastal flooding. Some groups of people are more vulnerable to hazards than others. Poverty, age, gender and disability are all factors that can affect vulnerability. For example, someone's poverty may mean that their house is badly built and unable to withstand flooding.

How can we reduce vulnerability?

To reduce vulnerability to a coastal hazard, organisations need to:

- Reduce the impact of the hazard. Examples include building flood walls and other coastal defences.
- Build **capacity** to cope with the hazard. Examples include educating people about what to do during a cyclone or storm surge so that they are more likely to survive, and creating a disaster plan so that emergency services, hospitals and local authorities all know what to do during and after a major disaster.
- Tackle the root causes of vulnerability. This means that governments need to reduce poverty and inequality in society so that everyone has the same opportunities and the same level of protection during a disaster.

Capacity

The opposite of vulnerability. It describes someone's ability to survive a hazard or recover from it quickly. Capacity is increased where individuals or communities have the resources they need to cope with the hazard. Resources can be financial or material, such as strong building materials. But capacity can also be improved by factors such as education, technology and disaster preparedness and planning.

▲ **Figure 1** Fishermen's stilt houses, Luzon, Philippines.

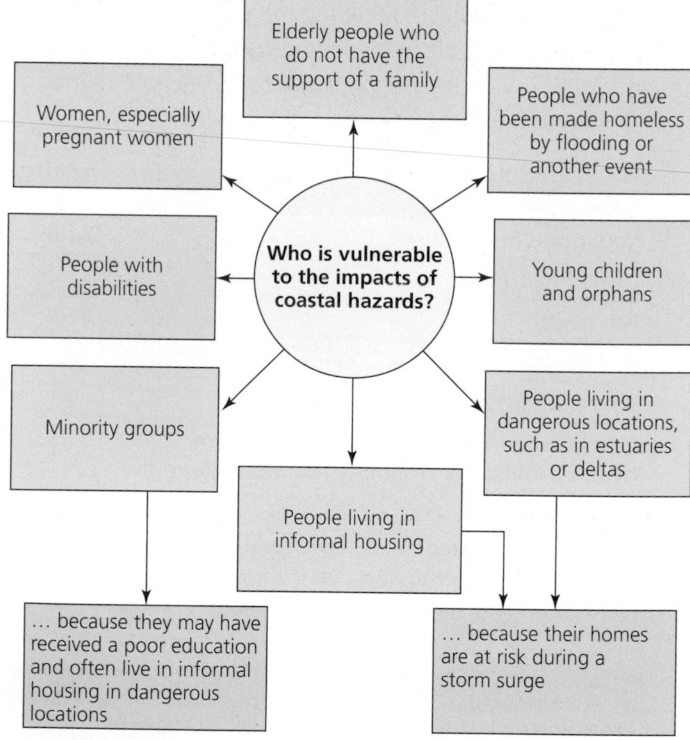

▲ **Figure 2** Why are some groups of people more vulnerable than others to coastal hazards?

Activities

1 a) Describe the housing in Figure 1.
  b) Why are people living here particularly vulnerable? What hazards do they face?
2 Choose four different groups of people in Figure 2.
  a) For each group explain why they may be more vulnerable to natural disasters.
  b) For one of these groups, suggest how their capacity could be improved to cope with coastal flooding.

Climate change

Climate change will make some coastal communities more vulnerable to coastal hazards. Rising sea levels will increase the rate of coastal erosion. More farmland will be lost and more expensive sea defences will be needed to 'hold the line' against erosion of our towns and cities. Climate change also means a warmer atmosphere, which means more storms like the devastating storm surges that flooded Jaywick in 1953 and 2013. In other words, climate change will increase the frequency and magnitude of extreme weather events, which are critical physical factors that affect the vulnerability of all coastal communities.

Year	2007	2032	2057	2082	2107
Rise in sea level (cm)	0	13	35	65	102

▲ **Figure 3** Predicted sea level rise at Jaywick, Essex.

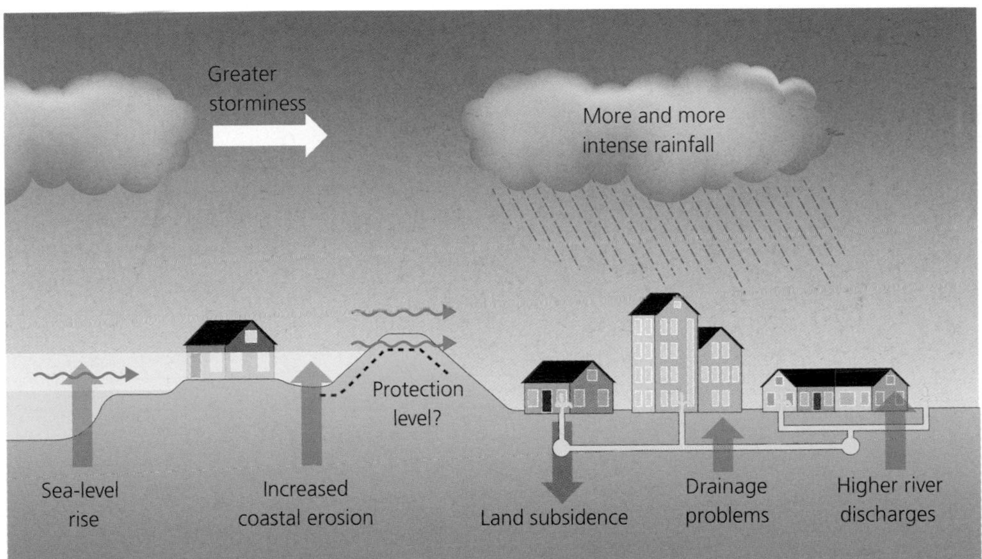

▲ **Figure 4** Some of the impacts of climate change on our coastline by 2050.

Activities

3 Use Figure 4 to describe five different impacts of sea-level rise on coastal communities in the UK.

4 Use Figure 5 and an atlas to name:
   a) five counties in England facing problems of extreme coastal erosion
   b) three counties in Wales facing very high rates of erosion.

5 a) Use Figure 3 to draw a graph of the predicted sea level rise at Jaywick.
   b) Give two reasons to explain why sea levels are rising here.

► **Figure 5** Coastal erosion if carbon dioxide emissions continue to increase and sea levels rise.

How could climate change affect London and the Thames Gateway?

One of the most vulnerable coastlines in the UK is the estuarine landscape of the River Thames to the east of London. This coastline, known as the Thames Gateway, is at risk from storm surges (like those in 1953 and 2013) that push sea into the narrow funnel-like coastline between Essex and Kent. This coastline has been sinking ever since the end of the ice age in the UK about 10,000 years ago – a process called **postglacial rebound**. As a result, the Thames Gateway is sinking at about 2 mm a year relative to current sea levels. Climate change means that sea levels in the Thames estuary are rising at about 3 mm a year. So the combined effect of sea level rise and postglacial rebound means that sea levels here are rising at 5–6 mm per year.

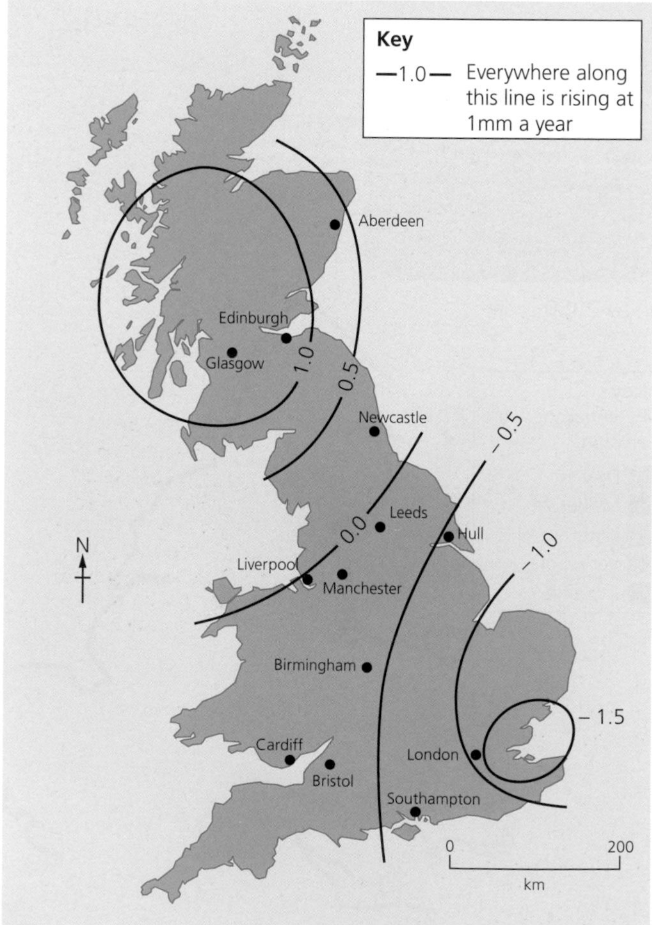

▲ **Figure 6** The amount of postglacial rebound (mm per year). Positive numbers mean the land is rising relative to the sea and negative numbers mean the land is sinking.

Year	Type of risk that caused closure		Total closures
	Tidal	River flood	
1983	1	0	1
1984	0	0	0
1985	0	0	0
1986	0	1	1
1987	1	0	1
1988	1	0	1
1989	0	0	0
1990	1	3	4
1991	2	0	2
1992	0	0	0
1993	4	0	4
1994	3	4	7
1995	2	2	4
1996	4	0	4
1997	1	0	1
1998	1	0	1
1999	2	0	2
2000	3	3	6
2001	16	8	24
2002	3	1	4
2003	8	12	20
2004	1	0	1
2005	4	0	4
2006	3	0	3
2007	8	0	8
2008	6	0	6
2009	1	4	5
2010	2	3	5
2011	0	0	0
2012	0	0	0
2013	0	5	5
2014	9	41	48
2015	1	0	1
2016	1	0	1
2017	2	0	2
2018	3	0	3

▲ **Figure 7** Closures of the Thames Barrier to protect against storm (tidal) surges (1983–2018).

Activities

1 Use Figure 6 to describe the parts of the UK where:
   a) land is rising fastest
   b) land is sinking fastest.
2 a) Use the data in Figure 7 to produce a graph of closures.
   b) Describe the trend of your graph.
   c) Explain why this graph could be seen to be more evidence for climate change.

Holding the line

The Thames flood barrier was completed in 1982. It is situated to the east of the City of London so protects large parts of London from tidal surges coming up the river from the North Sea. It protects 1.25 million people from tidal floods. However, it is now thought that the barrier is not large enough to protect London from future floods. The Thames Estuary 2100 Plan (TE2100) suggests that, by 2100, London needs to be protected from a possible flood that would be 2.7 m higher than current flood levels.

Factfile: Property at risk of tidal flooding on the Thames floodplain

- Over 500,000 homes
- 40,000 commercial and industrial properties
- 400 schools
- 16 hospitals
- 35 tube stations
- Over 300 km of roads

The TE2100 Plan uses three strategies to protect London and the Thames Gateway:

- Continue to renew and replace existing embankments, sea walls and sluices in the Thames Gateway.
- Increase the amount of inter-tidal habitat in the Thames estuary by 876 hectares. These salt marshes will help to store flood water as it moves up the estuary during a tidal surge. These storage areas would be created by managed realignment projects like the one at Tollesbury, Essex.
- Consider building a new, larger barrier at Long Reach to the east of the existing barrier. The construction of this new barrier would cost between £6 billion and £7 billion.

Activities

3 Use Figure 8 to describe:
   a) the distribution of breaches
   b) the amount and value of flooded land.

4 Explain why the cost of flood damage in Essex would be lower than that in London.

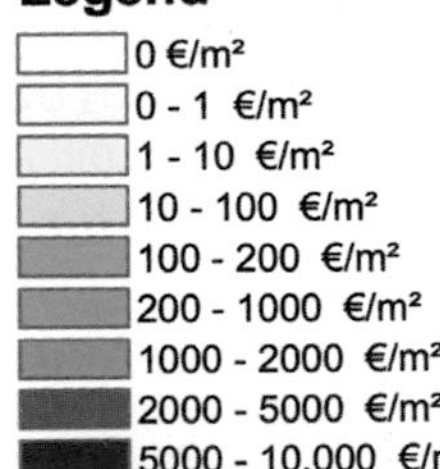

▲ **Figure 8** The cost of flood damage in 2050 after a flood similar to the 1953 storm surge if sea levels continue to rise and flood protection is not improved. Red dots show where coastal defences would be breached.

Enquiry

How should London and the Thames Gateway be protected in the future? Some people are sceptical about managed realignment. Justify why the TE2100 Plan proposes to combine a new flood barrier with managed realignment.

Why are some coastal communities more vulnerable than others?

A report by the Joseph Rowntree Foundation suggests that some of the most vulnerable people, living in isolated communities of the UK, will be the most affected by rising sea levels.

The report argues that poverty is a factor that makes some communities more vulnerable to sea level rise and coastal flooding than others. Poverty means that the local council has fewer resources available to reduce the threat of and the impacts of sea level rise.

The report suggests that vulnerability to coastal flooding increases where communities have:

- a higher proportion of people claiming benefits
- a fast turnover of people through economic migration
- a high proportion of poor-quality housing
- an over-reliance on tourism, resulting in seasonal employment and low incomes.

▲ **Figure 9** Coastal floods in Great Yarmouth, December 2013.

Skegness, Lincolnshire

Skegness is one of the better-known seaside resorts in England. It is situated within the largely rural area of East Lindsey and has poor road and rail links. Skegness has one of the largest concentrations of caravan parks in Europe. Some static caravans are the permanent homes for retired residents and people on low incomes.

Great Yarmouth, Norfolk

Great Yarmouth is a medium-sized port and is an important seaside resort. It has a high proportion of elderly and retired residents. Unemployment rates are higher in Great Yarmouth than in the rest of the East of England. The economy of the port is in decline and the nearby North Sea gas fields have begun to reduce production.

GEOGRAPHICAL SKILLS

Interpreting population pyramids

Population pyramids are a specialist form of bar graph. Each bar represents males or females in a specific age category. The bars may represent actual population numbers or percentage figures. Interpreting the shape of a pyramid can tell you a lot about the structure of a population. Pyramids with a wide base have a youthful population and those with a wide top have an ageing population. Interpreting the structure is the first step in analysing possible issues that may be facing the population. For example, does an ageing population have sufficient health care and social services suitable for this age group?

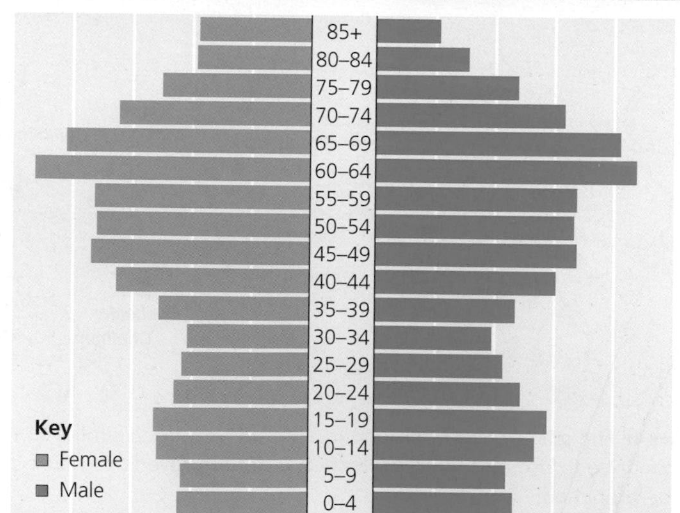

▲ **Figure 10** Population pyramid for East Lindsey, Lincolnshire (2011).

Weblink

www.ons.gov.uk/ons/interactive/uk-population-pyramid---dvc1/index.html

An interactive population pyramid for the UK showing change between 1971 and 2085.

Llanelli, Carmarthenshire

Unemployment in Llanelli is higher than the national average and the town has been particularly hard hit by the economic recession of 2008–09. Derelict industrial sites along the coastline are now the focus of several regeneration schemes. The town has a higher than average proportion of elderly residents. A high number of migrants from Eastern Europe have settled in the town.

Benbecula, Outer Hebrides

With a population of 1,200, the population and economy of this remote Scottish island have been in decline since the mid-1970s because of the closure of a military base. The lack of jobs for younger people has also contributed to the decline and to the ageing of the island's population. The island's coastal drainage system was built in the 1800s and is not very efficient. It struggles to cope with the combined effects of heavy rainfall and high tides.

Indicator	Skegness	Lincolnshire	England
% of working population in professional/ management jobs	14	18	23
% of working population claiming work-related benefits	18	13	13
% of adult population with no academic qualifications	33	26	22
% of the adult population who are in very good health	38	43	47

▲ **Figure 11** Socio-economic statistics for Skegness.

▲ **Figure 12** Skegness has a large concentration of static caravans.

In such communities, local people may lack the funds to make structural changes to their homes (e.g. to make them flood resistant). They may not be able to afford to move away. Coastal local authorities with areas of high deprivation may not be able to afford the resources for climate change preparation. Government policy means that there is an ever-increasing expectation that individuals and communities should help themselves to prepare for the likely increase in sea level rise. For disadvantaged residents, the threat of sea level rise is simply not a major issue at this moment in time. They have other things to focus on.

▲ **Figure 13** An extract (adapted) from the Joseph Rowntree Foundation Report, Summary of 'Impacts of climate change on disadvantaged UK coastal communities' (2011).

Activities

1. Compare the four communities described on these pages. Describe three similarities that link at least two of these communities.
2. a) Describe the structure of the population structure for East Lindsey.
   b) Use the Weblink to compare this population structure to that of the UK.
   c) What does this suggest is needed in East Lindsey?
3. Explain why every local authority may not be able to afford the resources needed to prepare for sea level rise.
4. a) Use Figure 11 to identify the overall pattern shown in the table.
   b) The figures in the table date from the National Census in 2011. The worst impact of sea level rise is not expected to happen until 2050. Should the local authority use the data to appeal for additional outside help when planning for sea level rise? Justify your answer.

Enquiry

Who should be responsible for protecting communities in the UK from the effects of sea level rise?

Government policy means that there is an ever-increasing expectation that individuals and communities should help themselves to prepare for the likely increase in sea level rise. How far do you agree with this statement? Discuss this in groups.

How might climate change affect coastal communities around the world?

By 2030 it is estimated that 950 million people around the world will live in the **low elevation coastal zone (LECZ)** – that is, coastal areas that are less than 10 m above sea level. Climate change presents a triple threat to people living in LECZs:

- Sea level rise increases the risk of coastal flooding at high tide.
- Heavier rainfall increases the risk of flash floods in urban areas with poor drainage.
- More violent storms and hurricanes increase the risk of coastal erosion and storm surges.

Small Island Developing States (SIDs) such as the Maldives in the Indian Ocean and the Marshall Islands in the Pacific are very low lying. A 1 m rise in sea level by 2100 would flood up to 75 per cent of the land in each of these two nations.

The worst-affected coastal communities could be those living on the world's major river deltas. People living here are affected by subsidence of the soft land as well as by sea-level rise. Millions of people live on deltas in Bangladesh, Egypt, Nigeria, Vietnam and Cambodia.

In 2013 the World Bank identified 136 of the world's coastal cities that were at greatest risk from climate change. Mumbai, home to 18.4 million people, is one. Built on a low-lying island, much of Mumbai is only 10 m above sea level. Other cities, like New York, Singapore and New Orleans, are also at risk. Many cities are in developing countries where the poorest members of society are at most risk of natural hazards. This is because the poorest neighbourhoods, like those in Figure 15, are often in low-lying areas and built alongside waterways or seafronts that are vulnerable to flooding. As sea levels rise, some people may have to leave their homes. They will become **environmental refugees**.

Low elevation coastal zone (LECZ)

Any area close to the sea that is less than 10 metres above sea level. It is estimated that 950 million people will live in the global LECZ by 2030.

Environmental refugee

Someone forced to leave their home permanently because of a change in the environment, such as coastal flooding or desertification.

Activities

1 Use Figure 14 to describe:
   a) the distribution of SIDS
   b) the location of:
      i) the Maldives
      ii) the Marshall Islands.
   c) Suggest why people living in isolated places like these are vulnerable to natural hazards.

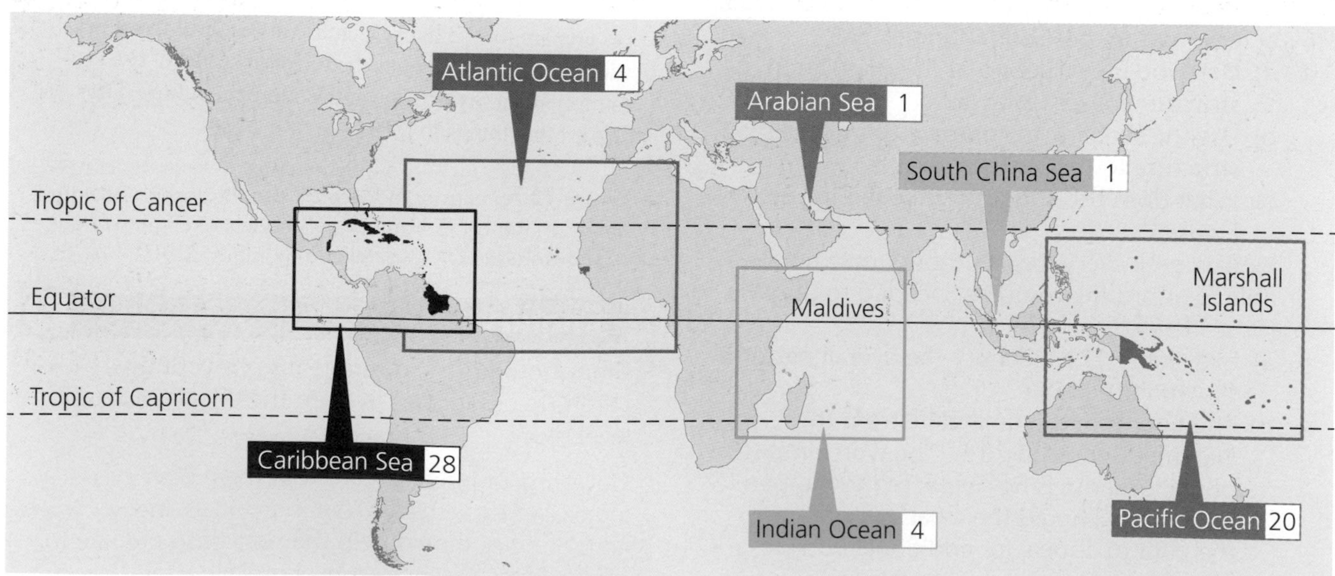

▲ **Figure 14** The locations of Small Island Developing States (SIDs).

City	Population (%) at risk in LECZ	Land (%) at risk in LECZ	Population 2015	Population 2030
Cotonou, Benin	94.7	85.4	682,000	979,000
Warri, Nigeria	90.8	92.0	663,000	1,298,000
Alexandria, Egypt	85.1	68.8	4,778,000	6,313,000
Port Harcourt, Nigeria	64.4	61.9	2,344,000	4,562,000
Dakar, Senegal	61.6	47.6	3,520,000	6,046,000

▲ **Figure 15** Selected African cities in the low elevation coastal zone (LECZ).

▲ **Figure 16** Poor neighbourhoods in Cotonou, Benin are built on stilts along the waterfront.

Activities

2 Study Figure 16.
   a) Describe the housing carefully.
   b) Suggest why the poorest members of society live in neighbourhoods like this.
   c) Suggest how this community will be affected by climate change.

3 Calculate the actual number of people expected to be living in the LECZ in each of the cities in Figure 15.

Enquiry

Analyse why it may be more difficult for the governments of SIDs to cope with climate change than a larger country like India.

Communities vulnerable to climate change and sea level rise in Egypt

The International Organization for Migration suggests that climate change will displace 200 million people by 2050. Rising sea levels, drought, food and water insecurity and increased health risks are the main reasons. Egypt is one country that could be affected.

Egypt has a GNI of US$2,800, which makes it one of the world's poorer middle income countries. Egypt is a desert country with a population of 85 million. Most urban areas and farmland are squashed into just 15 per cent of the nation's landmass – mainly along the length of the River Nile and in the Nile Delta. Climate change will create a number of challenges for Egypt:

- Rising temperatures and less frequent rain will increase the current problems of **water stress**. This will affect the urban poor and it will mean that farmers will have to find new efficient techniques to irrigate crops.
- Water-borne diseases and malaria will become more common. There is also likely to be a huge increase in parasitic diseases, skin cancer, eye cataracts, respiratory ailments and heat stroke.
- Sea level rise may erode and flood the Nile Delta, displacing as many as 8 million people.

Alexandria, in the Nile Delta, is Egypt's second largest city. This Mediterranean port handles 80 per cent of Egypt's imports and exports. The wealth of the city attracts migrants. Natural population increase adds to pressure on space. Houses and flats are often built without proper planning permission or building regulation. In fact, it is estimated that 50 per cent of Alexandria's population live in informal housing. The urban poor are perhaps at greater risk of climate change than others because they:

- have few savings so cannot afford to lose their jobs or their homes
- often rely on boreholes that are polluted by human waste so are at risk of water-borne disease or they have to buy water from street vendors at great cost
- often live in locations that are dangerous to human health, for example near to stagnant water where mosquitoes that carry malaria breed
- live in badly built multi-storey buildings that are at risk of collapse during earthquakes.

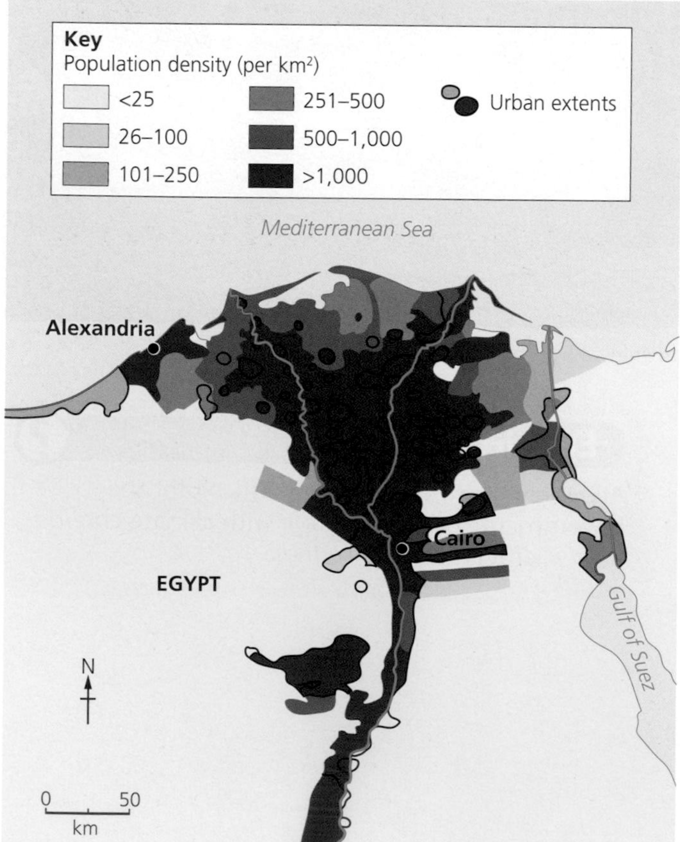

◀ **Figure 17** Population density of the Nile Delta.

Water stress

A lack of clean water that causes problems for people and the economy.

Year	Population
1950	1.04
1960	1.50
1970	1.99
1980	2.52
1990	3.06
2000	3.55
2010	4.33
2020	5.23
2030	6.31

▲ **Figure 18** The population of Alexandria (millions). Figures after 2010 are predictions.

Alexandria is built on Egypt's **low elevation coastal zone (LECZ)** so is vulnerable to permanent flooding if sea levels rise. If this happens, Egypt will find it difficult to re-house the huge number of environmental refugees.

▲ **Figure 19** Illegally built flats in Alexandria, Egypt.

Presumably, many people who currently live in informal housing in Alexandria will move into the poorest, most overcrowded districts of Cairo.

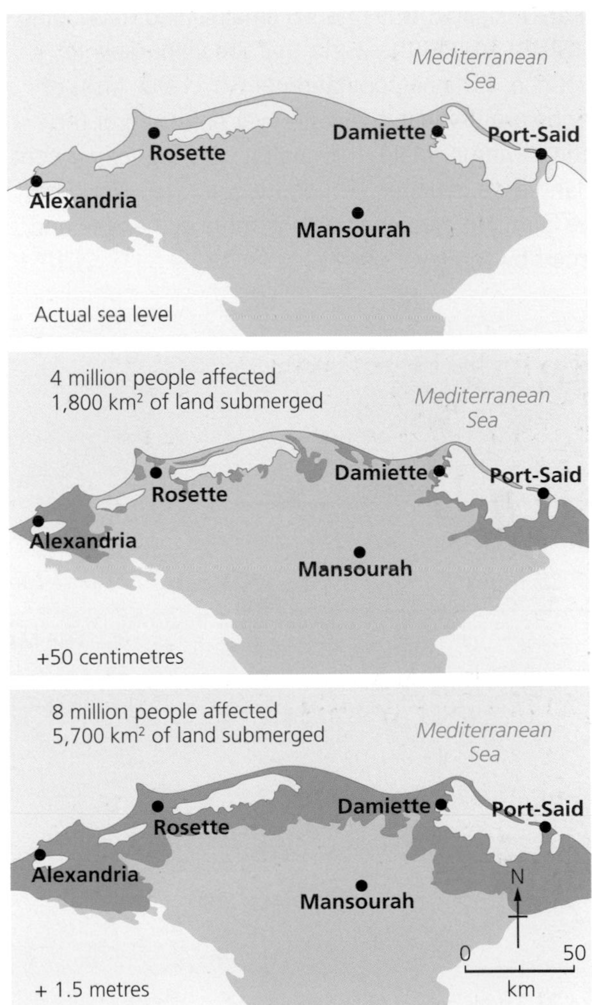

▲ **Figure 20** Predicted changes to Egypt's coast-line as the sea level rises.

Activities

1 Describe the location of Alexandria.
2 a) Use Figure 18 to draw a line graph of population growth.
   b) Describe the trend of your graph.
   c) If this trend continues, what might be the population of Alexandria in 2040?
3 a) Outline three different problems that will be created for the people of Egypt by climate change.
   b) Explain why the urban poor are most at risk. Give two different reasons.
4 Describe the possible loss of land in the Nile Delta. How big is the potential environmental refugee problem?
5 a) Use Figure 17 to describe the distribution and density of Egypt's population.
   b) Why is this a significant problem for the Egyptian government if sea levels rise as predicted?

Low elevation coastal zone (LECZ)

Any area close to the sea that is less than 10 metres above sea level. It is estimated that 950 million people will live in the global LECZ by 2030.

Enquiry

'Solving the problems caused by climate change refugees is the responsibility of all nations.' How far do you agree with this statement? Justify your answer.

Is it too late to save the Maldives?

The Republic of Maldives, in the Indian Ocean, is made up of 1,190 islands. It is one 58 small island developing states (SIDS) around the world that are vulnerable to sea level rise. It has a population of 350,000. Most of the islands are uninhabited, with over one-third of the population living in Malé, the capital city. Eighty per cent of the land area is under 1 m above sea level. Nowhere is above 3 m. No place on earth is more vulnerable and threatened by sea level rise.

The Maldives has a GNI of US$8,300 (2018), which makes it one of the wealthier middle income countries. The government is relying on hard engineering to hold the line against rising sea levels. It is building an artificial island close to Male called the City of Hope. The new island will be protected by 3-metre-high sea walls. It should be finished in 2023 and will house 130,000 people.

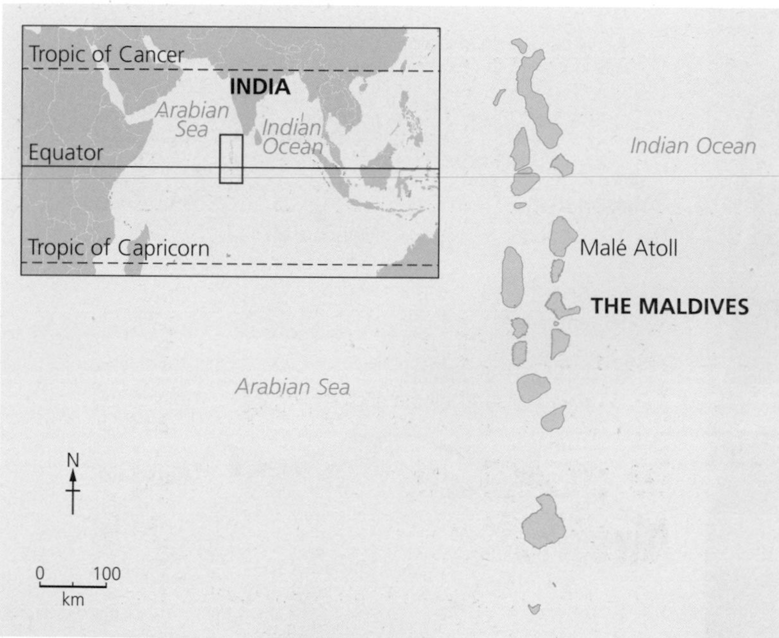

▲ **Figure 21** The location of the Maldives.

▲ **Figure 22** Is this the future for the Maldives?

▲ **Figure 23** The capital city of the Maldives, Malé, is surrounded by sea.

Factfile: The effects of sea level rise on the Maldives

- A rise in sea level of 0.5 m by the year 2100 would mean that 77 per cent of the land surface of the Maldives will be underwater.
- A rise in sea level of 1 m by the year 2100 would make the islands uninhabitable by 2085.

How is sea level rise already affecting the islands?

Frequent flooding of the islands over the past 30 years has caused ongoing problems for the Maldives. Most of the problems occur when the highest spring tides coincide with storms across the northern India Ocean.

The impact of flooding across the Maldives

Male – a capital city surrounded by sea walls:
In 2008 Japan offered $60 million in aid to fund a 3 m sea wall for Male. The sea wall has been completed and will hold back the advance of the sea for the medium term. All the other islands remain vulnerable. The sea wall needs constant repair and the cost has to be paid for through local tourist taxes.

Drinking water is in short supply:
Sea level rise is already beginning to put stress on the scarce freshwater resources of the Maldives; 87% of the population can currently be supplied by collecting rainwater. Groundwater sources across the chain of islands have been contaminated by salt-water intrusion and are now undrinkable. Bringing in supplies from abroad is unsustainable.

The tourist industry under threat:
Tourism is by far and away the most important industry. It accounts for 90% of government tax revenue. The damage caused by the 2004 tsunami-related sea surge destroyed many prize beaches and ruined some luxury resorts. For a year, tourist numbers dropped dramatically as the islands implemented a recovery programme. Tourist numbers have since recovered, but the industry may have been given a taste of what's to come.

▲ **Figure 24** The impact of flooding across the Maldives.

Islands that float

Floating islands will be moored to the seabed using cables to minimise environmental impact. This idea has been put forward by a Dutch company. One of the islands will be used to create an artificial golf course. Golfers will access the floating 'golf island' by a tunnel on the seabed. There will be a spectacular underwater clubhouse for golfers to relax in after their game. The artificial islands will be built in India or the Middle East and towed to the Maldives.

▲ **Figure 25** How might the Maldives develop tourism in the future?

Sea level rise forces castaways to move to Australia

The Maldivian President said his government was considering Australia as a possible new home if the Maldives disappear beneath rising seas. He explained that Maldivians wanted to stay but moving was an eventuality his government had to plan for. Australia may need to prepare for a mass wave of climate refugees, seeking a new place to live.

▲ **Figure 26** Sea level rise may create environmental refugees.

Activities

1 Describe the location of the Maldives.
2 Study the two newspaper headlines shown in Figures 25 and 26. Both were published in 2012. They offer very different radical solutions to the problems faced by the Maldives.
   a) Discuss how sustainable each idea would be.
   b) Discuss which is the most likely to be turned into reality.

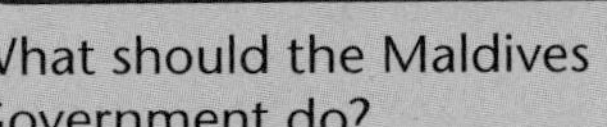

What should the Maldives Government do?

- List five different ideas for action – place your ideas in priority order.
- Justify why you think ideas one and two need to be put into action as soon as possible.

THEME 5 Weather, climate and ecosystems

Chapter 1 Climate change during the Quaternary

Why has climate changed?

The Quaternary, sometimes called the Pleistocene, is the most recent period of geological time. It is a period of Earth's history that has been dominated by cold climates and ice shaping the land. At the beginning of the Quaternary, the polar ice sheets were far bigger than they are today, as you can see in Figure 1. Throughout the 2.6 million years of the Quaternary the climate has changed constantly. There have been periods, known as **glacials**, when the polar ice has reached much further south, covering large parts of the Earth. At other times, known as **inter-glacials**, the polar ice retreated. Scientists have evidence of 60 different cycles of ice advance and retreat. The ice sheets retreated about 10,000 years ago and the Earth is currently experiencing an inter-glacial period. However, the ice has not totally disappeared. Technically, we are still living in an ice age!

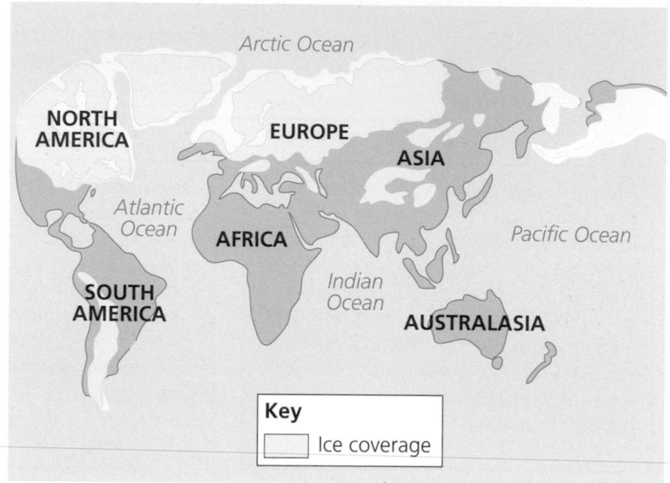

▲ **Figure 1** The extent of the ice during a colder period of the Quaternary period.

What are the natural causes of climate change?

There is much debate about the cause of the change from glacial periods to inter-glacial periods. The most commonly accepted theory is based on the work of a scientist called Milankovitch. He suggested that the warmer and cooler periods are caused by a combination of two things:

- The natural wobble of the Earth as it moves around the Sun. This affects the tilt of the Earth and the amount of energy it receives from the Sun.
- The fact that the Earth does not have a circular orbit around the Sun. The orbit is eccentric: sometimes it is closer to the Sun, and sometimes it is further away.

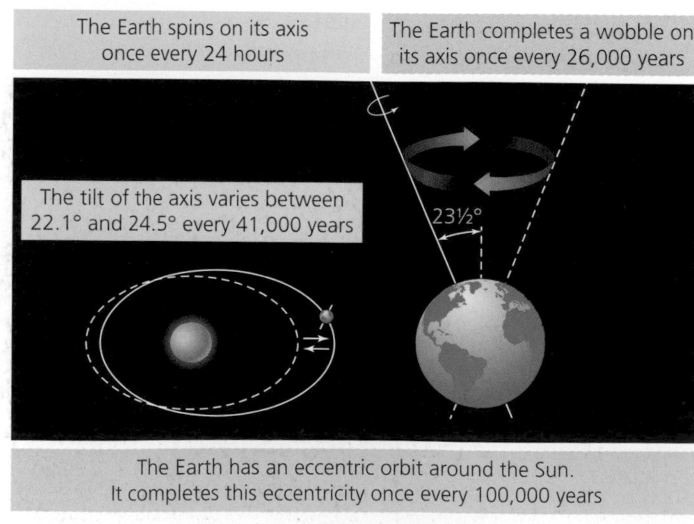

▲ **Figure 2** Why does the Earth's climate fluctuate?

Activities

1 Study Figure 1.
   a) Describe the amount and distribution of the ice sheets.
   b) Study the Mediterranean region carefully. Suggest why the Mediterranean Sea appears to be smaller than it is today.
   c) To what extent are we still living in an ice age? Use an atlas or the internet to compare the amount and distribution of permanent ice today with that shown in Figure 1.

2 Why does the surface temperature of the Earth change when:
   a) its tilt varies due to the natural 'wobble'
   b) due to its eccentric orbit?

How do volcanic eruptions affect climate?

Large volcanic eruptions can eject dust and sulphur dioxide (SO_2) into the lower stratosphere – a layer in the atmosphere that is 15 to 25 km above the Earth. At this altitude the jet stream is able to carry the volcanic material in a belt right around the globe. The mixture of ash and SO_2 form an **aerosol** – tiny droplets that scatter sunlight back into space. This can have the effect of reducing the amount of solar energy that reaches the Earth's surface, so average temperatures can be reduced. The largest volcanic eruption of the last 100 years was the eruption of Mount Pinatubo in the Philippines in 1991. This eruption had a magnitude of VEI-6. It ejected enough ash and SO_2 to block out about ten per cent of the sunlight reaching the northern hemisphere, reducing temperatures by 0.4 to 0.6°C for two years. However, the very largest volcanic eruptions could have much greater effects.

The eruption of Tambora in Indonesia in 1815 had a magnitude of VEI-7, making it ten times larger than Pinatubo's eruption. Temperatures in the northern hemisphere fell. Crops failed in North America and Europe during the cold, wet weather of 1816 and thousands died of starvation. Dust in the stratosphere scattered sunlight creating spectacular sunsets – an effect captured by the British artist JMW Turner as you can see in Figure 3.

▲ **Figure 3** Sunset over Chichester Canal, painted by JMW Turner.

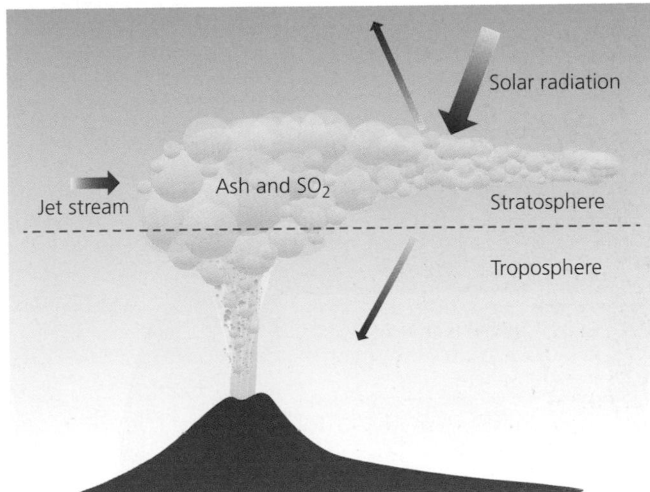

▲ **Figure 4** How volcanic eruptions can reduce global temperatures.

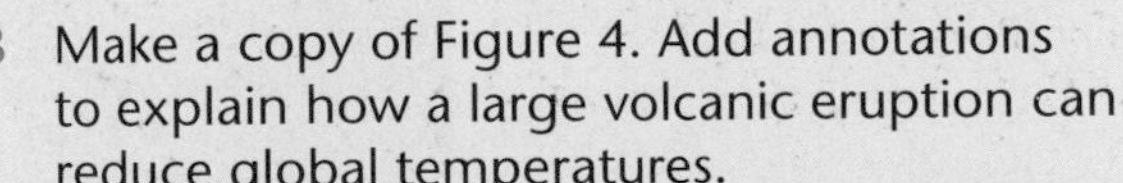

Activity

3 Make a copy of Figure 4. Add annotations to explain how a large volcanic eruption can reduce global temperatures.

Enquiry

How significant is the effect of large volcanic eruptions on climate?

Research the Toba eruption (75,000 years ago) – perhaps the largest eruption during the Quaternary period. What were its effects on climate? How might this compare to other natural reasons for warming or cooling?

▲ **Figure 5** Forests act as carbon stores. When they burn, the carbon is released as CO_2 into the atmosphere.

What is the greenhouse effect?

The **greenhouse effect** is a natural process of our atmosphere. Without it, the average surface temperature of the Earth would be –17° Celsius rather than the 15° Celsius we currently experience. At these temperatures, life would not have evolved on Earth in its present form and we probably wouldn't exist!

The greenhouse effect, shown in Figure 6, means that Earth's atmosphere acts like an insulating blanket. Light (short-wave) and heat (long-wave) energy from the Sun passes through the atmosphere quite easily. The Sun's energy heats the Earth and it radiates its own energy back into the atmosphere. The long-wave heat energy coming from the Earth is quite easily absorbed by naturally occurring gases in the atmosphere. These are known as greenhouse gases. They include carbon dioxide (CO_2), methane (CH_4) and water vapour (H_2O). Carbon dioxide is the fourth most common gas in the atmosphere. It occurs naturally in the atmosphere as a product of respiration from all living things. So carbon dioxide has existed in the atmosphere for as long as there has been life on Earth. Methane and water vapour have been in the atmosphere for even longer, so the greenhouse effect has been affecting our climate for thousands of millions of years.

Greenhouse effect

A natural process that traps heat in the atmosphere. This process has been enhanced by human activity.

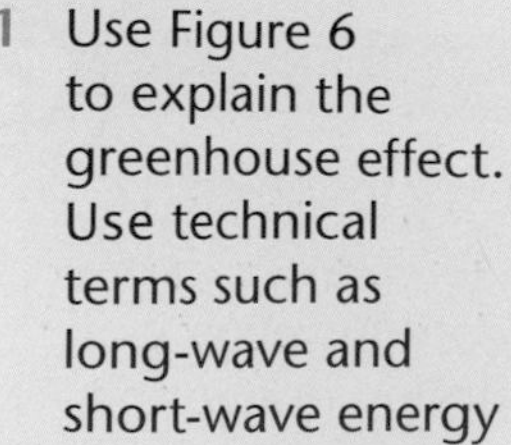

Activity

1 Use Figure 6 to explain the greenhouse effect. Use technical terms such as long-wave and short-wave energy in your answer.

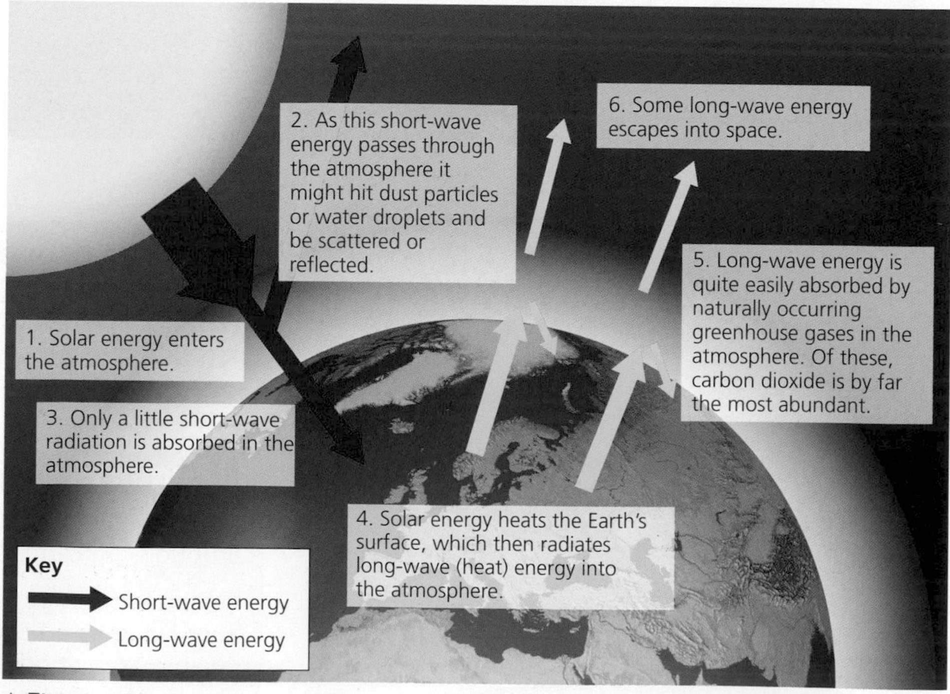

▲ **Figure 6** The greenhouse effect.

The carbon cycle

Carbon is one of the most common elements in the environment. It is present in:

- all organic substances, i.e. all living things
- simple compounds such as CO_2, which exists as a gas in the atmosphere and is dissolved in the oceans
- complex compounds, for example, hydrocarbons found in fossil fuels such as oil, coal and gas.

The **carbon cycle**, shown in Figures 7 and 8, contains stores and flows. Carbon is stored in many different parts of the environment, including forests, soils, oceans and the atmosphere. Carbon flows from one store to another through processes such as respiration and solution. Some stores, such as fossil fuels, store carbon for very long periods of time. These long-term stores are **carbon sinks**. Human activities, such as farming, clearing forests and burning fossil fuels, cause carbon to flow into the atmosphere.

At night photosynthesis stops. The tree continues to respire and it emits more CO_2 than it absorbs

Solar energy

While the tree is alive it absorbs more CO_2 from the atmosphere than it emits

When branches or leaves fall they transfer the carbon that is locked in the plant tissue into the soil

During the day the tree uses sunlight to convert carbon dioxide to plant sugars. This is **photosynthesis**

Organisms such as beetles and earthworms may digest the plant tissue. Their respiration adds CO_2 to air in the soil

Rainwater dissolves some of the carbon dioxide that has come from soil organisms. This water may carry the dissolved CO_2 into a river and eventually to the sea.

▲ **Figure 7** A simplified carbon cycle.

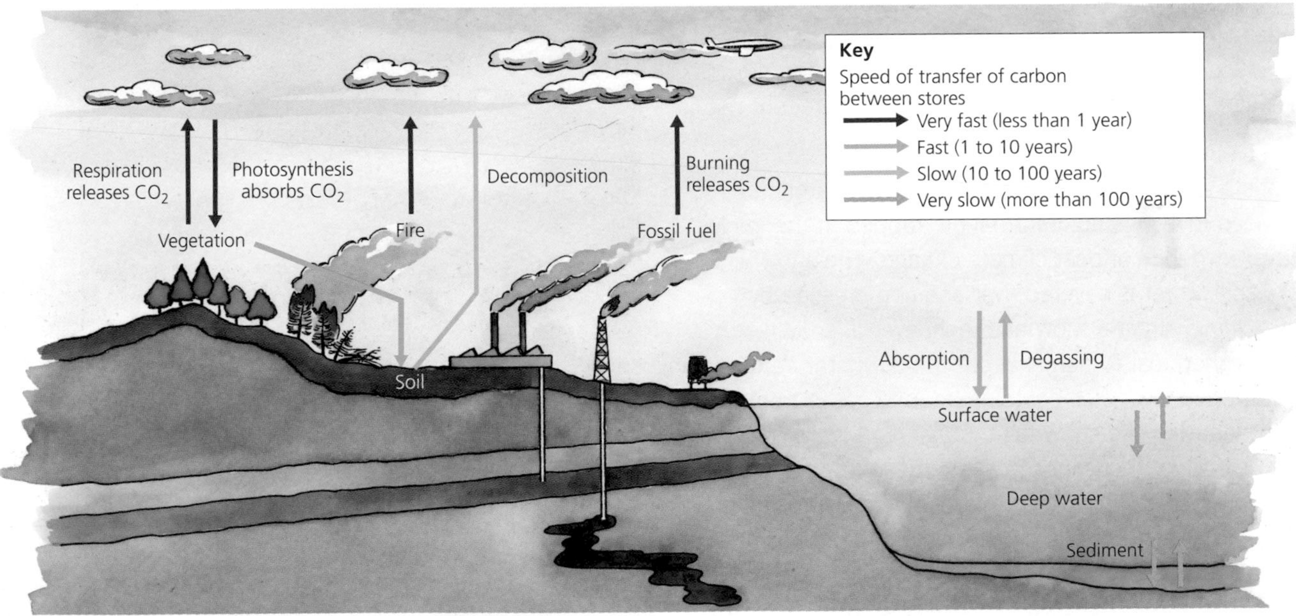

▲ **Figure 8** The carbon cycle, showing fast and slow transfers.

Activities

2 Study Figures 7 and 8.

a) Describe the human actions that release CO_2 into the atmosphere.

b) Explain the processes that allow forests to act as a carbon sink.

c) Give two reasons why the burning of tropical rainforests will increase the amount of CO_2 in the atmosphere.

3 Use Figure 8.

a) Describe the difference in the speed of transfer of carbon in the natural part of the cycle compared with the part of the cycle affected by human action.

b) Explain what difference this makes to the amount of carbon stored in the atmosphere compared with the long-lasting carbon sinks. Explain why this is alarming.

Evidence for climate change

Evidence for climate change comes from:

- historical accounts, such as descriptions of the cold weather of 1816 (see Figure 3)
- scientific measurements.

Scientists have been accurately measuring CO_2 levels in the atmosphere since 1958. The measurements are taken in Hawaii, where the atmosphere is unaffected by local emissions of CO_2 from traffic or industry. Figure 9, known as the Keeling Curve, shows how CO_2 levels have changed since 1958.

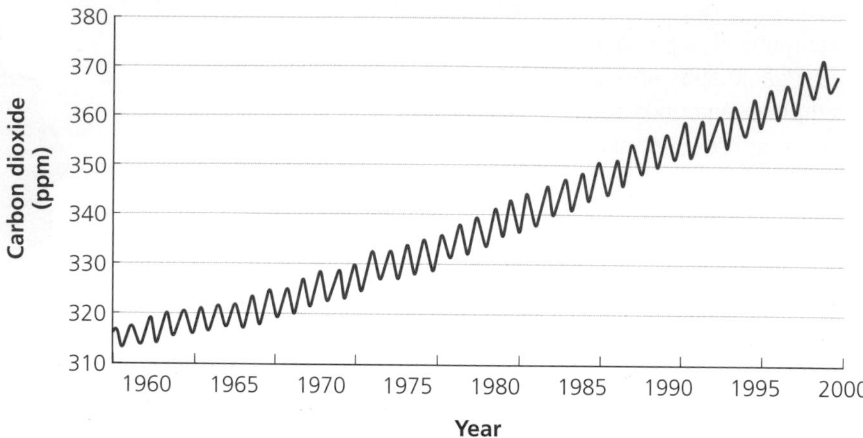

▲ **Figure 9** The Keeling Curve shows the rise of carbon dioxide in the atmosphere since monitoring began in 1958 (ppm = parts per million).

Evidence from the ice cores

We have already seen that scientific evidence from Hawaii proves that carbon dioxide levels have been rising steadily since 1958. However, can we be certain that this isn't part of a natural cycle? Perhaps carbon dioxide levels vary over long periods of time and the recent rise is part of one of those cycles.

Scientists working in both Greenland and Antarctica have been investigating information trapped in the ice to uncover evidence of past climate change. The snowfall from each winter is covered over and compressed by the following winter's snowfall. Each layer of snow contains chemical evidence about the temperature of the climate. Each layer also contains trapped gases from the atmosphere that the snow fell through. Gradually the layers turn to ice. Over thousands of years these layers have built up and are now thousands of metres thick. By drilling down into the ice, scientists can extract older and older **ice cores**. Chemical analysis of these ice layers and the gases they contain reveal a record of the climate over the last 420,000 years. This evidence suggests that the climate has indeed gone through natural cycles of colder and warmer periods known as glacials and inter-glacials. They also show that levels of carbon dioxide in the atmosphere have also gone up and down as part of a natural cycle.

▲ **Figure 10** Scientists taking ice core samples from the ice sheet in Iceland.

Ice core

A sample of ice from a glacier or ice sheet obtained by drilling into the ice. The deeper the drill goes, the older the ice. Gas trapped in the ice can be analysed to tell us about the atmosphere in the past.

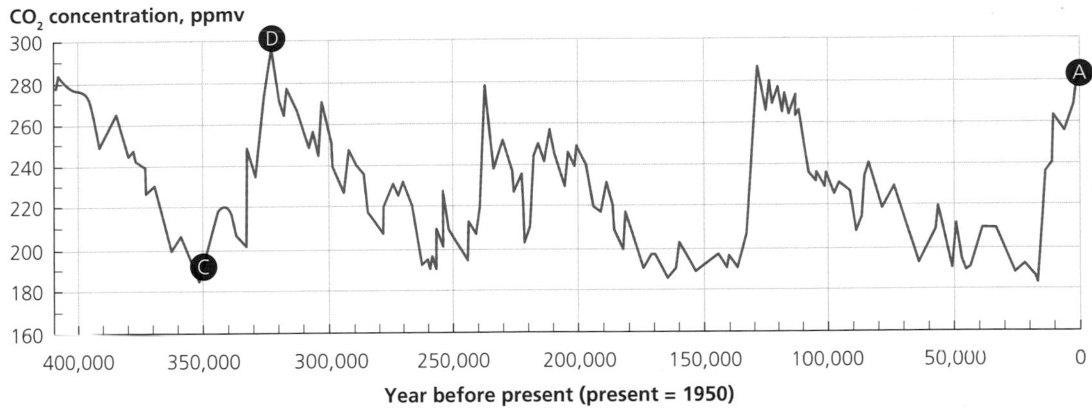

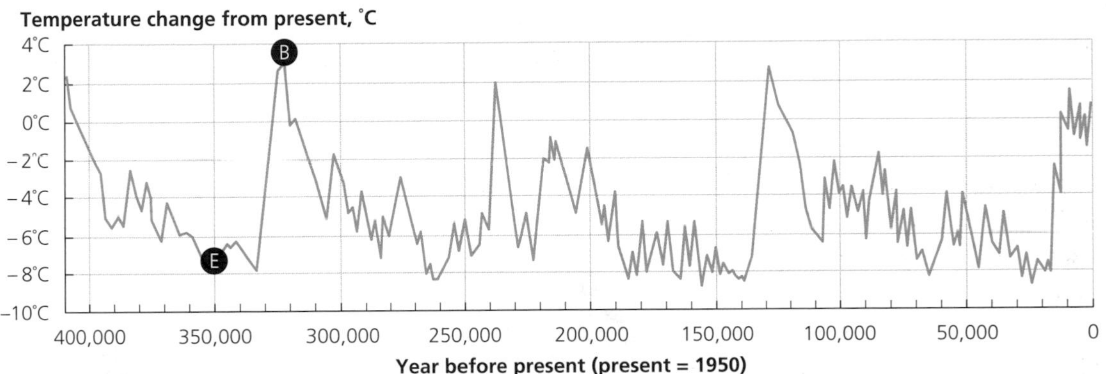

▲ **Figure 11** Temperature and CO_2 concentration (ppm) in the atmosphere over the past 420,000 years.

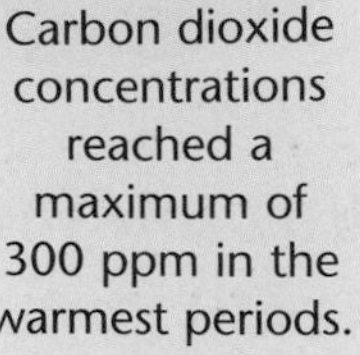

Carbon dioxide concentrations dropped to 180 ppm during the coldest periods

Narrow peaks in the temperature record represent short warm episodes (interglacials)

Broad dips in temperature represent glacial periods

Carbon dioxide concentrations were at about 280 ppm in 1950

Activities

1 Describe and explain the trend of the Keeling Curve (shown in Figure 9).

2 Use Figure 11.

a) Match the five statements to the correct place on the graph shown by the letters A, B, C, D and E.

b) Use the graph to copy and complete the following statement:
The graph shows natural cycles of .... periods and .... periods. Average temperatures were higher than present on *three/four/five* occasions. These are known as .... periods. The current interglacial period appears to have lasted much *longer than/shorter than* previous periods.

3 Use your understanding of the greenhouse effect to explain why reduced levels of carbon dioxide in the atmosphere might be linked to cooler periods of climate.

4 Compare Figure 11 with Figure 9.

a) How many times in the last 420,000 years have CO_2 levels been as high as in 2000?

b) Based on the ice core data, do you think that the Keeling Curve fits into a similar natural cycle of carbon dioxide concentrations? Explain your answer fully.

Enquiry

How conclusive do you find the evidence for:

- natural cycles of climate change over the last 420,000 years?
- an unusual rise in carbon dioxide levels since 1958?

Chapter 2 Weather hazards

Circulation of the atmosphere

Figure 1 shows regions of the world affected by violent tropical storms known as **cyclones** which are caused by low pressure in the atmosphere. The map also shows areas of drought caused by lack of rain. To understand these extreme weather events, we need to understand what causes the large-scale movement of air from one part of the globe to another – the global circulation of the atmosphere. Atmospheric circulation is driven by heat at the Equator.

Activities

1 Study Figure 1.

a) Describe the location of cyclones that affect:
   i) Australasia
   ii) North and South America
   iii) South East Asia.

b) Compare the direction of storm tracks in the northern and southern hemispheres.

c) Name two countries that are at risk of both cyclones and drought.

d) Describe the location of areas affected by drought in the northern hemisphere. How are these areas different to those that are vulnerable to cyclones?

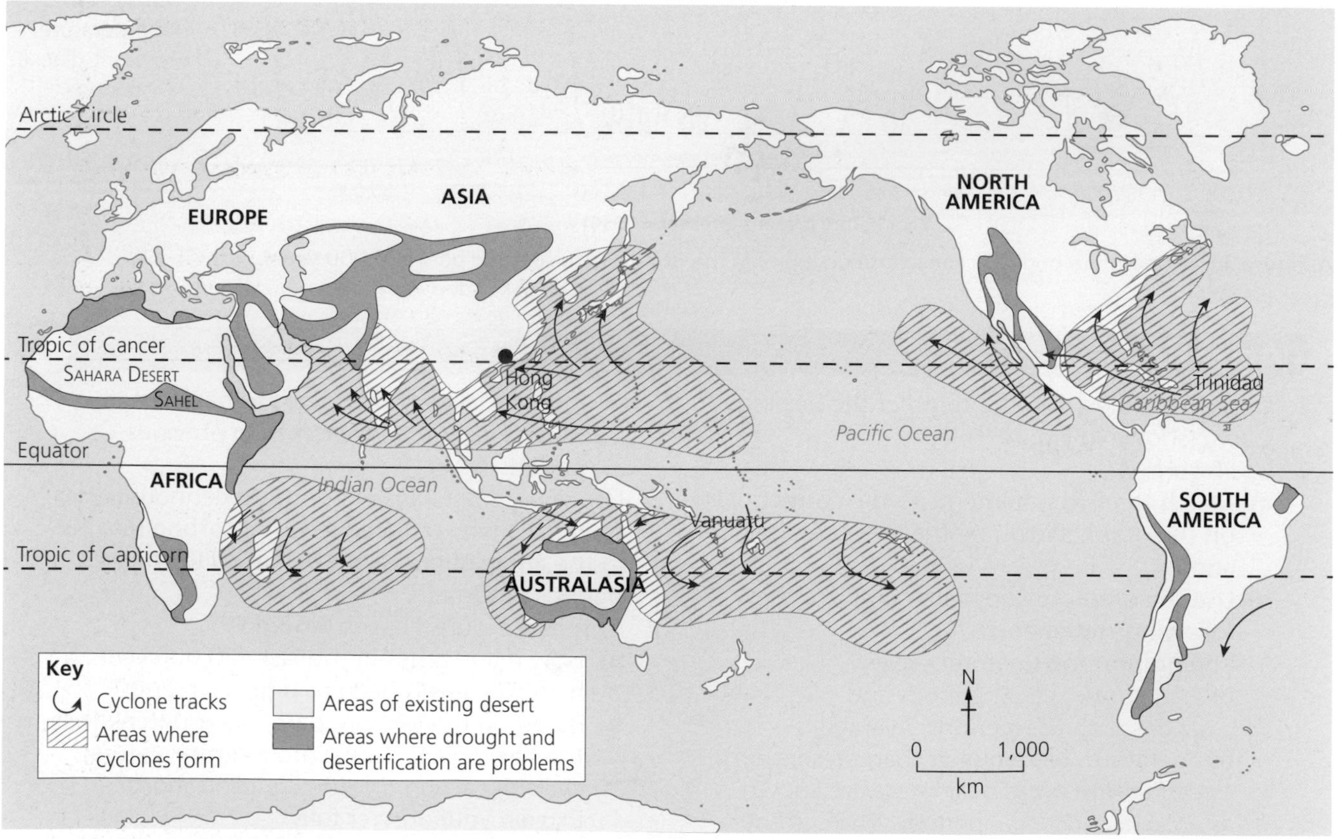

Arctic Circle
EUROPE
ASIA
NORTH AMERICA
Tropic of Cancer
SAHARA DESERT
SAHEL
Hong Kong
Trinidad
Caribbean Sea
Pacific Ocean
Equator
AFRICA
Indian Ocean
SOUTH AMERICA
Vanuatu
Tropic of Capricorn
AUSTRALASIA
Key
Cyclone tracks
Areas where cyclones form
Areas of existing desert
Areas where drought and desertification are problems
N
0 1,000
km

▲ **Figure 1** The global distribution and location of areas affected by cyclones and drought.

Activities

2 a) Make a copy of Figure 2 on page 161. Use the text boxes to annotate your diagram. Place your annotations in appropriate places on the diagram.

b) Use your completed diagram to explain why there is rainforest at the Equator and desert at latitudes 30° to the north and south.

What is happening in our atmosphere to cause weather hazards?

We can begin to understand extreme weather by considering the intensity of the Sun's heat on the ground at different latitudes on the Earth. The climate close to the Equator (within 5° of latitude) is hot throughout the year. The Sun heats the Earth and the Earth heats the air above, which becomes **unstable** and rises. This creates a band of low pressure in the atmosphere, known as the **intertropical convergence zone (ITCZ)**, which circles the equatorial region of the Earth. The position of the ITCZ is shown in Figure 3. Notice that its position varies throughout the year. This is because of the tilt of the Earth's axis. The northern hemisphere leans towards the Sun in June and July so the ITCZ is slightly north of the Equator. The ITCZ migrates to south of the Equator in December and January when the southern hemisphere leans towards the Sun.

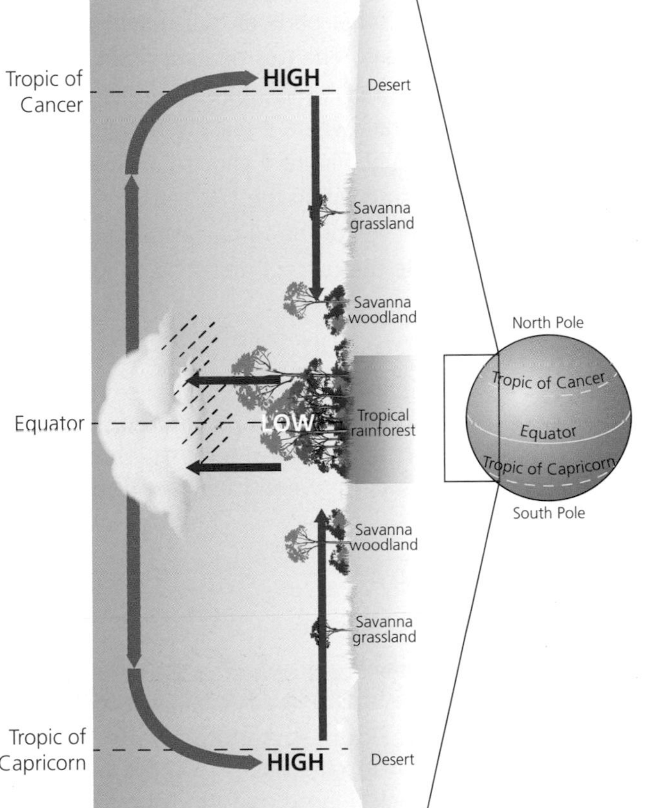

The Sun heats the Earth and the Earth heats the air above, which becomes unstable and rises creating an area of low pressure.

The air circulates back towards the equator in the lower atmosphere creating the trade winds.

At about 30 °N and 30 °S the air descends creating an area of high pressure. This air is dry. It seldom rains.

The air reaches a boundary layer in the atmosphere called the tropopause which is about 17 km above the equator. The air spreads outwards towards the poles.

◀ **Figure 2** How solar heating at the Equator creates the ITCZ.

Activities

3 a) Using Figure 3, describe the position of the ITCZ:
   i) over the Pacific Ocean in January and July
   ii) over Central America in July.

b) Compare the location and distribution of cyclones on Figure 1 to the position of the ITCZ in Figure 3. What conclusion do you come to?

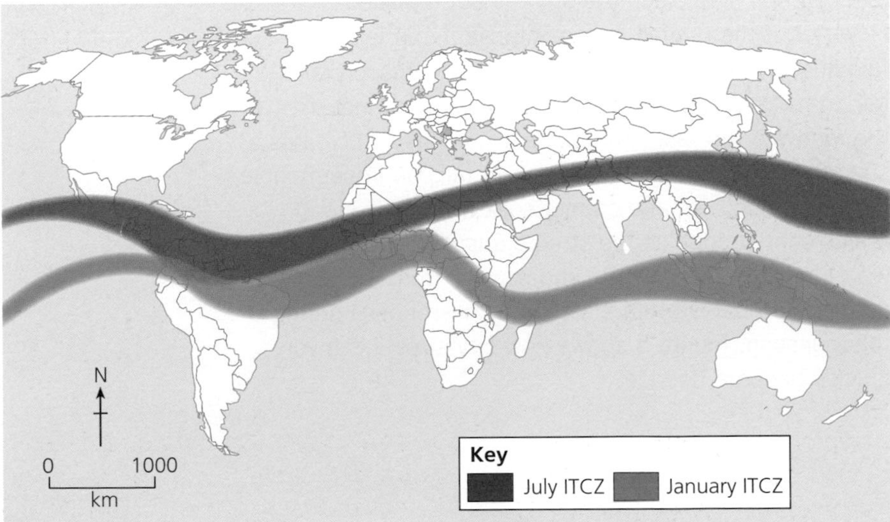

▲ **Figure 3** The location of equatorial and tropical low-pressure systems (ITCZ).

Low-pressure weather hazards

Figure 3 (page 161) shows the regions of the world that are affected by the ITCZ. The ITCZ is a large-scale feature in the atmosphere that wraps around the globe. Within this zone, the air is warm and rising. This creates huge areas of low pressure in the atmosphere. Air moves towards the centre of areas of low pressure, creating strong winds. The rising air cools and condenses, forming towering rain clouds. In this way, the ITCZ creates the right conditions for intense rainfall, which sometimes leads to flooding.

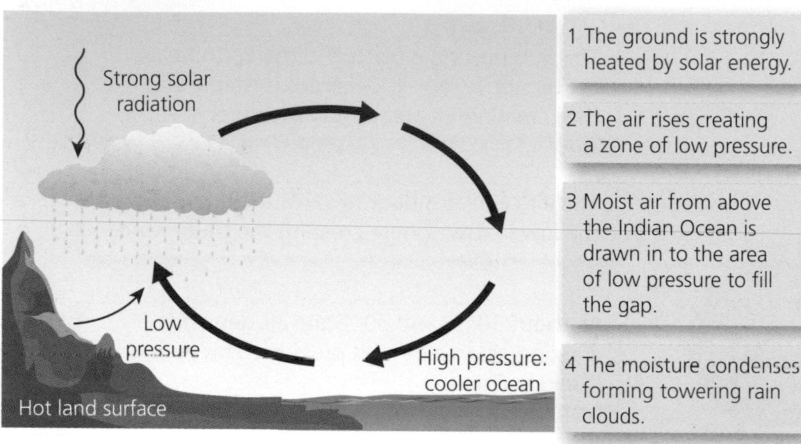

▲ **Figure 4** Circulation of the atmosphere over South Asia during June and July.

The South Asian monsoon

Low air pressure within the belt of the ITCZ brings seasonal rainfall to tropical regions called the **monsoon**. One region of the world that experiences the monsoon is South Asia. The position of the ITCZ moves during the year. From June to August, the ITCZ is north of the Equator and, during this period, its leading edge is moving steadily northwards across South Asia. This means that the southern tip of India experiences the monsoon rains before other parts of India each year, usually during early June. The wet weather gradually moves northwards through India and into Pakistan, where it usually arrives in mid-July. Figure 4 shows how low pressure within the ITCZ and the closeness of the Himalayan Mountains create the perfect monsoon conditions.

Monsoon

A seasonal period of intense rainfall that occurs in tropical regions. In the northern hemisphere, the monsoon occurs between June and August.

Consequences of the monsoon

In many parts of India, groundwater (water stored in rocks in the ground) is used to supply water for drinking and irrigation. Some rain is useful during the monsoon because it replaces this store of water. If the monsoon rains are too light, then the store is not replaced. However, if the rainfall is very intense it can lead to flooding. Flooding can also occur if the monsoon rains fall on hard, dry, baked earth. The ground cannot soak the rainwater up fast enough, so it runs off and creates a sudden flash flood. In cities the rain falls on impermeable tarmac or concrete and cannot soak away. The storm drains cannot cope so the streets quickly become flooded with a mixture of rainwater and sewage from the foul drains. Mumbai is often affected by flash floods during the monsoon. Figure 5 shows how city streets can very quickly become flooded.

Activities

1 Make a copy of Figure 4. Annotate your diagram in appropriate places to show why the ITCZ creates such a large quantity of rainfall during the monsoon in South Asia.
2 Study Figure 5. Describe how floods of this sort could affect everyday life or people's health and well-being.

The severity of the monsoon floods in South Asia varies from year to year. In 2018, the southern state of Kerala suffered its worst flooding for over 100 years. 450 people died in these floods. In 2019, monsoon floods affected parts of the southern states of Kerala and Karnataka, and the western states of Maharashtra and Gujarat. At least 244 people died and more that 1.2 million people were forced to evacuate their homes because of the flooding.

▲ **Figure 5** Flooding caused by monsoon rains in Kolkata (August 2019)

The amount of rain that falls during the monsoon varies significantly during the three-month season and also from one location to another. In 2018, the state of Kerala suffered extreme flooding. However, Figure 6 shows that the rainfall in the early part of the season was close to average. The most intense rain fell in the last two to three weeks. Figure 7 shows that a few districts in Kerala had higher than average rainfall. These are the districts that experienced flooding. Other districts received below average rainfall.

Month	Actual rainfall 2018 (mm)	Average rainfall June–August (mm)	% departure from average
June, 2018	749.6	649.8	+15
July, 2018	857.4	726.1	
August, 2018	821.9	419.5	

▲ **Figure 6** Actual rainfall in Kerala's monsoon (2018) compared with average rainfall for June–August.

District	Actual rainfall 2018 (mm)	Average rainfall June–August (mm)	% departure from average
Lakshadweep	991	811	+22
Alappuzha	1,322	1,398	
Idukki	1,741	2,185	
Kozhikode	2,724	2,249	
Palakkad	1,613	1,314	
Mahe	2,328	2,203	

▲ **Figure 7** Actual rainfall in selected districts of Kerala (June–August 2018)

Activities

3 Explain why the monsoon rains can have both positive and negative consequences for India.

4 Study Figures 6 and 7.

a) Use Figure 6 to calculate the total amount of rainfall in the 2018 monsoon. How does this compare with the average?

b) The first row in each table shows the percentage departure from the average. Calculate the other values. See page 44 for help with this skill.

Cyclones

Cyclones are another example of a low pressure weather hazard. These violent storms affect tropical regions of the world, as you can see in Figure 1 (page 160). Cyclones form over warm oceans. Sea temperatures need to be at 27°C or above for several weeks before a cyclone will form. The warm water heats the air above it which rises rapidly, creating an area of very low pressure in the atmosphere. This causes towering clouds to form and torrential rain to fall. At sea level, the rising air is replaced by more warm moist air coming in from the outside. As the air moves toward the centre of the low pressure, it spirals upwards into the atmosphere. The spiral effect comes from the rotation of the Earth, a process known as the Coriolis Effect.

Cyclones are seasonal events. They occur north of the Equator between June and October. This is the time of year that the northern hemisphere is leaning towards the Sun so the ITCZ is overhead. During December to March, the southern hemisphere leans towards the Sun so the ITCZ moves south of the Equator and cyclones occur in the southern hemisphere.

Cyclone

A large-scale weather hazard characterised by strong wind and heavy rain. Cyclones are caused by extreme low pressure in the atmosphere. They are known as hurricanes in America and typhoons in Asia.

	Jan	Feb	Mar	Apr	May	Jun	Jul	Aug	Sep	Oct	Nov	Dec
Vanuatu	29.1	29.5	29.5	29.3	28.4	27.7	27.0	26.5	26.6	27.0	27.4	28.2
Trinidad	27.6	27.3	27.3	27.6	28.1	28.2	28.3	29.3	29.4	29.0	28.4	27.9
Hong Kong	19.1	18.9	20.1	22.7	26.8	28.5	29.1	28.6	28.3	26.5	24.3	21.0
Cornwall, UK	10.4	10.0	9.7	10.6	12.0	14.2	16.5	17.1	16.5	14.8	13.2	11.7

▲ **Figure 8** Average sea temperatures for selected locations (°C).

Consequences and responses

Cyclones can have devastating consequences for people, especially in coastal regions of tropical countries. The very low air pressure causes a temporary rise in sea level beneath the cyclone, known as a **storm surge**. The strong winds create large waves. The combination of a storm surge, large waves and powerful winds often causes coastal flooding. Cyclones also create heavy rainfall. When a cyclone is slow moving, the rain falls for several hours in the same location, leading to river flooding and landslides. Flooding may cause death by drowning and the destruction of roads and bridges, which may delay the distribution of aid after the cyclone. Cyclones also have very strong winds, which can bring down powerlines.

Storm surge

A local rise in sea levels that occurs during a cyclone, which may cause flooding. Sea levels rise because the air pressure is low, so the atmosphere is not pressing down on the ocean as much as usual.

Activity

1 Use Figure 8 to draw two graphs to represent average sea temperatures in Hong Kong and Cornwall. Use your graphs and Figure 1 to explain why Hong Kong is at risk of cyclones but Cornwall is not.

Enquiry

Predict when cyclones are most likely to threaten each of the following islands. Use the information in Figures 1, 3 and 8 to justify your answer:

- Trinidad
- Vanuatu.

Cyclone Idai, March 2019

Cyclone Idai was one of the deadliest storms ever recorded in the southern hemisphere. This cyclone affected Mozambique, Malawi and Zimbabwe. The worst affected area was the port city of Beria in Mozambique, where the cyclone came onshore. The city was affected by coastal flooding and power cuts. Heavy rainfall to the north of the city caused further flooding and landslides. The lack of clean water made people vulnerable to diseases such as cholera. Cyclone Idai was particularly destructive because:

- the low-lying coastline was vulnerable to flooding during the storm surge
- a lot of housing in Beria was in the informal sector, so it was badly built
- the cyclone was very slow moving and released a lot of rainfall once it came onshore.

Mozambique and Malawi are both low-income countries (LICs). The poorest countries find it hard to respond to natural disasters because they are unable to provide sufficient aid. As a consequence, a number of foreign governments and NGOs provided short-term aid to help people affected by the cyclone. These included UNICEF, the World Food Programme (WFP) and Oxfam.

▲ **Figure 9** Cyclone Idai in numbers, April 2019.

Activities

2 Compare Figures 10 and 11.
   a) Describe the similarities and differences.
   b) Suggest how this might have affected the people of Beria and the rescue workers.

3 Write a newspaper report describing the consequences of Cyclone Idai.

4 Explain why Beria was particularly vulnerable. Give human and physical reasons.

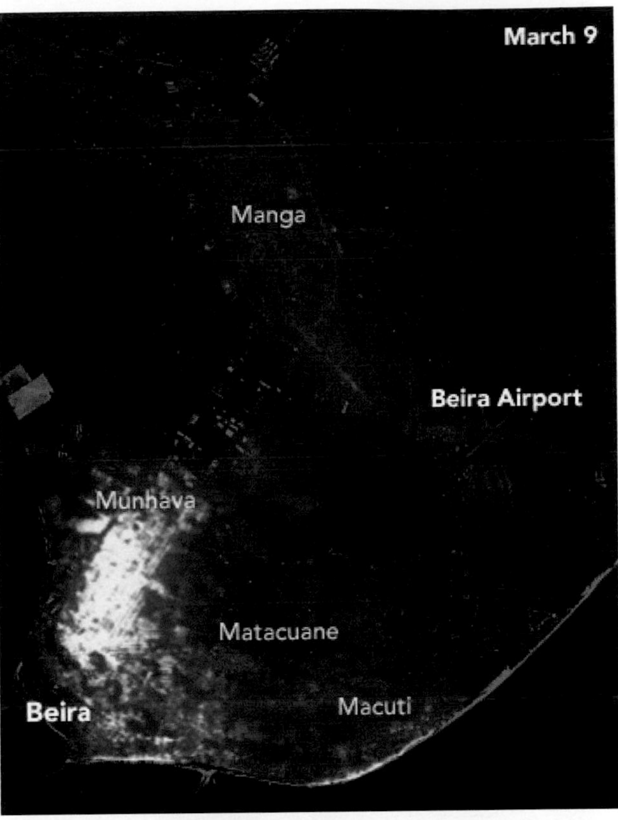

▲ **Figure 10** Night-time satellite image of Beria taken before Cyclone Idai.

▲ **Figure 11** Night-time satellite image of Beria taken three days after Cyclone Idai came ashore.

High-pressure weather hazards

High-pressure systems are large-scale weather systems. They occur where air in the lower atmosphere is descending and pressing down on the Earth. High pressure brings dry weather with very low wind speeds. When high pressure lasts for long periods of time it can cause weather hazards such as **drought** and **heatwaves**. Drought occurs most frequently in the regions marked dark orange in Figure 1 (page 160).

Drought

When significantly less rain than usual falls over a prolonged period of time, leading to a shortage of water.

Heatwave

When temperatures are at least 4.5°C above the mean temperature for two or more days.

California drought (2012–19)

Between 2012 and March 2019 California experienced a severe drought. The seven-year drought caused water shortages and wildfires, like the one shown in Figure 12. The drought was caused by high pressure centred over California and the eastern Pacific Ocean, as can be seen in Figure 14. In normal conditions, the western states of the USA receive low-pressure systems from the Pacific Ocean during the winter months that bring snow to the mountains. In a normal year, California, Arizona and Utah get 75 per cent of their water from snow melt. During 2012–19 the high pressure blocked the movement of low pressure across the western states, so the amount of winter snow was much less than usual. The summer months, however, were are usually warm and dry as usual, so any moisture in the soil quickly evaporated.

▲ **Figure 12** A helicopter fighting a wildfire near Burbank, California (2017).

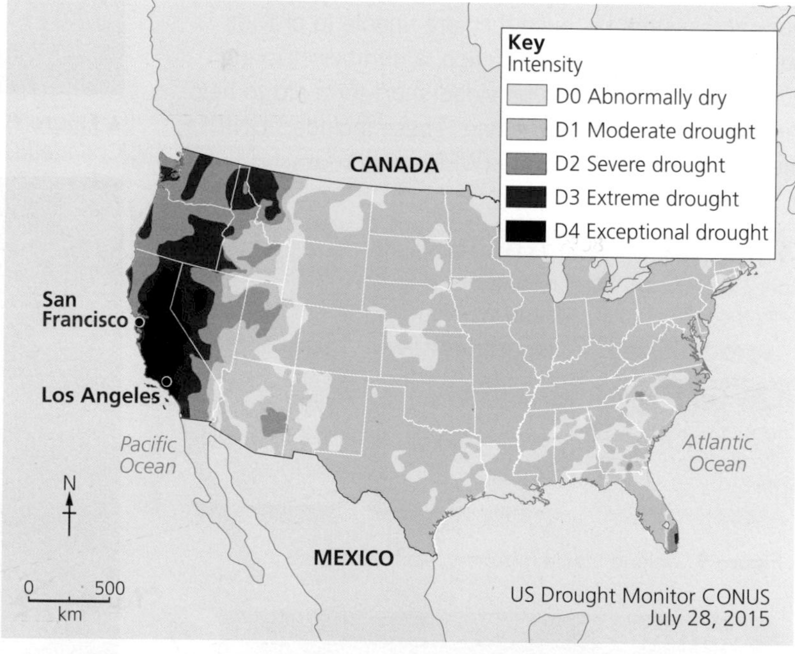

▲ **Figure 13** The extent of the drought across the USA in July 2015.

Activities

1 Use Figure 13 to describe:
   a) the location of Los Angeles
   b) the distribution of states affected by extreme and exceptional drought conditions.

2 Suggest how the drought and wild fires might have affected farmers, consumers, homeowners and firefighters.

3 Using Figure 14 and an atlas, forecast the weather in each of the following places. You should be able to comment on the temperature and precipitation.
   a) Los Angeles
   b) Ottowa
   c) Mexico City

Weblink

http://droughtmonitor.unl.edu/ – website of the US drought monitoring service.

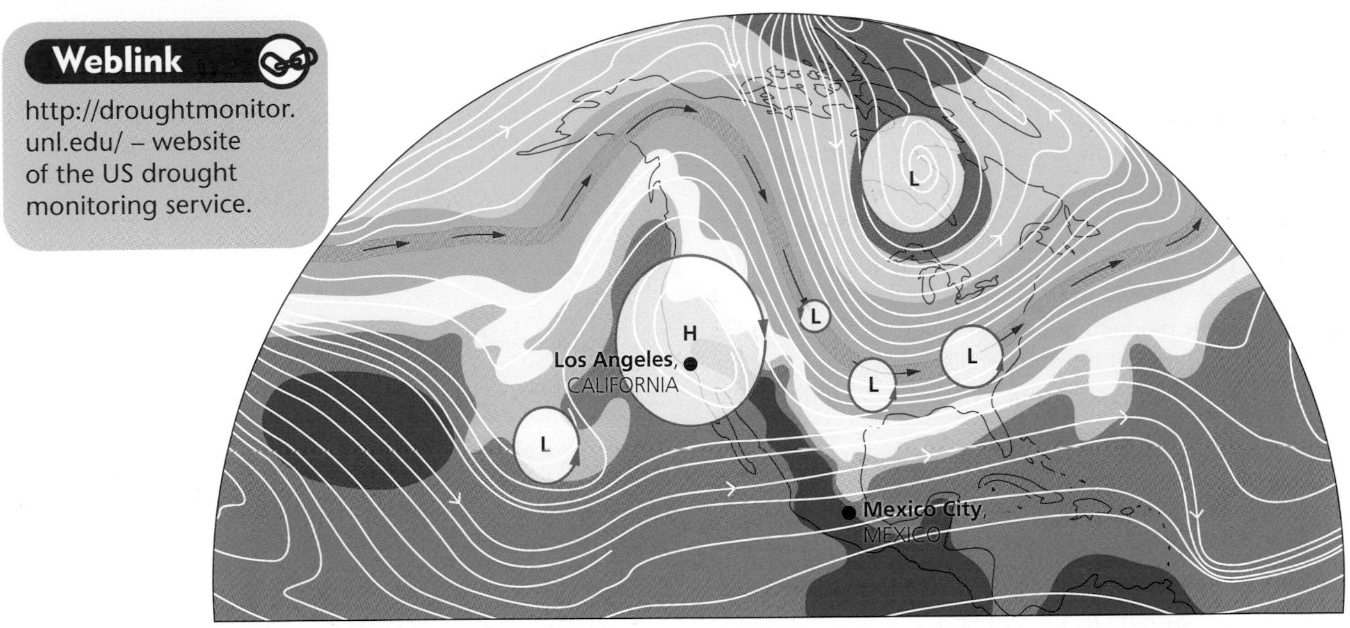

▲ **Figure 14** A weather chart during the drought (February 2015), showing high pressure over California.

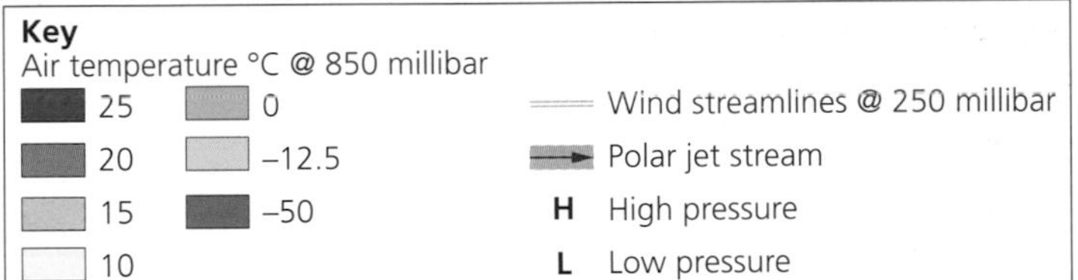

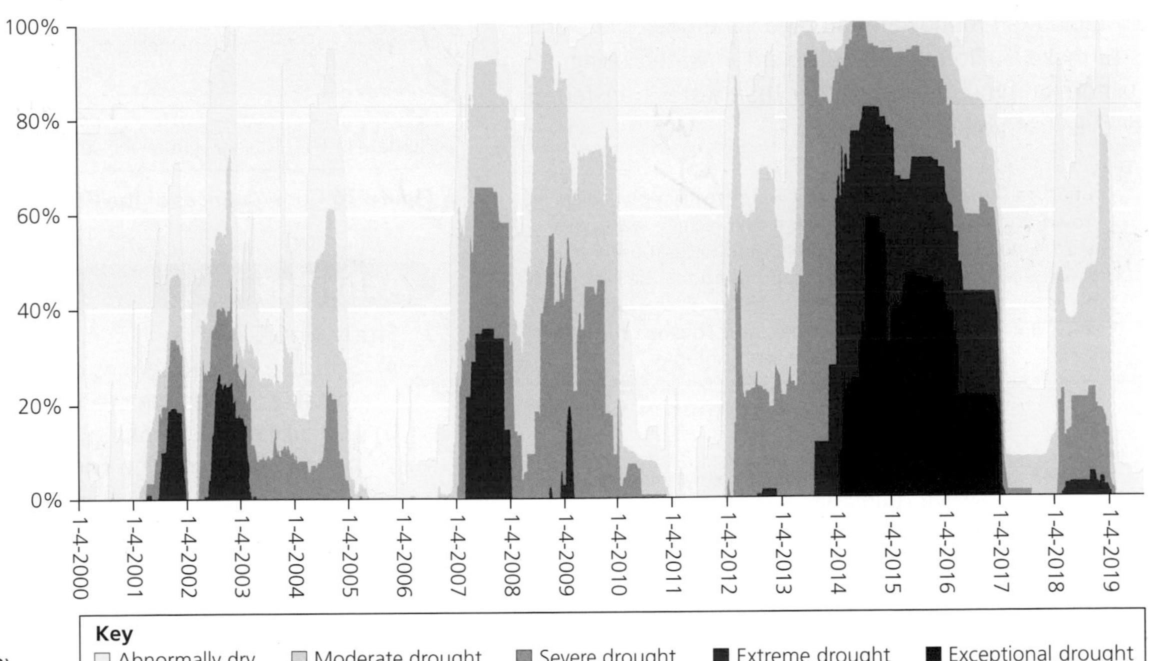

▶ **Figure 15** The percentage of California experiencing drought (2000–19).

Activity

4 Study Figure 15. Describe how drought conditions changed across California in the period 2000–19.

Consequences of the California drought

The drought had consequences for the people, environment and economy of California. The lack of water meant that many homeowners suffered water shortages. Some homes had no running water for five months and had to rely on water delivered by road. Farmers, who produce two-thirds of the USA's fruit and nuts, also faced shortages of water to irrigate their crops. It is estimated that lost farm production cost as much as $2 billion. As moisture evaporated from the soil the soil dried leading to dust storms, while dry vegetation led to wildfires.

Responses to the California drought

During drought, when snow melt is lower than normal, farmers cannot take as much water from rivers and reservoirs as they would like. Instead, they rely on water stored in the ground to irrigate crops. If groundwater is abstracted quicker than it is replaced the water table drops. In the long term, this is unsustainable. The Governor of California, Jerry Brown, responded to the drought by passing laws that will make the use of groundwater in California sustainable by 2022. $2.5 billion will be invested in new reservoirs and $5 billion in groundwater storage schemes. The plan is to make California self-sufficient in water, using Californian ground water rather than water transferred by river from other states.

I had to introduce compulsory water restrictions. Every town and city must show how it will reduce water use by 25% compared to 2013 levels. Agricultural users must provide the State with frequent reports about their medium term plans to reduce water consumption. Those who install toilets, domestic washing machines and showers must only use modern, low-water technologies.

Jerry Brown, Governor of California

We have introduced a mixture of voluntary and compulsory water conservation programmes. Watering gardens, cars and driveways is banned. We will offer subsidies to people who want to change their toilets and washing machines for newer models. We have sent out millions of leaflets offering advice on how to conserve water in the home.

Harlan Kelly, General Manager of the San Francisco Water Utility

▲ **Figure 17** Short-term responses to the drought.

Farmers in Central Valley lost $810 million during 2015.

Homeowners were told to stop using water to wash down their driveways, or water gardens. Using a hose to wash cars was banned. Californians caught ignoring water restrictions were shamed on Twitter using the hashtag #DroughtShaming.

Most HEP dams stopped producing electricity.

Cracks appeared in buildings and roads due to subsidence. This happened because water was being pumped out of the ground faster than it was being replaced naturally.

California usually produces nearly half of the fruits and vegetables grown in the USA. Shortages meant that prices rose by 6% in the shops. More food was imported.

There was a 36% increase in wildfires. Property was damaged and wildlife killed. 31,000 acres of oak habitat burned.

The state government paid $687 million of its savings to compensate farmers and homeowners who lost earnings or property. This money could have been used for other much needed projects.

Salmon and trout died in the San Joaquin River Delta. An increase in river temperature means the water carries less oxygen for fish.

The state lost 17,100 agricultural jobs due to the drought.

▲ **Figure 16** Consequences of the California drought.

Activities

1 Study Figure 16.
   a) Sort the consequences of the drought into social, economic and environmental effects.
   b) Use a diamond ranking technique (see page 83) to rank the consequences of the drought.
   c) Choose the three most serious consequences. Explain why you have chosen each one.

Long-term drought patterns

Research by US scientists suggests that California's climate has changed a lot over longer periods of time. It seems likely that drought in California used to be common. Large dams, such as the Hoover Dam, were built and reservoirs created in the early part of the twentieth century when rainfall was higher. Water storage schemes and cheap hydro-electric power (HEP) allowed the population of California to increase. However, the wetter climate of the twentieth century may not have been 'normal' and scientists believe that global climate change will increase the frequency and severity of droughts in the future.

Figure 18 shows annual values for the Palmer Drought Severity Index (PDSI). This Index is used to measure levels of drought by the US Weather Bureau. Zero PDSI means normal climate conditions. Positive PDSI numbers indicate above average precipitation. Negative PDSI numbers indicate drought: –2 is moderate drought, –3 is severe drought and –4 or lower is extreme drought.

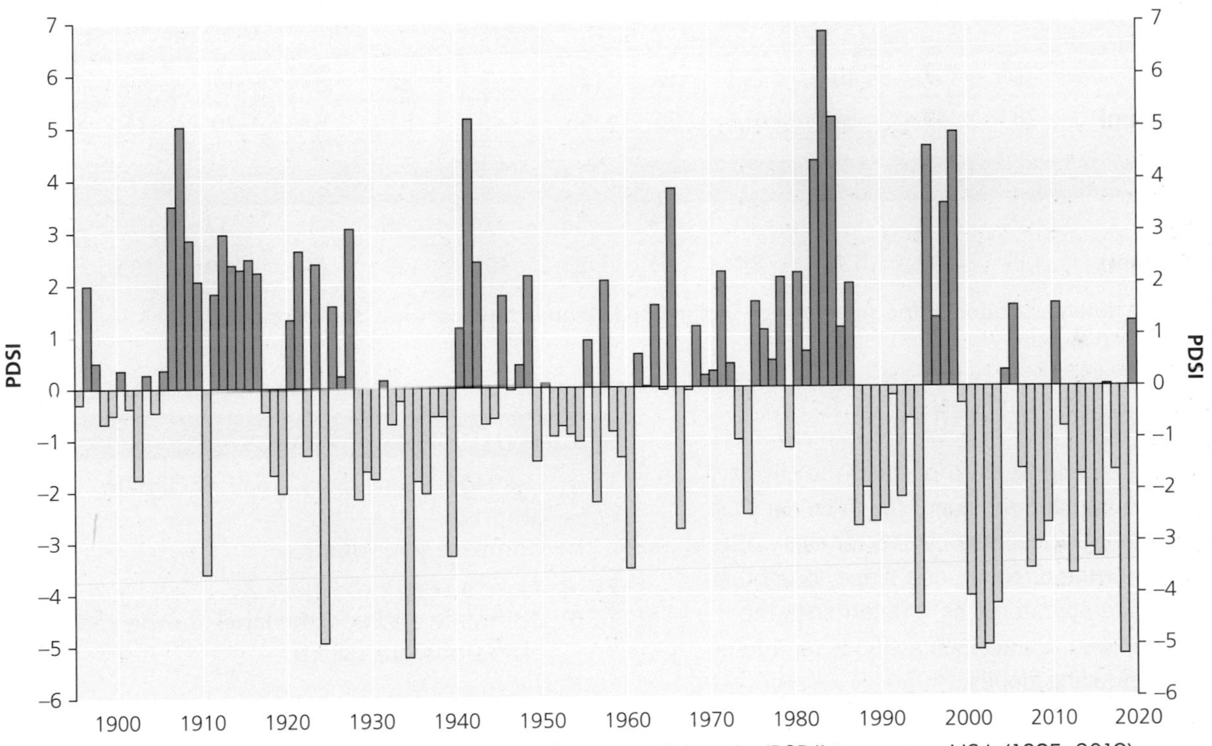

▲ **Figure 18** Long-term changes to the frequency and severity of drought (PSDI) in western USA (1895–2019).

Activities

2 Describe the short-term and long-term responses to the drought. Identify the main groups of people who have to make these responses happen.

3 Using Figure 18:

a) Describe how patterns of drought have changed since 1895.

b) Use evidence from the graph to explain why building lots of dams in the 1920s may have been a mistake.

Enquiry

'California must limit the future growth of farming.'

To what extend do you agree with this statement? Use evidence from the figures on pages 168–169.

▲ **Figure 19** A roadside poster in California during the drought.

THEME 5

Weather, climate and ecosystems

Chapter 3 UK weather and climate

The temperate maritime climate of the UK

The UK has a **maritime climate** that is mild, without extremes of temperature. The climate is strongly influenced by moist air masses and warm ocean currents crossing the Atlantic Ocean. A feature of the climate is its variability throughout the year. The UK has four distinct seasons: spring, summer, autumn and winter.

The main factors that affect the UK's climate are:

- latitude
- the track of the jet stream and its effect on the movement of air masses
- the effect of ocean currents
- altitude and aspect.

London	Jan	Feb	Mar	Apr	May	Jun	Jul	Aug	Sep	Oct	Nov	Dec
Temperature °C	6	7	10	13	17	21	23	22	19	14	10	7
Precipitation (mm)	78	59	61	51	55	56	45	51	63	70	75	79

Oban	Jan	Feb	Mar	Apr	May	Jun	Jul	Aug	Sep	Oct	Nov	Dec
Temperature °C	8	8	8	10	13	14	16	16	15	12	10	8
Precipitation (mm)	195	142	155	84	69	83	105	123	174	189	185	197

▲ **Figure 1** Climate data for London in the southeast of England and Oban in the northwest of Scotland show how the UK's climate varies across the country.

Effects of aspect and altitude

The northern, central and western parts of the UK have upland landscapes. Upland areas are much colder than lowlands. Temperatures decrease by 1°C for every 100 m in height. **Aspect**, or the direction of a slope, is another factor that affects temperatures as it determines the amount of sun received. South-facing slopes tend to be warmer than north-facing slopes.

Activities

1 a) Use Figure 1 to draw a pair of climate graphs.
b) Compare your graphs.

2 a) Make a sketch of Figure 2.
b) Complete and add the labels to the correct places on your sketch.

▲ **Figure 2** A valley in the Lake District shows how aspect can affect temperature.

Maritime climate

A mild and wet climate influenced by the nearby ocean.

Flat valley floor	... will remain in shadow for most of the day in winter.
North-west facing slope	... is warmed by early morning sunshine.
South-east facing slope	... will remain in shadow until the Sun is higher in the sky so frost remains longer in the morning.

The maritime climate

Oceans contain flowing currents of sea water. These **ocean currents** circulate around the globe. They are able to transfer heat from warm latitudes to cooler ones. The Gulf Stream, and its extension the North Atlantic Drift, is one of these currents. It carries warm water from the Gulf of Mexico across the Atlantic, towards Europe. This warm water transfers heat and moisture to the air above it and influences the climate of the UK. It gives the UK a maritime climate which is warmer and wetter than other places at similar latitudes in continental parts of Europe.

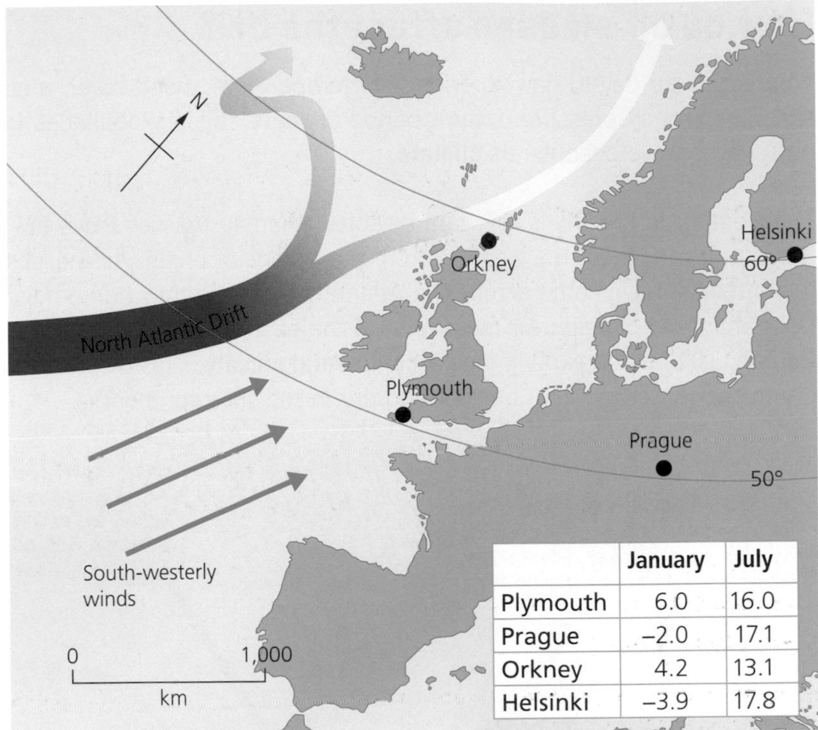

	January	July
Plymouth	6.0	16.0
Prague	–2.0	17.1
Orkney	4.2	13.1
Helsinki	–3.9	17.8

▲ **Figure 3** Comparing temperatures in the UK with the more continental parts of Europe.

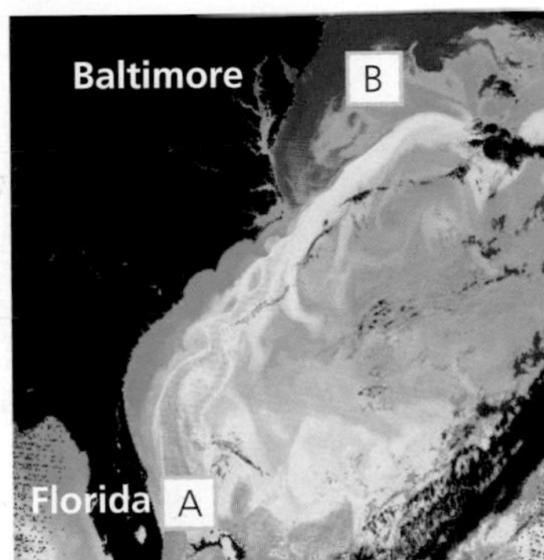

▶ **Figure 5** Satellite image of the Gulf Stream. The orange colours show warm water. Cold water is blue. Land is black.

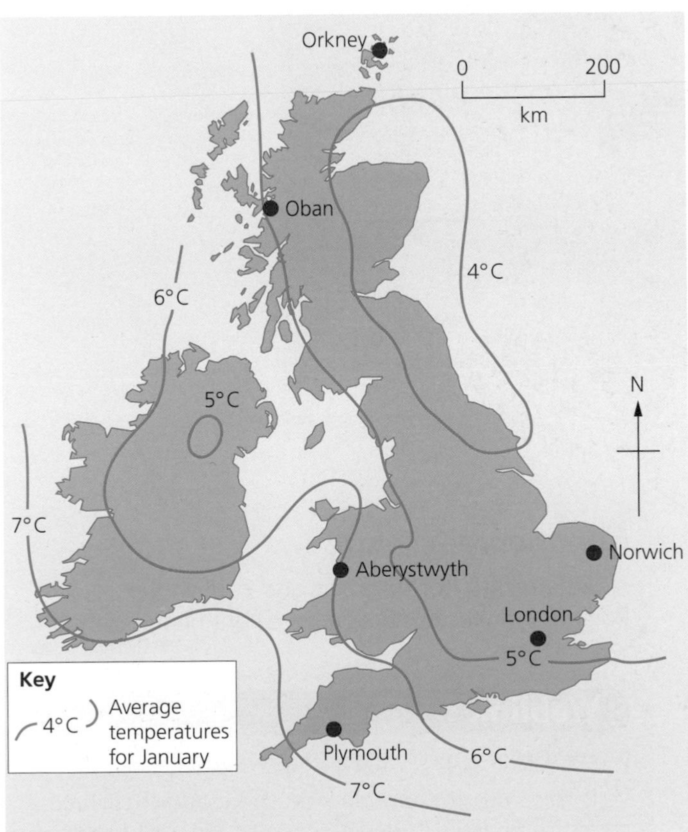

▲ **Figure 4** Average temperatures for January. Lines of equal temperature are known as isotherms.

Activities

3 Compare the temperatures that are at the same line of latitude in Figure 3. Explain why this pattern occurs.

4 a) Use Figure 4 to describe the January temperature in:
   i) Aberystwyth
   ii) Norwich
   iii) Orkney.
  b) Which parts of the UK have the lowest January temperatures?
  c) Draw a cross-section to represent the temperature change between Plymouth and Orkney.

5 a) Make a sketch of Figure 5. Label what is happening at A and B.
  b) Suggest how the ocean current at B will affect the climate of Baltimore.

How do air masses affect the UK?

Weather is our day-to-day experience of temperature, cloud cover, precipitation, wind, sunshine and air pressure. Over a period of years, these experiences form normal patterns that we describe as **climate**.

The weather in the UK is very changeable. When air masses move towards the UK, they bring with them the weather from their place of origin. Air masses from the west or southwest have come across the Atlantic Ocean, so they bring moisture, giving the UK a maritime climate. Air masses from the east bring weather from northeastern Europe. This region experiences a **continental climate**. This means the air is cold and dry in the winter months, but hot and dry in the summer months.

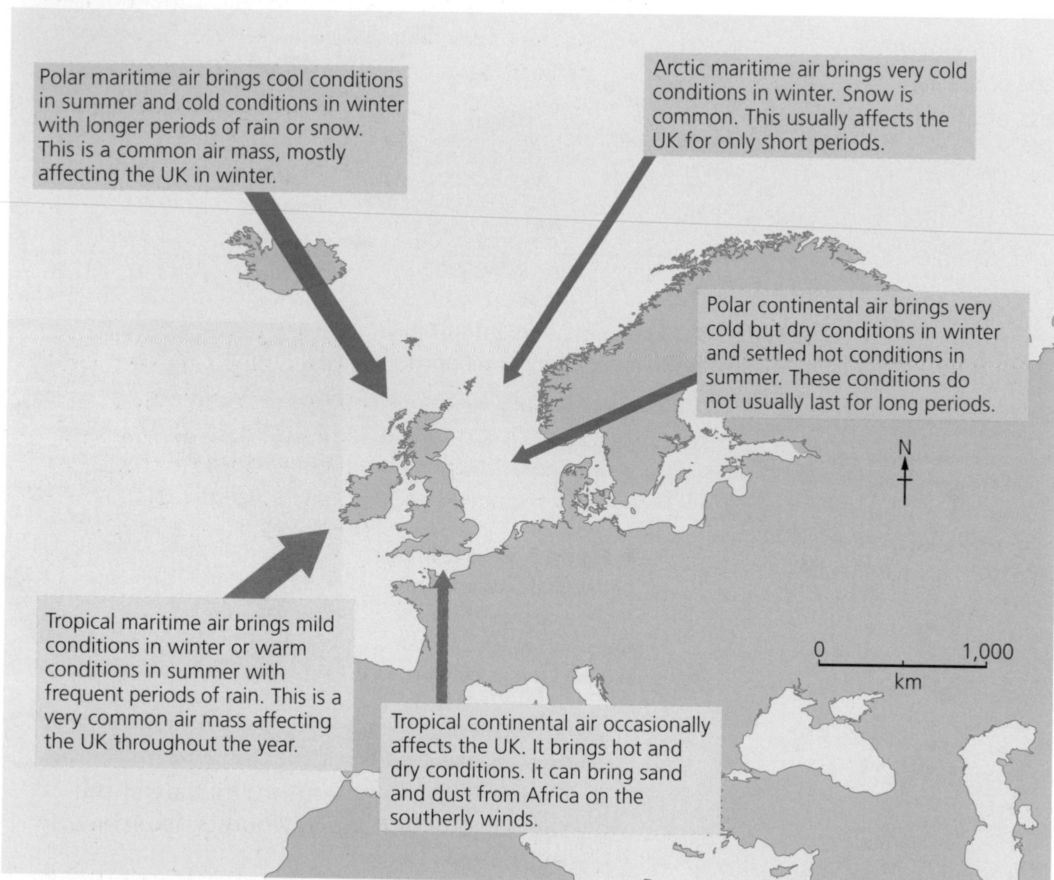

▲ **Figure 6** How do air masses affect the UK?

Enquiry

Can you predict the weather?

Use the Weblink on this page to view the surface pressure chart for the next four days. Use the position of the high and low pressures to predict places across Europe that will have:

- dry, settled weather
- wet weather
- the strongest winds.

Continental climate

A climate with hot summers and cold winters. Poland and western Russia have continental climates.

Weblink

www.metoffice.gov.uk/public/weather

Visit this website and click on the icon for surface pressure charts to see the current areas of high and low pressure across the UK.

How does the jet stream affect the UK's weather?

The jet stream is a strong ribbon of wind that circles the globe between 9 and 16 km above the surface of the Earth. It crosses over the UK, taking a sinuous course (as shown in Figure 7). It separates the cold polar air masses to the north from the warmer tropical air masses to the south.

Key
1. Track that typically allows high pressure to settle over the UK like in summer 2003
2. Track that will bring unsettled weather conditions
3. Track that will bring a series of depressions across the UK like in winter 2013–14

▲ **Figure 7** Typical paths for the jet stream across the UK.

When the jet stream takes a northerly track to the west of the UK, it tends to drag **high pressure** over the UK from the south. Areas of high pressure, also known as **anticyclones**, bring periods of dry, settled weather. The track taken by the jet stream tends to shift slightly over time. But if it stays in the same position for several weeks, then the UK will experience a long spell of similar weather. When high pressure becomes fixed over the UK in winter, the weather is sunny and dry but cold, and especially cold at night. During the summer an anticyclone brings hot dry weather. If the jet stream becomes fixed in this position it can cause problems such as heatwaves or drought. The UK's worst drought in recent years was in 2003 when an anticyclone stayed over Western Europe for several weeks.

Activities

1 Use Figures 6 and 7 to explain:
   a) Why the more northerly route (track 1) of the jet stream will bring settled weather.
   b) Why the more southerly route (track 3) of the jet stream will bring unsettled weather.

2 Use Figure 8 and an atlas to describe:
   a) the location of the three zones of high pressure
   b) the cold front.

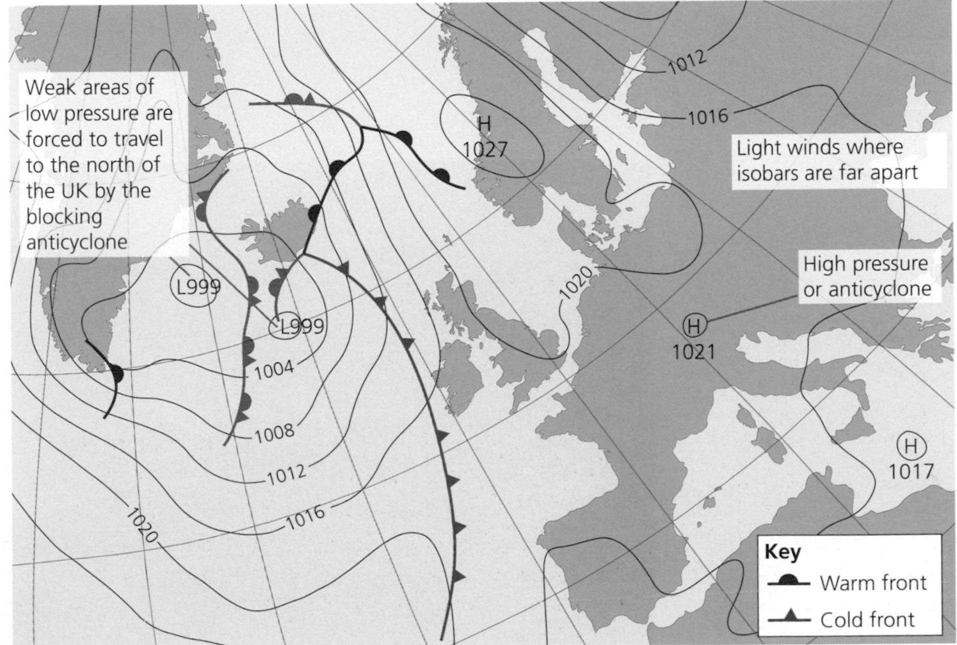

◄ **Figure 8** A weather map showing an anticyclone in August 2003.

High-pressure system (anticyclone)

A large-scale weather system that occurs where air in the lower atmosphere is descending and pressing down on the Earth.

The effects of low pressure

Regions of low pressure in the atmosphere are formed when air lifts off the Earth's surface. It is common for several cells of **low pressure**, also known as **depressions**, to form in the North Atlantic at any one time. They then track eastwards towards Europe bringing changeable weather characterised by wind, cloud and rain. Depressions are more likely to be deeper (have lower pressure) in the winter months. These weather systems can bring damaging gusts of wind and large waves on to the coast, as well as heavy rain leading to floods such as those of winter 2015–16 (see pages 41 to 45). However, low pressure in the summer months is also common: depressions during July 2012 caused flooding across large parts of England.

The winter storms of 2014

Between December 2013 and February 2014 the UK suffered the stormiest period for twenty years. Storms, created by extreme low pressure, battered the North Sea coastline during December 2013 (see pages 130 and 131). The coasts of Devon and Cornwall were badly damaged by a combination of fierce winds and huge waves. Southwesterly winds reached speeds of 146 kmph on the exposed coast of Devon. The storms caused a major incident in Dawlish, Devon, on 5 February 2014 when the sea wall collapsed causing closure of the main London-to-Cornwall rail line. The line was reopened on 4 April 2014 after repairs costing £35 million.

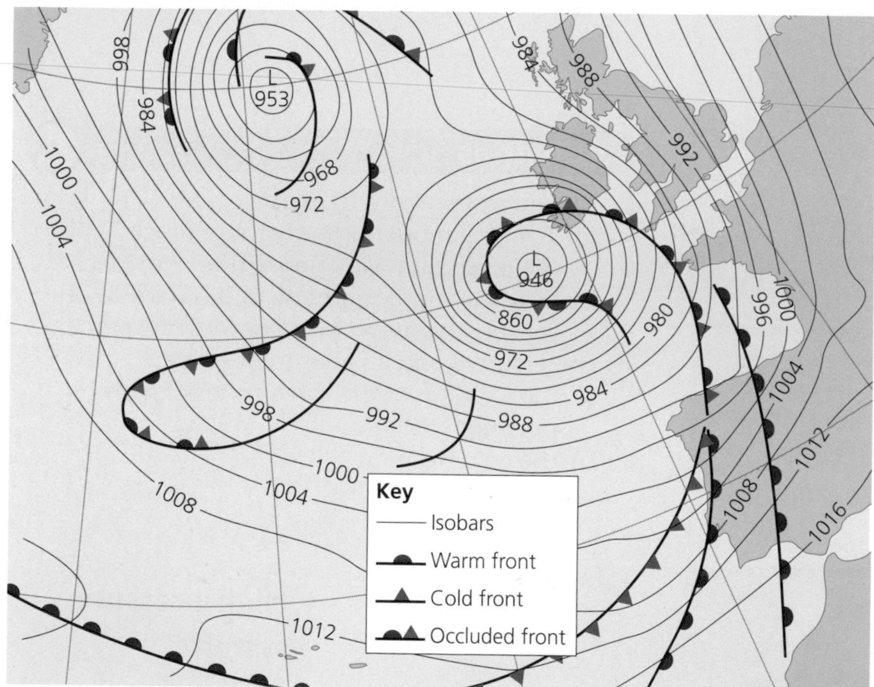

◀ **Figure 9** The weather map for 4 February 2014.

Low pressure system (depressions)

A large-scale weather system that occurs where air is rising in the lower atmosphere. Low pressure systems bring wet, windy weather.

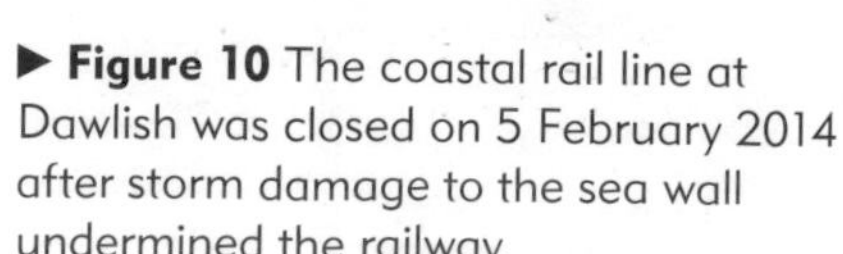

▶ **Figure 10** The coastal rail line at Dawlish was closed on 5 February 2014 after storm damage to the sea wall undermined the railway.

The passage of a depression

Inside the depression there is a battle between huge masses of warmer and colder air. These air masses revolve slowly around each other in an anticlockwise direction (in the northern hemisphere) as the whole system tracks eastward. Huge banks of cloud form along the fronts where cold and warm air meet. Rain fall is heaviest along these fronts.

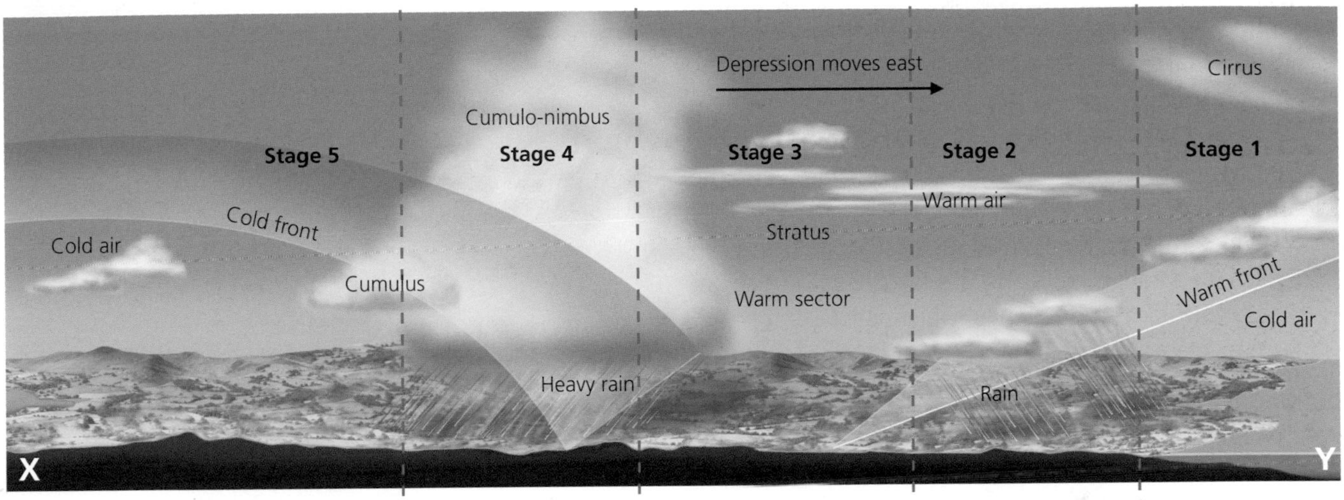

	Stage 5	Stage 4	Stage 3	Stage 2	Stage 1
Air mass	Cold	Cold			
Temperature (°C)		7	11	6	5
Wind strength		Very strong	Strong		
Wind direction	SSW	S	SSE	SE	E
Cloud/rain		Thick, low cloud and heavy rain	Some high cloud and clear skies; no rain		

▲ **Figure 11** Weather that would have been associated with the easterly progress of the depression shown in Figure 9.

Feature	Cyclones or depressions	Anticyclones
Air pressure		High, usually above 1,020 millibars (mb)
Air movement		Sinking
Wind strength		Light
Wind circulation		Clockwise
Typical winter weather		Cold and dry; clear skies in the daytime; frost at night
Typical summer weather		Sunny and warm

▲ **Figure 12** Comparing cyclones and anticyclones.

Activities

1 Study Figure 9.
   a) Describe the location of the areas of low pressure.
   b) What do the isolines on the area of map to the west of the UK tell you about the strength and direction of the winds?
2 Use evidence from Figure 9 to write a weather forecast for the southern half of the UK on 5 February 2014.
3 Make a copy of the table in Figure 11. Use the evidence in Figure 9 to complete the missing sections.
4 Make a large copy of Figure 12 and use the information on these pages to complete the blank spaces.

Microclimates

The buildings and traffic in a large city influence the local climate, an affect known as an **urban microclimate**. One of the main impacts of a city on the local climate is to create temperatures that are warmer than in the surrounding rural area. This is known as the **urban heat island**. Basically, the city acts rather like a huge storage heater, transferring heat from buildings and cars to the dome of air that covers the city:

- Concrete, brick and tarmac all absorb heat from the Sun during the day. This heat is then radiated into the atmosphere during the evening and at night.
- Buildings that are badly insulated lose heat energy, especially through roofs and windows. Heat is also created by cars and factories, which is lost to the air from exhausts and chimneys.

Parks and green spaces in our cities do not heat up as much as buildings and roads. So, open spaces help to keep our urban areas cool in summer. Vegetation helps to remove dust from the air, such as the particulates emitted by diesel cars, and trees and hedges reduce noise from roads. Parks and gardens soak up rain water and help to prevent the flash floods that might otherwise occur due to all of the impermeable surfaces in our towns and cities.

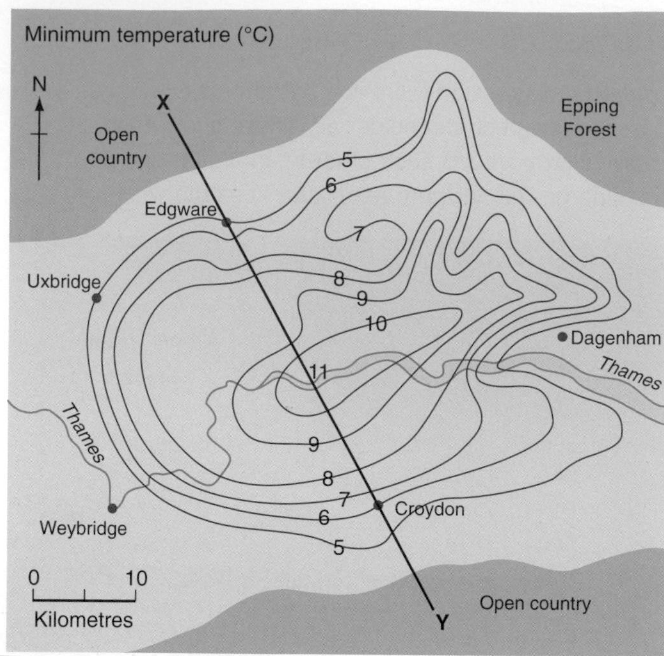

▲ **Figure 13** London's urban heat island.

▲ **Figure 14** Llandaff Fields, a large urban park in Cardiff.

Weather features	Urban microclimate
Sunshine duration	5–15% less
Annual mean temperature	1–2 degrees warmer
Temperatures on sunny days	2–6 degrees higher
Occurrence of frosts	2–3 weeks fewer
Total precipitation	5–30% more
Number of rain days	10% more
Number of days with snow	14% fewer
Cloud cover	5–10% more
Occurrence of fog in winter	100% more
Amount of condensation nuclei	10 times more

▲ **Figure 15** The effects of the urban climate in the UK.

Activities

1 Using Figure 13:
   a) Describe the location of the area of highest temperatures in London.
   b) Describe the distribution of places with lower temperatures.
   c) Suggest reasons for the pattern shown on the map.
2 Make copies of Figures 16 and 17. Sort the labels and add them to appropriate places on your diagrams.
3 Study Figure 15. Outline the advantages and disadvantages of the urban microclimate for the people who live in cities.
4 Give three reasons why planners need to protect urban parks from new homes or other development.

How do cities affect patterns of wind and rainfall?

Tall buildings in a city affect local patterns of wind. In the open countryside farmers plant trees to reduce wind speeds and the damage that wind can do to their crops. The trees act as a barrier, forcing the wind to rise over the treetops and therefore creating a shelter belt on the leeward side. Tall buildings in cities have an even greater sheltering effect, so average wind speeds in cities are lower than in the surrounding countryside. However, rows of tall buildings can also funnel the wind into the canyon-like streets between them. As the wind is forced to flow around or over tall buildings, the wind can suddenly gust at speeds that are two to three times the average wind speed. This may cause hazards for pedestrians and, in some extreme weather conditions, has led to the collapse of scaffolding.

During the summer months, the extra heat due to the urban heat island causes air to rise over larger cities. This can lead to convectional rainstorms – usually on hot afternoons. The dust in urban air is another factor that leads to higher amounts of rainfall in urban areas. When water vapour condenses in the air to form water droplets it does so by attaching itself to a dust particle. Urban air has ten times more dust particles in it than in rural air. The dust comes from car exhausts, heating systems in people's homes, industry and building sites.

Wind is funnelled between tall buildings

Gusty winds may be two to three times stronger than average

Prevailing wind

Less wind in leeward side of buildings

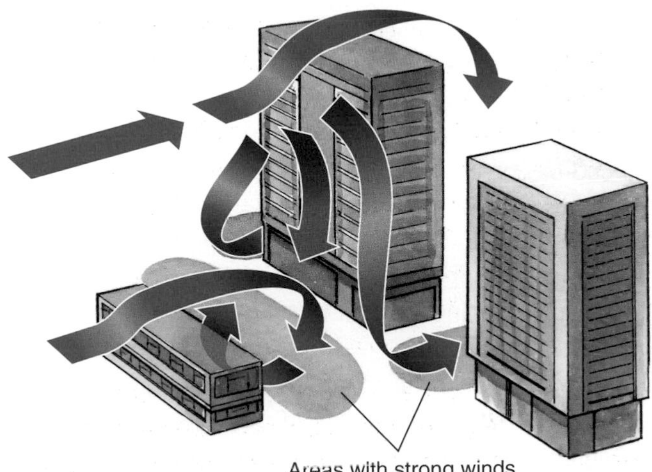

▲ **Figure 16** Urban wind patterns.

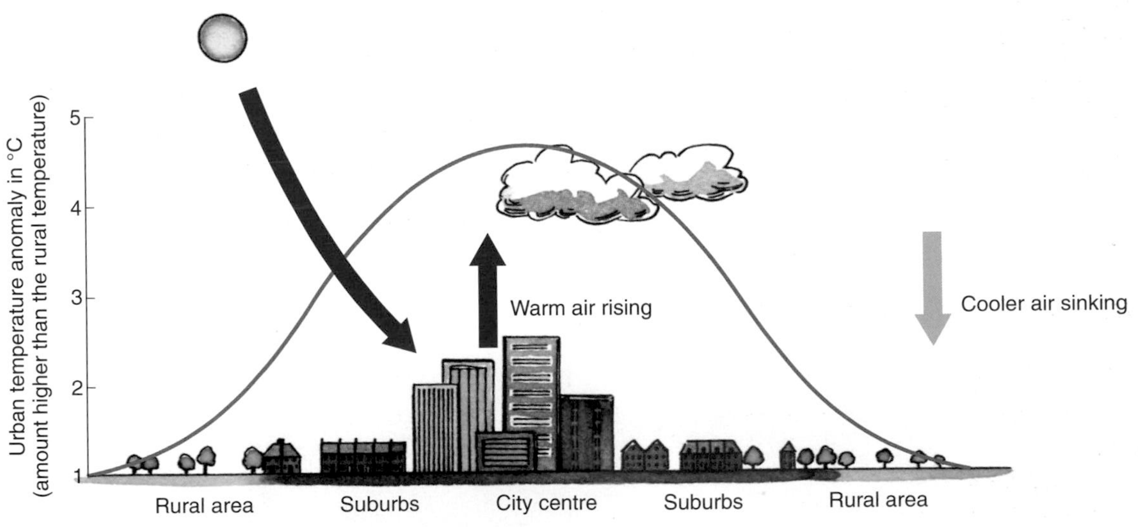

Buildings radiate heat

Condensation

Sun's energy is absorbed by building materials

Cumulus clouds form where rising warm air meets colder air

▲ **Figure 17** Why it rains more often in cities.

Chapter 4 Ecosystems

How does climate affect the global distribution of biomes?

An **ecosystem** is a community of plants and animals and the environment in which they live. Ecosystems contain both living and non-living parts. The living part includes such things as insects and birds, which depend on each other for food. It also includes plants, which may also depend on insects and birds for pollination and seed dispersal. The non-living part of an ecosystem includes such things as the climate, soils and rocks. This non-living environment provides nutrients, warmth, water and shelter for the living parts of the ecosystem.

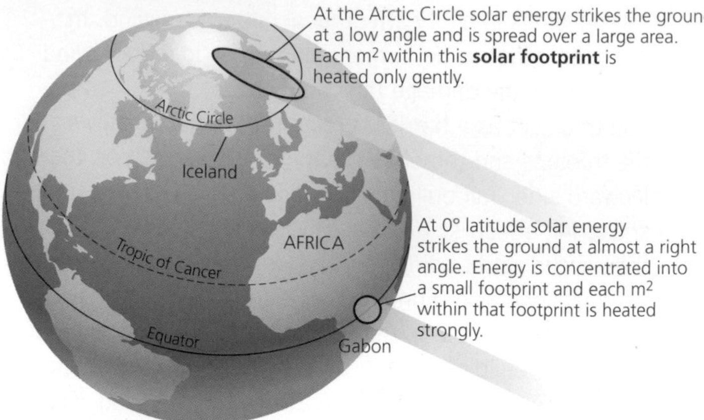

▲ **Figure 1** Solar heating of the Earth varies with latitude.

Climate is such an important factor in influencing the natural vegetation and wildlife of a region that **biomes** (the largest-scale ecosystems) broadly match the world's climate zones. **Tropical rainforests** grow in a band around the Equator where the Equatorial climate is hot and wet. Treeless **tundra** and forested **taiga** ecosystems exist where winters are cold and summers are short. The effect of latitude on temperature is explained by Figure 1.

Arctic climates and ecosystems

The Arctic region of northern Scandinavia and Iceland has cold winters and short, cool summers. These conditions have a major impact on plant growth. Plants have to survive the long, dark winters when temperatures fall well below freezing and when strong winds or snowfall can damage the branches of trees. In the summer, plants benefit from long hours of daylight but the growing season is very short. Plants therefore grow slowly.

The further north you go in northern Norway and Finland, the smaller the plants become. South of the Arctic Circle, the ecosystem is taiga. This is a forest ecosystem of conifer trees and birches. As you travel north, the trees become shorter and grow further apart. Eventually, a little north of the Arctic Circle, the climate becomes too extreme for trees to grow and the treeless Arctic tundra takes over.

◀ **Figure 2** Short coniferous trees growing in the taiga (or boreal forest) of northern Norway.

Tropical climates and ecosystems

Tropical climates are hot and wet. In the equatorial region of the Amazon Basin (within 5° of the Equator) there is between 1,500 mm and 2,000 mm of rainfall a year. London, by comparison, has an average of 593 mm of rainfall each year. The heat and abundant rainfall allow rapid plant growth and trees can reach a height of 40 m or more. This contrasts greatly with the slow growing plants of the tundra, which never grow more than a few centimetres high.

In tropical regions further from the Equator, temperatures remain high. Rainfall, however, becomes seasonal, with wet and dry seasons. Water shortages during the dry season mean that trees compete for water in the soil. Trees grow further apart and the savanna ecosystem of sparse trees and grasslands develops (see pages 190–1)

▲ **Figure 3** Reindeer grazing on lichen in the tundra region of Arctic Norway.

1. Temperatures are only above 10° C (the temperature at which most plants grow) for two or three months …		… plants grow close to the ground where they are less likely to be damaged.
2. Precipitation in the winter months falls as snow …	… so …	… plants have a short growing season.
3. Rocks weather (break down) slowly in the cold conditions which means soils have few nutrients …		… plants are extremely slow growing.
4. With few trees around there is little shelter from wind …		… plants have small leaves and so don't lose any moisture.

▲ **Figure 4** How the Arctic climate affects plant growth.

▲ **Figure 5** Tropical rainforest plants in Tobago.

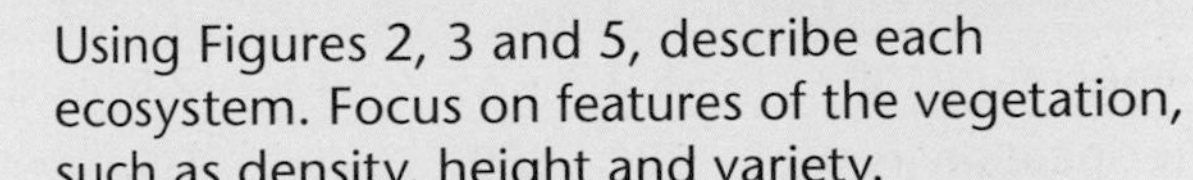

Activities

1. Using Figures 2, 3 and 5, describe each ecosystem. Focus on features of the vegetation, such as density, height and variety.
2. Pair up the phrases in Figure 4 to make four sentences that explain the features of Arctic ecosystems.
3. Use Figure 4 and the labels around Figure 5 to explain the differences you noted in your answer to activity 1.
4. Use Figure 1 to help explain the importance of latitude in the development of different biomes.

Nutrient cycles depend on climate

Plants need minerals containing nitrogen and phosphates. These nutrients exist in rocks, water and the atmosphere. The plants take them from the soil and release them back into the soil when the plant dies. This process forms a continuous **nutrient cycle**.

Figure 10 represents nutrient stores and flows in the tropical rainforest ecosystem. The circles represent **nutrient stores**. The size of each circle is in proportion to the amount of nutrients kept in that part of the ecosystem. The arrows represent **nutrient flows** as minerals move from one store to another. The thickness of each arrow is in proportion to the size of the flow, so large flows of nutrients are shown with thick arrows while smaller flows are shown with narrow arrows.

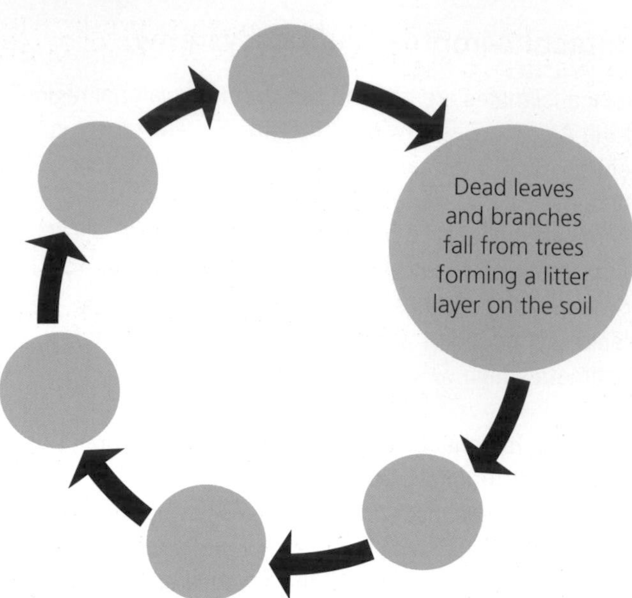

Leaf litter breaks down slowly in the cold conditions

Roots are shallow so they can take in nutrients near the surface

Decomposers such as beetles and fungi grow in the litter

Nutrients from leaf litter return to the soil

Plants use nutrients from the soil to help growth

▲ **Figure 6** Nutrient cycles in the taiga.

▲ **Figure 7** Lichens growing on a rotting tree stump in the taiga ecosystem, Norway.

Activities

1 Make a copy of Figure 6 and add the labels to the correct places to make a complete cycle.

2 a) Use Figures 8 and 9 to draw two climate graphs.

b) Describe three main differences between these two climates.

	Jan	Feb	Mar	Apr	May	Jun	Jul	Aug	Sep	Oct	Nov	Dec
Temperature (°C)	−8.0	−7.5	−4.5	2.5	8.5	14.0	17.0	15.5	10.5	5.5	0	−4.0
Precipitation (mm)	38	30	25	35	42	48	76	75	57	57	49	41

▲ **Figure 8** Climate data for the taiga forest, northern Norway (65° latitude).

	Jan	Feb	Mar	Apr	May	Jun	Jul	Aug	Sep	Oct	Nov	Dec
Temperature (°C)	27.0	26.5	27.5	27.5	26.5	25.0	24.0	25.0	25.5	26.0	26.0	27.5
Precipitation (mm)	249	236	335	340	244	13	3	18	104	345	373	249

▲ **Figure 9** Climate data for the tropical rainforest, Gabon (0° latitude).

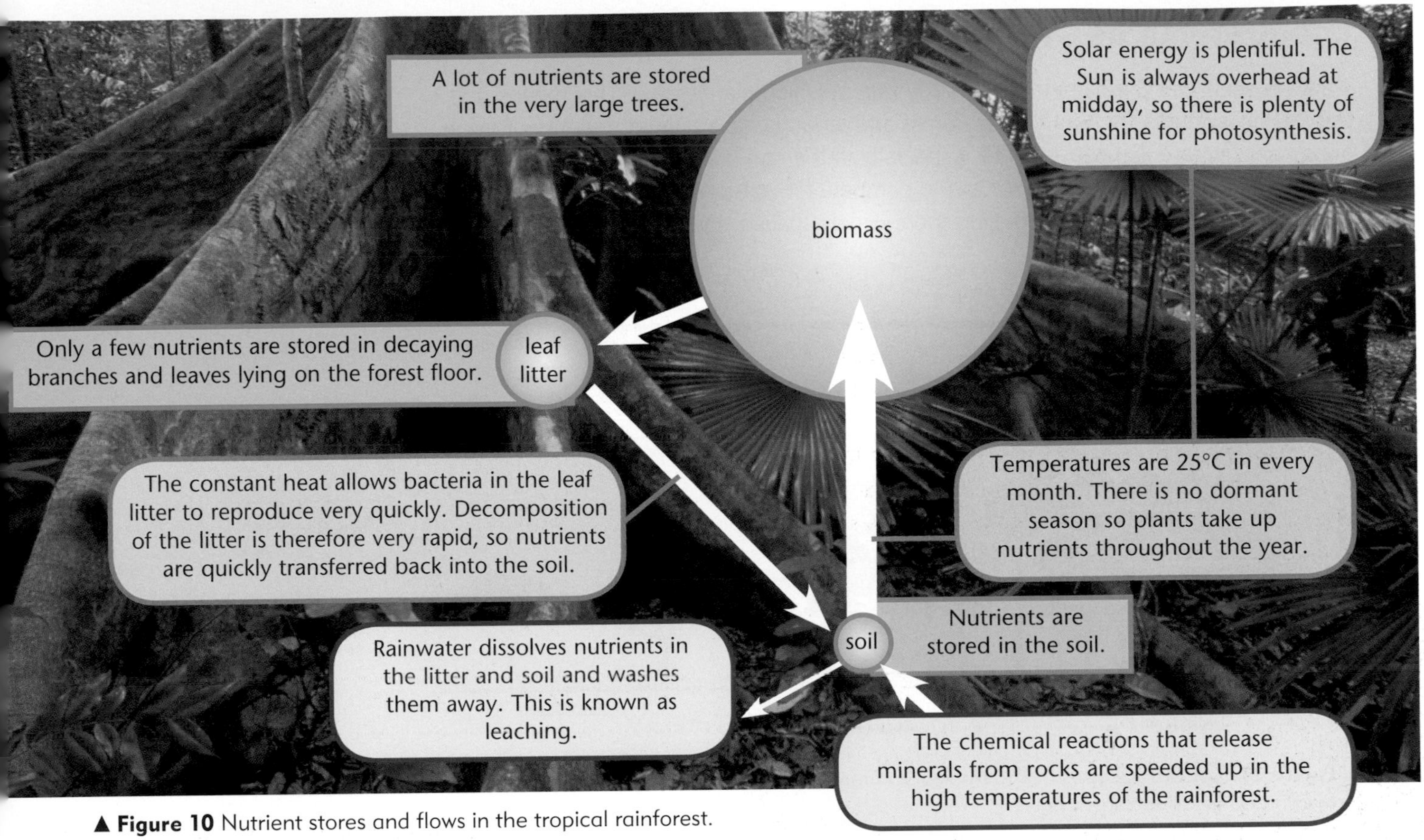

A lot of nutrients are stored in the very large trees.

Solar energy is plentiful. The Sun is always overhead at midday, so there is plenty of sunshine for photosynthesis.

biomass

Only a few nutrients are stored in decaying branches and leaves lying on the forest floor.

leaf litter

The constant heat allows bacteria in the leaf litter to reproduce very quickly. Decomposition of the litter is therefore very rapid, so nutrients are quickly transferred back into the soil.

Temperatures are 25°C in every month. There is no dormant season so plants take up nutrients throughout the year.

soil

Nutrients are stored in the soil.

Rainwater dissolves nutrients in the litter and soil and washes them away. This is known as leaching.

The chemical reactions that release minerals from rocks are speeded up in the high temperatures of the rainforest.

▲ **Figure 10** Nutrient stores and flows in the tropical rainforest.

Activities

3 a) Define what is meant by nutrient stores and nutrient flows.
   b) Describe three places where nutrients are stored in an ecosystem.

4 Study Figure 10.
   a) Describe two ways that nutrients can enter the soil.
   b) Explain why these two nutrient flows are rapid in the rainforest.
   c) Explain why these nutrient flows are likely to be much slower in the taiga.

5 Study Figures 6 and 10. Explain why nutrient cycle diagrams for the taiga would have:
   a) a larger circle for leaf litter than in the rainforest
   b) a thinner arrow for leaching
   c) a thinner arrow showing nutrient flows into the biomass.

Nutrient cycle

The flow and storage of minerals within an ecosystem. Nutrients are stored in plants, soil and rocks. Nutrient flows occur through processes such as weathering and leaching.

Enquiry

How is the climate of the Arctic region affected by latitude? Do temperatures become more extreme as you travel north?

- Research the climate and record the latitude of each of the following locations:
  a) Reykjavik, Iceland
  b) Oulu, Finland
  c) Murmansk, Russia
  d) Churchill, Canada.
- Apart from latitude, suggest one other factor that may explain these differences.

Ecosystems as stores

We have seen that the climate affects the rate at which plants grow. This means that tropical ecosystems such as the rainforest and savanna have a great quantity of living things per hectare. Growth is rapid, especially in the rainforest, where the amount of rainfall means that plants do not have to compete for water. Tropical regions that have a long dry season tend to have smaller, more widely spaced trees. These differences are shown in Figure 11.

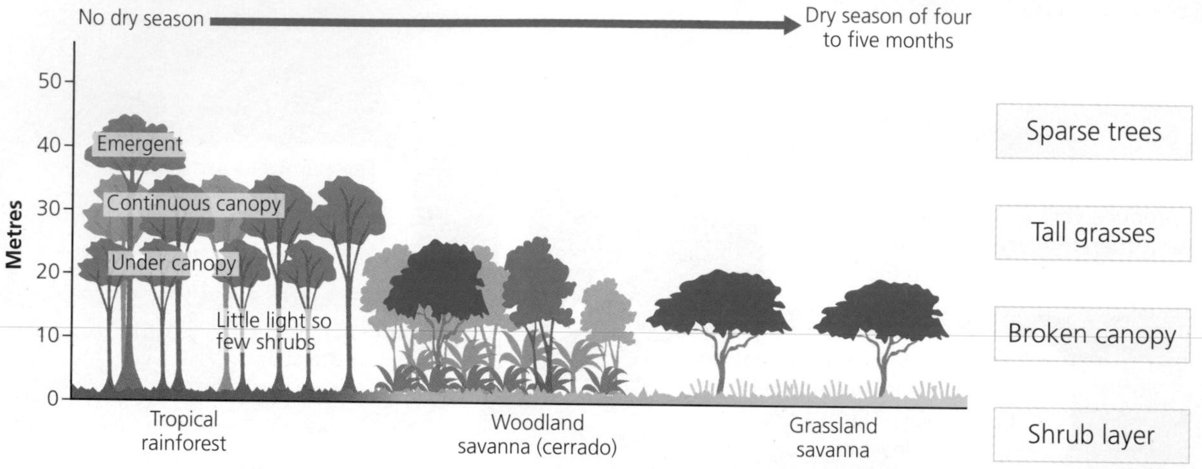

▲ **Figure 11** Typical structures of the tropical rainforest, woodland savanna (known as cerrado in Brazil) and grassland savanna.

Activities

1 a) Make a copy of Figure 11.
b) Complete your diagram by adding the labels to suitable places.
c) Describe how the increasing length of the dry season affects the structure of these ecosystems.

2 a) Make a copy of Figure 12.
b) Add the annotations to appropriate places on your diagram. You can use page 157 to help you.

3 a) Make a copy of Figure 14. Add labels to the correct places to make a complete cycle.
b) Using evidence in Figures 13, 15 and 16, explain why the tundra provides an important function at a global scale.

4 Using Figure 16:
a) Calculate the percentage of the global total amount of carbon stored in each ecosystem. Tundra has been done for you.
b) Represent the percentages using a suitable graphical technique.

Ecosystems store carbon

Ecosystems contain huge amounts of carbon – stored in plants and other organisms. Some is stored above ground in trees, plants, birds, insects and animals. A lot of carbon is also stored in the soil. Some is in living things, such as roots, beetles and earthworms. The rest is stored in dead organic material, such as leaf litter, fallen trees and decaying wood lying in the soil.

Because climate affects rates of plant growth, it also affects how much carbon is stored – and where it is stored. For example, the treeless tundra has a very low amount of carbon stored above ground. However, decomposition is slow because of the cold climate, so this ecosystem stores huge amounts of carbon in the peat-rich soil and in sediment in ponds and wetlands. The tundra is an example of a **carbon sink** (see page 157) where carbon is stored for long periods of time. If the Arctic warms due to climate change there is a concern that some of this carbon will be released into the atmosphere as carbon dioxide (CO_2) and methane (CH_4). This could have an undesirable impact on the greenhouse effect – leading to even more global warming.

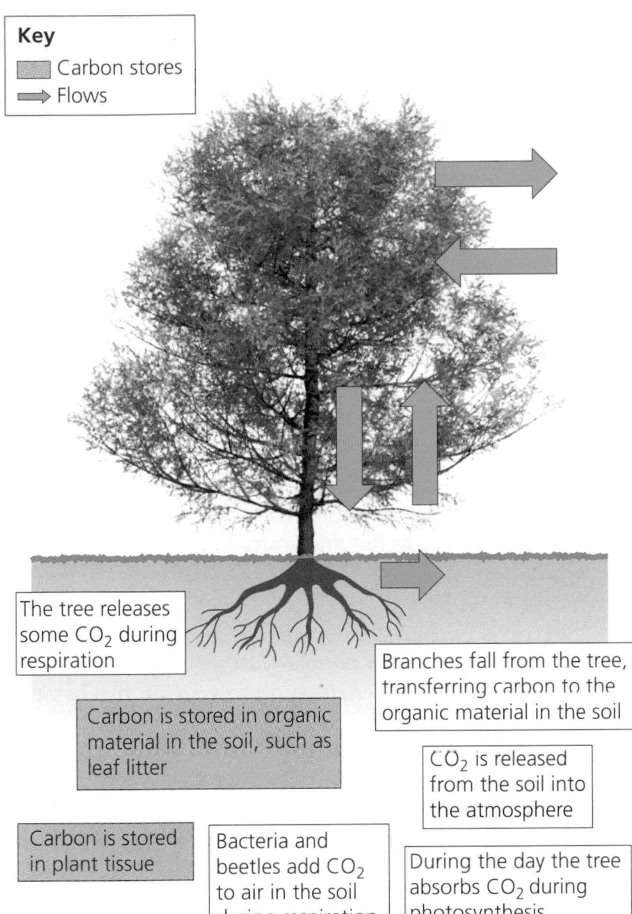

▲ **Figure 12** A diagram showing carbon stores and flows in a typical ecosystem.

▲ **Figure 13** Tundra ecosystem of an upland area of Arctic Norway; there is little carbon stored above ground but the bogs and peaty soils contain huge stores.

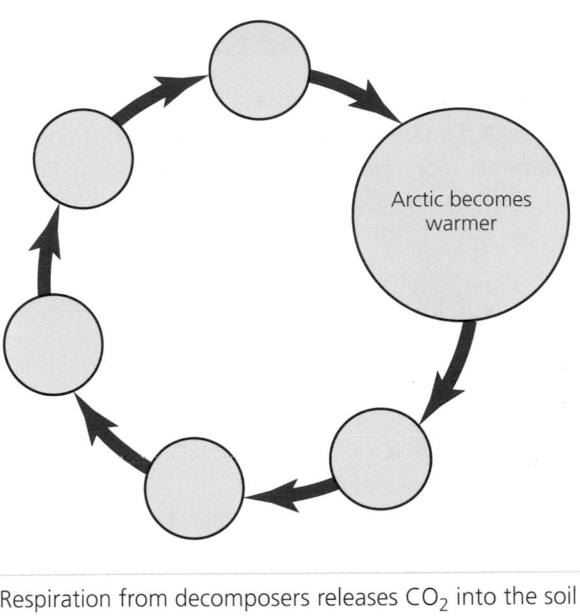

Respiration from decomposers releases CO_2 into the soil

Decomposers in the soil become active

The greenhouse effect becomes stronger

CO_2 and CH_4 in the soil is released into the atmosphere

Frozen soil thaws

▲ **Figure 14** The impacts of a warming Arctic on carbon stored in the soil.

Ecosystem	Carbon stored above ground	Carbon stored below ground
Tundra	Less than 1	500+
Taiga	60–100	120–340
Savanna grasslands	2	7–54
Savanna woodlands	30+	7–54
Tropical rainforest	160–200	90–200

▲ **Figure 15** Estimated carbon stores in selected ecosystems (tonnes per hectare).

Ecosystem	Total carbon store	Per cent
Tundra	155.4	7.6
Taiga	384.2	
Temperate forest (e.g. oak)	314.9	
Temperate grasslands	183.7	
Deserts and dry shrublands	178.0	
Savanna grasslands and woodlands	285.3	
Tropical rainforest	547.8	
Global total stored in ecosystems	2,049.3	

▲ **Figure 16** Total amount of carbon stored in each biome (gigatonnes).

Ecosystems provide key services

Sadly logging, oil exploration, intensive farming and over-fishing are all damaging natural ecosystems. But does it really matter if there are fewer forests and less wildlife? After all, farming and fishing provide us with food, jobs and wealth.

Scientists argue that ecosystems should be protected and not just for their scientific value. They argue that ecosystems provide people with a number of essential services which they describe as **key services**. Furthermore, they say that these key services have financial value. They include:

- maintaining a steady supply of clean water to rivers
- preventing soil erosion
- reducing the risk of river floods
- providing natural materials such as timber for building, or plants for medicinal use; 75 per cent of the world's population still rely on plant extracts to provide them with medication
- providing foodstuffs such as honey, fruit and nuts.

▲ **Figure 17** Bees provide a service to humans by pollinating our crops. Beetles also provide a key service. They digest waste materials such as leaf litter and dung.

Provide a safe environment for fish to spawn and juvenile fish to mature, so helping to maintain fish stocks

Tropical rainforests

Provide people with the opportunity to develop recreation or tourism businesses

Coniferous (boreal or taiga) forests

Support thousands of plants and wild animals that contain chemicals that may be useful to agriculture or medicine

Mangrove forests

Inspire a sense of awe and wonder in human beings

Peat bogs/moors

Act as natural coastal defences against storm surges, strong winds and coastal floods

Tropical coral reefs

Soak up rainwater and release it slowly, therefore reducing the risk of flooding downstream

Sand dunes

Act as huge stores of carbon dioxide, so helping to regulate the greenhouse effect

▲ **Figure 18** Key services provided by ecosystems.

Activities

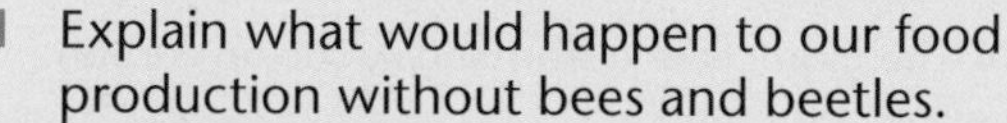

1. Explain what would happen to our food production without bees and beetles.
2. Using Figure 19:
   a) List the places where water is stored in the rainforest.
   b) Explain how water flows from the atmosphere to the forest and back again.
3. a) Describe how tropical rainforests maintain a steady supply of water for local communities.
   b) Describe how damaging the structure of the rainforest could affect local people, and people in the wider region.
4. Discuss the six ecosystems in Figure 18. For each ecosystem identify at least one key service (the yellow boxes) that it provides.

Enquiry

'It is more important to protect the tropical rainforest than the tundra.'

To what extent do you agree with this statement? Use evidence from this page, and your research, to decide which of these two ecosystems has the greatest environmental value. You should consider factors such as biodiversity and carbon stores as part of your argument.

Tropical rainforests regulate water supply

Figure 19 shows how rainforests play an essential role in the regional **water cycle** of tropical areas. The forest acts as a store for water in between rainfall events. After a rainstorm it is thought that about 80 per cent of the rainfall is transferred back to the atmosphere by evaporation and transpiration. This moisture condenses, forming rain clouds for the next rainstorm. So rainforests are a source of moisture for future rainfall events.

At least 200 million people live in the world's tropical rainforests. This includes the tribal groups, or indigenous peoples, of the rainforest. Many more people live downstream of the rivers that leave these forests. The forest maintains a constant and even supply of water to these rivers. If the rainforest water cycle were to be broken, then the water supply of many millions of people could be put at risk. The total amount of water flowing in the rivers would be reduced and the supply would become more uneven, with periods of low water supply punctuated by sudden flooding.

Conservationists argue that we need to place a greater value on these key services than on the value of the tropical timber alone. The benefit of a clean and regular water supply can be measured in financial terms. Rebuilding homes after a river flood can also be measured financially. The conservationists argue that these key services are more valuable in the long term than the short-term profits gained from logging.

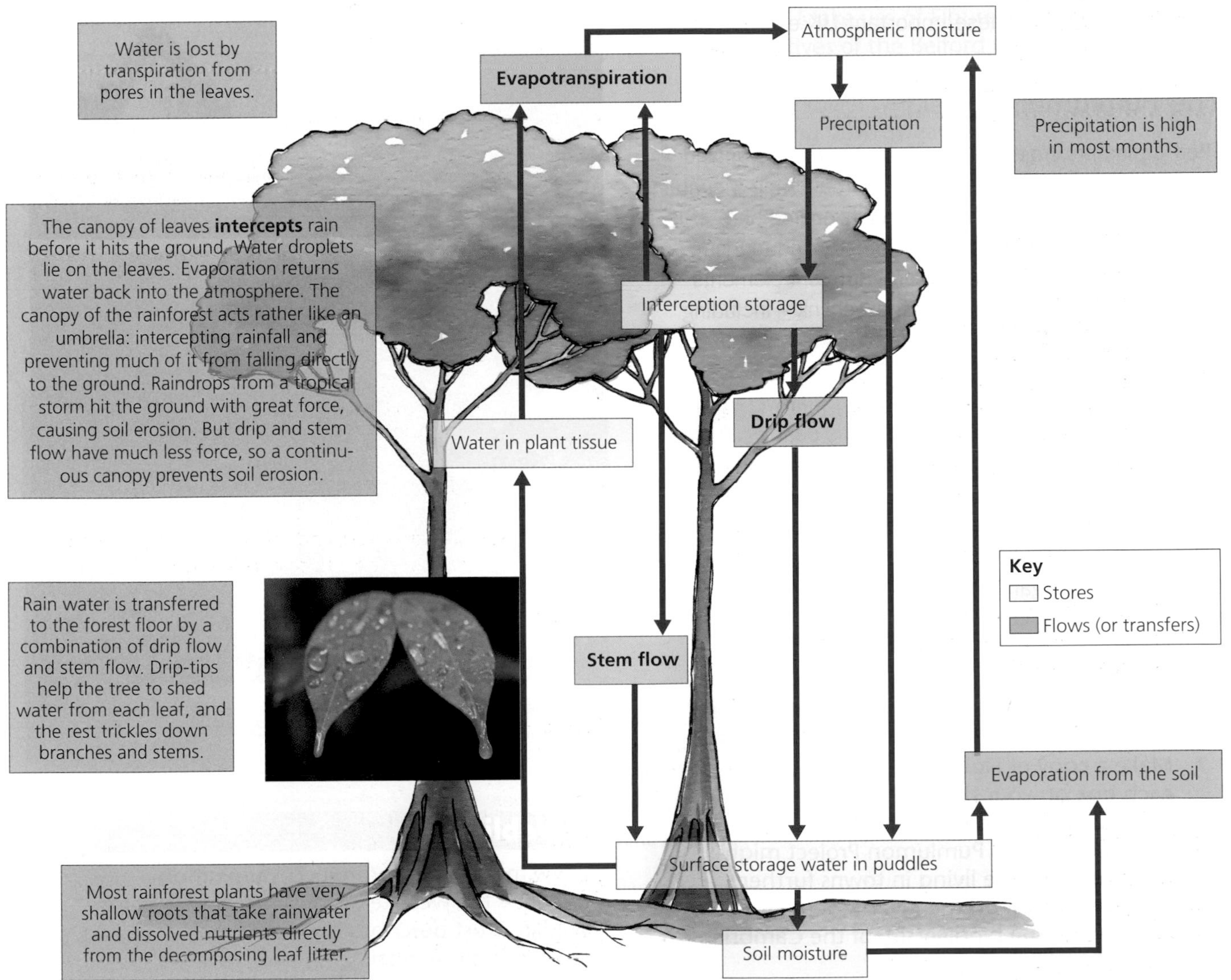

▲ **Figure 19** The water cycle in a tropical rainforest.

Investigating sand dunes

The UK has a number of small-scale ecosystems that you can investigate through fieldwork. You could set up an enquiry to investigate structures or stores and flows in a moorland, woodland or sand dune ecosystem. Data for such an enquiry is usually collected along a transect. A transect is an imaginary straight line drawn across or through a geographical feature. Observations can be taken at regular (systematic), random or stratified intervals along the length of the transect (see page 9 for these sampling strategies).

Step 1: Design your enquiry

A fieldtrip to the sand dunes allows you to see how the plant life and growing conditions vary as you walk through this ecosystem from the strand line to the dune slacks. How quickly do the zones change? What factors cause these changes? Does your sand dune system follow the typical pattern? You could use a hypothesis to organise your enquiry, for example:

The variety of plants increases with distance from the strand line.

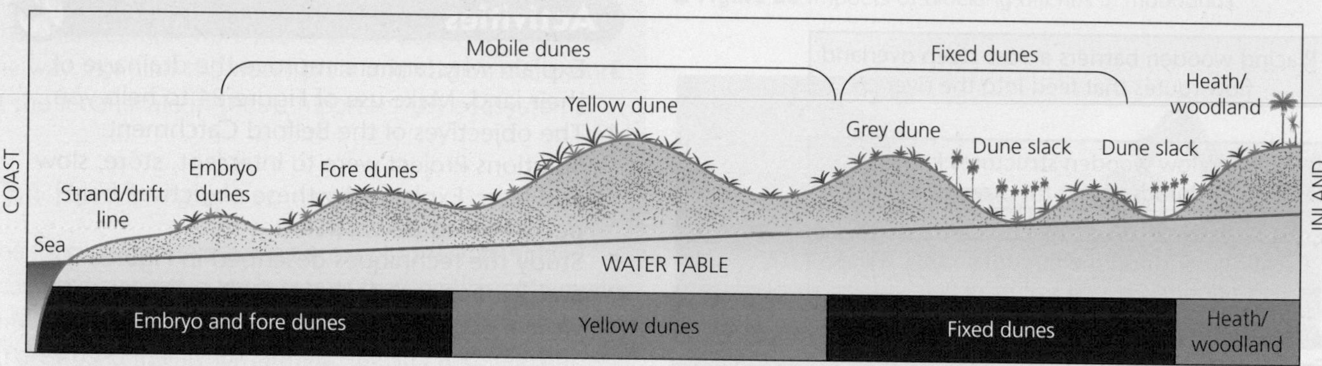

▲ **Figure 25** Cross-section through the zones of a sand dune ecosystem.

▲ **Figure 26** Sea rocket is a pioneer plant. It grows in the embryo dunes. The waxy outer surface of the leaf helps to reduce water loss by evapotranspiration. Shallow roots take nutrients from decaying material on the strand line.

▲ **Figure 27** Marram grass is the main plant of the yellow dunes. Strong leaves bend in the wind without breaking. Long tap roots find water. Burial by sand stimulates new shoots to grow from surface roots.

▲ **Figure 28** Restharrow is a low-growing flowering plant of the yellow dunes. Sticky hairs on leaves reduce water loss by evapotranspiration. Nodules on the long roots are able to take nitrogen from air and turn it into nutrients that are useful for plant growth.

Step 2: Starting and ending your transect

Sand dunes are formed when the wind frequently blows in the same direction. The dunes make ridges that are roughly parallel to the coast and at right angles to the prevailing wind direction. So, your transect should be at right angles to the beach so that it samples each zone within the sand dune system – like in Figure 25. You will need to start on the beach at the strand line. It will end when you have sampled data in each zone. If the dunes are managed you may be asked to use a specific line for your transect.

Collect your data

If your enquiry is about zonation then you will need to record the percentage of plants at each sample point along the transect using a quadrat. You may want to record other variables that could affect the growing conditions in each zone. If so, you could record:

- wind speed
- soil pH
- soil colour (as an indication of the amount of organic material providing nutrients)
- evidence of trampling or management.

	Distance (m) from embryo dunes											
	0	50	100	150	200	250	300	350	400	450	500	550
Sea rocket	20	10	0	0	0	0	0	0	0	0	0	0
Sea spurge	0	10	10	10	10	0	0	10	0	0	0	0
Marram	0	40	60	70	60	70	50	30	30	20	0	0
Restharrow	0	0	0	0	10	10	20	10	0	0	0	0
Fescue	0	0	0	0	0	10	30	50	60	50	30	70
Bramble	0	0	0	0	0	0	0	0	0	0	70	0
Others	0	0	0	0	0	0	0	0	10	30	0	30
Bare sand	80	40	30	20	20	10	0	0	0	0	0	0

▲ **Figure 29** Percentage of each type of vegetation in each quadrat.

Present the data

The percentage of each plant type along the transect is best shown using a **kite diagram** like Figure 30. The axis of this graph represents the length of the transect. The vertical axis is divided by two so that half of the total percentage is displayed on each side of the horizontal axis.

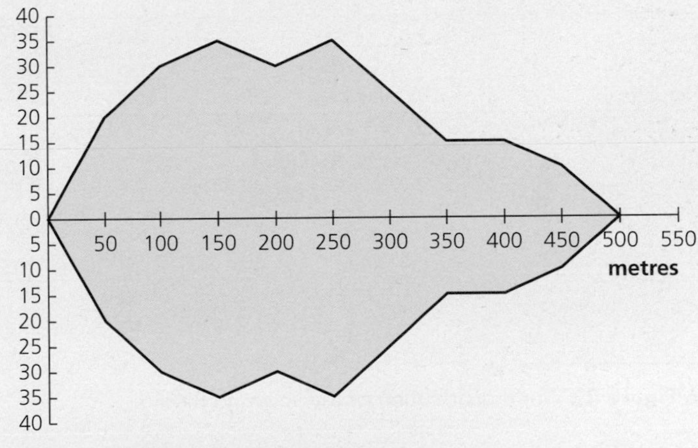

▲ **Figure 30** Kite diagram for marram.

Enquiry

How might you design an enquiry to investigate the effects of wind speeds on sand dune zonation? Use Figure 25 when you consider the following points.

- What questions can you pose?
- How should you sample along your transect? Would regular, random or stratified sampling be best? How many sample points should you use? Do you need a control?
- How would you design your data collection sheets?

Activities

1 Study Figures 26, 27 and 28. Describe two adaptations to:
   a) strong winds experienced in the yellow dunes
   b) the lack of nutrients in the sandy soils
2 a) Use the information in Figure 29 to draw a series of kite diagrams.
   b) What conclusions can you reach about?
      i) The plants that commonly grow in each zone.
      ii) How nutrient levels must change as you travel along the transect.

Characteristics of the savanna biome

The global distribution of the savanna biome (sometimes described as hot semi-arid grass and shrub land) is shown in Figure 32. This biome forms a transition zone between tropical forests and deserts. This biome occurs in regions which have a tropical semi-arid climate. The climate pattern is one of marked wet and dry seasons with the rainfall concentrated in 5–6 months of the year, often in the form of heavy storms and high humidity. This is followed by months of drought with clear skies and fine sunny weather.

	Jan	Feb	Mar	Apr	May	Jun	Jul	Aug	Sep	Oct	Nov	Dec
Average Temp in °C	27	28	27	25	23	23	22	23	25	27	26	26
Precipitation in mm	71	68	151	289	122	27	11	13	12	33	149	106

▲ **Figure 31** Climate data for Arusha, Tanzania.

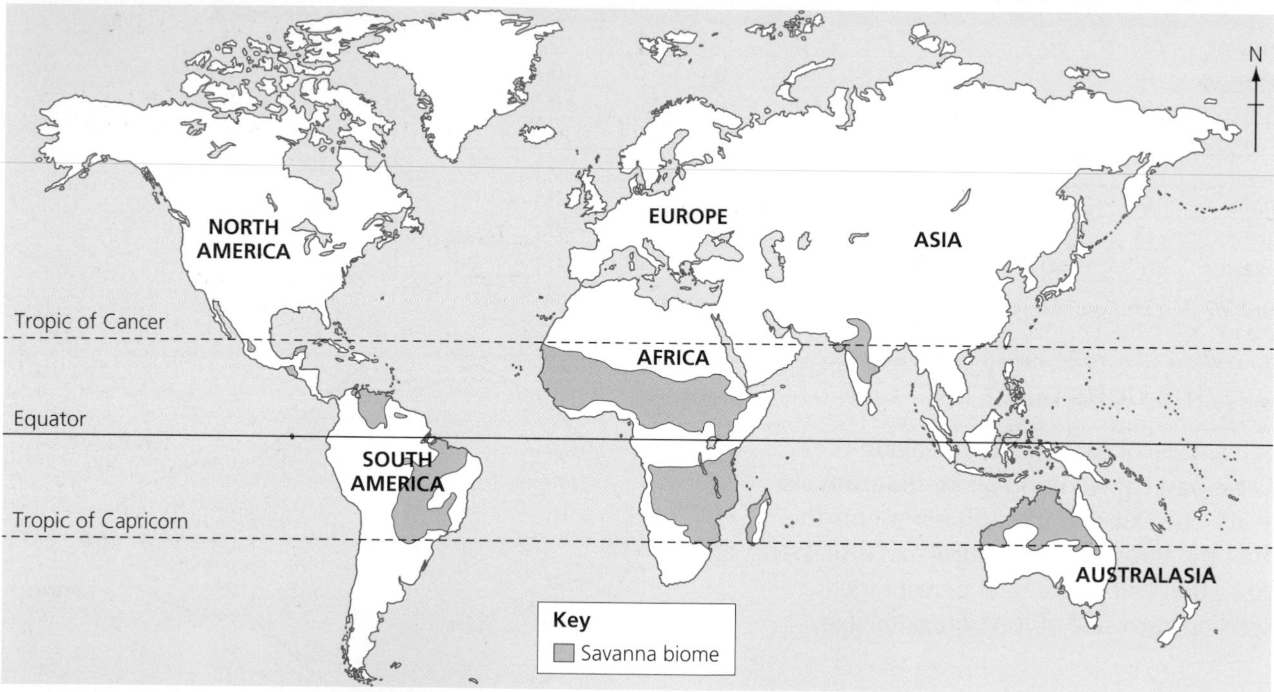

▲ **Figure 32** World distribution of the savanna biome

▲ **Figure 33** Semi-arid grassland in Tanzania.

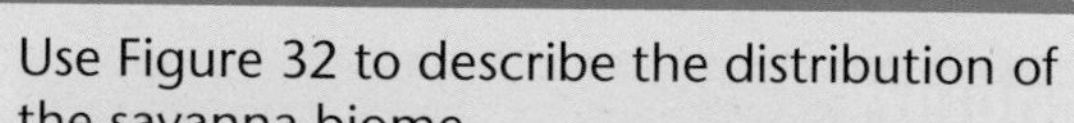

Activities

1 Use Figure 32 to describe the distribution of the savanna biome.

2 a) Use the data in Figure 31 to draw a climate graph for Arusha in Tanzania.
  b) Add labels to your graph to show the wet season and dry season.
  c) What is the total annual rainfall in millimetres?
  d) Describe the annual range of temperature.
  e) Many people visit Tanzania to go on a safari holiday. What time of year would be the best time to go?

How does vegetation adapt to the hot, dry climate?

In savanna climates, the soils are porous which means they drain rapidly. The thin humus layer provides nutrients for plants. The typical vegetation is of scattered trees and drought-resistant bushes. The climate is too dry for thick forest to form because trees need a lot of water to grow and survive. Between the widely spaced trees and bushes there are also grasses that grow rapidly to 3–4 m in height in the wet season. In the dry season they turn yellow and die back, leaving the ground vulnerable to soil erosion. The baobab and acacia are examples of xerophytic (drought-resistant) trees found in this biome – this means they can survive long periods with very little rainfall during the **dry season** of the year. The baobab and acacia are adapted to survive drought in a number of ways – these are shown in Figure 34.

Dry season

A period of the year in which rainfall is very low leading to a shortage of water. In the hot semi-arid grassland the dry season lasts between 5 and 8 months.

▶ **Figure 34** Adaptations of plants to the climate of the savanna.

ACACIA TREE

Broad flat canopy reduces water loss. It provides shade for animals.

Thorns on branches deter animals from eating them.

Long tap roots reach ground water deep underground.

Small leaves with waxy skins reduce the amount of water lost through transpiration.

Grows up to 20 m in height and 2 m in diameter with whitish bark.

BAOBAB TREE

Grows over 30 m in height and 7 m in diameter. Can live for thousands of years.

Lots of shallow roots spread out from the tree. They collect water as soon as it rains. It also has long roots to collect water deep in the soil.

Thick bark is fire-resistant.

Few leaves reduce water lost by transpiration.

Large barrel-like trunk stores up to 500 litres of water.

Activities

3 Study Figures 33 and 34.

a) Describe the vegetation shown in each photograph.

b) At what time of year do you think each photograph was taken? Give reasons for your answer.

c) Draw a table like the one below. Use the information in Figure 34 to describe how these plants have adapted to the climate conditions.

	High rates of transpiration	Long periods of drought and high temperatures	Animals such as zebra eat the leaves
Acacia tree			
Baobab tree			

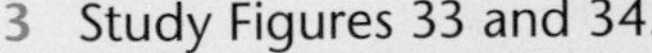

Enquiry

Australia has a large area of savanna. Research the climate and ecosystems of East Kimberley, Australia. Hall's Creek is one of the only large towns in this region.

How do the climate and ecosystem of East Kimberley compare with the savanna of Africa that are introduced on these pages? What are the main similarities and differences?

How do ecosystem processes operate within the savanna?

The savanna has a complex food web. The trees, shrubs and grasses are the producers in this food web. They use the sun's energy, water and carbon dioxide in the atmosphere to make plant tissue. The producers are eaten by a wide variety of insects, birds and grazing animals. The grazing animals include zebras, antelopes, gazelles, wildebeests, giraffes, rhinoceros and elephants. Elephants are the world's largest living land animals. These grazing animals are the prey of secondary and tertiary consumers such as lions and leopards.

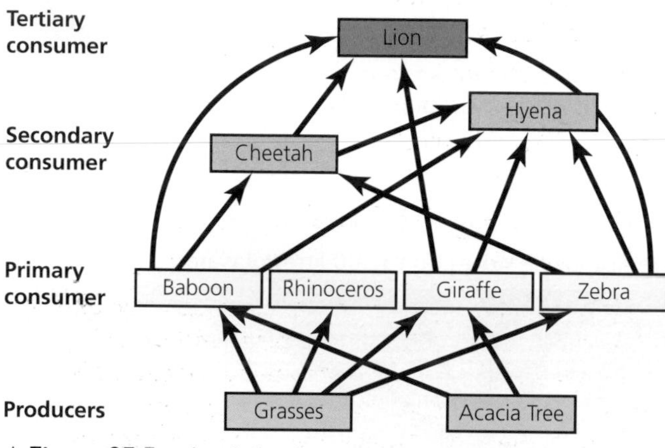

▲ **Figure 35** Food web for the savanna ecosystem in Africa.

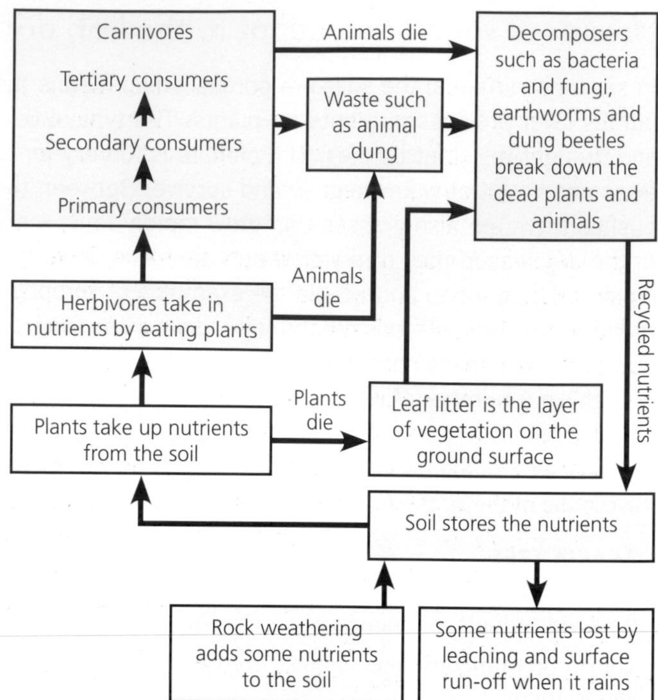

▲ **Figure 36** How nutrients are recycled in the savanna ecosystem.

Nutrient cycles in the savanna ecosystem

The savanna has a dry season where there is very little rain. Nutrients are recycled rapidly during the dry season. Dead plants and animals decompose quickly in the tropical heat. Termites, a type of insect that chews through leaves and the stems of plants, are common in tropical regions. It is thought that termites cause the decomposition of about 20 per cent of organic matter in the tropics. Fire is another common way that nutrients are recycled in this ecosystem, especially where shrubs and trees are growing closer together. Vegetation becomes very dry during the dry season and fires are easily lit by lightning strikes. Nutrients that had been stored in the wood are returned to the soil as ash.

Activities

1 a) Draw a food chain by putting the following into the correct sequence.
lion   grass   energy from the Sun   zebra
b) Explain the difference between a food chain and a food web.
c) Draw a table like the one below and add examples from the savanna food web in Figure 35.

Biotic (living) part of the ecosystem	Examples
Producer	
Primary consumer	
Secondary consumer	
Tertiary consumer	

d) Explain why the hyena is both a secondary and tertiary consumer.
e) What would be the effect on the savanna ecosystem if most of the lions were killed by people?

2 Give two reasons why nutrient recycling is rapid in this ecosystem.

3 Explain why grassland plants need shallow roots to get the nutrients they need.

Why is the biodiversity of Africa's grasslands under threat?

The savanna is under threat from both human activity and natural processes. During the dry season, fires are caused both by lightning strikes and by local farmers burning the grass to encourage new growth when the rains arrive. Grass can survive these fires, but young trees are destroyed.

This ecosystem contains a huge variety of plants, insects, birds, reptiles and mammals. This **biodiversity** – or range of living things – makes the grasslands of Africa a popular tourist destination. People come to see endangered species such as elephant, cheetah, lion and rhino. The safari industry is very important to the economy of countries like Kenya and Tanzania but it is difficult to create a sustainable balance between allowing visitor access to the natural scenery and wildlife, while protecting and conserving it. One of the greatest threats is illegal hunting and poaching. There were once over 100,000 black rhinos in Africa. By 1995, the population had fallen to less than 2,500. Rhinos are killed for their horns which are used in traditional Asian medicines and sell for over five times the price of gold. WWF estimate that 20,000 African elephants are killed every year for the ivory in their tusks. But it is not all bad news. Conservation projects mean that there are now a little over 5,000 black rhino in Africa. However, the species is still critically endangered.

▲ **Figure 37** Conservationists remove the horn from a black rhino to discourage poachers from killing the rhino for its horn.

Biodiversity

Biodiversity is a measure of the richness of wildlife – the variety of plants, insects, birds and other animals that live in any ecosystem. For example, one hectare of tropical rainforest supports a greater variety of wildlife than the same area of hot desert. Some plants and animals are only found in certain places. These are endemic species. Places with a high biodiversity, which also have a variety of endemic plants or animals, are especially worthy of conservation.

Enquiry

Poaching of wild animals is a serious problem in many parts of Africa. Rhino and elephant are key targets for poachers in the savanna environments of Africa. Debate the following enquiry questions. You may want to do some research first.

- Why should we conserve elephant and rhino?
- How do these animals benefit the environment and people?
- What is the best way to protect these animals? Which of the following options might work best?
  - i) Dehorning the wild rhino
  - ii) Moving (translocating) wild rhino and elephant to National Parks
  - iii) Lifting the trade ban and exporting 'farmed' rhino horn and elephant ivory that has been reared sustainably.

How is the marine environment used for energy production?

With 1,200 km of coastline, Wales has huge potential to generate renewable energy using wind, wave and tidal technologies. The Gwynt y Môr offshore wind farm opened in June 2015. It is the second largest wind farm in the world and consists of 160 tubines. Gwynt y Môr is located off the coast alongside the smaller Rhyl Flats and North Hoyle wind farms. Another very large offshore wind farm has been proposed for the Bristol Channel. The proposed wind farm is called the Atlantic Array and would consist of up to 400 wind turbines. Each turbine would be 140 meters tall, or almost three times the height of Nelson's Column. It would be visible from Devon and some local residents are expected to object. If it is built, it will generate enough energy for 900,000 homes.

Pembrokeshire probably provides the greatest potential for wave energy in Wales. Swansea-based Marine Power Systems has developed a technology called WaveSub. The device is to be tested in the water at Milford Haven for between six and twelve months. If tests are successful they will build a full-scale version (between 35 and 40 m long) which will have an output of 1.5 MW.

Technology	Target capacity (GW)	Target date(s)
Onshore wind	2.0	2015–17
Offshore wind	6.0	2015–16
Biomass	1.0	2020
Tidal range	8.5	2022
Tidal stream/wave	4.0	2025

▲ **Figure 1** Renewable energy potential for Wales (1 gigawatt equals 1,000 megawatts).

◀ **Figure 2** The Gwynt y Môr offshore wind farm near Llandudno.

Tidal energy

The Skerries are a group of small rocky islands lying close to the northwest coast of Anglesey. The area between the Skerries and Carmel Head has strong tidal currents and will be the site of Wales' first commercial tidal energy farm – the Skerries Tidal Stream Array. The scheme will generate electricity for up to 10,000 homes in Anglesey. It should deliver other benefits to the local community also, such as long-term jobs in maintenance and environmental monitoring.

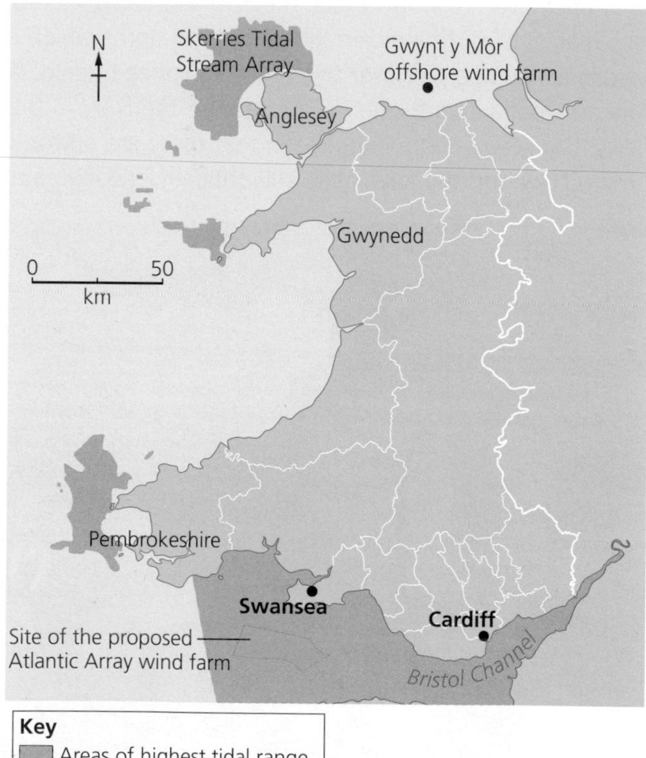

▲ **Figure 3** Coastal areas with greatest potential for tidal and wave power and the location of the Gwynt y Môr offshore wind farm.

The building of a tidal lagoon in Swansea Bay has also been proposed. The tidal lagoon on the eastern side of Swansea Bay, between the docks and the new university campus, would use the flow and ebb of the tide to generate electricity. A 9.5 km long sea wall would be built between 5 m and 20 m in height. It would house sixteen underwater turbines. Enough electricity could be produced for 155,000 homes for 120 years. Figure 4 shows the possible size of the proposed lagoon.

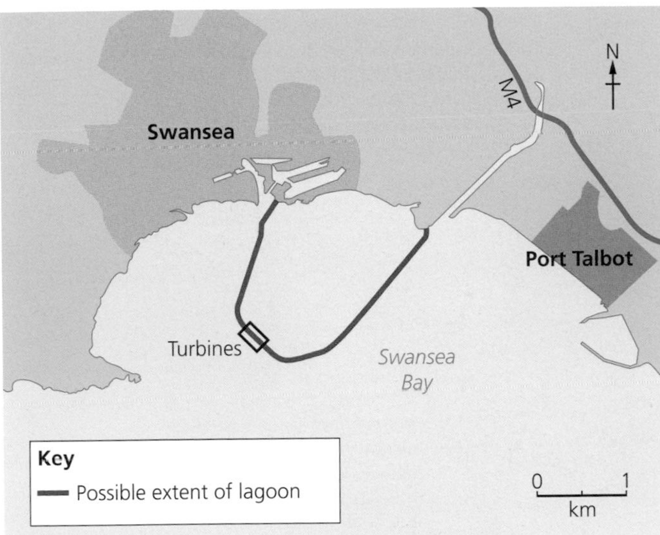

▲ **Figure 4** The possible extent of the Swansea Bay Tidal Lagoon.

Activities

1 What is the difference between renewable and non-renewable types of energy? Give examples of each.
2 Discuss the arguments for and against building more offshore wind farms.
3 Use Figure 3 to describe the distribution of those areas in Wales with high potential for developing marine energy.
4 Use Figure 5 to identify the arguments for and against building a tidal lagoon at Swansea.

Enquiry

Should the Swansea Bay Tidal Lagoon be built?

Prepare a speech so that the issue can be debated in class. Make sure that you consider:

- the environmental, social and economic issues
- the arguments for and against.

Welsh Assembly Government spokesperson

We need more clean and home-grown energy which reduces our imports of fossil fuels. Wales is well-placed to develop marine energy and the Swansea Bay Tidal Lagoon project could create thousands of jobs in construction and maintenance.

Swansea councillor

The project is hugely encouraging for Swansea as it will boost leisure and tourism in an area which has already seen much regeneration.

Friends of the Earth Cymru

Instead of relying on climate-changing fossil fuels, it could help us build a clean and safer energy future and improve our energy mix. However, there are environmental concerns about the impact on migratory fish as the lagoon would be between the Tawe and Neath estuaries. There are also worries regarding silting and sand dredging.

Spokesperson for the nuclear energy sector

The estimated cost of the project has almost doubled to £1 billion. A government subsidy for the power generated – a strike price – has to be agreed. The company is asking for a higher subsidy than solar or nuclear power. The energy output would be a third of an average-sized power station.

▲ **Figure 5** Opinions about Swansea Bay Tidal Lagoon.

How and why do we use tropical ecosystems?

The Amazon rainforest is the largest remaining rainforest in the world. It is a vast lowland rainforest covering much of the drainage basin of the River Amazon and its tributaries. On the southern edge of the Amazon rainforest is the cerrado – another huge forest that grows in a region that has a tropical wet and dry climate.

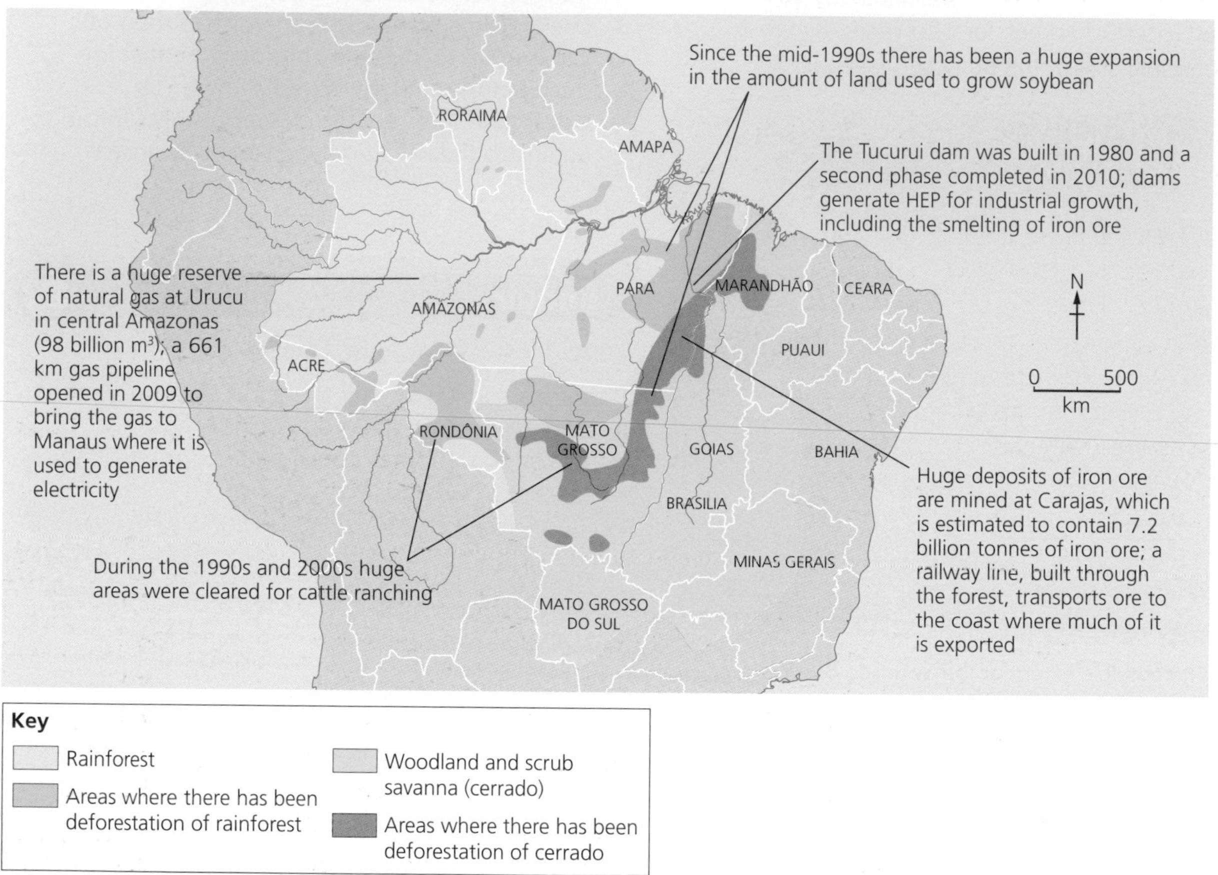

▲ **Figure 6** Deforestation of the tropical rainforest and woodland savanna (cerrado) ecosystems of Brazil.

Year	2009	2010	2011	2012	2013	2014	2015	2016	2017	2018
Deforestation (km²)	1,049	871	1,120	757	1,139	1,075	1,601	1,489	1,561	1,749

▲ **Figure 7** Deforestation of tropical rainforest (km² per year) in the state of Mato Grosso.

Activities

1 a) Describe the distribution of areas where there has been most deforestation.
b) Use map evidence to explain why so much forest has been destroyed.

2 a) Use Figure 7 to draw a line graph showing deforestation rates in Mato Grosso.
b) Describe the trend shown by your graph.
c) Brazil's government has been trying to protect the Amazon rainforest since 2009. How successful do they seem to have been?

Why has soybean farming expanded so rapidly?

The agricultural sector is a huge boost to Brazil's economy. Brazil's largest export is soybeans. In 2017, it exported US$25.7 billion of soybeans. It is creating jobs, wealth for agricultural businesses, and tax revenue for the government. The government needs this money to repay debts and to invest in welfare services, such as education and health.

Soybean, or soya, is a very useful product. It is high in protein so can be used as a meat substitute in vegetarian cooking. It is also used widely in agriculture as a source of protein in the feed given to pigs and chickens. Soybean can also be crushed to extract oil that is used in the manufacture of cooking oils.

Soybean will not grow in the hottest regions of the tropical rainforest ecosystem, but varieties have been created that will grow in the transition zone between the rainforest and savanna woodland (cerrado). Huge plantations now produce soybeans, which are exported to Europe and China. The rich biodiversity of the rainforest has been replaced by a single plant or **monoculture**. This change in habitat greatly reduces the variety of insects, birds and mammals that can continue to live here. The growth of soybean plantations has, therefore, been opposed by environmental pressure groups.

Year	Metric tonnes (million)
2003	51.0
2004	53.0
2005	57.0
2006	59.0
2007	61.0
2008	57.8
2009	69.0
2010	75.3
2011	66.5
2012	82.0
2013	86.7
2014	97.2
2015	96.5
2016	114.6
2017	120.8
2018	117.0

▲ **Figure 8** Production of soybean in Brazil.

Country	Metric tonnes (million)
USA	123.7
Brazil	117.0
Argentina	55.0
China	15.9
India	11.0
Paraguay	9.5
Canada	7.3
Other	21.6
Total	361.0

▲ **Figure 9** Global production of soybeans, 2018.

Source	Metric tonnes (million)
Soybeans	235.4
Rapeseed	39.1
Sunflower	20.7
Cottonseed	15.7
Palm kernel	10.4
Peanut	6.8
Other	6.9
Total	335.0

Figure 10 World protein meal consumption, 2018.

Source	Metric tonnes (million)
Palm	70.7
Soybean	56.3
Rapeseed	28.1
Sunflower	17.8
Palm kernel	8.1
Peanut	5.5
Cottonseed	5.1
Coconut oil	3.4
Olive	3.1
Total	198.1

Figure 11 World vegetable oil consumption, 2018.

Monoculture

When farms grow a single crop across a large area of land. Soybean and palm oil are examples of monoculture crops grown in tropical regions.

Activities

3 Suggest why biodiversity is lower on soybean plantations than in rainforest or cerrado. Use pages 182–4 to help you.

4 Explain why the growth of agribusiness has benefitted the economy of Brazil. Use evidence from Figures 8 to 11 to support your statement.

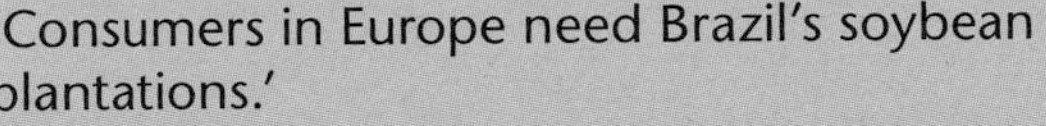

Enquiry

'Consumers in Europe need Brazil's soybean plantations.'

- Discuss this statement. Do consumers have any options?
- Suggest whether European consumers should be concerned about the production of soybean used in products that they buy in the supermarket. Use evidence from this page, and your own research, to support your argument.

How do human activities change processes in the tropical rainforest ecosystem?

Study Figure 12. It shows the continuous **canopy** of the rainforest which acts rather like a giant umbrella. Raindrops from a tropical storm can hit the ground with great force, causing soil erosion. By intercepting rainfall, the canopy prevents much of it from falling directly to the ground. Water then drips from the leaves with much less force.

In Figures 13 and 16 the canopy has been removed. Where the canopy has been removed by human activities, such as logging or forest clearance for palm oil plantations or other agriculture, there is an increased risk of soil erosion. The soil is washed into local rivers where it reduces the capacity of the river channel. This can lead to flooding problems.

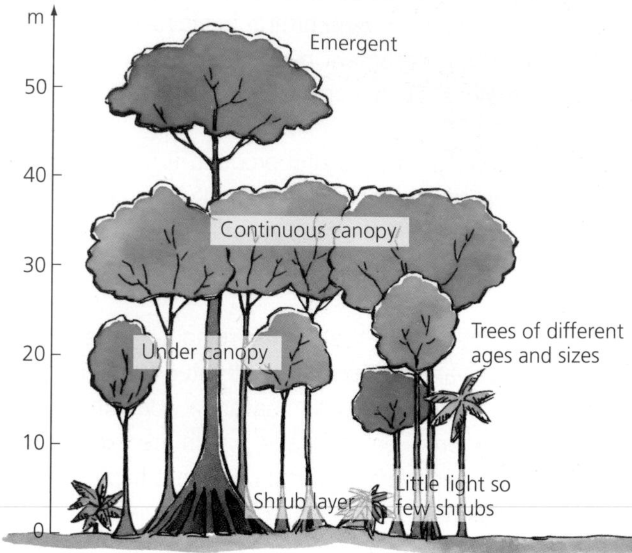

▲ **Figure 12** Typical structure of the tropical rainforest.

▲ **Figure 13** Tropical rainforest cleared in Madagascar.

Forest type	Location of study	Percentage intercepted and evaporated from canopy
Sitka spruce (conifer)	Scotland	28
Douglas fir (conifer)	Oregon, USA	19
Beech (deciduous)	England	15
Tropical rainforest	Indonesia	21
Tropical rainforest	Dominica	27
Tropical rainforest	Malaysia	27

▲ **Figure 14** Interception and evaporation of water from different forest canopies.

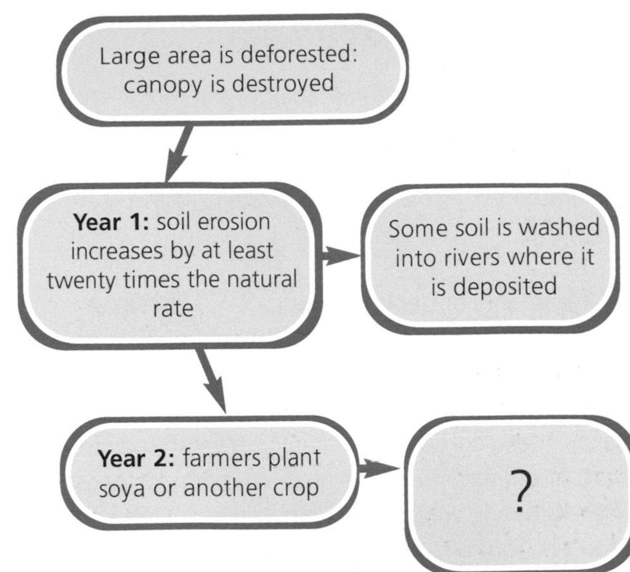

▲ **Figure 15** The consequences of deforestation.

▲ **Figure 16** Logging next to a river in Borneo.

Activities

1 Explain how the canopy of the rainforest reduces the risk of soil erosion.
2 Study Figure 14.
   a) Choose a suitable method to create a graph to represent this data.
   b) Calculate the mean interception rates for
      i) coniferous forest
      ii) deciduous forest
      iii) tropical rainforest.
   c) Suggest why the deciduous forest has significantly lower figures.

Enquiry

How does human activity affect water stores and flows in the tropical rainforest?

- Using Figure 15 as a starting point, create a flowchart or spider diagram to explain the links between deforestation, soil erosion, silt deposition in rivers, and river floods.
- Write suitable annotations to fill the boxes on Figure 16.

Rainforest management

Deforestation causes fragmentation of the forest and this is a major problem for wildlife. As clearings get bigger the wildlife is restricted to isolated fragments of forest that are separated by farmland. You can see this in Figure 13 on page 198.

The governments of Central America (also known as Mesoamerica) are co-operating with each other in an ambitious conservation project. They want to create a continuous **wildlife corridor** through the length of Central America. These corridors are created by planting strips of forest to connect the remaining fragments of forest together. Creating wildlife corridors allows animals to move freely from one area of forest to another without coming into conflict with people. The project is called the Mesoamerican Biological Corridor (known by its Spanish initials, CBM) and involves all seven governments of Central America, plus Mexico.

Some rainforest countries, including Brazil and Costa Rica, have used debt-for-nature swaps to help fund rainforest conservation. These are voluntary agreements in which a donor of aid (such as the USA, Canada or Germany) cancels part of the debt of the developing country. In return, the developing country agrees to invest more money in conservation projects.

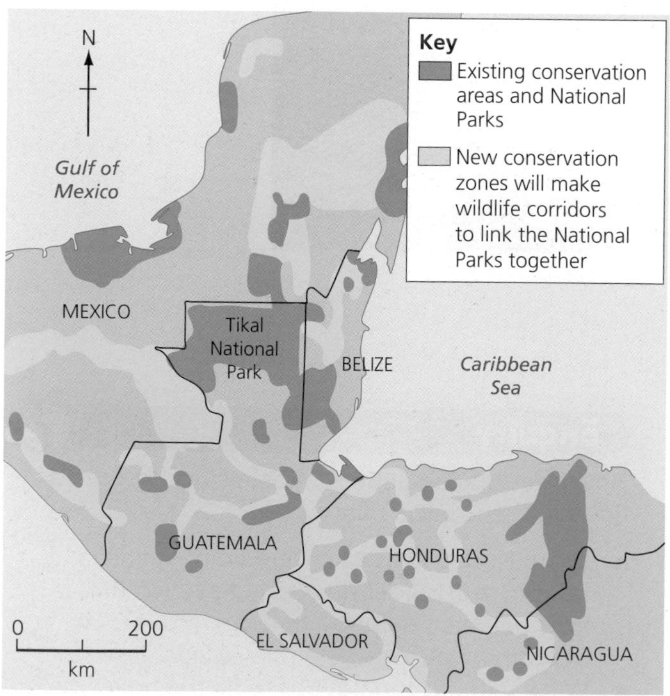

▲ **Figure 17** Protected areas, including forest reserves, in Central America and Mexico, and the proposed wildlife corridors.

Country	1996	2000	2018
Belize	18.0	35.2	37.7
Costa Rica	20.7	23.9	27.6
El Salvador	0.4	0.1	8.8
Guatemala	28.4	29.5	20.0
Honduras	15.0	21.0	23.9
Mexico	4.1	7.7	14.5
Nicaragua	29.3	30.5	37.2
Panama	17.7	19.0	20.9

▲ **Figure 18** Protected areas in Central America and Mexico as a percentage of total land area.

Activities

1 Using Figure 17:
   a) Describe the location of Tikal National Park.
   b) Describe how the pattern of conservation areas in Honduras will change if new wildlife corridors are created.
2 Working in pairs, draw a spider diagram to show how fragmentation of the rainforest affects wildlife. Consider the likely impacts of fragmentation on:
   - food chains
   - success of mating
   - predator–prey relationships
   - pollination and seed dispersal.
3 Explain how the new wildlife corridors will help to conserve wildlife.

Conservation zones and buffer zones

It is difficult to conserve tropical rainforests even if the forest is protected by National Park status. In reality, many protected forests suffer from some illegal logging, clearance of farmland or poaching of wild animals. Forests are often large and remote, so it is difficult for forest rangers to catch poachers or loggers or prevent clearance by burning. Besides, many of the people doing the damage are very poor. They have few ways of making a living and are tempted by opportunities for poaching or farming. Sustainable development of a forest should take into account the needs of local communities, who could make a living by using the forests' resources. At the same time, the overall biodiversity of the forest needs to be conserved for future generations of local people. One way to achieve this balance between the needs of local people and the desire to conserve the forest is to create **buffer zones**.

▲ **Figure 19** The motmot lives in the rainforests of Central America; it eats insects, small lizards and forest fruit. Bird-watching holidays bring valuable income to local communities.

Local people are encouraged to use the buffer zone sustainably – to understand what can be done without doing lasting damage to the forest. For example, by planting food, fruit and nut crops among the trees people can create a diverse environment for insects and birds, and protect the soil from erosion. This type of farming is known as **agro-forestry**. Ecotourism is another permitted activity – with local people becoming guides. Hunting may also be permitted in the buffer zone – as long as animals are only killed for meat when the populations are high and can easily recover. Meanwhile, few if any activities are permitted in the central conservation zone, except perhaps scientific study.

The use of buffer zones may be a more realistic approach to forest management than banning all use of the forest entirely. If managed well, they involve local communities in decision making. They also help people to make a living and educate them about the value of conservation. However, buffer zones may also encourage migration – people see economic opportunities in agro-forestry or ecotourism and move into the buffer zone, putting pressure on local resources such as food, water and firewood.

Buffer zone

An area in which some human use of the forest is permitted. Buffer zones are used to protect a central conservation zone.

The Talamanca Highlands buffer zone

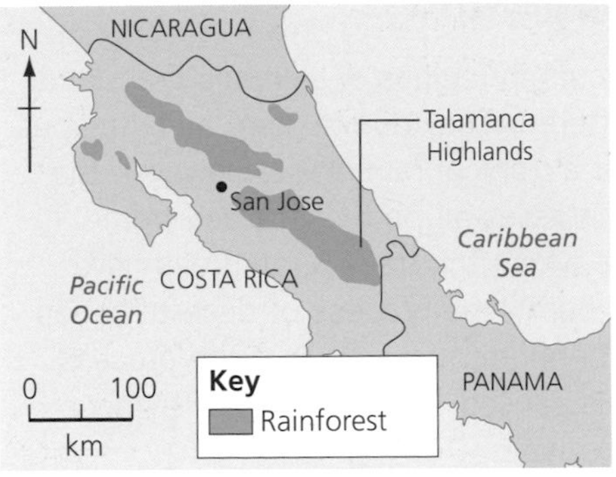

◀ **Figure 20** The location of the Talamanca Highlands.

The Talamanca Highlands are a mountainous area of Costa Rica and Panama. The region contains large areas of unspoilt tropical rainforest that are conserved as a World Heritage Site. The diversity of plants and birds is nationally and globally important. Rainfall in the mountains feeds into rivers that supply fresh drinking water for communities in lowland Costa Rica. In order to protect the central conservation zone, a buffer zone has been created.

The buffer zone has not prevented damage to the environment, however. Land has been cleared for smallholdings, banana plantations and cattle ranching, which has led to soil erosion. The rivers have washed the sediment out to sea where it has damaged coral reefs. There are now plans for HEP dams and roads that could cause further damage to the rivers and forests of the buffer zone.

Activities

4 Describe the location of the Talamanca Highlands.
5 Describe the benefits of creating buffer zones for:
   a) wildlife
   b) local communities.
6 Do you think buffer zones are a sustainable strategy for a tropical rainforest? Explain your answer.

Enquiry

How good is Costa Rica's record on conservation compared with that of its neighbours?

- Study Figure 18. Calculate the average amount of land that is protected in Central America and Mexico.
- Present the data in graphical form – include a bar for the mean.
- What conclusions can you reach?

How does human activity affect the savanna grassland?

Desertification occurs in regions of savanna grassland. The trees are scattered. They do not form a continuous canopy like that of a tropical rainforest. However, the trees, shrubs and grasses all protect the soil from erosion. In regions where the trees and shrubs have been cut down or burnt, the process of desertification has been rapid. Therefore, it seems that the process of desertification is caused, at least in part, by poor management of the land:

- Vegetation is an important regulator of the water cycle. In more heavily forested areas, as much as 80 per cent of rainfall is recycled back into the atmosphere by a combination of evaporation and transpiration from the leaves. Slash and burn of savanna trees and bushes to make space for farming significantly reduces evapo-transpiration and so eventually leads to reduced rainfall totals. This in turn leads to a reduction in water for people, who rely on rivers for water supply.
- The removal of vegetation means that leaf litter can no longer fall into the soil. The nutrient cycle is broken and shrubs no longer replace nutrients or help to maintain a healthy soil structure by adding organic material to the soil.
- The destruction of the tree canopy exposes the soil to rain splash erosion. During heavy rainfall the water flows over the surface of the ground in sheets, eroding all the organic material from the upper layers of the soil. On steeper slopes the power of the water picks up and carries soil particles and smaller rocks. It uses these to erode downwards into the soil in a process known as gulley erosion.

▲ **Figure 21** Interactions between vegetation, soils and climate.

Desertification

A process that turns areas of semi-arid grassland into desert. Desertification is caused by drought, the removal of shrubs and trees, and/or the use of unsuitable farming techniques.

▲ **Figure 22** Gulley erosion is a common problem in areas suffering from desertification. This farmer is placing large stones across the width of the gulley.

Farmers allow their goats to overgraze shrubs, and vegetation is killed

Annual rainfall totals are gradually falling

Trees and shrubs are burnt to clear land for farming or urbanisation

Trees are cut down for firewood for cooking

The rain in the wet season is unpredictable and can be very heavy, causing soil erosion

Commercial farms use the land so intensively that the soil is quickly worn out

Less vegetation means less water is returned to the atmosphere by evapotranspiration

▲ **Figure 23** Physical and human factors that may cause desertification.

Activities

1 Sort the causes of desertification listed in Figure 23 into physical and human factors.

2 Study Figures 21 and 22. Use the information to write an explanation of what will happen if…

Farmers allow goats to overgraze	Effect on vegetation	
	Effect on soils	
	Effect on climate	

3 Explain why the farmer is placing rocks across the gulley in Figure 22.

4 a) Use page 37, and the Glossary, to make sure you understand the following key terms:
- infiltration
- interception
- overland flow
- evaporation of soil moisture
- transpiration
- gulley erosion

b) Predict whether each of the processes in question 4a will increase or decrease during desertification.

c) Write a short news report about the issue of desertification. Make sure you use each of the key terms from question 4a in your report.

How has food production affected the savanna?

Farming in the savanna grasslands of Africa is a mixture of arable (crop-growing) and pastoral farming (animal grazing). Farmers keep goats and cattle for both their milk and meat. Crops are grown using a traditional bush fallow system. Scrub vegetation is removed by slashing and burning. Crops such as maize, root crops and vegetables are grown for between one and three years. The land is then abandoned for between eight and fifteen years. This is known as the **fallow period**. During this fallow period, the natural shrubs grow back. Leaves from the shrubs decompose in the soil, replacing organic fibre and nutrients that have been taken out by farming. This system is sustainable as long as the fallow period remains long enough. However, in some villages the fallow period is now only two to three years. This does not give the soil enough time to recover. It loses its organic content and its structure becomes dusty. This means that the soil is at risk of erosion from both wind and rainfall.

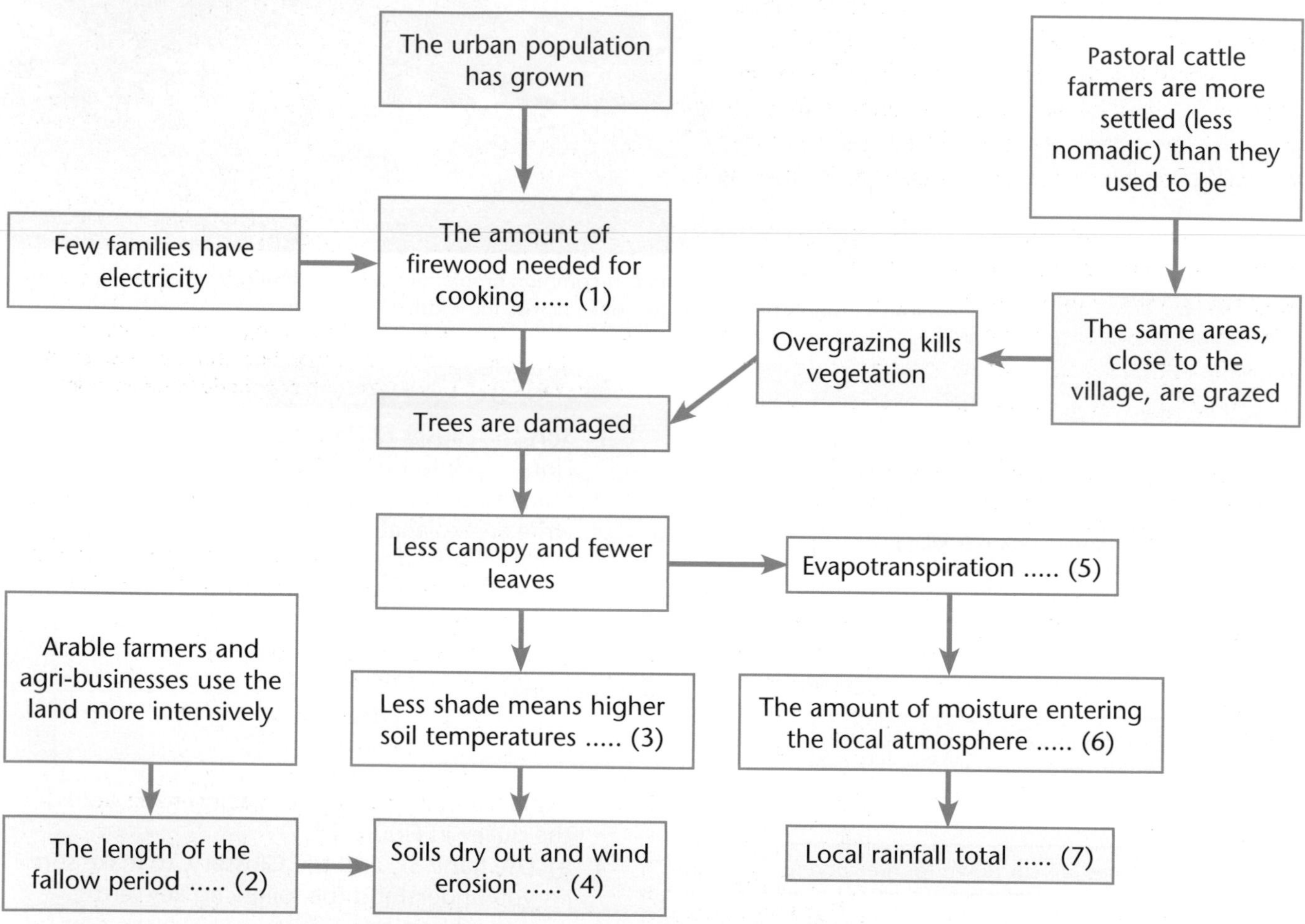

▲ **Figure 24** How poor land management leads to desertification.

Activities

1 Copy and complete the following description of the bush fallow system:
Natural vegetation is cleared using slash and .... techniques. Crops are grown for one to .... years. The land is then allowed to rest for at least .... years in between crops. This is known as the .... period. During this time .... are returned to the soil.

2 Explain why:
  a) the traditional bush fallow system is sustainable
  b) reducing the length of the fallow period has degraded the soil.

3 Make a copy of Figure 24. Complete the boxes numbered 1–7 by adding the word *decreases* or *increases*.

Can drought resistant crops help solve the problem?

Ghana is a tropical country in West Africa. Northern Ghana has a hot semi-arid climate with a dry season that can last up to eight months of the year. Trees are used for firewood and shrubs are over-grazed so soil erosion has become a serious issue in the Northern, Upper West and Upper East regions. Farmers are also concerned about the impacts of climate change. Rainfall patterns seem to be increasingly unpredictable. Crop failures and the death of livestock lead to economic losses for farmers, food shortages and higher food prices. According to UNICEF, one in five children in Ghana is stunted because they are suffering from chronic malnutrition. In the Northern region it is thought that 37 per cent of children are stunted due to malnourishment.

Drought-resistant crops may help this region produce food even when rains are poor. These are crops that can grow well even if rainfall totals are low. Crops like chickpea, pigeonpea, groundnut, millet and sorghum can all grow in hot semi-arid regions and are all suitable for growing on smallholdings. The Ghanaian government is encouraging farmers to use four new varieties of maize that have been developed with the help of the Nippon Foundation which is a non-government organisation (NGO). Local farmers, however, have complained that the new seeds are more expensive than the usual varieties that they grow.

▲ **Figure 26** Clementia Talata cooking banku, a staple made from maize, at a local 'chop bar' (simple restaurant) using water from a Safe Water Network distribution point.

Natural vegetation is savanna grassland. The risk of soil erosion in the Northern and Upper West regions is classified as Moderate to Very Severe.

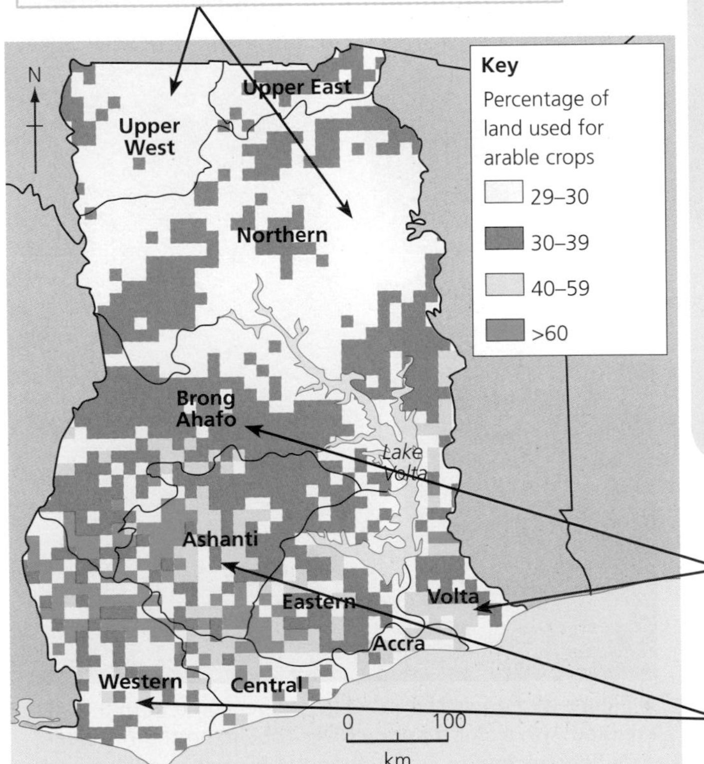

Natural vegetation is savanna woodland. The risk of soil erosion in Brong Ahafo and Volta is classified as Slight to Moderate.

Natural vegetation is tropical rainforest. The risk of soil erosion in Ashanti and Western regions is classified as Moderate to Severe.

▲ **Figure 25** Arable farming and the risk of soil erosion.

Activities

4 Use Figure 25.
   a) Estimate the amount of the Upper West and Upper East regions that is used for arable farming.
   b) Compare the arable farming in Ashanti to the Northern region.
   c) Suggest why the risk of soil erosion is highest in regions that have:
      i) high rainfall totals, like Ashanti
      ii) a long dry season, like Upper West.

5 Describe one benefit and one disadvantage of growing drought-resistant crops in Ghana.

6 Explain the links between climate change, desertification and poverty (including health issues such as malnutrition).

How can intensive farming affect water cycles and climates?

Lake Chad is in the **Sahel** region of Africa. In the last 50 years, Lake Chad has shrunk dramatically. The reasons are complex. They include overgrazing and deforestation in the lake's **drainage basin**, which have resulted in a drier climate. However, a more significant reason may be the over-abstraction of water from the rivers that supply Lake Chad. Since 1970 a large number of dams have been built in both the Komodougou-Yobe river basin and Chari-Logone river basin. Some water is abstracted for domestic supply in cities such as Kano. The rest is used for intensive farming in irrigation projects that grow crops such as onions, tomatoes, chilli peppers and rice. It is thought that only five to ten per cent of the water in the Chari-Logone river now flows into Lake Chad. The rest evaporates or is abstracted and used. Twenty dams have been built in northeastern Nigeria since the construction of the Tiga Dam in 1974. This leaves only about two per cent of the water in the Komodougou-Yobe river basin to flow into Lake Chad.

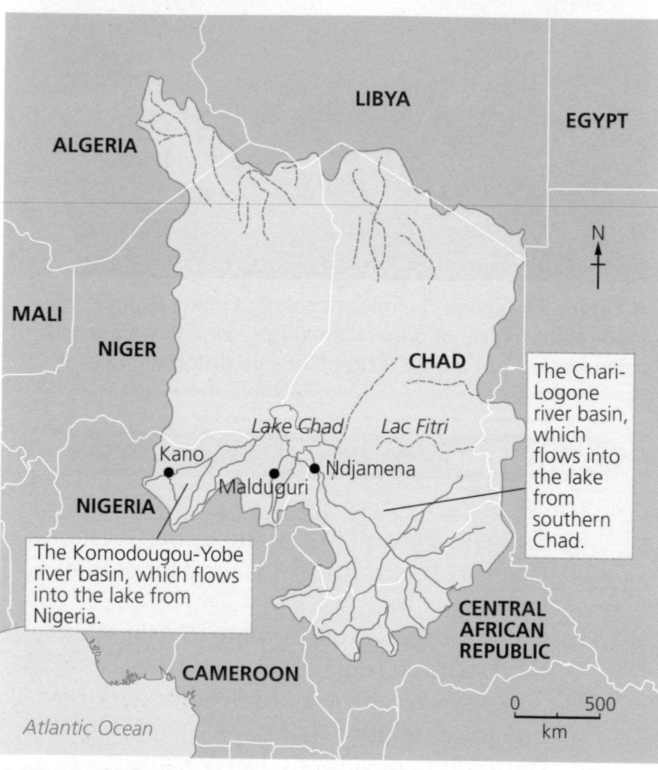

▲ **Figure 27** The drainage basin of Lake Chad is shown by the pale green area.

Country	Total area (km²)	Area within the basin (km²)
Nigeria	923,770	179,282
Niger	1,267,000	691,473
Algeria	2,381,740	93,451
Sudan	2,505,810	101,048
Central Africa	622,980	219,410
Chad	1,284,000	1,046,196
Cameroon	475,440	50,775
Total area of Lake Chad basin		2,381,635

▲ **Figure 28** The drainage basin of Lake Chad.

▲ **Figure 29** People harvest chilli peppers on a farm that is irrigated with water from the River Yobe in northern Nigeria; chilli peppers are an essential part of Nigerian cooking and fetch a good price, but have a very high water footprint.

Activities

1 Study Figure 27.
   a) Describe the location of Kano.
   b) Describe the drainage pattern within the Lake Chad drainage basin.
   c) Use your understanding of tropical climates to explain why Lake Chad has this drainage pattern.

2 Study Figure 28.
   a) How could this information be adapted to make it easier to use?
   b) What percentage of Nigeria is within the Lake Chad drainage basin?

How has the shrinking lake affected people and the environment?

This is a very poor part of Africa. There are 40 million people living in the drainage basin of Lake Chad and 60 per cent of them live on less than $2 a day. Over-abstraction means that:

- Some people rely on the lake for their water supply; this is not safe for human health and is a source of cholera and polio.
- Soils have suffered from salinisation. This process occurs when too much water is used to irrigate crops. Water evaporating from the soil leaves behind harmful minerals.
- The wetland ecosystems surrounding the lake have dried out. Wetland bird populations and fish stocks have declined. Less lake fish has meant declining incomes for fishermen.
- Poverty has increased, leading to increased migration to cities such as Kano. Increased poverty has also led to a rise in support for extremist groups such as Boko Haram.

Can Lake Chad be saved?

Thirty years ago an ambitious plan was announced to save Lake Chad. It was proposed that the Transaqua Project would transfer water from the Democratic Republic of Congo (DRC). A 2,400 km canal would transfer 100 billion m^3 of water every year from the River Congo to the River Chari. Few people have taken the project seriously until now. The governments of India, China and Brazil see this project as an opportunity to invest in Africa – a continent that is rich in resources that would help support their rapidly growing NIC economies. This infrastructure project would provide water for agriculture, industry and electricity production. It would create job opportunities and water security. This peace-through-development approach to tackling poverty may also reduce extremism and violence in the region.

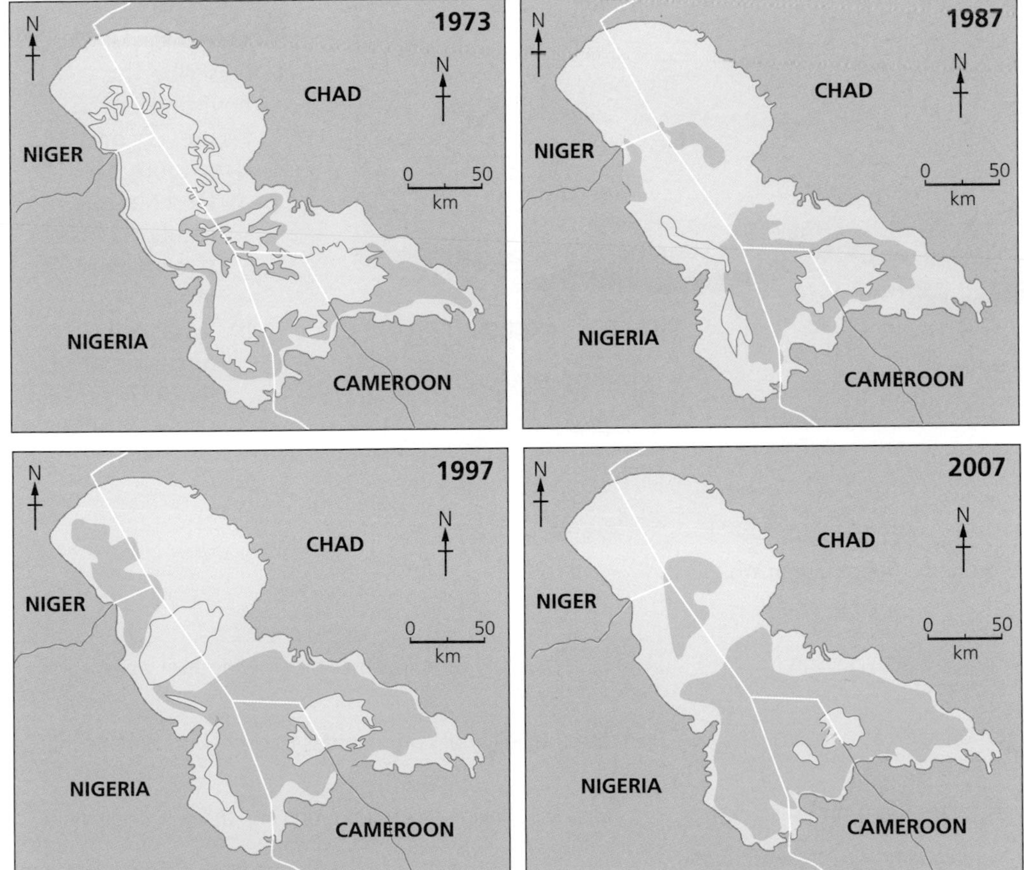

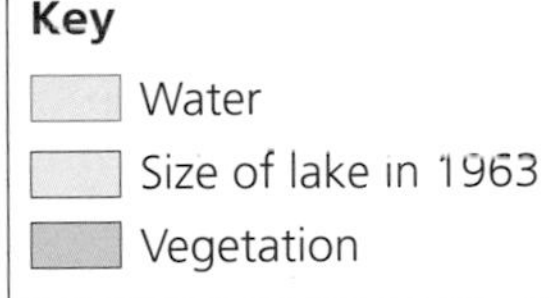

▲ **Figure 30** The shrinking of Lake Chad.

Activities

3 Use Figure 30 to describe the changes in the shape and size of Lake Chad between:
   a) 1963 and 2007
   b) 1987 and 1997.
   c) Suggest reasons why Lake Chad has reduced in size so erratically.

4 Outline the impacts of over-abstraction on the people, economy and environment of the region.

Enquiry

Should the Transaqua Project be built? Justify your decision by considering the possible social, environmental and economic benefits.

The Great Green Wall of Africa

The Great Green Wall (GGW) is one example of an initiative where countries are working in partnership with one another to try to tackle the issue of desertification. Eleven countries signed an agreement in 2010 to begin planting this 'wall'. The plan is to plant a 15 km-wide strip of land with trees and shrubs across the width of Africa. It is hoped that this wall of vegetation will help prevent further soil erosion from the Sahel and improve incomes.

The plan is to encourage local communities to plant a mixture of native trees. These will include fruit and nut trees. Small fields between the trees can be planted with food and cash crops – a type of farming called agro-forestry because it combines farming and forestry.

▲ **Figure 31** Fields of millet growing between native shrubs and trees in Zinder, Niger.

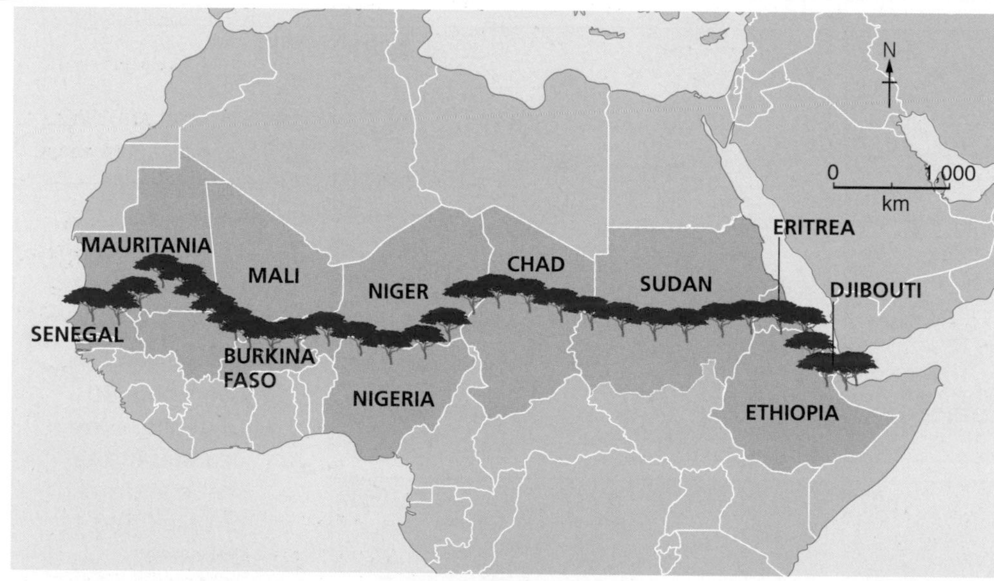

▲ **Figure 32** The proposed location of the Great Green Wall.

Activities

1. Describe the environment in Figures 31 and 34.
2. Use Figure 32.
   a) Describe the location of the Great Green Wall.
   b) What is its approximate length?

Reduce soil erosion during the rainy season	Diversify farm incomes by growing fruit trees
Improve soil fertility by using leaves as a mulch	Increase the ability of communities to cope with climate change
Increase the amount of fodder (plant food) for livestock	Trees will provide shade for crops and increase their yield
Reduce the amount of time women spend collecting firewood	Grow medicinal plants
Increase biodiversity	

▲ **Figure 33** Benefits of the Great Green Wall.

How successful is the GGW?

Huge progress has been made by Niger. But Niger had a head start. It began its tree-planting programme 25 years before the 2010 international agreement was signed. Five million hectares of land in the Zinder region of Niger have been planted with trees since the mid-1980s. Senegal has also made good progress. Eleven million trees have been planted across 27,000 hectares of land. The Senegal Government wants local communities to develop ecotourism in the newly planted areas to take advantage of the larger number of bird species that live in these new forests.

▲ **Figure 34** The Project Eden Research station in Niger. Project Eden is a Swedish NGO. It funds research into plants that will grow in semi-desert conditions without the use of fertilisers or irrigation.

Enquiry

How do changes in climate over time affect spatial patterns (i.e. change over a map) in the environment?

Figure 35 uses isohyets to show spatial patterns of rainfall. Every place along an isohyet has an equal amount of rainfall over the year.

- Use the isohyets on Figure 35 to explain the change in ecosystems as you travel south to north through Niger.
- Use evidence from Figure 35 to explain why relatively small changes in rainfall can have significant impacts on people and the environment.

Activities

3 a) Discuss the benefits of the Great Green Wall shown in Figure 33.
 b) Classify each benefit as economic, social or environmental. Do any of the benefits fit into more than one category? Explain why.
 c) Make a copy of the diamond nine ranking diagram (page 83) and place each of the benefits into your diagram.
 d) Explain why you have chosen your top three benefits.

4 Suggest why some people may be suspicious of top-down development.

The Food and Agriculture Organization (FAO) of the United Nations claims that tree planting in these two countries has been a success:

- crop yields have increased
- livestock is better fed
- the trees are providing medicines and firewood.

Progress has been slow in the other nine countries who signed the agreement. This may be because some local communities do not feel as though they have been involved in the decision-making process and they feel suspicious. This is an example of 'top-down development' and some communities are disappointed because they have not been consulted. They cannot imagine how their own community might benefit.

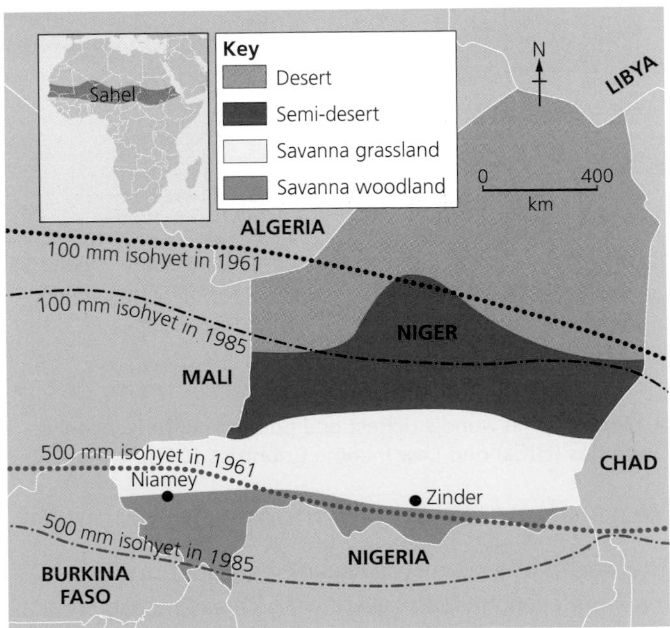

▲ **Figure 35** Ecosystems and the location of the Great Green Wall in Niger.

THEME 6 Development and resource issues

Chapter 1 Global inequalities

Where is the development gap?

Figure 1 shows the economic gap that exists between the wealthiest and poorest countries of the world today. This gap was first identified in a report written in 1980 by a German politician, Willy Brandt. His report drew a line, shown as the Brandt Line on Figure 1, separating richer countries from poorer ones.

- The line showed that most wealthy countries were in the northern hemisphere, so these became known as the 'global north'. The line loops around Australia and New Zealand in the southern hemisphere so that they could be included in the 'north'.
- Poorer countries became known as the 'global south' because many were in the southern hemisphere.

The line dividing these groups became known as the 'North–South divide'. Other people call it the 'development gap'. Forty years later, what, if any, progress has been made in closing the gap?

Activities

1 Study Figure 1. Describe the distribution of:
   a) high income countries (HICs).
   b) low income countries (LICs).
2 Identify any countries that do not fit the Brandt 'development gap' from 1980.

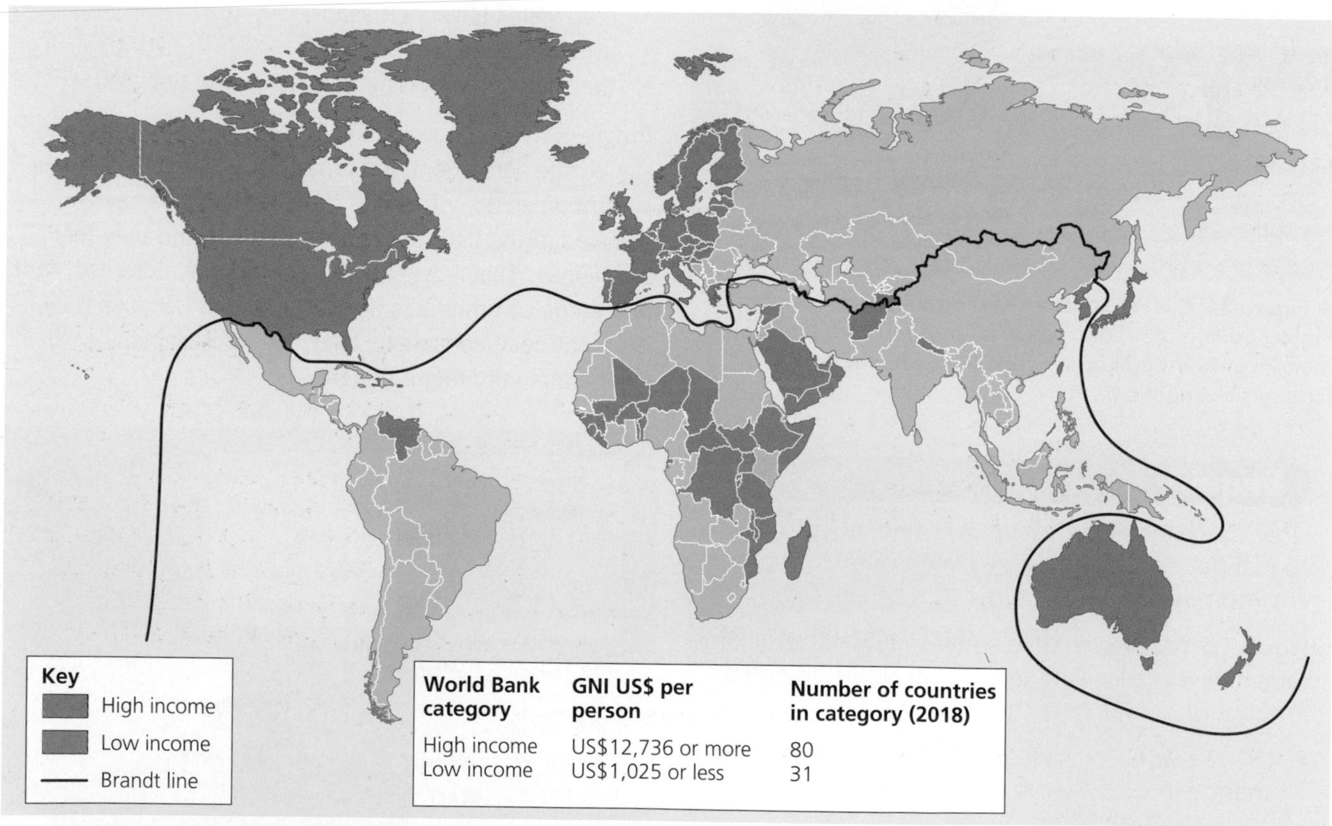

World Bank category	GNI US$ per person	Number of countries in category (2018)
High income	US$12,736 or more	80
Low income	US$1,025 or less	31

▲ **Figure 1** The world's richest and poorest countries; colours on this map indicate GNI per capita for High Income Countries (HICs) and Low Income Countries (LICs).

Ways of measuring economic development

The wealth of a country is usually measured by how much its economy earns each year. This is measured in a variety of ways, but the most common are:

- gross domestic product (GDP), which is the total value of the economy each year
- **gross national income (GNI) per person**, which is a little like the average wage
- poverty line – this measures the percentage of the population who earn less than a certain amount per day; the World Bank used a figure of US$1.90 in 2019.

These ways of measuring wealth have limitations. For example, GNI is an average for the whole country. The GNI in Malawi is around US$360, but some people in Malawi earn much more than this while others earn less. Also, comparing GNI for different countries can be a bit misleading. Food is cheaper in Malawi than in many countries, so US$360 will buy more there than in, for example, the UK. To take account of this, the World Bank changes its data to **Purchasing Power Parity (PPP)**. This converts GNI into a figure that describes what that money will buy in local prices. For Malawi, the GNI per capita figure in PPP in 2018 was US$1310, over three and a half times more than GNI, because prices there are three and a half times cheaper.

Using GNI to measure development

The GNI per person of a country is calculated as follows:

1 Add up the value of goods and services produced by people living in that country and by people overseas who are still citizens of that country. For example, the total earnings for Malawi in 2018 were US$7.1 billion.
2 Divide this figure by the number of citizens of that country. For example, the total number of citizens in Malawi in 2018 was 19.6 million. By dividing Malawi's earnings by its population we can see that the average per capita GNI was about US$360.

Rich and poor?

In 2016 Oxfam published a report on global inequality. In this report they point out that the richest 62 people in the world own the same amount of wealth as the poorest 3.5 billion people (which is 50 per cent of the world's population). This is a shocking fact. It reminds us that wealth is still very unevenly distributed. However, unlike the Brandt Line, which drew our attention to the gap between rich and poor *countries*, this report highlights the gap between rich and poor *people*.

▲ **Figure 2** How do you see the world?

The gap between richest and poorest is widening

Oxfam is calling for governments to take urgent action to tackle the inequality crisis. It says there is a widening pay gap in most countries between the poorest and highest earners. Oxfam is urging world leaders to take three forms of action:

1 Crack down on large companies that avoid paying tax.
2 Increase investment in public services such as health care and education.
3 Boost the income of the lowest paid workers in society – especially women.

Oxfam points out that low pay particularly affects women, who make up the majority of low-paid workers around the world. Those on low wages do not seem to be benefitting from national economic growth. By contrast, people who are already wealthy benefit most from banking and tax systems that reward further investment.

▲ **Figure 3** An article based on Oxfam's 2016 report.

Activities

3 Describe two different limitations of using GNI as a measure of economic development.
4 Study the cartoon in Figure 2.
   a) Describe the two characters – what each is doing and how each is dressed?
   b) Suggest who the two characters represent, and why they have been represented in this way.

Enquiry

What should be done to tackle the gap between rich and poor?

The Oxfam's 2016 report suggests that there are three forms of action that would help reduce the gap between rich and poor.

- Describe how each action would help.
- Which of these three would you prioritise and why?

A continuum of economic development

It is very simplistic to divide the world into rich and poor as is shown in Figure 1 on page 210. In reality, every country fits somewhere along a line, or continuum, between very poor countries such as Malawi and super-rich ones such as the UK. If we rank countries by GNI, it is possible to see this continuum. The World Bank does this, then divides the world's 215 countries into four categories using GNI per capita:

- High Income Countries (HICs) have a GNI greater than US$12,736
- Upper Middle Income Countries have a GNI between US$3,996 and US$12,735
- Lower Middle Income Countries have a GNI between US$1,026 and US$3,995
- Low Income Countries (LICs) have a GNI less than US$1,025.

Figure 4 shows the distribution of Middle Income Countries. This large group, which the World Bank divides into two, includes **newly industrialised countries (NICs)** such as Brazil, India, Mexico and China.

Newly industrialised country (NIC)

A country with a large percentage of the workforce employed in manufacturing. Most NICs are middle income countries and their economies are strong and growing. They often attract investment from multinational companies.

Activities

1 Use Figure 4 to describe the distribution of:
   a) Upper Middle Income Countries
   b) Lower Middle Income Countries.
2 Study Figures 6 and 8.
   a) Classify each country in 2015 by its World Bank category.
   b) Use a suitable graph to represent the data.
   c) How much progress has each country made in closing the development gap?
   d) What conclusion can you make about the progress of these two groups of countries in closing the development gap?

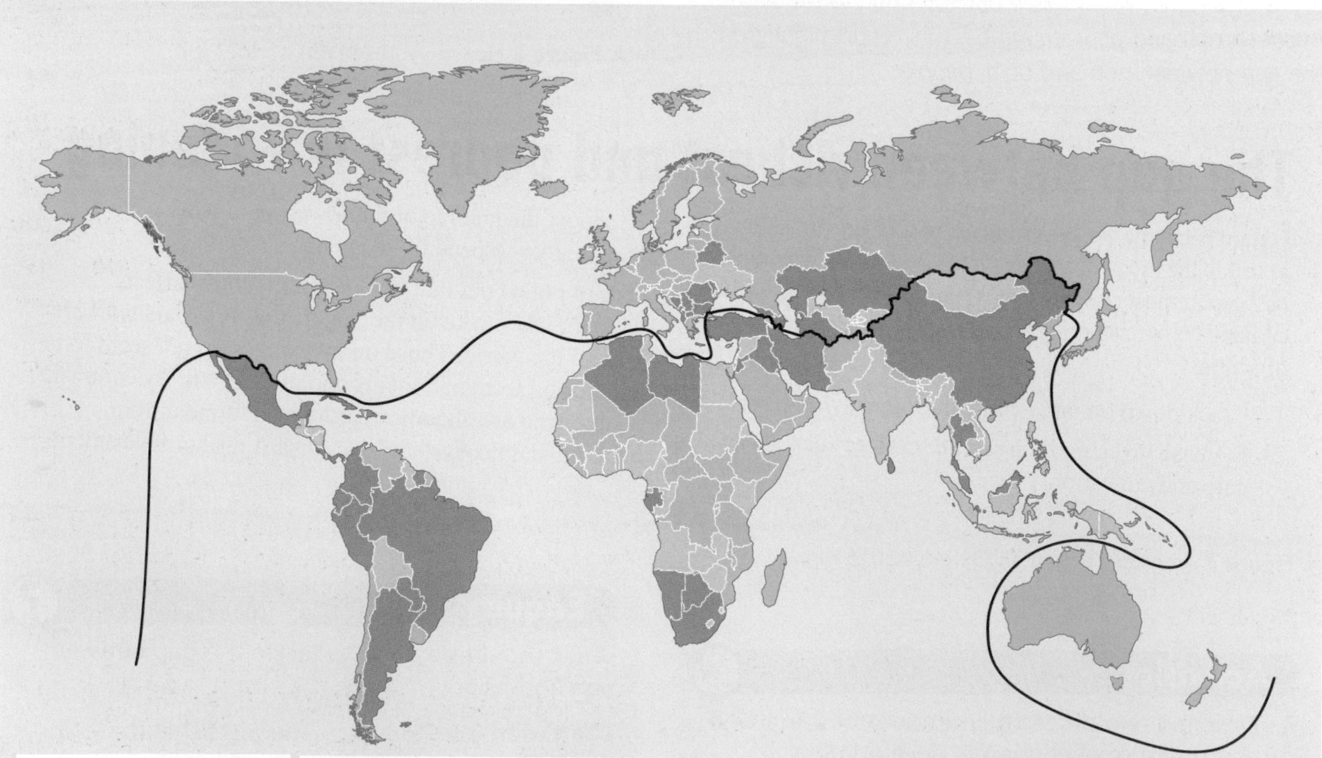

Key
Upper middle income
Lower middle income
Brandt line

World Bank category	GNI US$ per person	Number of countries in category (2018)
Upper middle income	US$3,996 to US$12,375	60
Lower middle income	US$1,026 to US$3,995	47

▲ **Figure 4** The world's Upper Middle and Lower Middle Income Countries; colours on this map indicate GNI per capita.

Is the development gap closing?

Figure 1 on page 210 shows the distribution of HICs and LICs. They clearly lie on either side of the Brandt Line – does this mean nothing has changed in 40 years? Figure 4 shows a more complex pattern. NICs such as India have experienced rapid economic growth since 1980. Investment in education has created a skilled workforce. This has encouraged investment by foreign-owned **multinational companies (MNCs)**, creating jobs in the manufacturing industry. India's manufacturing sector now employs 22 per cent of the total workforce. Chemicals, electronics and engineering are all growth industries that employ millions of highly skilled and well-paid workers. The development of global cities in India, such as Mumbai (see pages 90–7) has connected India to the global economy – encouraging further investment. Foreign MNCs such as Siemens, IKEA, Samsung and Suzuki have all recently invested in factories in India. Indian-owned MNCs, such as Tata, are now investing Indian money in other countries – including the UK. Modern India has a complex economy – it earns foreign income by exporting a wide range of goods and providing a range of services.

▲ **Figure 5** Wealth and poverty in India.

	1985	1990	1995	2000	2005	2010	2018
China	290	330	540	930	1,750	4,300	9,470
India	300	390	380	450	730	1,260	2,020
Indonesia	520	610	990	560	1,220	2,530	3,840
Mexico	2,130	2,750	4,570	5,750	7,720	8,720	9,180

▲ **Figure 6** GNI per capita for selected newly industrialised countries (NICs).

However, other countries have made much less economic progress. Malawi, an LIC in southern Africa, is one example. Malawi is a **landlocked** country – without a coastline it has been unable to develop a port and its global connections with the world economy are much fewer than those of India. Malawi's economy depends on the export of a very small range of farm products including tobacco, tea, cotton and sugar. Unprocessed farm products like these have a low value compared to the processed products that Malawi imports, such as oil and farm machinery.

▲ **Figure 7** A tea picker working in Malawi.

	1985	1990	1995	2000	2005	2010	2018
Gambia	310	320	740	670	410	580	700
Kenya	310	380	270	420	530	1,000	1,620
Mali	150	270	250	230	400	610	830
Malawi	160	180	160	160	220	350	370

▲ **Figure 8** GNI per capita for selected sub-Saharan African countries.

Activity

3 Use the information on this page to suggest why Malawi has failed to make as much economic progress as India.

Multinational companies (MNCs)

Businesses that have branches, mines, factories or offices in several different countries. MNCs take advantage of cheaper labour costs in countries that have lower GNIs. Having branches in different countries makes it easier to expand sales of their products into those countries.

What is globalisation?

People and places around the world are linked by the process of globalisaton. Close links are made between countries through trade, the movement of people, and the investment of MNCs. Constantly improving communication technologies is one factor that helps to make these links. Globalisation helps to create economic growth. But do some people and places benefit more than others from this process?

Example: Avocados grown in Mexico will be flown to a UK supermarket. This improves customer choice but food air miles have an impact on carbon emissions.

Example: Local people demonstrating about the falling water levels in their wells after this soft-drinks company opened a bottling plant in Kerala, India.

The factors that drive globalisation

Multinational companies (MNCs): Large companies open branches in several different countries throughout the world.

Communication: The growth of fibre and satellite communication technology has increased global connections via smart phones, the internet and satellite television.

Trade: Improved technology and cheap aviation fuel mean that fresh food can be flown from distant places to our supermarkets.

Culture: Western styles of music, TV and film are released throughout the world at the same time.

Example: Films made in Hollywood, USA, advertised in an Asian street. But will local styles of music and entertainment survive?

▲ **Figure 9** Factors that drive globalisation.

Activities

1. Explain why globalisation is more rapid now than ever before.
2. Use evidence on this spread to make a list of the advantages and disadvantages of globalisation for ordinary people and for businesses.
3. To what extent has India benefited from globalisation? Use evidence from Figure 10.

Globalisation

A complex process that links people and places across the globe through business.

How do newly industrialised countries benefit from globalisation?

MNCs create economic benefits. Newly industrialised countries (NICs) such as India and China have benefited from globalisation with rapid economic growth. In the 1990s, they offered cheap labour so their manufacturing industries boomed. Their economies have benefited from changes in technology (e.g. container shipping) and from investment of capital by their governments and by multinational companies (MNCs). Now they are becoming centres of the world's largest companies – just look at how rapidly the growth of companies in Asia now puts them in the Forbes 'rich list' of the top 2,000 companies.

Flows of people: Indian migrants work in many parts of the world, earning money which is often sent home to the family in India. In 2017, 829,000 people who were born in India lived and worked in the UK.

Rank	Country of birth of resident UK population (2017)	Population (1,000s)
1	Poland	922
2	India	829
3	Pakistan	522
4	Romania	390
5	Republic of Ireland	390
6	Germany	318
7	Bangladesh	263
8	Italy	232
9	South Africa	228
10	China	216

Forbes Top 2000 Companies List

Rank in 2015	Rank in 2008	Country	Number of companies
1	1	United States	525
2	4	China (with Hong Kong)	301
3	2	Japan	210
4	3	United Kingdom	93
5	8	South Korea	67
6	10	India	58
7	5	France	57
8	7	Germany	54
9	6	Canada	51
10	-	Taiwan	47

Foreign investments: Indian companies, such as Tata, are very successful in the world economy. In 2018, India's companies had taken it from 10th to 6th in the Forbes Top 2,000 companies.

Flows of ideas and culture: The Hindi movie industry based in Mumbai (known as Bollywood) produced over 1,986 films in 2016. These films are extremely popular in South Asia and, with the growth of satellite TV, are easily accessible in other parts of the world.

	Mobile subscriptions (millions)	Mobile subscriptions (% population)	3G/4G subscriptions (% population)
World	6587.4	93	27
China	1246.3	92	33
India	772.6	62	3
United States	345.2	110	92
Indonesia	285.0	115	18
Brazil	272.6	137	55
Russia	237.1	165	29
Japan	137.1	108	85
Nigeria	128.6	76	8
Vietnam	127.7	144	20
Pakistan	126.1	70	-
Bangladesh	116.0	75	22
Germany	113.6	139	56
Philippines	109.5	113	17
Mexico	102.7	95	16

Improved communication technologies: India and China dominate the global communications market with over 2 billion customers. India produces thousands of IT and software graduates from its universities each year.

▲ **Figure 10** Examples of the growth of NICs and their influence in the global economy.

Globalisation: Nike are just doing it

Every year, the Forbes Directory researches and puts a value on every major company in the world. It publishes a Top 100 called 'The World's Most Valuable Brands', which tries to put a value on brand names. It also groups them by category, such as technology or clothing. In 2019, the most valuable clothing company in the world was Nike. Forbes valued its 'brand', that is to say the value of the name if it were to be sold, at US$36.8 billion. Most people immediately associate the 'swoosh' logo as Nike, and their 'Just do it' slogan is also an asset. Nike's global sales in 2018 were US$36.7 billion, that's bigger than the gross national income (GNI) of some countries! US$9 out of every US$10 the company earned came from merchandise with Nike logos.

Nike is a multinational company (MNC) because it operates and sells in over 140 countries around the world. 41 of these countries manufacture Nike products, as Figure 11 shows. In 2018, it employed 73,100 people directly worldwide. It provided jobs for 20 times this number in factories which are under contract to make Nike products. This process is known as **outsourcing**, where companies work for Nike for a period of time under a signed contract. It gives Nike one big advantage – that it can negotiate on price. In 2012, China was Nike's biggest source of outsourced workers. By 2015 China had lost its place to Vietnam, because China's currency has increased in value, which made its products more expensive.

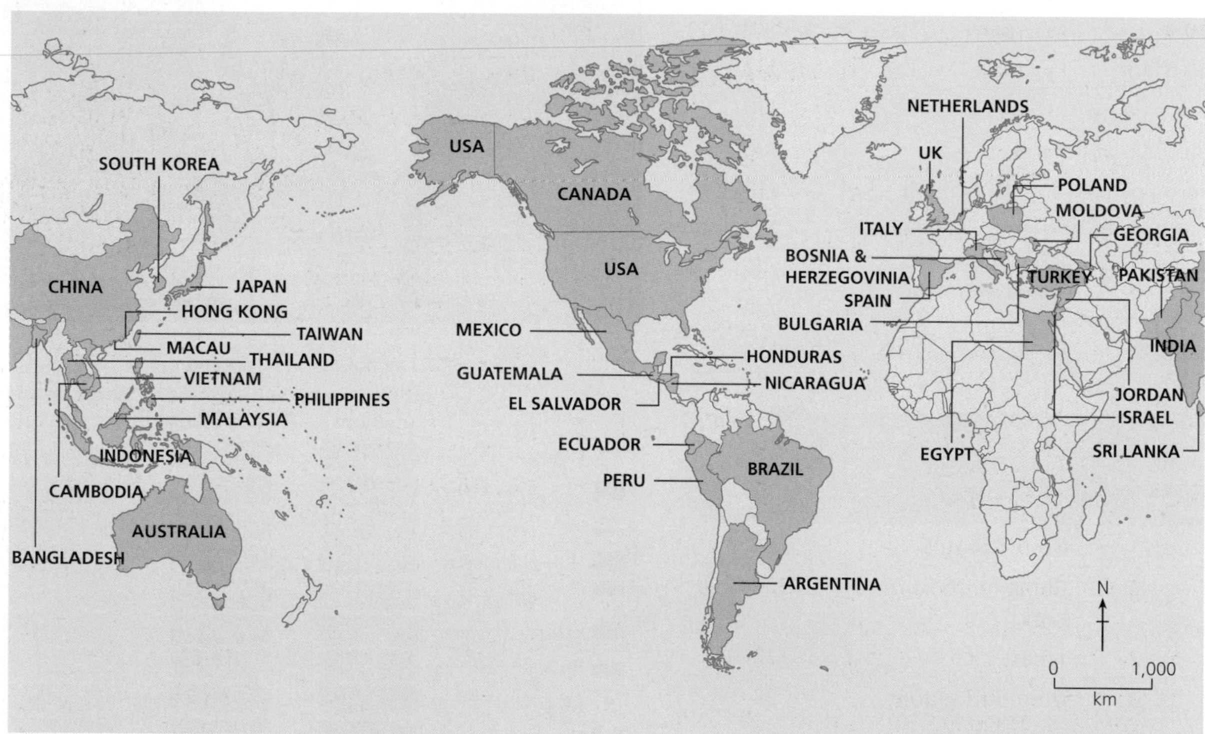

Japan: 1960s A factory making training shoes opened in Japan because labour was so cheap at the time. It was probably one of the world's first outsourced factories.

Thailand and Indonesia: late 1980s South Korean-owned companies making Nike products moved their manufacturing into Thailand and Indonesia in search of cheaper labour.

Taiwan and South Korea: 1970s Nike opened up its manufacturing here, using local companies who were contracted to make sportswear.

Vietnam: 2010s Now that China's currency is worth more, it is cheaper to make many items in Vietnam, so Vietnam has become the largest producer. Companies working for Nike in Vietnam employed over 140,000 more people between 2012 and 2015. China lost 43,000 jobs in the same period.

China: 1980s Nike began production here, taking advantage of cheaper labour in China as labour in South Korea and Japan became more expensive.

USA: today The location of Nike Head Office: Nike World Campus in Beaverton, Oregon. All research and development, and new product design is done in the USA.

▲ **Figure 11** World map showing the countries in which Nike manufactures its products.

Although small amounts of Nike clothing are made in the USA and Europe, almost all factories are located in Asia (Figure 12). Most factories employ a majority of women because pay is rarely equal for women compared to men. In Vietnam, the largest producer, many workers are migrants from rural areas. Workers live in hostels owned by the factory.

Manufacturing country	Manufacturing workers 2015	Workers + or -
1 Vietnam	341,204	Up 143,000
2 China	228,732	Down 43,000
3 Indonesia	186,425	Up 60,000
4 Sri Lanka	33,587	Up 13,000
5 Thailand	31,770	Down 20,000
6 India	28,165	Down 4,000
7 Honduras	26,090	Up 16,000
8 Brazil	20,935	No data
9 Bangladesh	15,090	No data
10 Pakistan	14,899	Up 4,400
Global total	1,016,657	

▲ **Figure 12** Nike's ten largest manufacturing countries by number of employees.

Outsourcing

When a business gets some of its goods, components or services from an outside supplier.

Activities

1 a) Represent the information in Figure 12 using a suitable map or graph.
  b) What are the strengths and limitations of your chosen technique?
2 Suggest why Nike contracts companies to do its work instead of opening its own factories.
3 Explain the risks for Nike manufacturing overseas.
4 Why do many countries such as Vietnam want to attract MNCs such as Nike?
5 a) Using Figure 13, calculate the percentage of a US$65 pair of trainers that goes to
    i) the country of manufacture
    ii) the USA
    iii) the retailer.
  b) Calculate the percentage of US$65 that consists of profit.

Nike and outsourcing

The company we know as Nike began in 1964 when Phillip Knight, a former athlete, began manufacturing training shoes in Japan, because of its cheap labour, and importing them to the USA. But all management operations and decisions – from design to finance and marketing – operate from the USA. The jobs in Nike's Asian and Central American companies are low-paid and unskilled. You can calculate from Figure 13 just how much of a US$65 pair of trainers goes to the manufacturing country, and how much goes to the headquarters in the USA.

	Cost
Production labour	$2.50
Materials	$9.00
Factory costs	$3.25
Supplier's operating profit	$1.00
Shipping costs	$0.50
Cost to Nike	$16.25
Nike costs (Research and development, promotion and advertising, distribution, admin.)	$10.00
Nike's operating profit	$6.25
Cost to retailer	$32.50
Retailer's costs (rent, labour, etc.)	$22.50
Retailer's operating profit	$10.00
Cost to consumer	$65.00

▲ **Figure 13** Where the money goes: breakdown of a US$65 pair of Nike training shoes.

Enquiry

What are the common features of MNCs?

Compare Nike with a company of your choosing, such as Apple, L'Oréal or Coca-Cola. For your chosen company, research:

- where it manufactures in different parts of the world
- their methods of outsourcing
- who they use as outsourced workers
- the benefits and problems that each company gets from outsourcing.

What are the benefits and problems brought by MNCs?

MNCs create economic benefits. If labour costs are low, then a product can be made more cheaply and the price of the product falls. American, European and Asian consumers gain through these cheaper goods.

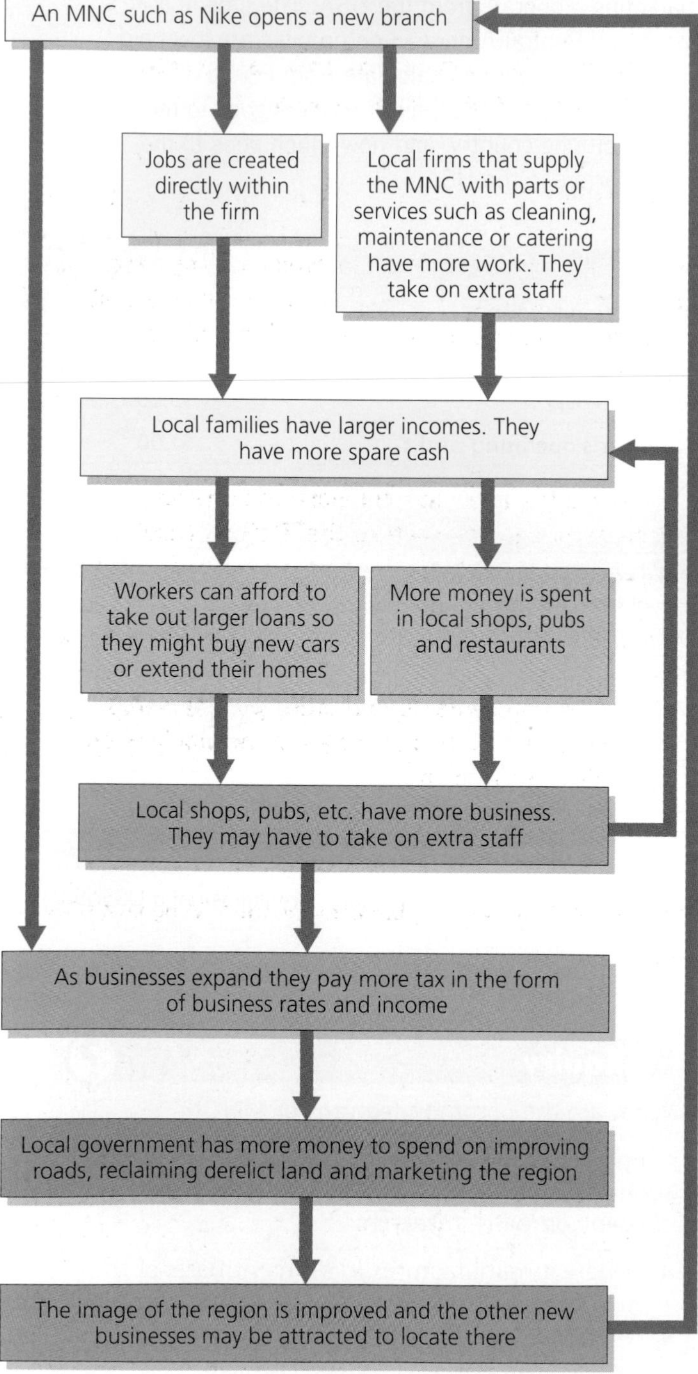

▲ **Figure 14** The multiplier effect – how investment brings growth that builds on itself.

The countries, like Vietnam, in which MNCs such as Nike operate also benefit as factories are built, creating jobs and further investment. This creates a demand for housing, which in turn leads to demand for shops and services to cater for the working population. These have to be built, which creates a huge increase in employment in the building industry. New networks develop for transport, energy and water supplies, known as infrastructure. The process creates an upward spiral, known as the **multiplier effect**.

Multiplier effect

When economic change creates a chain reaction of cause and effect that ripples through the economy. Multiplier effects are positive after money is invested. They can be negative if a business has to close.

The downside to MNC investment

Many NICs have few workers' rights (e.g. maximum working hours) and no minimum wage. The governments of these countries think a minimum wage might put companies off from investing there. Thailand introduced a minimum wage of US$10 a day in 2013. Within two years, Nike had cut the number of jobs there by a third. As wages increase in one country, an MNC might decide to move production to another country where wages are lower. Wage levels in 2015 were as low as US$1 a day in Indonesia.

Activities

1 Copy and complete the following table to show the advantages and disadvantages brought by MNCs such as Nike for:
   a) the host countries
   b) the MNC.

	Advantages	Disadvantages
For the host country (e.g. Vietnam)		
For the MNC (e.g. Nike)		

2 Explain why, in spite of disadvantages, an NIC might decide to attract investment from MNCs such as Nike.

3 In pairs, decide whether you think Nike should have invested in Vietnam in 1995. Explain your reasons.

Nike in Vietnam

Nike began producing sportswear in Vietnam in 1995. Vietnam was one of the world's poorest countries. But its people had a reputation for hard work from other companies who had already invested there. Costs were much lower:

- There were no trade unions, or strikes for higher pay, so wages were low.
- The Communist government was keen to develop exports.

Getting to know Vietnam

Vietnam has become one of South East Asia's most rapidly industrialising and growing countries. It has a Communist government which is keen to raise the standard of living through the growth of manufacturing and services. The government began to open up opportunities for investment by overseas companies in the mid-1990s. It began by welcoming travel companies and overseas tourists. More recently a programme of industrialisation has rapidly increased Vietnam's GNI and other economic indicators (shown in Figure 15).

	Vietnam 2000	Vietnam 2018
GNI per person (in US$) PPP	2,070	6,220
Where GNI comes from (%)	Agriculture: 26 Industry: 33 Services: 41	Agriculture: 15 Industry: 34 Services: 51
Percentage of people by occupation	Agriculture: 67 Industry and Services: 33	Agriculture: 40 Industry: 26 Services: 34
Value of exports (US$)	11.5 billion	214 billion
Export goods (in rank order	Crude oil, seafood, rice, coffee, rubber, tea, garments, shoes	Clothing, shoes, electronics, seafood, crude oil, wooden products, machinery, rice

▲ **Figure 15** Comparing the economy of Vietnam in 2000 and 2018.

Enquiry

What should Nike do? Write a letter to the chief executive of Nike to show how you feel about reports of abuses in NICs such as Vietnam or Indonesia.

Nike became Vietnam's largest foreign employer with ten factories employing 40,000 people. In 2015, it employed 300,000 people. But Nike made mistakes in the late 1990s:

- All its sub-contracted companies in Vietnam were South Korean and Taiwanese, so all profits went overseas, instead of Vietnam.
- Factories gained reputations for sweatshop conditions. There were stories of abuse of workers.

The Vietnamese Government attacked Nike for its practices. Nike found itself in the newspapers every day. Some stories were later found to be false, but by then the stories had reached overseas. The harm was done. Nike told the Vietnamese Government that if attacks continued, it would leave. The criticism stopped.

Many pressure groups have tried to persuade Nike to improve workers' conditions. Many consumer groups want to see fair treatment for workers overseas. Pressure from groups such as the 'Boycott Nike' campaign in the early twenty-first century also came from US Congress. Nike now argues they have improved factory conditions, working hours and wages. Their workers live in hostels which are high quality, and there are agreements to restrict working hours. Nike now publishes data about all of its suppliers on its website, nikebiz.com.

▼ **Figure 16** Photo showing protestors in the USA against Nike.

Trade

Speaking in the 1960s, Martin Luther King (seen in Figure 25) reminded us that countries rely on each other for the **import** and **export** of goods and services. Since he spoke, larger, faster and more fuel-efficient aircraft and container ships move goods around the world. Globalisation brings everything from strawberries in December to the latest smart phones from China. Increasingly, countries rely on goods traded with others. Trade is made easier where partnerships have been agreed between countries. These trading partnerships are known as **trade blocs**. The European Union (EU), G20 and APEC are all examples.

Before you finish eating breakfast this morning, you've depended on more than half the world.

▲ **Figure 25** Martin Luther King, the American civil rights leader.

Figure 26 describes the trade of Malawi and Mexico:

- Malawi is an LIC in sub-Saharan Africa, whereas Mexico is an NIC.
- The value of Mexico's exports is 480 times greater than that of Malawi.
- Malawi's exports are mainly tobacco, tea and other raw materials. Malawi's imports are mainly manufactured goods, such as chemicals or machines.
- Mexico exports and imports a wide variety of manufactured goods.

Manufactured goods are worth more than raw materials. Value has been added by designing, making and marketing products that consumers and other manufacturers need. This is why countries that export a lot of manufactured goods earn more than those who export mainly raw materials.

Imports

Goods that are bought from other countries.

Exports

Goods that are sold to other countries.

Trade bloc

A group of countries that agree to buy and sell goods without any restrictions or tariffs (taxes).

Malawi		
Exports (total US$869 million)	**US$ (m)**	**Per cent**
Tobacco	620	71
Tea	90.8	
Other foodstuffs	28.3	
Plastics and rubber	8.77	
Wood products	8.47	
Imports (total US$1,440 million)	**US$ (m)**	**Per cent**
Chemical products	334	23
Machines	290	
Metals (mainly iron)	128	
Foodstuffs	108	
Vehicles (mainly trucks and cars)	102	

Mexico		
Exports (total US$419 billion)	**US$ (m)**	**Per cent**
Machines (including computers)	156	37
Vehicles	112	
Minerals (including crude oil)	27.9	
Instruments (including medical)	15.5	
Foodstuffs (including beer)	13.8	
Imports (total US$356 billion)	**US$ (m)**	**per cent**
Machines	116	33
Vehicles	43.9	
Metals	34.3	
Plastics and rubber	29.4	
Chemical products	28.3	

▲ **Figure 26** Comparing trade data for Malawi and Mexico (2017).

Free trade

Developing countries rely on trade to increase their GNI. Since the 1980s, the world has moved towards **free trade**, that is, trade without limits, duties or controls. Each country within a trade bloc has a free trade agreement with other countries in the bloc, or is working towards a free trade agreement. Countries can export as many goods as they wish. This is good for producers who export goods and services. The disadvantage of free trade is that a country can find itself swamped by cheap imports from other countries, with job losses in its own producing industries.

To avoid this, some countries use **protectionist policies**. These include:

- Placing **quotas** that restrict the amount of imports each year.
- Placing **tariffs** on imports. This is a tax on imports to make them more expensive.
- Paying **subsidies** to businesses so that their own goods can be sold at lower prices.

Ghana joined the World Trade Organization (WTO) in 1995. The WTO exists to promote free trade between its members. Until it joined, Ghana's government had paid farmers a subsidy to encourage them to grow food for Ghana's cities.

It sounds fine. But WTO rules are that farmers cannot be subsidised, even though the USA and EU (who are members of WTO) pay subsidies to their own farmers. These subsidies make American and European food cheaper. Farmers in Ghana suffer because of imports of subsidised EU food.

- Ghana's tomato farmers find it hard to sell their produce, because imported EU tomatoes are cheaper, and factories canning Ghanaian tomatoes have closed.
- Ghana's rice growers have been affected by cheap imported rice from the USA.

Protectionist policies

Schemes that are used by countries to protect their own industries from cheap imports. For example, by placing a tariff on imports, the imported goods become more expensive, so people are more likely to buy the cheaper product that has been made in their own country.

▲ **Figure 27** Most Ghanaian farmers sell at street markets. Farmers find it difficult to compete with the low prices of these imported tomatoes.

Activities

1 In pairs, research the internet to find out where breakfast items come from, e.g. cereal, orange juice, tea, coffee.
2 Use Figure 26 to calculate the value of each import and export as a percentage of the total. The first row in each table has been done for you.
3 Explain the advantages and disadvantages of farm subsidies for:
   a) consumers in Europe
   b) tomato growers in Ghana.

Enquiry

Are subsidies fair? Write a 300-word letter to the WTO, arguing why either **a)** subsidies to EU and US farmers should be allowed to continue, or **b)** subsidies should be abolished.

How important is tourism to Mexico's economy?

Mexico is an NIC. It has a large and complex economy with exports in a wide variety of manufactured goods. Around 39 million foreign tourists visited Mexico in 2017. Most tourists came from the USA (61%), Canada (11%) and the UK (3%). About 3.6 million Mexicans work directly in travel and tourism, and this is expected to rise to 4.8 million by 2025. The growth of tourism creates jobs in construction and leads to improved infrastructure, such as airports and ports. The industry itself creates **direct employment** in travel (for example in airports and on aircraft) and in hotels and restaurants. It also creates **indirect employment** – these are existing jobs that benefit because of the arrival of tourists. For example, demand for food in hotels supports jobs in agriculture and fishing. In 2017, the average long-stay tourist in Mexico spent US$909 while visiting the country. This is money that can create an economic multiplier.

Activities

1 Working with a partner, make a list of the kinds of jobs that are created by tourism:
 a) directly
 b) indirectly.

2 Describe the location of Mexico's tourist resorts/regions of:
 a) Cancun
 b) Acapulco
 c) Baja California.

3 Use Figure 39.
 a) Use the data to draw a suitable graph.
 b) Describe how the number of tourists visiting Mexico has changed.

▲ **Figure 37** A street vendor selling fruit, Cancun. One example of how tourism can boost employment indirectly.

Direct employment

Jobs created within the industry itself, in this case, jobs in hotels, tourism and airports.

Indirect employment

Jobs created within businesses that supply goods and services to the industry. In this case, the tourism industry in Mexico is supplied by taxi drivers, food suppliers, fishermen and farmers.

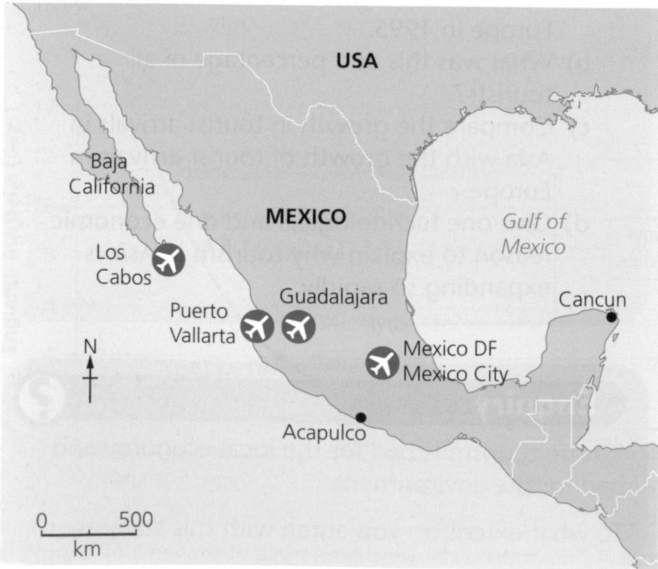

◀ **Figure 38** The location of Mexico's busiest international airports and the mass tourist resorts at Cancun and Acapulco.

Year	Million
2010	23.3
2011	23.4
2012	23.4
2013	24.2
2014	29.3
2015	32.1
2016	35.1
2017	39.3

▲ **Figure 39** International tourist arrivals to Mexico.

The growth of mass tourism in Cancun, Mexico

In the 1970s Cancun was a tiny fishing village. Cancun is now the largest mass-tourist resort in the Caribbean. Cancun's hotel zone has over 150 hotels with a total of 35,000 rooms. This is expected to grow to 46,000 by 2030. Three million visitors to Cancun add US$4.36 billion to Mexico's economy each year.

Cancun's hotel zone is separate from the local community. It was planned like this as a **tourist enclave**. While the tourists stay in air-conditioned comfort in the hotel zone, the hotel workers live in the separate city of Cancun. Local people aren't even allowed on the beaches. The beaches are owned by the state but access is controlled by the hotels. They keep locals away to prevent tourists from being hassled by vendors who try to sell food, drinks or their services as guides.

Enclave tourism

Hotels often sell 'all-inclusive' packages. This means that tourists pay one price and get all their food, drink and entertainment from the hotel. Cruise ships offer the same kind of deal. In either case, their customers have no real need to leave the hotel or cruise ship. A consequence of enclave tourism is that tourists do not visit or spend money in locally owned bars or restaurants. If they did, less money spent by tourists would leak away from the local economy.

Activities

4 Study Figure 40.
   a) Suggest a new caption for this cartoon.
   b) Justify the view that enclaves are bad for both tourists and local people.
   c) Suggest why the Mexican government developed Cancun as an enclave.

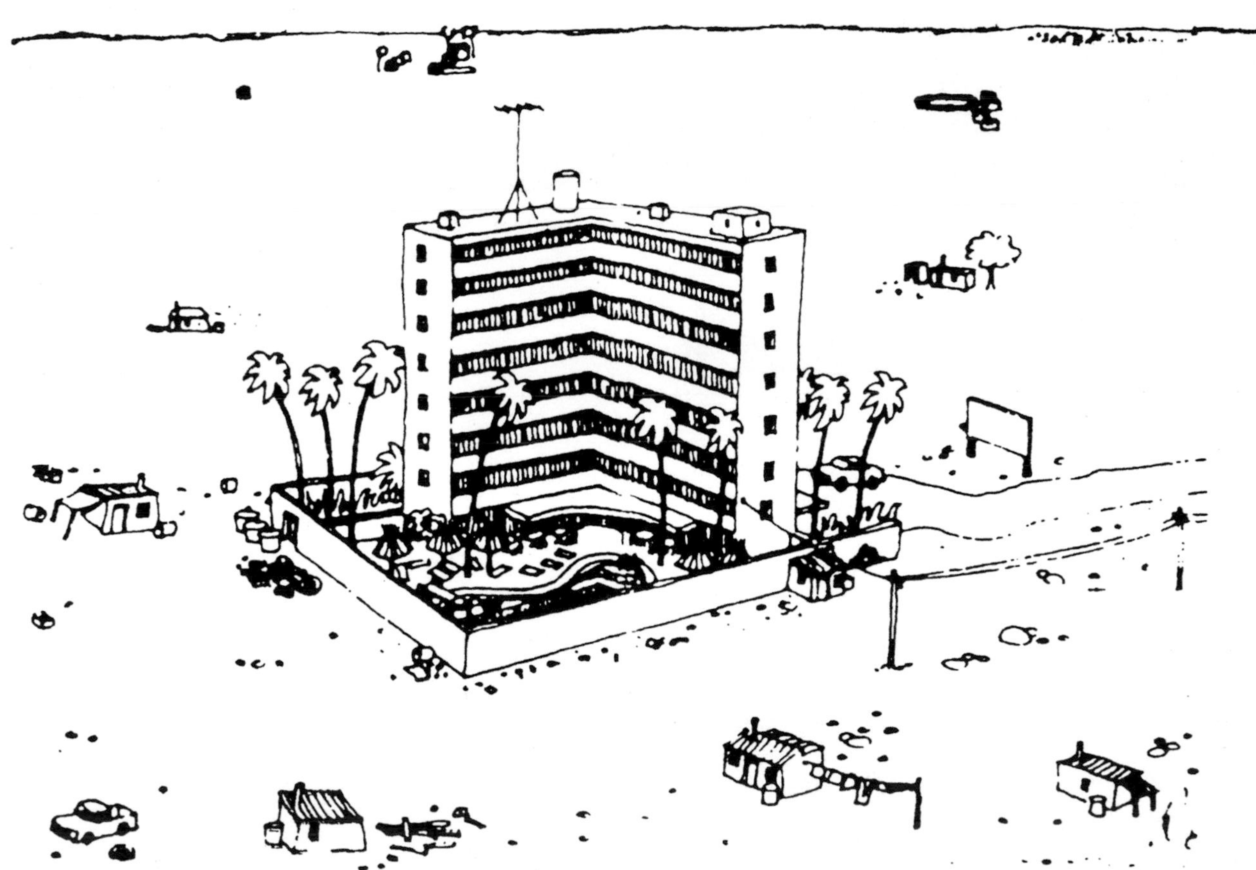

"Dear George, here we are in the middle of things having a great time. We feel we're really getting to know this exotic country . . . "

▲ **Figure 40** Enclave tourism.

Has tourism been good or bad for Cancun?

Tourism creates direct employment for 52,000 people in Cancun and as many as 175,000 jobs indirectly. However, many workers in the large hotels are only offered short, temporary contracts. This means that workers can be laid off at the end of a one- or three-month contract if the seasonal pattern of visitors means there is insufficient work.

▲ **Figure 41** Workers in Cancun's tourist industry in front of their home in an informal settlement on the outskirts of Cancun.

> I work long hours but only earn about US$5 a day. I can't afford a decent home. I live in a rented shack. It costs US$80 a month. I share the outside toilet with neighbours and the tin roof leaks when it rains heavily. The tourists have everything they need, but we have no space or leisure facilities. We're not even allowed on the beach!

Hotel worker

> Many people who work in the hotels of Cancun are migrants from other parts of Mexico. They suffer because they are separated from their families and original communities. Most of them are badly paid and rely on tips to make up their wages. Many work long hours and have stressful working conditions. Some suffer from alcohol or drug abuse.

Social worker

> We use security guards to keep locals off the beach. The problem is that some 'beach boys' hassle the tourists. They try to sell fast food or souvenirs. Sometimes there have been problems with drug dealers or even muggings.

Hotel manager

▲ **Figure 42** Views on the development of tourism in Cancun.

Activities

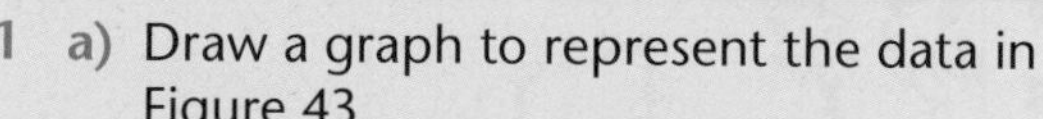

1 a) Draw a graph to represent the data in Figure 43.
  b) Using Figure 41, and your understanding of other global cities, suggest the problems that rapid urban growth may cause in Cancun.
2 Discuss the points of view in Figure 42.
  a) Identify three different issues created by tourism in Cancun.
  b) What are the main causes of these issues?
  c) Suggest possible solutions to one of these issues.
3 Summarise what you consider to be the advantages and disadvantages of mass tourism for the local people and economy of Cancun.

Year	Cancun	Acapulco
1970	0	178
1980	34	304
1990	192	658
2000	438	794
2010	680	864
2020	993	944
2030	1,161	1,075

▲ **Figure 43** Population of Cancun and Acapulco (1,000s).

The impacts of tourism on Cancun's environment

Cancun is built on a long thin barrier of sand – the Cancun–Nizuc barrier. The lagoon behind this barrier was largely filled with mangrove forest. Offshore is another important habitat – a tropical coral reef. Scientists are concerned that all three parts of this environment have been damaged by the growth of tourism:

- Sewage from the resort has caused the spread of disease in the coral reef.
- Sand was taken from the beach during the 1970s and 1980s to be used in the construction of the hotels.
- Mangrove trees have been cleared away and the lagoon has become polluted.
- Plants that naturally grow along the top of the beach, which help to hold the sand together, have been stripped away. This makes the beaches more attractive to tourists but it means that sand is more easily eroded during storms.
- The high-rise hotels are too heavy for the soft sediments beneath them. Some hotels have had structural problems due to subsidence.

Coral reefs, beaches and mangroves all act as natural buffer zones during a tropical storm. Scientists believe that damage to these environments has left Cancun at greater risk to natural disasters. Cancun has been hit by several storms. The worst was Hurricane Wilma (2005), which caused US$19 million damage to Cancun's hotel zone. Wilma eroded all of the sand from the 20 m wide beach. The beach was replenished in 2016 at a cost of US$19 million.

Activities

4 Describe three ways that the development of Cancun has damaged the environment.

5 a) Make a sketch of Figure 44.
  b) Label the tombolo (the thin spit of sand), the beach and the lagoon.
  c) Add an annotation to explain why the hotel zone is at risk of tropical storms.

Enquiry

'The resort of Cancun has no long-term future.'

Make use of evidence from these pages to decide whether or not you agree with this statement. Explain your decision carefully.

▲ **Figure 44** Cancun's hotel zone.

Tourism in the Gambia

The Gambia is a low income country (LIC) in West Africa. Poverty, poor health care and education place this small country among the poorest 25 in the world. The UN estimates that 40 per cent of the population live in poverty.

About 75 per cent of the population are employed in agriculture, although farm products only contribute about 20 per cent of Gambia's wealth. Peanuts are the main export and tourism is Gambia's second highest earner. The Gambia has few natural resources and only a very limited manufacturing sector. So, with golden beaches and a warm climate, Gambia has become dependent on tourism. Tourism contributes about 40 per cent of all national wealth. It is estimated that between 20,000 and 35,000 people are employed directly in tourism. At least another 35,000 are employed indirectly. The Gambia relies on the money it earns from tourism to import basic food stuffs such as rice.

The first tour operator began flying tourists to the Gambia, from Sweden, in 1965. Today, most tourists come from the UK, Germany, Belgium, the Netherlands, Denmark, Sweden and Norway. The Gambia is only six hours flight time from most European cities. It is in the same time zone as London, so tourists don't get jet-lagged.

Dependency

When a country's economy relies on a particular industry. For example, the economies of most Small Island Developing States (SIDS) rely heavily on tourism. Many of these small countries have few exports, so money earned by tourism represents more than 30 per cent of their total exports. By comparison, the average amount of money earned by tourism for the world is just over five per cent. The Maldives, Seychelles and Jamaica are particularly dependent. Some LICs, such as the Gambia, also rely very heavily on tourism compared with other ways of earning foreign income. Dependency on tourism is risky for an economy. A natural disaster such as a cyclone, earthquake or tsunami can severely reduce visitor numbers. Political instability, conflict, disease and economic recession can also affect visitor numbers.

Enquiry

How dependent is the economy of the small island developing states (SIDS) on tourism?

- Research the location of the SIDS.
- Choose one SID to research further. How many tourists visit? What are the positive and negative impacts of tourism here?

▲ **Figure 45** Informal work is common in the Gambia; here, a vendor tries to sell souvenirs to tourists.

Is tourism good for the Gambia?

Gambia has a seasonal climate with a rainy season during the summer months. Tourism is also seasonal, with most European tourists visiting during the winter from October to March. This means that many tourist workers are under-employed or unemployed for several months of the year and they consequently rely on work in the **informal economy**. Figures for unemployment and informal employment are unreliable, but both are very high.

The Gambia cannot rely on earning the same income from tourism each year. Tourists want to feel safe and secure when they are on holiday. They are fearful of problems such as disease, terrorism and political unrest. If a destination seems to be unsafe, it is very simple to book a flight to another destination. In recent years some tourists have been put off visiting the Gambia by the country's political unrest. The country's leader has been very outspoken on some topics, including the rights of gay and lesbian people. His views have triggered protests that have been reported in European news broadcasts. In 2014, parts of West Africa were affected by Ebola. There were no cases in Gambia but tourist numbers were exceptionally low in 2014.

Hotel owner in the Gambia

Our hotel is usually 80–95 per cent full during the winter months. But during the winter of 2014 it was only 30 per cent full. I think it will take at least three years for our business to recover.

Tourist from the UK

I think tourists are frightened to come here. They know that Ebola has affected West Africa, but they don't realise that West Africa is twelve times the size of the UK and only some parts are have been affected.

Hotel cleaner

Things are much worse when we have a bad tourist season. Obviously there is less work, so I have less money to buy food. The price of rice and oil, which I use for cooking, also goes up!

▲ **Figure 46** Views on the Gambia's 2014 tourist season.

Year	Tourists
2000	79,000
2001	57,000
2002	81,000
2003	89,000
2004	90,000
2005	108,000
2006	125,000
2007	143,000
2008	147,000
2009	142,000
2010	91,000
2011	106,000
2012	157,000
2013	171,000
2014	156,000
2015	135,000
2016	161,000
2017	162,000

▲ **Figure 47** Annual visitor numbers to the Gambia.

Informal economy

Part of the economy that is not regulated by the state. It includes irregular jobs in tourism such as street sellers, informal guides and sex workers.

Activities

1 a) Use an atlas to find the location of the Gambia.
  b) Draw a sketch map and label it to show why it is popular with tourists from Europe.
2 a) Explain why the tourist industry in Gambia is seasonal.
  b) What problems does this create?
3 a) Compare tourism in the Gambia and Cancun. Include facts and figures.
  b) Explain why the Gambia is more heavily dependent on tourism than Mexico.
  c) Explain why this is a problem for:
    i) the Gambian government
    ii) ordinary people in the Gambia.
4 a) Choose a suitable technique to represent the data in Figure 47.
  b) Describe the trend shown on your graph.
  c) Explain why visitor numbers fell so sharply in 2014.
  d) Give one other reason why visitors from Europe may stop visiting the Gambia.

Investigating long-term aid

We are familiar with the TV appeals after there has been a natural disaster such as a major earthquake. Such aid is **emergency aid**. However, most aid is actually given as **long-term aid**. One LIC that receives long-term aid is Malawi.

Emergency aid

Help given urgently after a natural disaster or conflict.

Long-term aid

Aid planned over long periods of time to tackle poverty and improve health and education.

Getting to know Malawi

Malawi remains one of sub-Saharan Africa's poorest countries. Although keen to develop manufacturing and services, there are several barriers:

- It is landlocked.
- It exports little compared to other countries.
- It has suffered major impacts from HIV/AIDS.

	Malawi 2000	Malawi 2018
GNI per person (in US$) PPP	490	1,310
Population living below poverty line (%)	54	53
Where GNI comes from (%)	Agriculture: 37 Industry: 29 Services: 34	Agriculture: 29 Industry: 15 Services: 56
Percentage of people by occupation	Agriculture: 86 Industry and services: 14	Agriculture: 77 Industry and services: 23
Value of exports (US$	0.5 billion	1.42 billion
Export goods (in rank order)	Tobacco, tea, sugar, cotton, coffee, peanuts, wood products	Tobacco (55%), tea, sugar, cotton, coffee, peanuts, wood products

▲ **Figure 52** The economy of Malawi in 2000 and 2018.

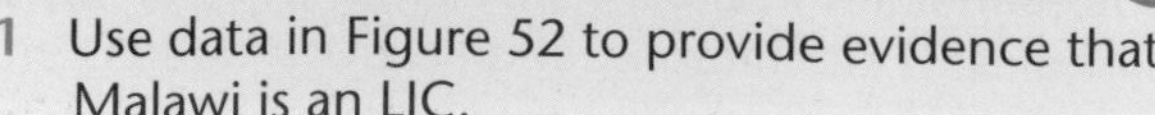

Activity

1 Use data in Figure 52 to provide evidence that Malawi is an LIC.

COVAMS – saving the soil

Middle Shire is a district in southern Malawi named after a local river, the Shire. The river is vital to this part of Malawi, because it provides most of Malawi's hydro-electric power (HEP). It's an agricultural region. Most families are subsistence farmers, growing crops such as maize or cassava. Others work on large sugar cane and cotton plantations.

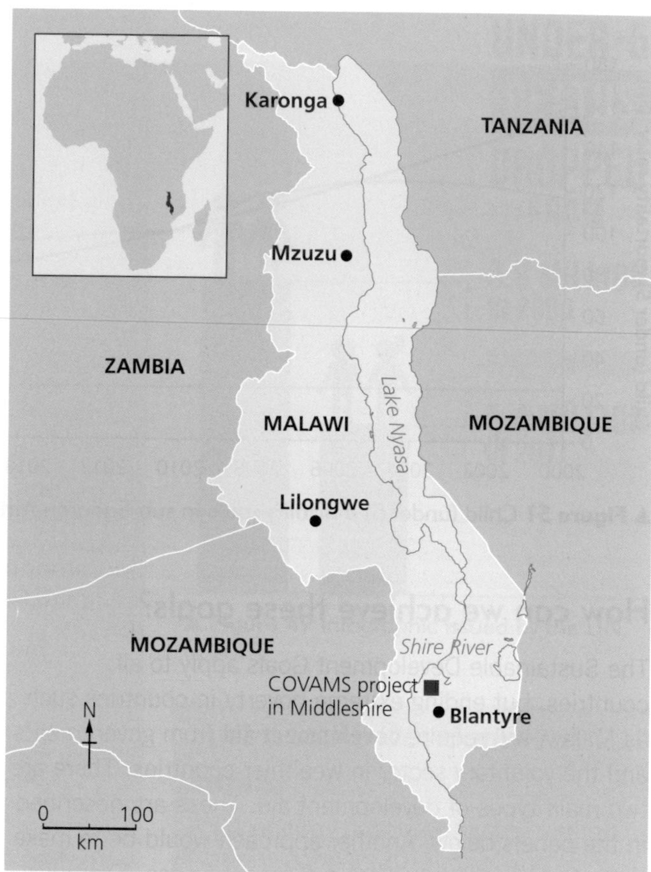

▲ **Figure 53** Map showing the location of the COVAMS project in Middle Shire.

Middle Shire is affected by soil erosion during the intense rainy season each year. Between 1990 and 2005, Malawi lost 13 per cent of its forest due to land clearance for farming. Two pressures have caused this:

- A rapidly rising population, so farmers have cut down forest to grow more food.
- Tobacco, Malawi's most valuable export, has suffered falling prices because of global over-production and falling consumption in HICs of the world. So, if farmers are to earn the same income, they must grow more tobacco.

▲ **Figure 54** An agricultural training scheme in Malawi.

A project in Middle Shire aims to save soils and re-plant trees. It's known as Community Vitalization and Afforestation in Middle Shire (COVAMS), a ten-year aid project funded by the Japanese government. Its aim is to allow Malawi to increase food production and jobs. The project uses several methods to prevent social erosion:

- Explain to communities what causes soil erosion.
- Build rock, wood and bamboo barriers across streams to prevent soil loss during rainy seasons.
- Train villagers how to conserve soil and plant trees.
- Train farmers to plough around hillsides, following the contours rather than ploughing up and down the hills, which increases surface run-off.
- Build terraces into hillsides to reduce surface run-off.
- Supply fast-growing tree species from local nurseries to speed up reafforestation.

This approach uses **intermediate technology**, teaching local people skills in using local materials to provide low-cost solutions to problems. It has had three impacts:

- By 2011, 75 per cent of households in Middle Shire had taken part.
- Crop yields improved dramatically.
- With less soil erosion, water quality improved in the dams, so Malawi produces more HEP.

Intermediate technology

Simple, practical ways of solving problems that do not require expensive machinery or high levels of skill for maintenance.

Why do countries give aid?

Malawi benefits from aid, especially long-term projects such as COVAMS. But Japan also benefits for the following reasons:

- **Diplomacy and good relations.** Aid improves relations between countries, and enables the Japanese to influence politicians in southern Africa.
- **Tactics.** Japan is keen to become a member of the UN Security Council, and needs the support of African countries.
- **Economics.** Sub-Saharan Africa countries have averaged 5 per cent annual economic growth during the 21st century. Japan sees opportunities to sell more Japanese products in Malawi if it invests in Malawi.

Enquiry 

Analyse the impacts of COVAMS. Create a table to show the economic, social and environmental impacts of COVAMS on Middle Shire.

Activities 

2 In pairs, design a flow diagram to show causes, processes and effects of soil erosion.
3 Explain why it is important to train Malawian farmers about how to prevent soil erosion.
4 Write a five-minute script for local radio in Middle Shire to tell people about the COVAMS project.

Is trade fair?

All countries have to export goods to earn money. An issue faced by many low-income countries (LICs) is that they rely heavily on the sale of just a few exports. For example, cotton makes up 35.5 per cent of Mali's exports. If countries are dependent on the export of just a few products, they find it hard to become wealthier.

Study Figure 55. The price of sugar traded on the world market in 2018 was about US$0.28 per kg, which is almost the same as its value in 1996. Meanwhile, the prices of items that farmers need to produce sugar, like pesticides, herbicides and tractors, have risen. This means that sugar cane farmers are, in real terms, poorer than they were 22 years ago.

Prices of primary commodities (like tea, coffee, cocoa, sugar and cotton) go up and down on a daily basis. Sometimes, if worldwide production is low, or worldwide demand is high, prices rise. You can see this in Figure 55 when the price of cotton rose to a peak of US$3.33 in 2011. However, in 2012, when production increased, the price fell again. The fluctuating price of primary commodities makes it almost impossible for poor farmers to plan ahead or invest in their farms.

The Fairtrade Foundation

The Fairtrade Foundation was established in 1992. It is an example of a **non-government organisation (NGO)**. The main aim of the Fairtrade Foundation is to help lift farmers and farm workers out of poverty by helping them to sell their products to consumers at a fair price. The Fairtrade Foundation works in a variety of different ways:

- It sets standards for the companies and farmers it works with. For example, making sure that workers' rights are protected and helping to protect the environment by making sure that farmers use sustainable methods.
- It uses the FAIRTRADE Mark on products to show consumers that products have met the social, economic and environmental standards that the Foundation demands.
- It raises awareness about the need for fair trade and better working conditions. It lobbies the UK government and informs the public about the issues facing farmers in the developing world.

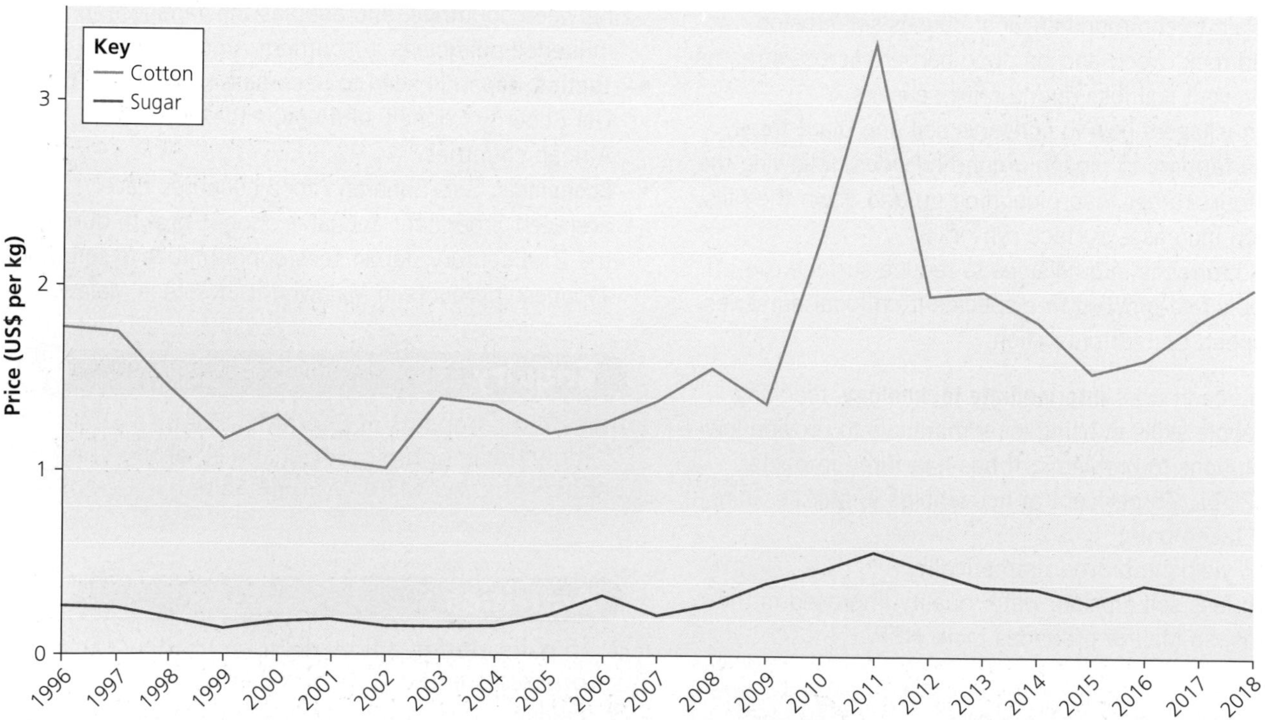

Figure 55 Price of cotton and sugar on the world commodity markets (1996–2018).

What are the benefits of Fairtrade?

Fairtrade gives farm workers and farming communities a number of benefits as well as helping to protect the environment:

- Farmers receive a Fairtrade minimum price for their product that is agreed and stable – not like the variable prices in Figure 55. This means farmers face less uncertainty, so they can plan ahead.
- Farmers get technical advice on how best to grow their crops.
- Farmers are taught how to minimise the use of pesticides and herbicides. They are taught about biological pest control and how to improve soil fertility and reduce soil erosion.
- Farming communities receive an additional payment called a Fairtrade Premium to be used for local projects, for example providing clean water or new school buildings.

Figure 56 A new village well provided by the Fairtrade Premium scheme, Mali

Fairtrade sugar cane in Malawi

Malawi is an LIC in sub-Saharan Africa. Around 80 per cent of Malawi's population live in rural areas and many poor families rely on farming to make a living. Tea, sugar, cotton and groundnuts are all grown on farms supported by Fairtrade projects. Sugar is the country's third most important export commodity after tobacco and tea. Larger farms (plantations) only grow sugar cane, whereas smallholders (small farms) grow sugar as well as rice, cotton and maize.

The Kasinthula Cane Growers Association (KCG) is a Fairtrade project in southern Malawi that supports 800 farm workers who grow sugar cane on smallholdings (very small farms). The raw sugar cane is processed in a local mill and then exported. KCG gets a Fairtrade Premium of US$60 for every tonne of sugar it sells. Since 1996, the Premium has been used to:

- build a new, local primary school
- supply computers and printers to local high schools
- dig 18 boreholes to supply clean water
- pay for drugs to treat bilharzia, an infection caused by a tropical parasitic worm
- buy four bicycle ambulances.

Activities

1. Study Figure 55.
   a) Describe how the price of cotton changed on the world market between 1996 and 2018.
   b) Calculate the value of cotton in 2016 (when it was US$2.01) as a percentage of its value in 1996 (when it was US$1.77).
   c) Explain why the evidence shown in Figure 55 means that cotton and sugar cane farmers are likely to remain poor.
2. Draw and complete a table to show the economic, social and environmental benefits of Fairtrade.
3. Describe how KCG has benefitted the farming community in the short term and long term.

Enquiry

'All trade should be fair trade.'

To what extent do you agree with this statement? Make use of the evidence on this page to support your answer.

THEME 6 Development and resource issues

Chapter 2 Water resources and their management

How much water do we use?

Everyone needs water. A supply of clean water to our homes is essential for healthy life. We also use vast quantities of water in industrial processes and even more in agriculture. In fact, growing the food we need consumes 70 per cent of the world's water. In 1900, there were 1.6 billion people on Earth. By 1999, the world's population had risen to 6 billion and it is now around 7.5 billion. The increased population largely explains the growth in water consumption shown in Figure 1. As the world's population has grown, and more food needs to be grown, the consumption of water has increased.

Another reason for the growing demand for water is economic growth. As countries become wealthier, more water is consumed. This may be because:

- water **abstraction** is expensive – it requires huge investments to build dams and water-transfer schemes
- wealthier people tend to use more water in non-essential ways, such as watering gardens, washing cars or filling swimming pools
- as countries become wealthier, they invest in modern farming technologies, for example:
  - growing higher-yielding crops that need more water
  - using high-tech irrigation systems to grow food in arid regions, like in Figure 2.

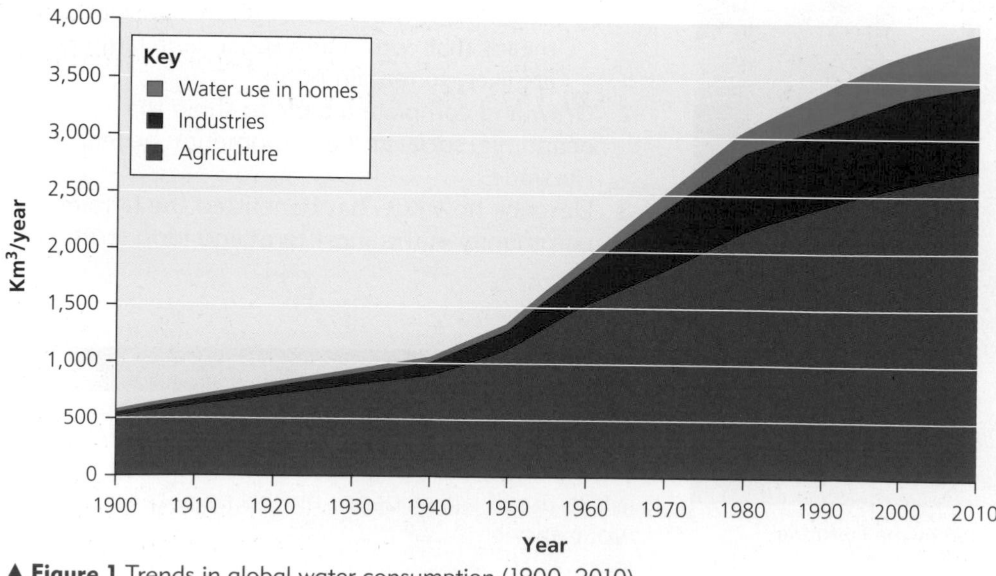

▲ **Figure 1** Trends in global water consumption (1900–2010)

Abstraction

The removal of water from rivers or aquifers, where water is stored underground in porous rocks. Over-abstraction means to remove water more quickly than it is naturally replaced by rainfall.

▲ **Figure 2** The amount of water used by agriculture has risen as technology has improved. This circular field, in California's Mojave Desert, is irrigated from a pipe which sweeps around like the hand of a giant clock.

Activities

1 a) Working with a partner, make a list of all the ways that you use water every day.
  b) How many of these uses are essential and how many could you live without?

2 Study Figure 1.
  a) Describe how the total amount of water consumption has changed since 1900.
  b) How much water was used by agriculture in i) 1900 and ii) 2010?
  c) Explain why global water consumption has increased.

Some groups of people have very poor access to water. Many people in rural regions of sub-Saharan Africa have to collect water and carry it some distance from their home. Collecting water is time-consuming and heavy work. In Africa, 90 per cent of the work of collecting water and firewood is done by women and children. A piped supply to the home would not only be safer, it would save time. A study in Tanzania showed that reducing the distance to a source of water from 30 to 15 minutes increased girls' attendance at school by 12 per cent. Having piped water also gives families privacy and dignity. Imagine how unhappy children must feel if they have nowhere private to wash.

▲ **Figure 3** Street vendor selling water in Nigeria.

In sub-Saharan Africa, people living in urban areas are twice as likely to have access to safe, piped water as people living in rural areas. However, even within urban areas, there are massive differences between the way that rich and poor have access to water. In the informal settlements of Africa's cities, many people do not have access to piped water and cannot afford to drill a borehole. They are forced to buy water from street vendors like in Figure 3. As a result, people who live in the shanty towns of some African cities can pay up to 50 times the amount for water as people living in European cities.

▲ **Figure 4** African women of the Sahel (see page 286) spend a lot of time doing work that does not contribute directly to family or state income.

Long working days are the norm for women in the Sahel. Women work up to a total of 16 hours per day in the growing season, of which about half is spent on agricultural work. Time allocation studies from Burkina Faso and Mali show women working one to three hours a day more than men. In rural areas, the lack of basic services such as reliable water supplies, health centres, stores (shops) and transport adds considerably to the time women must also spend on household chores. Shortage of time constrains women's attendance at activities to benefit them, the time and attention they can pay to productive activities, and visits to health facilities.

▲ **Figure 5** Extract from a World Bank report.

Enquiry

How close is the connection between a nation's wealth and its water use?

- Study Figure 1. Suggest a hypothesis that links wealth and water use.
- Draw a scattergraph to investigate the possible relationship between these sets of data.
- What conclusions can you draw?
- How could you improve this enquiry?

Activities

3 Study Figures 4 and 5.
   a) Give three reasons why women in Sahel countries such as Mali have such long working days.
   b) Suggest a number of ways in which the lives of rural African women and children would be improved if they had access to a clean and safe water supply close to their home.
   c) Suggest how women in Mali might use four extra hours a day.

What is your water footprint?

Every day each of us drinks between 2 and 4 litres of water a day. You use a lot more water through washing, bathing and flushing the toilet. For example, you will use about 95 litres in a five-minute shower. However, even this is only a small fraction of the total amount of water you will use in a day. Our food and clothing contain **embedded water**. Each of us uses 2,000–5,000 litres of embedded water every day. So, as consumers, we each have a **water footprint** – the impact of our water use on the planet.

Embedded water

The amount of water that has been used to grow, manufacture and transport any product through its life cycle.

Water footprint

A measure of human use of water and our impact on natural water resources.

Other footprints	Litres
T-shirt and pair of jeans	10,000
Pizza	1,260
100 g chocolate	1,700
A dozen bananas	1,920
1 kg beef	13,500

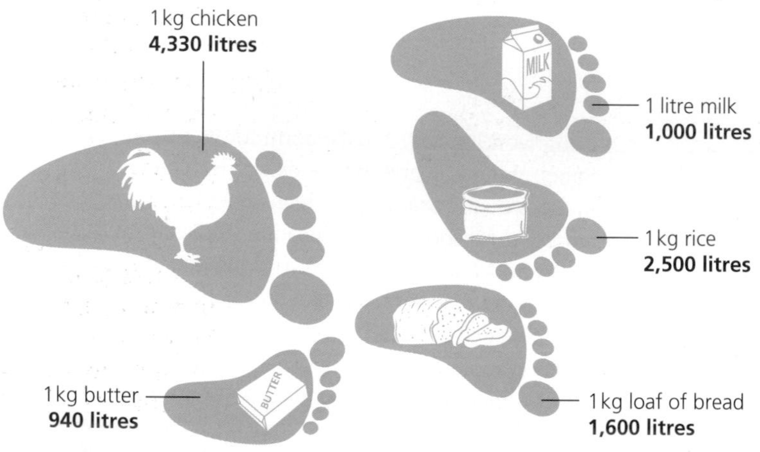

▼ **Figure 6** Water footprints for selected items.

Activities

1 Present the data for other footprints in Figure 6 as five pictograms. Try to make your pictograms roughly proportional so that larger water footprints are shown using larger pictures.
2 Explain why drip irrigation is more suitable for farming in hot semi-arid climates than other forms of irrigation.

◀ **Figure 7** Drip irrigation reduces the amount of water lost by evaporation. The system was invented in Israel where water supply is an issue.

Water for food

Globally we use about 70 per cent of our water to grow food, 20 per cent to supply industry and just 10 per cent to give us a safe domestic water supply. There are 7 billion people in the world today and the world's population is expected to rise to 9 billion by 2050. To feed these extra people, and overcome poverty and malnutrition, we need to produce 60 per cent more food by 2050.

This won't be a problem in regions which have an equatorial climate where high quantities of rainfall provide water for farming without the need for irrigation. However, in regions where rainfall totals are low, water is withdrawn or abstracted from the ground or reservoirs so that it can be used to irrigate crops. In these regions farmers will need to adopt techniques that use less water to produce more food. Figure 7 shows one example of these techniques.

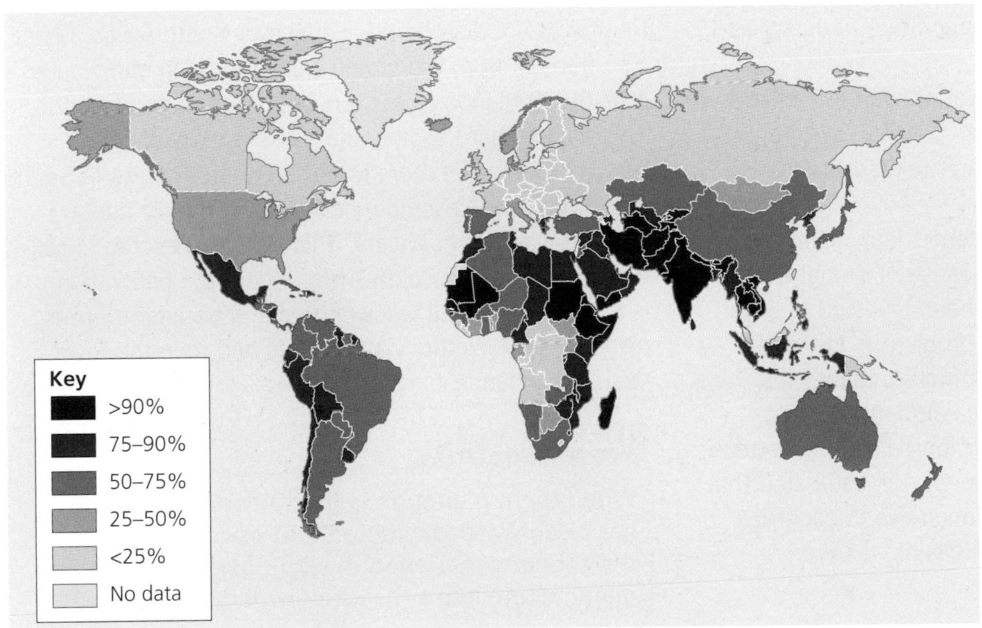

▲ **Figure 8** The percentage of water withdrawn for use by agriculture.

Activities

3 Use the data in Figure 9 to compare global water use in 2000 and 2050.

4 Study Figure 8.

a) Describe the location and distribution of countries that use more than 90 per cent of water withdrawals for agriculture.

b) Suggest one way that this style of map could be improved to make it more effective.

6,000
5,000
4,000
3,000
2,000
1,000
0
km^3
2000
2050
World

Key
Irrigation
Domestic
Livestock
Manufacturing
Electricity

▲ **Figure 9** Global water demand in 2000 and 2050.

	Agriculture	Industry	Domestic
Cambodia	94.0	1.5	4.5
Egypt	86.4	5.9	7.8
Ghana	66.4	9.7	23.9
India	90.4	2.2	7.4
Malawi	85.9	3.5	10.6
Pakistan	94.0	0.8	5.3
UK	9.2	32.4	58.5
USA	40.2	46.1	13.7

▲ **Figure 10** The percentage of water, withdrawn for each sector, in selected countries.

Maintaining water security

We have seen that we need fresh water to provide drinking water, grow food and supply industry. Without enough fresh water our health and the development of our economies could suffer. So, **water security**, which means making sure that sufficient fresh water supplies are always available, is essential for any country. Water security means that:

- people have enough safe and affordable water to stay healthy
- there is sufficient water for agriculture, industry and energy
- ecosystems that supply our water, such as wetlands, are conserved
- people are protected from water-related hazards such as drought.

Some parts of the world suffer from a lack of water, or water insecurity. This may be because of drought or because insufficient money has been invested in water resources. One way to solve the problem is to transfer water from places that have too much to places that have too little. This is a more common solution than you might think. Rivers transfer water long distances, often from one country to another. Indeed, it is estimated that 90 per cent of the world's population live in countries that take water from shared river basins.

Investigating water supply in South Africa

Around 91 per cent of South Africa's population has access to fresh piped water. That means that about 4 million South African's do not. Many of these live in rural places. In urban areas, the theft of water is becoming a serious issue. People make illegal connections to water supplies because they cannot afford a legal connection. So, improving water security is a high priority for the South African government.

Rainfall is not distributed evenly over South Africa. Moist air comes in from the Indian Ocean forming rain clouds over the highlands of eastern South Africa and Lesotho. This means that Lesotho and eastern parts of South Africa receive a lot more rain than western parts of South Africa. Parts of Lesotho receive 1,200 mm of rain a year (similar to mid Wales). This enables Lesotho to sell its excess water to South Africa where the population is higher but rainfall is lower. Water is transferred from reservoirs in Lesotho, via pipelines and then the River Vaal into Johannesburg in Guateng.

Water security

When there is enough water to ensure everyone has clean water, sanitation and good health. Also, the economy has enough water to grow food and manufacture items that are needed.

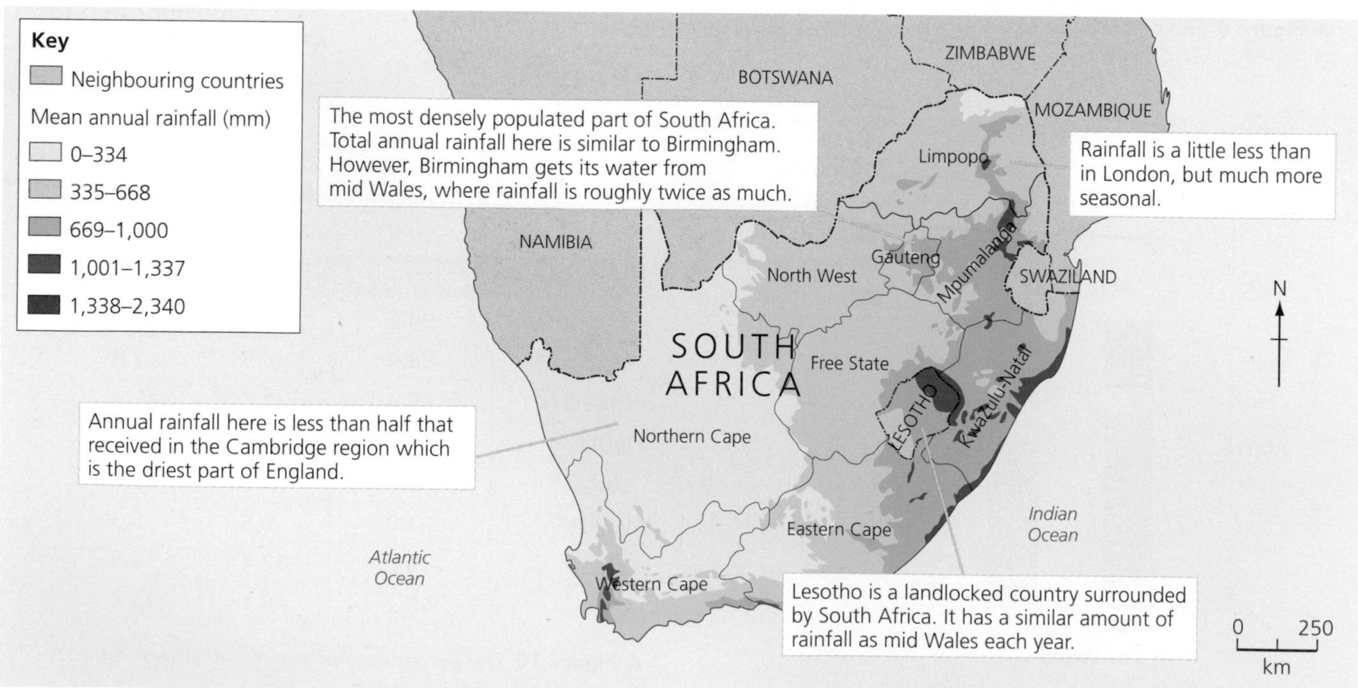

▲ **Figure 11** Average annual rainfall in South Africa and Lesotho.

	Jan	Feb	Mar	Apr	May	Jun	Jul	Aug	Sep	Oct	Nov	Dec	Total
Western Cape	8	4	11	24	40	41	47	45	24	12	12	10	278
Gauteng	125	90	91	54	13	9	4	6	27	72	117	105	713

▲ **Figure 12** Rainfall (mm) for selected regions of South Africa.

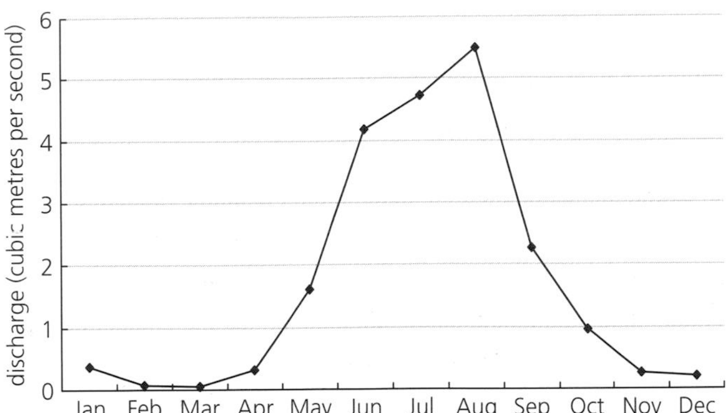

▲ **Figure 13** Hydrograph for the River Dorling, Western Cape Province, South Africa. The catchment area of the River Dorling before this station is 6,900 km².

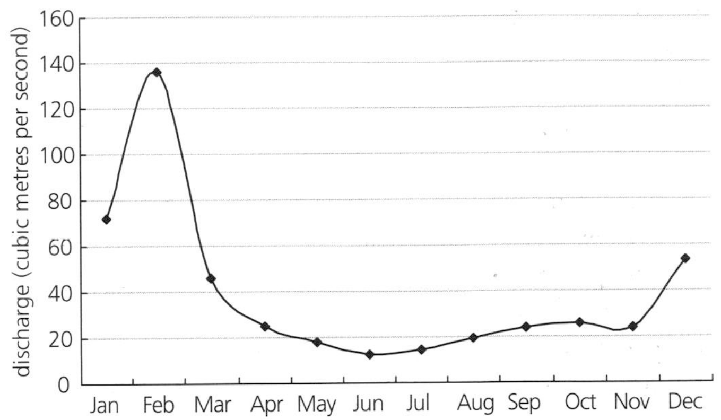

▲ **Figure 14** Hydrograph for the River Vaal, Gauteng Province, South Africa. The catchment area of the River Vaal before this station is 38,560 km².

▲ **Figure 15** Relief rainfall patterns in South Africa and Lesotho.

Activities

1 Use Figure 11 to describe the distribution of rainfall in South Africa.
2 Explain why Lesotho is able to sell water to South Africa and why South Africa wants to buy it.
3 a) Use Figure 12 to draw a pair of rainfall graphs.
  b) Compare your rainfall graphs with Figures 13 and 14.
  c) At which times of year do these two regions face possible water shortages?
4 Use Figure 15 to explain why Lesotho has so much more rainfall than Free State in South Africa. Describe what is happening at each point (1–6) on the diagram.

Investigating a major water-transfer scheme

The Lesotho Highlands Water Project (LHWP) is an example of a large-scale water-transfer scheme. Dams collect water in the mountainous areas of Lesotho to supply water to areas of water deficit in the neighbouring country of South Africa. It is the largest water-transfer scheme in Africa, diverting 40 per cent of the water in the Senqu river basin in Lesotho to the Vaal river system in the Orange Free State of South Africa via 200 km of tunnel systems. The River Vaal then carries the water into Gauteng Province.

The first phase of the project, started in 1984, is complete. Two dams, the Katse and the Mohale, have been built. The LHWP transfers 24.6 m³ of water per second to South Africa. A second phase of construction began in 2019 with the building of the Polihali Dam. This development will deliver an extra 45.5 m³ per second of water to South Africa. The new dam will also generate 1,000 MW of electricity for Lesotho. Phase two is expected to cost US$1.3 billion and be completed by 2023. There are plans for two more dams to be built in later phases of the LHWP.

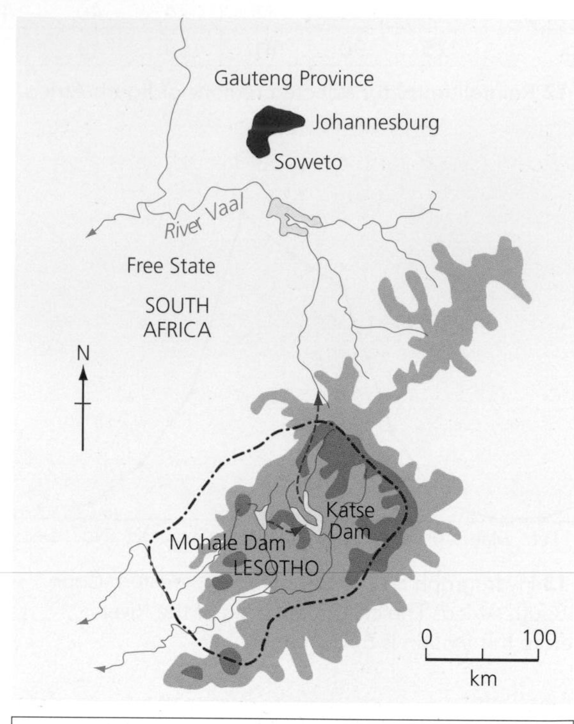

Key

Pipe transferring water into the River Vaal catchment

Land over 3,000 m   Land over 2,000 m

► **Figure 16** A map of the Lesotho Highlands Water Project water management and transfer scheme.

Year	Urban areas			Rural areas		
	Improved supplies			Improved supplies		
	Piped in home (%)	Other improved (%)	Unimproved (%)	Piped in home (%)	Other improved (%)	Unimproved (%)
1990	85.9	12.2	1.9	24.0	42.3	33.7
2000	87.5	11.0	1.5	27.9	42.7	29.4
2010	90.3	8.9	0.8	34.5	43.3	22.2
2015	91.7	7.9	0.4	37.7	43.7	18.6

▲ **Figure 17** Improved and unimproved water supplies in urban and rural areas of South Africa.

Year	Urban areas			Rural areas		
	Improved supplies			Improved supplies		
	Piped in home (%)	Other improved (%)	Unimproved (%)	Piped in home (%)	Other improved (%)	Unimproved (%)
1990	26.3	66.3	7.4	2.3	72.6	25.1
2000	39.4	53.7	6.8	2.9	72.6	24.5
2010	61.3	32.8	5.9	3.9	72.5	23.5
2015	70.0	24.5	5.4	4.3	72.6	23.0

▲ **Figure 18** Improved and unimproved water supplies in urban and rural areas of Lesotho.

What are the advantages and disadvantages of the LHWP?

Lesotho is one of the world's poorest countries. It has few natural resources so selling water to South Africa provides 75 per cent of Lesotho's income. However, many people in Lesotho do not have access to improved water sources. A new dam, the Metlong Dam is being constructed to supply lowland areas of Lesotho with water. The project is costing US$31.80 million – money that Lesotho has borrowed and will have to pay back. The dam is finished and the pipes and tunnels to transfer the water should be completed by 2020. Until then, many people in rural areas of Lesotho will spend many hours fetching and carrying water.

▼ **Figure 19** Impacts of the LHWP.

Roads have been constructed in the highlands to gain access to the dam sites.

Farmland is flooded to create the reservoirs. About 20,000 people were displaced when the Katse Dam was built.

Compensation is paid to families who lose land during construction of the dams. Many families have complained that the compensation is too little and too late.

Phase two dams will generate 1,000 megawatts of electricity for Lesotho.

Phase one created around 20,000 jobs. Many workers moved to shanty towns on the construction sites. Alcoholism and HIV/AIDs became significant problems in the shanty towns.

▲ **Figure 20** A family in Lesotho collecting water.

Activities

1 Summarise the aims of the Lesotho Highlands Water Project.

2 Use the text on these pages to complete a copy of the following table. You should find more to write in some boxes than in others.

Country	Short-term advantages (+) and disadvantages (–) of LHWP	Long-term advantages (+) and disadvantages (–) of LHWP
Lesotho	+ –	+ –
South Africa	+ –	+ –

3 Summarise what each of the following groups of people might think about the LHWP:
   a) a farmer in the Lesotho Highlands
   b) a government minister in Lesotho
   c) residents in Johannesburg, South Africa.

Enquiry

How much progress have Lesotho and South Africa made in providing improved water supplies?

Use the evidence in Figures 17 and 18 to:

a) Compare progress in urban and rural areas.
b) Compare progress in South Africa and Lesotho.
c) What conclusions can you draw about the benefits of the LHWP?

Alternative ways to manage water

South Africa has 539 large dams, which is almost half of all the dams in Africa. But despite this, there are still a large number of South Africans without access to clean drinking water. Many of these people live in rural, remote parts of South Africa; they are too isolated to become part of the big projects such as the LHWP and they are too poor to drill boreholes to tap into groundwater supplies. Instead they have to rely on cheap, small-scale methods of **rainwater harvesting**.

Ma Tshepo Khumbane is a South African farmer who teaches rainwater harvesting techniques. Her management strategies (shown in Figure 21) are affordable and practical for families, no matter how small the farm is or how little money they have. They use ways that are cheap, practical and easy to maintain using appropriate technology. They are commonly used in rural areas of Limpopo that do not have piped water. They are designed to:

- collect and use rainwater, for example, by collecting water from the roof of the farm
- maintain soil moisture by encouraging as much infiltration as possible. In this way groundwater stores are recharged.

Rainwater harvesting techniques and soil moisture conservation are examples of sustainable water development. They benefit people now without doing any lasting damage to the environment or using up valuable resources. These methods of water management are not usually big enough to have negative impacts on the surrounding drainage basin – unlike the big dam building schemes used by governments.

Rainwater harvesting

Where rainwater is collected and stored, for example, from the roof of a house.

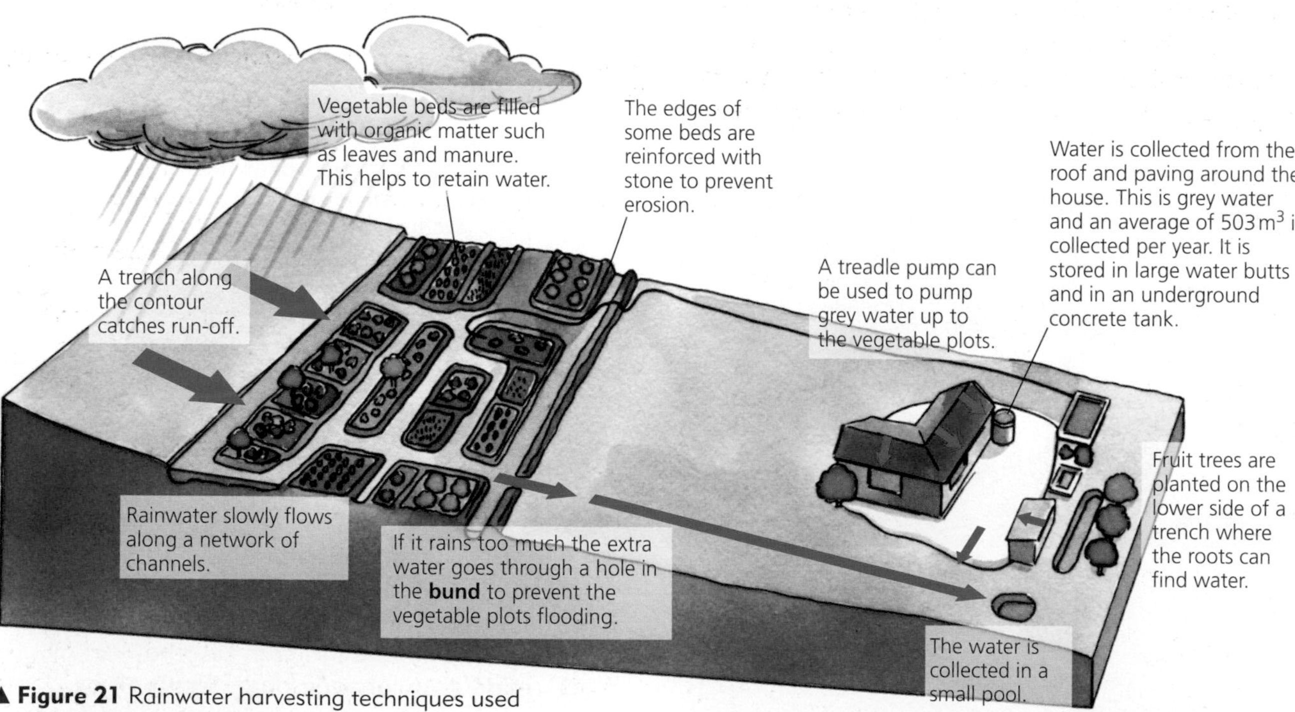

▲ **Figure 21** Rainwater harvesting techniques used by Ma Tshepo Khumbane.

Activities

1 Choose five techniques shown in Figure 21. For each technique explain how it either collects rainwater or recharges groundwater.
2 Explain why the type of management shown in Figure 21 is sustainable.

Enquiry

Should South Africa invest in big schemes like LHWP or small-scale sustainable water management schemes?

- Contrast the impacts of this type of water management with the building of large dam and water-transfer schemes like the LHWP.
- Which would you prioritise? Explain your answer.

Could South Africa harvest water from fog?

Fog has been 'harvested' to provide clean drinking water to isolated rural communities since 1987 when a scheme was set up in Chungungo, Chile. Since then, similar systems have been used successfully in Peru, Ecuador, Tenerife, Ethiopia and South Africa. It is another form of appropriate technology because it is relatively cheap to install and maintain. To collect water from fog, a simple system of fine-mesh nylon nets are suspended vertically between tall poles. The fog condenses on the net and drips into a gutter below. It then passes through sand before being piped to where it is needed.

Fog harvesting works best in upland regions (at least 400 m above sea level) that experience moist air being blown from the coast. As the air rises it condenses to form fog. So would fog harvesting work in Limpopo in South Africa? Most of Limpopo is over 1,000 m above sea level. Moist air from the Indian Ocean is blown inland by prevailing easterly winds. The first fog harvesting scheme was built at Tshanowa Primary School in Limpopo. All 130 school children used to bring bottled water to school every day. Now they drink pure water collected from fog.

▲ **Figure 22** Fog collectors in Tenerife. The nets, which look like giant volley ball nets, trap moisture as it condenses from the air.

It is not foggy every day.
The nets and poles are relatively cheap to buy.
Repairs are essential. Nets are easily torn in the wind.
Repairs are easy to make and require little training.
Ground water is contaminated.
Fog harvesting technology does not need any electrical energy.
Many rural areas do not have a piped supply from a reservoir.
Some of the foggiest sites are some distance from rural communities.

▲ **Figure 23** Advantages and disadvantages of fog harvesting in Limpopo.

Activities

3 a) Use Figure 23 to sort the advantages and disadvantages of fog harvesting.

b) Based on the evidence in Figure 23, explain why fog harvesting may be considered to be an appropriate technology for a poor rural community.

Enquiry

Research the fog harvesting systems in Chungungo and Tshanowa by typing the names into an internet search engine. To what extent has each scheme been a success?

Weblink

http://www.weathersa.co.za

This is the website of the South African weather service. Click on the Limpopo link to find out how foggy it will be in the next five days. Or, scroll to the bottom of the home page to find recent rainfall maps.

Over-abstraction of groundwater

Some layers of porous rock, known as **aquifers**, store huge quantities of water under the ground. Water from rain and melting snow percolates down into the ground where it is stored in these aquifers – this process is known as **recharge**. To use this water, all we have to do is dig or drill into the ground and the water can be abstracted. The abstraction of water from the ground is a major source of water supply in India. In fact, groundwater provides 65 per cent of all water used in agriculture and 85 per cent of the domestic supply. But in some areas, **over-abstraction** is taking place. This is when water is taken from the ground faster than it can be recharged by natural processes. In many places the water table is 4 m lower than it was in 2002. There are concerns that, as demand for water in India continues to increase, water shortages will become more common.

	Jan	Feb	Mar	Apr	May	Jun	Jul	Aug	Sep	Oct	Nov	Dec
Temperature (°C)	19	20	24	27.5	30	31	30	29.5	28	27.5	24	20
Rainfall (mm)	13	10	7	3	3	18	81	40	13	3	3	7

▲ **Figure 24** Climate data for West Gujarat.

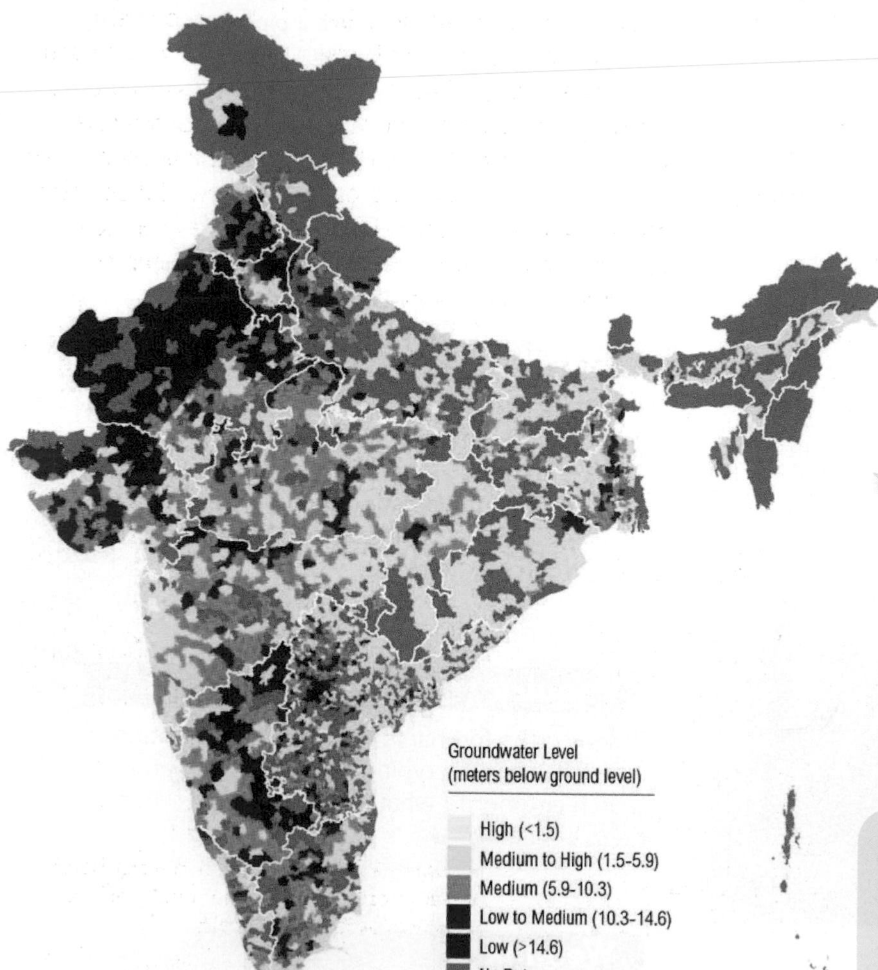

▲ **Figure 25** Groundwater level, India.

I own three and a half hectares of farmland and two hectares have been submerged. The compensation given to us is not enough. Unless they visit this place, and listen to our complaint, how can government officials understand our plight? We depend on our land for our livelihood and we will die with our land!

Local farmer

The farmers have received compensation. The water will not enter their homes – just flood a few fields. The extra water in the Omkareshwar Dam is needed by other farmers. It will provide water for 60,000 hectares of farmland.

Government official

▲ **Figure 26** Points of view on the raising of the Omkareshwar Dam.

Weblink

www.indiawatertool.in – an interactive map that allows you to investigate India's water issues.

Activities

1 a) Use Figure 24 to draw a climate graph for West Gujarat.
b) Describe the annual rainfall pattern.
c) Explain why farmers in West Gujarat need to abstract groundwater.

2 In which parts of India is the level of groundwater lowest? Use Figure 25 to help your answer.

Why is so much water abstracted?

There are a number of reasons for India's reliance on groundwater. There are physical and human factors:

1 India has a seasonal rainfall pattern. States in the northwest, such as Gujarat, have a long dry season. Surface stores such as lakes and rivers can be dry for long periods of time.
2 Surface stores are often polluted by human waste. Groundwater is considered to be cleaner – although groundwater can be also be polluted, especially by pesticides used on farms and shrimp farms.
3 Urban water supplies are poor. Many people buy water from water vendors at very high prices. Drilling a well is seen by many as a cheaper source in the long term.
4 The Indian government has encouraged farmers to grow more food to feed the growing population. This 'green revolution' has used crops that tend to need more water than traditional crops.
5 Cheap electricity has encouraged farmers to drill very deep wells and install pumps to extract water. The water table falls and the wells of neighbouring farms dry out, so they too have to drill deeper.
6 India's monsoon rainfall (see pages 162–3) is unpredictable. In seasons with lower than average rainfall (like 2012) the aquifers are not fully recharged.

Who should solve this problem?

India has over 3,200 major and medium-sized dams, which provide water for farms and industry. The Indian government could borrow money to build even more dams as the demand for clean water increases. This type of project is known as **top-down** development. However, dam construction has sometimes been very unpopular in India. Farmers often complain that they are not offered enough compensation for their land when it is flooded. Figure 27 shows farmers protesting about the Omkareshwar Dam. The height of this dam was increased by 91 m, flooding an extra 1,100 hectares of farmland.

An alternative approach would be for farmers to work together on water conservation techniques – like those in South Africa shown on page 250. One successful approach is for groups of farmers to build low stone and earth check dams. These create small pools of water for a few months after the rainy season. The water soaks into the ground and naturally recharges the aquifers. This type of **self-help** scheme doesn't need powerful decision makers or lots of money like a top-down development project. It comes from grassroots level – sometimes called a bottom-up approach to development.

Top-down

An approach to development in which decisions are made by politicians, government officials or international organisations.

Self-help

When decisions about development are made by local people.

◄ **Figure 27** Farmers stand in waist-height water to protest about the raising of the Omkareshwar Dam in 2015.

Enquiry

Who should solve India's water shortage?

Should India use top-down development solutions or self-help schemes to solve the over-abstraction problem? Justify your decision.

Activities

3 Why is so much groundwater abstracted in India? Give two physical and three economic reasons.
4 Explain why the Indian Government will find it difficult to solve the problem of over abstraction.

THEME 6 Development and resource issues

Chapter 3 Regional economic development

Inequality in India

India is making great economic progress, but wealth is not spread evenly. Some people are very rich. Oxfam estimates that there are 119 billionaires in India. Although the proportion of people living in poverty is falling, the most recent reports by the World Bank estimate that there are still 175 million people in India earning under US$1.90 a day. That is more people than the population of the UK and Germany put together.

Causes of inequality

A number of factors cause poverty and inequality in India. Some groups of people remain poor because they are excluded from part or all of their education and from better jobs. Women, people with disabilities, Muslims, **Dalits** and tribal groups all suffer from some degree of discrimination in Indian society. These are social and cultural factors that cause inequality.

There are also economic reasons for inequality. Huge numbers of Indians work in very low paid jobs in construction work, farming and the informal sector. Pay in these sectors of the economy remains very low, while growth in manufacturing and service industries has created well-paid jobs and personal wealth for many Indians.

There are environmental factors too. Extreme weather events, such as flooding, force 1.5 million Indian people from their homes every year. Climate change is creating environmental refugees. Many of these homeless people move to slums in India's cities.

There are also political reasons for regional inequality across India. Each state in India has its own government that is responsible for providing health care and education. The state government of Kerala, for example, has always been generous in funding public education and health care. This means that the poverty rate in Kerala is lower than in almost any other Indian state. By comparison, Bihar's state government has only recently begun to invest effectively in public health and education. Bihar is India's poorest state – average incomes are four times lower than in Kerala or Maharashtra.

Consequences of inequality

Inequality can have serious economic and social consequences. The economic consequences can be seen in the huge gap between rich and poor in India. Around 10 per cent of the population have 77 per cent of India's total wealth.

Some of these consequences can be seen in Figures 2, 3 and 4. Inequalities have social consequences for health and education. In Kerala, for example, life expectancy is high and the birth rate is low. Almost 92 per cent of the population can read and write. By contrast, in Bihar, India's poorest state, almost 58 per cent of the population are below the age of 25 and only 47 per cent can read and write. Bihar is also extremely rural, with 85 per cent of the population living in rural areas with few opportunities for better-paid work in manufacturing or service industries.

Dalits

The lowest social group in Indian society, sometimes referred to as 'untouchables'. There are 200 million Dalits in India.

Figure 1 Dalit women working in a stone quarry in West Bengal.

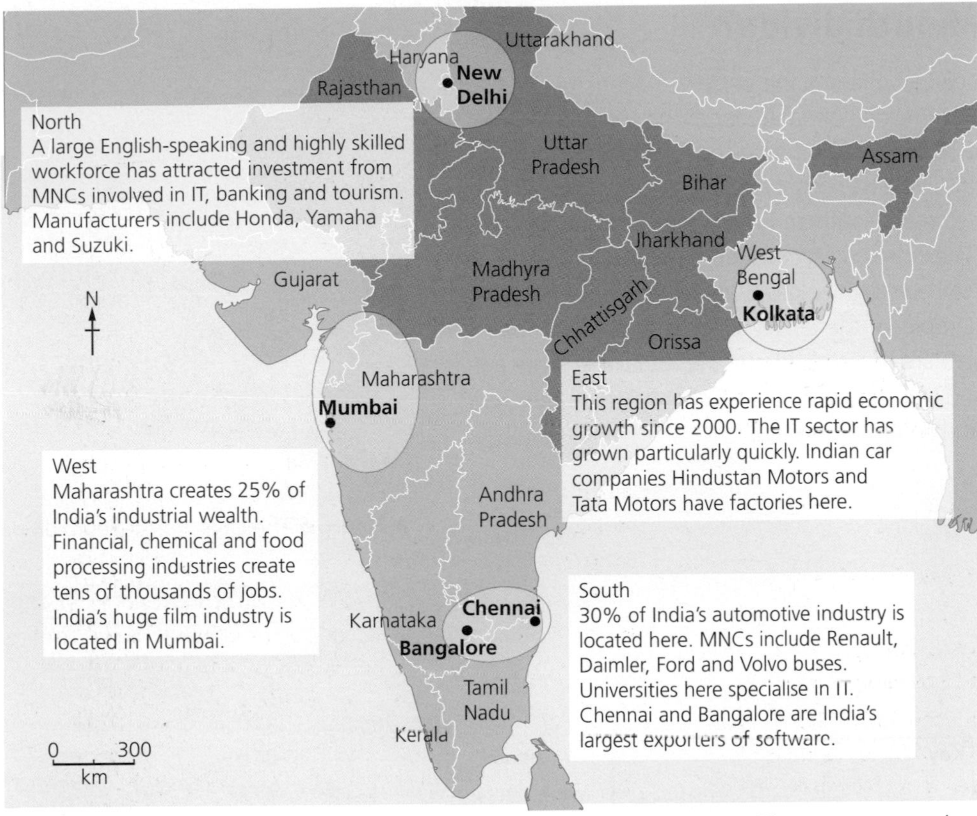

▲ **Figure 2** India has four regions that are experiencing industrialisation. The states in purple have 50% of India's population but 71% of infant deaths (before the age of 1), 72% of child deaths (before the age of 5) and 62% of maternal deaths.

	1974	1984	1994	2004	2013
Bihar	62	62	55	41	34
Kerala	60	40	25	15	7
Maharashtra	53	43	37	31	17

▲ **Figure 4** Percentage of the population living in poverty in three selected states.

Activities

1. Identify four different reasons for inequality in India.
2. Explain why inequality can have serious consequences for some groups of people.
3. Use Figure 2 to make your own labelled sketch map of India. Use it to show the reasons why some states are wealthier than others.
4. Study Figure 3.
   a) Describe the distribution of states that have the lowest levels of poverty.
   b) Use Figure 2 to help explain this pattern.
5. Suggest two reasons why Maharashtra still has higher levels of poverty than Kerala, despite having more industrial growth.

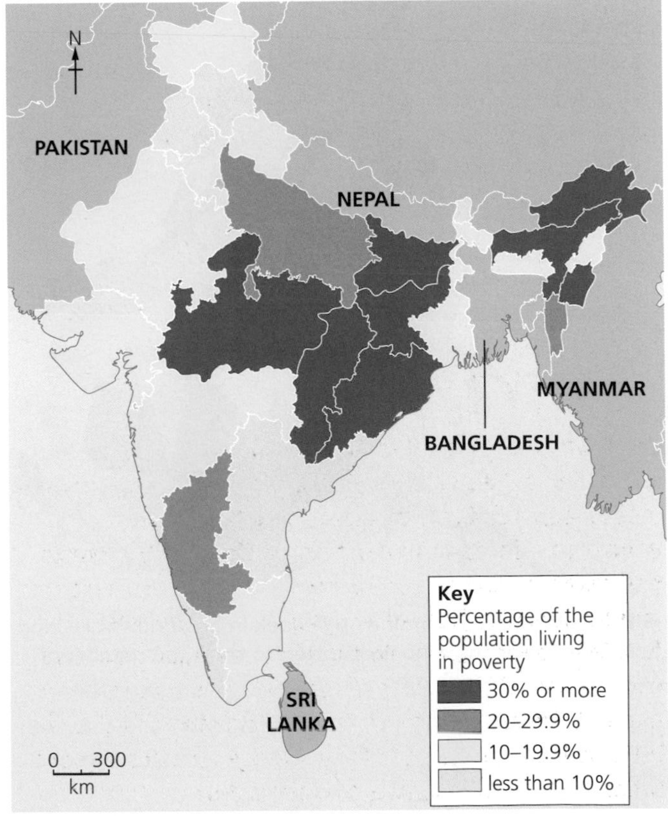

▲ **Figure 3** Percentage of the population of India living in poverty, 2013.

Is there a UK North–South divide?

In the UK the term 'North–South divide' refers to the perceived economic and social differences between southern England and the rest of the UK. The divide is usually shown as a diagonal line across England from the Severn Estuary to the Humber Estuary. The North–South divide has economic and social consequences. In general, people who live in London and southern England have higher incomes than those in the rest of the UK. There are social consequences too. People in the southern part of the UK tend to have slightly longer life expectancy, and fewer long-term health problems. For example, of the ten local authorities where male life expectancy is below 75 years, seven are in Scotland and the remaining three are in the North West region of England. Almost every local authority where male life expectancy is over 81 years is in the south of England.

▲ **Figure 6** The UK's North–South divide.

▲ **Figure 5** Average earnings per week in England, Wales and Scotland, 2015; the map is distorted to show the number of jobs available in each area.

	Earnings per week (£)	
Local authority	Men	Women
Aberdeen	657	411
South Tyneside	480	324
Hull	470	299
Liverpool	478	369
Solihull	575	328
Gwynedd	367	300
Powys	426	296
Cardiff	479	344
Cornwall	402	266
Bristol	533	370
Mid Devon	468	254
North Norfolk	436	221
Winchester	575	363
Thanet	404	273
Hounslow	681	464
Westminster	706	565
City of London	1,045	780

North
South

▲ **Figure 7** Average weekly earnings, 2015.

Enquiry

To what extent does the evidence in Figure 7 reveal a gap in earnings between the north and south of the UK?

You might want to consider whether the pay gap between men and women or between urban and rural areas is greater than the North–South divide.

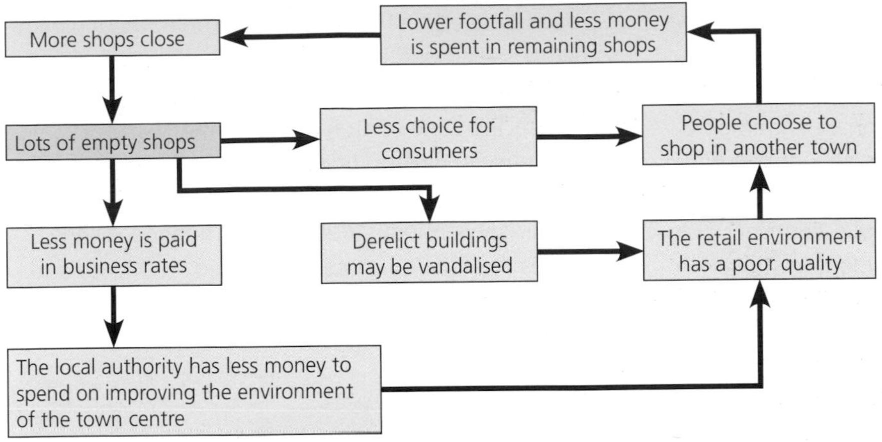

▲ **Figure 8** The impact of empty shops on the high street.

Activities

1 a) Use Figure 5 to describe the average wage and the relative number of jobs in:
   i) London
   ii) Birmingham
   iii) the North East region.
 b) To what extent does this map confirm that there is a north–south divide?

2 a) Put the earnings of men in Figure 7 in rank order.
 b) What is the median wage?
 c) How many of the local authorities that have higher than median wages are in the south?

3 Using Figure 9, describe the distribution of towns that have:
 a) high vacancy rates
 b) low vacancy rates.
 To what extent is it true that there is a North–South divide in the high street?

4 What are the impacts of high vacancy rates on our town centres? Study Figure 8 and divide the impacts into the following categories:
 a) social
 b) economic
 c) environmental.

5 a) Explain why a few empty shops can lead to other shops closing.
 b) Suggest three ways that local authorities might try to solve the problem of shop vacancies.

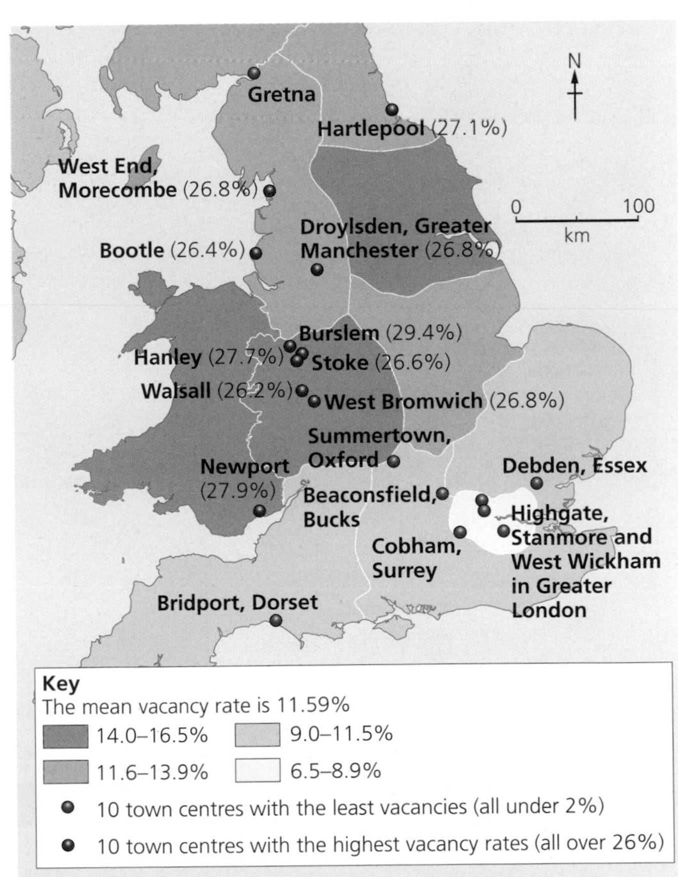

▲ **Figure 9** Empty shops in the UK's high streets, 2015; vacancy rates with the highest and lowest percentages of empty shops.

Factors that influence economic growth and decline

The UK's northern cities were the traditional centres of UK manufacturing industry. But, as it became cheaper to import electrical goods, textiles and steel from abroad, these industries declined. Highly skilled and relatively well-paid jobs have been lost. Unemployment is still a problem in some of these regions of the UK. Where jobs have been created, in retailing for example, the jobs are less highly skilled and less well paid.

In contrast, the zone stretching along the M4/M11 corridor between Bristol, London and Cambridge, has attracted modern high-tech manufacturing industries and jobs in service industries such as finance. These businesses are located close to a skilled workforce and with great transport links to their market.

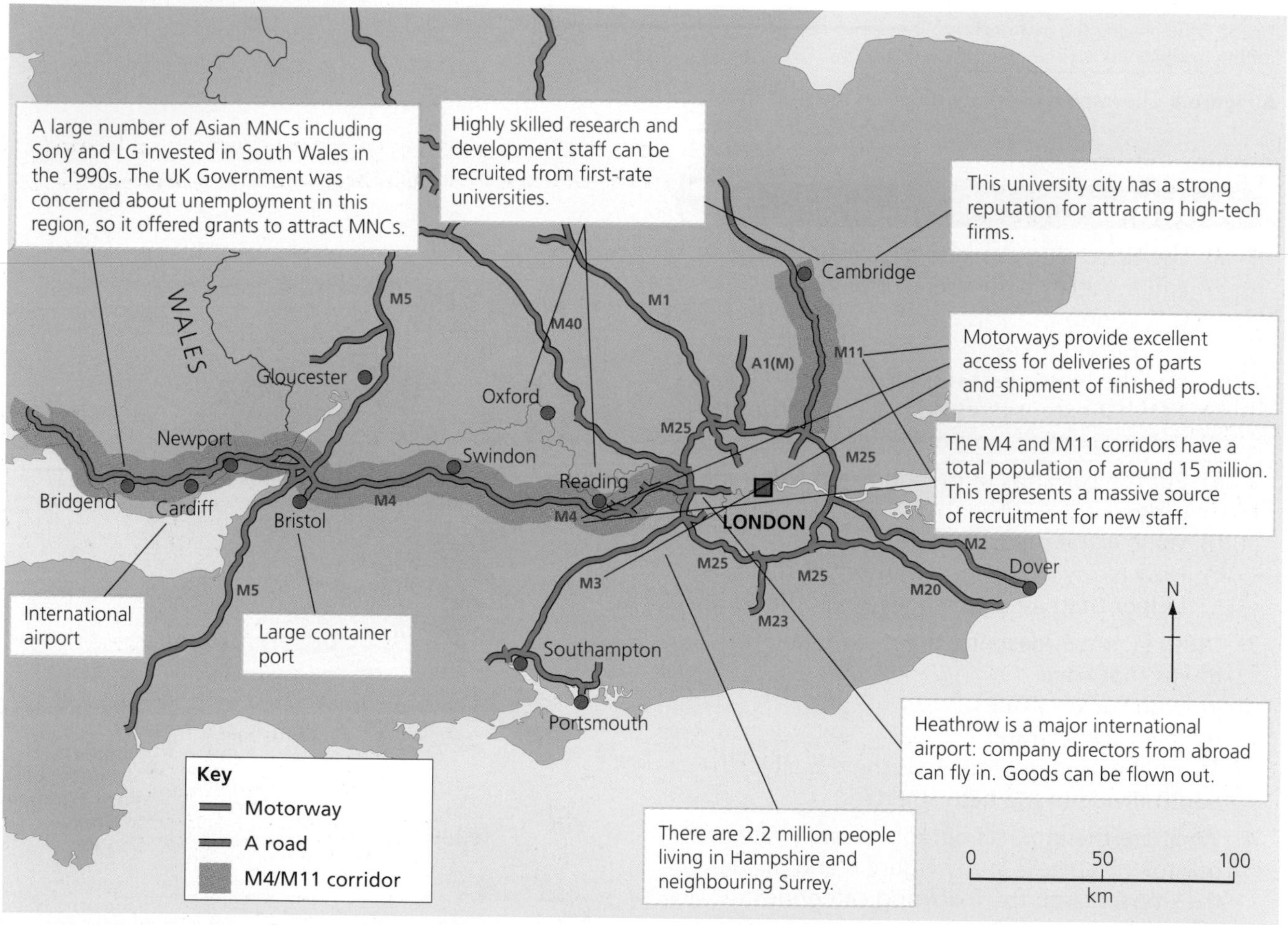

A large number of Asian MNCs including Sony and LG invested in South Wales in the 1990s. The UK Government was concerned about unemployment in this region, so it offered grants to attract MNCs.

Highly skilled research and development staff can be recruited from first-rate universities.

This university city has a strong reputation for attracting high-tech firms.

Motorways provide excellent access for deliveries of parts and shipment of finished products.

The M4 and M11 corridors have a total population of around 15 million. This represents a massive source of recruitment for new staff.

International airport

Large container port

Heathrow is a major international airport: company directors from abroad can fly in. Goods can be flown out.

There are 2.2 million people living in Hampshire and neighbouring Surrey.

WALES

Gloucester

Oxford

Swindon

Reading

Newport

Bridgend

Cardiff

Bristol

Cambridge

LONDON

Southampton

Portsmouth

Dover

M5

M1

M40

A1(M)

M11

M25

M4

M3

M23

M20

M2

N

0 50 100

km

Key

Motorway

A road

M4/M11 corridor

▲ **Figure 10** The location of the M4/M11 corridor, shown in lilac.

Activities

1 Study Figure 10.
   a) Identify five specific reasons why MNCs might wish to locate in the M4/M11 corridor.
   b) Describe the location of Newport.
   c) Suggest why businesses might be more attracted to locations to the east of the M4/M11 corridor rather than the west.

2 Give two reasons why Chinese steel is cheaper than UK steel.

3 Study Figure 11.
   a) Give four-figure grid references for the two grid squares that contain the steel works.
   b) Compare the relief of the area to the south and north of the railway line.
   c) Use map evidence to suggest three reasons why the steel works were located here.

When regions decline

In the winter of 2015-16 steel manufacturers reached a crisis point. Tata, the Indian owned multinational company (MNC), announced that it would have to lay off workers. 900 jobs were lost at Scunthorpe in North East England, 750 at Port Talbot and another 250 at Llanwern, both in South Wales. SSI, a Korean MNC was also forced to close its steel plant at Redcar, in Teesside, with the loss of 2,200 jobs. All of these job losses were in regions already affected by low pay and higher than average levels of unemployment. The job losses were blamed on cheap steel, imported from China, which was being sold in Europe for less money than it cost to make it in the UK. China is able to make steel more cheaply than in Europe. China has lower labour costs and the Chinese government keeps the cost of energy low.

GEOGRAPHICAL SKILLS

Locating places on an OS map

The vertical lines on an OS map are eastings. The horizontal ones are northings. Each grid square is known by the intersection in its lower left corner. When you locate a place on an OS map you must give the easting first, then the northing.

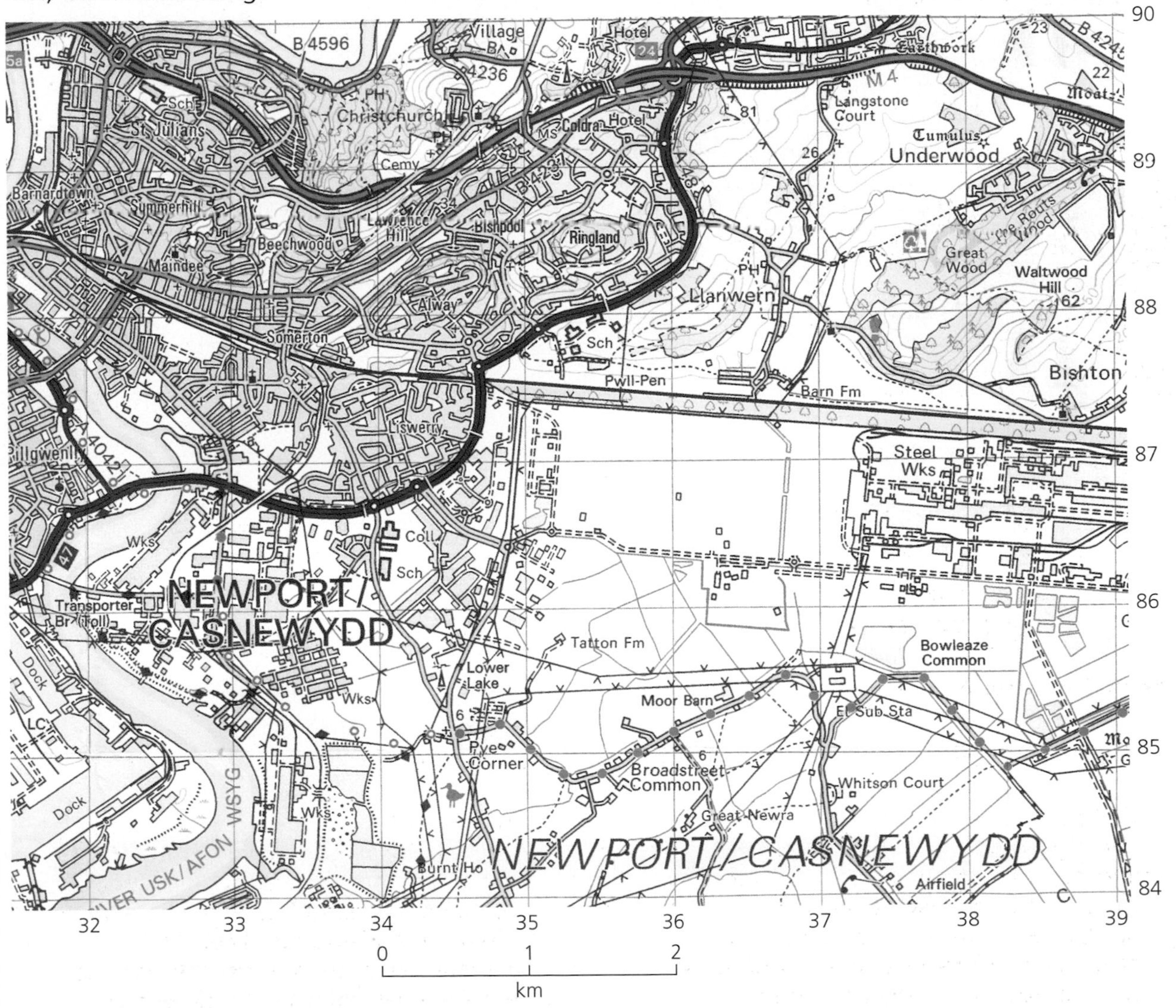

▲ **Figure 11** The location of the Llanwern Steelworks, owned by Tata, near Newport; scale 1 : 50,000.

How can regional inequalities in the UK be reduced?

One way to tackle the lack of jobs in the UK's regions is through investment in major infrastructure projects, such as road and rail schemes. Better transport links should attract new investment from business and create economic multiplier effects in the region (see pages 218–9).

The main transport route that runs through southeast Wales is the M4 motorway, which connects this region to London. However, between Newport and Cardiff the motorway narrows to two lanes as it crosses the River Usk and through the Brynglas Tunnels. This causes congestion and traffic hold ups, especially at peak times. In 2014 the Welsh government confirmed its intention to build a £1.6 billion relief road: a 23 km, three-lane motorway would be built around Newport to bypass the congested section of the M4. Figure 12 shows the proposed route. This would have been the largest capital investment programme ever made by the Welsh government. However, in June 2019, the Welsh government announced that the project would be scrapped because of its high cost and impact on the environment.

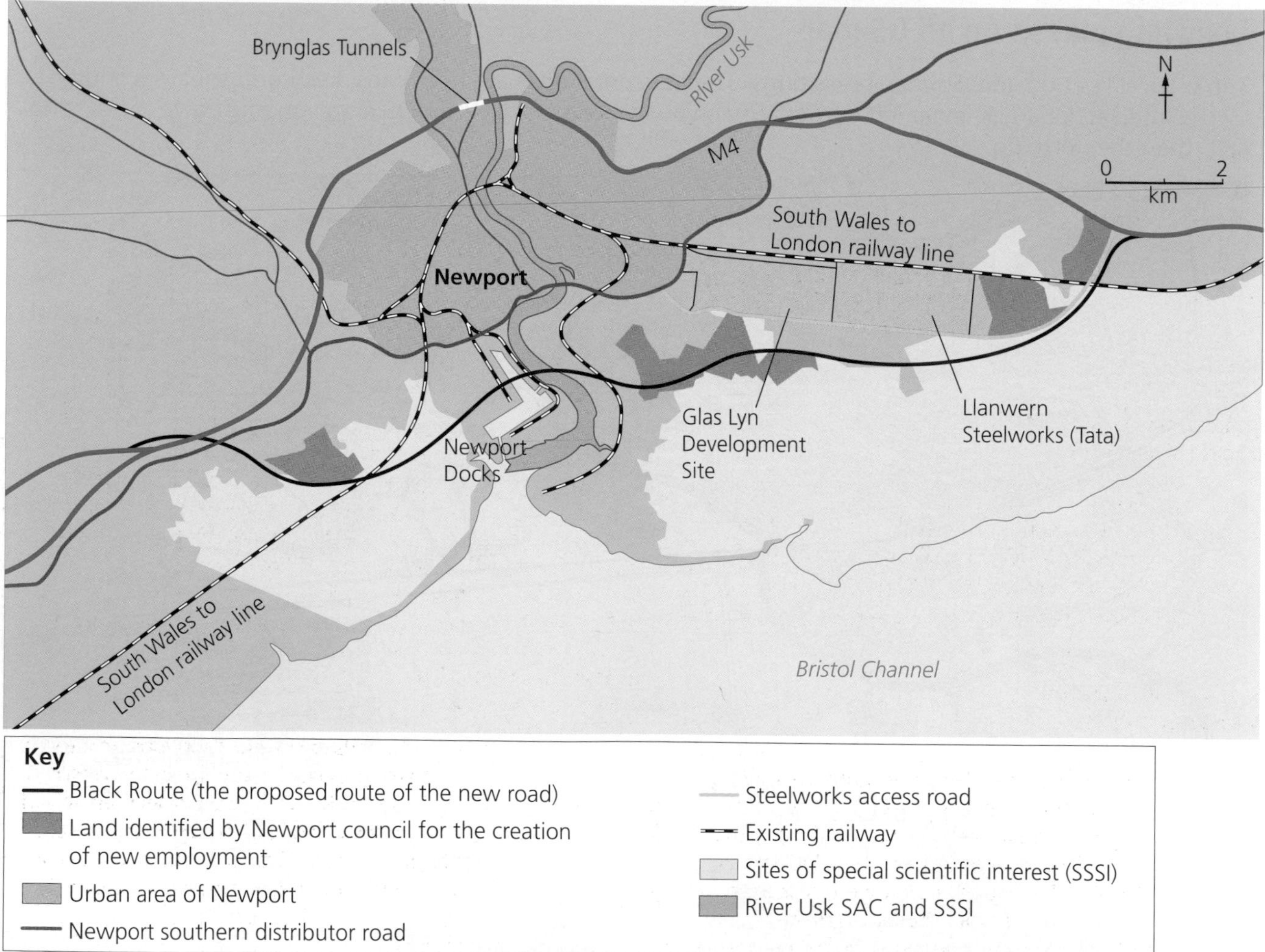

▲ **Figure 12** The proposed M4 relief road.

Activities

1 Describe the route of the proposed M4 relief road.
2 Use evidence from Figure 12 to suggest:
   a) one major cost of taking this route
   b) one environmental disadvantage of this route.
3 Use evidence from Figure 11 on page 259 and Figure 13 to suggest why this route is better than one to the north of the railway line.
4 Create a flow diagram to suggest the likely positive multiplier effects for South East Wales if this road had been built.

▲ **Figure 13** Llanwern Steelworks: the camera is pointing east; the new road would have been built to the south of the steelworks.

Spokesperson for British Associated Ports

These plans will jeopardise future investment in Newport. I don't think it is a good idea to put the route through the centre of Wales' most important general cargo port.

Director, Friends of the Earth Cymru

The decision will see the motorway plough through a part of one of Wales' most protected and environmentally sensitive landscapes – the Caldicot and Wentloog Levels. This hugely expensive transport route is the worst possible choice.

Spokesperson for Confederation of British Industry (CBI) Wales

The M4 around Newport is simply not fit for the twenty-first century. It is already congested and predicted to get worse. We need a transport system that will ease congestion and improve our economic competitiveness as well as providing jobs and growth.

Spokesperson for Sustrans Cymru

We should be investing in clean forms of public transport – like the electrification of the main railway line between London and Swansea. Electric trains are cheaper to operate than diesels. They require less maintenance and have lower energy costs. They are also lighter and do less damage to the tracks.

Spokesperson for Newport Council

Newport needs investment and jobs. The town is the third most deprived place in the whole of Wales. The construction of the relief road would create better access to land to the southwest and east of the steelworks – land that we have set aside for future development for industrial estates.

▲ **Figure 14** Conflicting opinions about the M4 relief road.

Enquiry

Do you think the Welsh government made the correct decision to scrap this project? Justify your answer using evidence from this page.

THEME 7

Social development

Chapter 1 Measuring social development

What is meant by social development?

Development means 'change'. Wealth is one way to measure development, and we examined this in Theme 6 Chapter 1. However, measuring the state of the economy is not the only way to find out whether society is changing for the better. Figure 1 lists a range of ways in which development could be defined.

Development is ...

- reducing levels of poverty
- increasing levels of wealth
- bringing benefits to all, not just to the wealthiest in a society
- reducing the gap between richest and poorest
- creating equal status and rights for men and women
- creating justice, freedom of speech and the right to vote (democracy) for everyone
- making everyone safe from conflict or terrorism
- making sure that everyone has their basic needs of food, water and shelter
- making sure every child has the right to a good standard of education and actually gets it.

▲ **Figure 1** Different ways of seeing development.

▲ **Figure 2** Development is ...

▲ **Figure 3** Development is ...

Activities

1 Work in pairs to discuss Figure 1.
   a) What are the advantages and disadvantages of each of these statements?
   b) Choose the five statements that you think are the best definitions of development. Join with another pair and justify your choice.
2 Working on your own, explain which aspects of human development are illustrated by Figures 2 and 3. Complete the caption for each figure.
3 Describe the pattern shown on:
   a) Figure 4
   b) Figure 5.
4 Which of these two maps is more useful to a geographer who is studying patterns of education across South Africa? Explain your answer.

Enquiry

How much regional inequality is there in South Africa?

Use the weblink to investigate access to clean water supply across South Africa. Remember that water is needed for good health. How similar or different are Figures 4 and 5 to the maps for water? What conclusions can you reach about regional inequality?

How do we measure social development?

There are many ways to measure social development. Many measures relate to health and education, and most can be easily quantified. This is useful because numbers allow us to identify the size of the social development gap. For example, one simple measure is the percentage of adults who had no formal education as a child. In an HIC like the UK this number will be close to zero. But in South Africa, an NIC, this figure was 8.7 per cent at the last census (2011).

We can also use numbers to measure progress. In South Africa, progress in education is good: the percentage of the population aged twenty years and older who had no schooling decreased from 19.1 per cent in 1996 to 8.7 per cent in 2011. However, figures like this can be misleading. There are wide variations in education across South Africa, as shown in Figures 4 and 5.

GEOGRAPHICAL SKILLS

Using choropleth maps

A **choropleth** is a coloured or shaded map. Darker colours or darker shading represent higher values on the map. Choropleth maps are useful for showing patterns. You don't need to know pin-point locations for your data. A choropleth may be drawn when you have data for each area or district on a map. There are two factors to consider when you draw a choropleth:

1 How many colours or shades should you use? Too many colours and patterns will be difficult to see. However, if you have too few then any small differences between areas will be lost.
2 If the data is available for different-sized areas, which data set should you choose? This issue is illustrated by Figures 4 and 5.

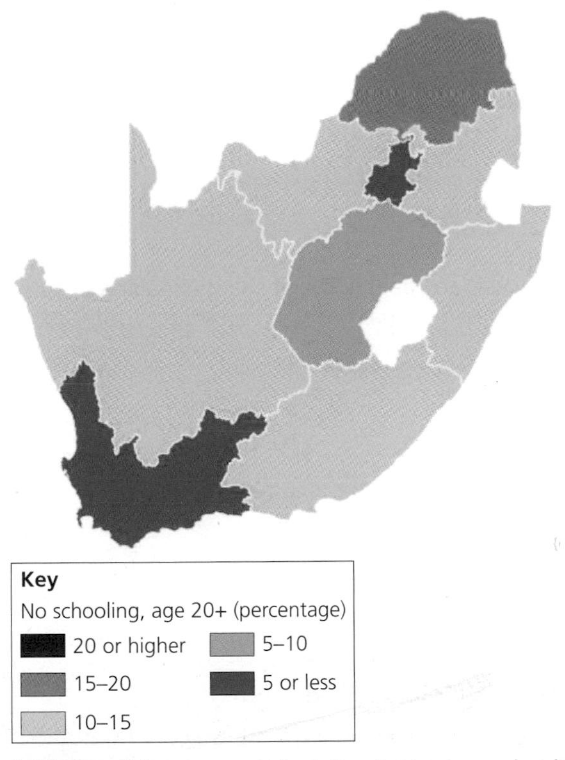

▲ **Figure 4** Percentage of adults who had no schooling using data from each state of South Africa.

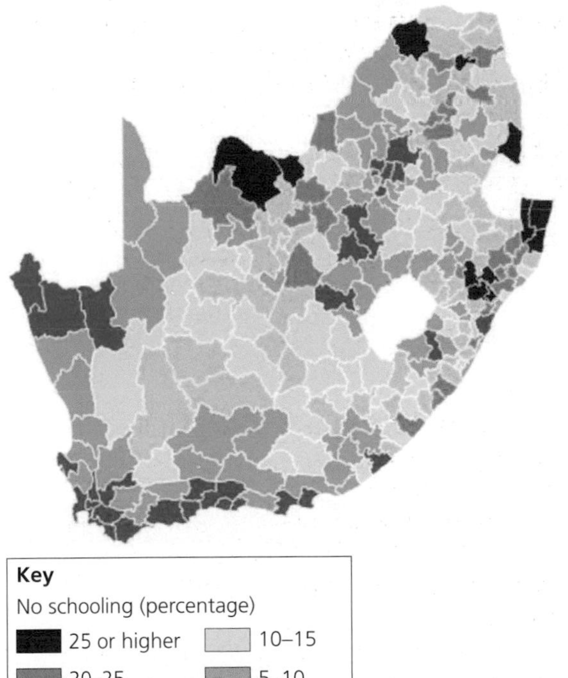

▲ **Figure 5** Percentage of adults who had no schooling using data from each municipality of South Africa.

Weblink 

http://tellmaps.com/mapsalive – this website uses a geographical information system (GIS) to display an online atlas of South Africa.

Using health data as a measure of development

Health data is often used to describe a country's level of development. Two commonly used measures are **infant mortality rate (IMR)** and **average life expectancy.** A number of factors contribute to improved life expectancy and lower IMR. Increased government spending on health care, better access to clean water and improved sanitation (sewage disposal) will all have a beneficial impact on health. That's why health data is a useful indication of development.

Just as it would be simplistic to assume that there is a gap between rich and poor countries (pages 210–13), so it would be too simple to divide the world into healthy and unhealthy countries. There is, in fact, a continuum of social development from countries like Central African Republic (which has an IMR of 88) to Sweden, Japan and Singapore (which have an IMR of around 2). Furthermore, Figures 6 and 7 indicate that there has been much progress, since the Brandt Report in 1980, to close the development gap.

	1985	1990	1995	2000	2005	2010	2017
Gambia	91	80	71	63	57	52	41
Kenya	63	66	72	67	54	42	34
Mali	144	131	125	116	97	83	66
Malawi	147	143	122	104	71	58	39

▲ **Figure 6** Infant mortality rate (IMR) for selected sub-Saharan African countries.

	1985	1990	1995	2000	2005	2010	2017
Bangladesh	118	100	81	64	51	39	27
India	101	88	78	66	56	46	32
Pakistan	115	106	97	88	80	74	61
Sri Lanka	25	18	17	14	12	9	8

▲ **Figure 7** Infant mortality rate (IMR) for selected countries in South Asia.

	High life expectancy and low IMR might indicate that ...	Low life expectancy and high IMR might indicate that ...
Government spending on hospitals and clinic	A wealthy government has prioritised health spending so ...	
Diet		
Access to safe drinking water		
State of sanitation	Most people have clean water piped to their homes so ...	Many people live in shanty housing or rural homes that have no sewerage system so ...
Standards of personal and social education		

▲ **Figure 8** What health data may be telling us about a country's development.

Activities

1 Study Figures 6 and 7.
  a) Classify each country by its World Bank category (see pages 210–13).
  b) Use a suitable graph to represent the data.
  c) What conclusion can you make about the progress of these two groups of countries in closing the development gap?

2 Make a copy of Figure 8.
  a) Complete the three statements by adding elaboration at the end of the sentence.
  b) Complete the blank spaces in the table.

Infant mortality rate (IMR)

The number of children who die before the age of one for every 1,000 who are born.

Average life expectancy

The average age to which a person can expect to live.

Changes to average life expectancy

Some of the biggest improvements in life expectancy since 1980 have been in the poorest countries of sub-Saharan Africa. The biggest improvement (24 years) has been in Central African Republic.

Since 1980, life expectancy has increased in almost every country of the world, and this trend is expected to continue, as shown in Figure 9. The map represents changes to life expectancy. The greater the expected increase in life expectancy, the bigger the country appears:

- Countries with large populations and the biggest improvements to life expectancy are the largest on this map.
- Countries with an already large life expectancy appear smaller, because they have less room for improvement.

Population and health

In most wealthy countries primary health care is generally very good and people have long life expectancy. However, as a country's population ages, causes of death change:

- Deaths in countries with low life expectancy tend to be caused by infectious disease (e.g. HIV/AIDS, malaria), and conditions caused by dirty water (e.g. diarrhoea, which particularly affects children). These deaths are reducing quickly, as global organisations such as UNICEF vaccinate against infectious disease, and as campaigns to get rid of malaria have some success.
- Deaths in wealthier countries with the longest life expectancy are most commonly caused by diet and lifestyle. The main causes of death are heart disease (deaths from which could be reduced by improved diet and exercise), cancer (caused by smoking, poor diet and environmental factors such as poor air quality) which together account for over 80 per cent of deaths.

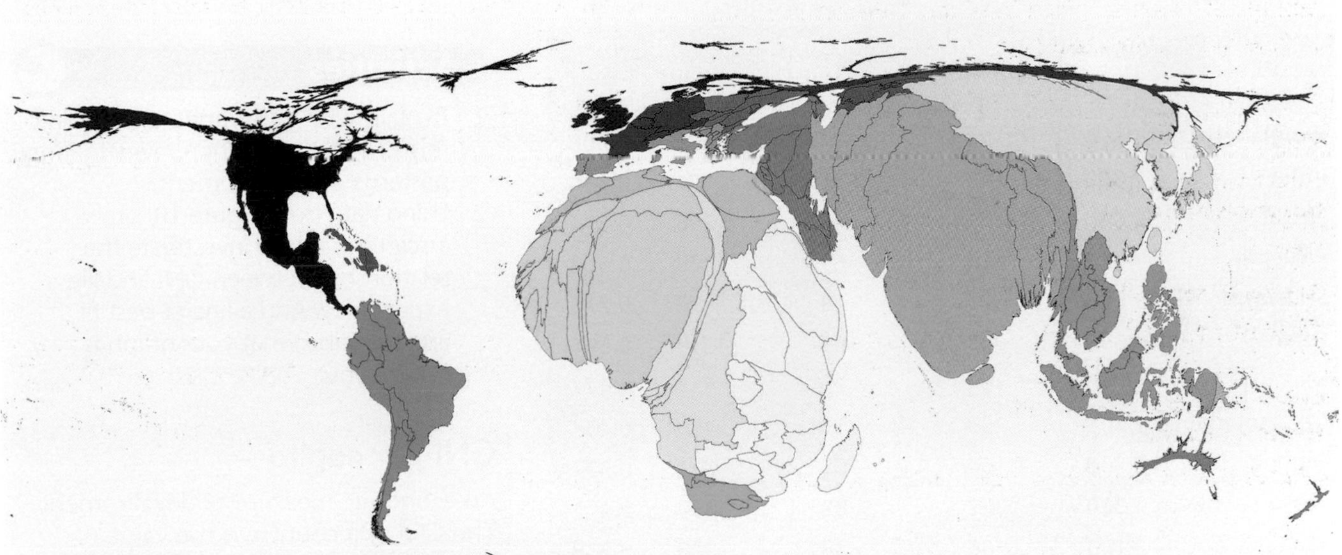

▲ **Figure 9** Expected increase in global life expectancy, 2015–50.

Activities

3 Study Figure 9. Describe the distribution of countries which will improve their life expectancy most.

4 Suggest reasons why life expectancy has increased in some countries more than others.

5 a) Explain why not all people in wealthy countries have healthy diets.
  b) Explain what effect these diets might have on health data.

Enquiry

How and why does health vary across the globe?

- Research the phrase 'main causes of death in the world's poorest and wealthiest countries'.
- Choose one major cause of death that is preventable and explain how the number of deaths could be reduced.

Human Development Index

Economic development usually brings social change. So, as GNI per capita increases, a number of social indicators tend to change as well.

- There is more money to spend on health, for example, fighting disease or spending on doctors and hospitals. As health improves, so death rates fall, as do infant and maternal mortality rates. Overall, life expectancy increases.
- As more money is spent on education, more children spend longer in school and so literacy rates increase.

Human Development Index (HDI)

A way of measuring development that takes into account both economic and social measures. HDI gives a single figure per country, between 0 and 1. The closer the figure is to 1, the better.

The UN uses the **Human Development Index (HDI)** as a way of showing a country's level of development. HDI is calculated using an average of four development indicators:

- education – average length of schooling in years
- education – literacy (as a percentage of the adult population)
- **GNI per capita (PPP)** in US dollars
- life expectancy in years.

The advantage of using HDI rather than just GNI is that it identifies those countries which have not used their wealth to benefit people's health or education. For example, in Figure 10, Nigeria has a relatively high GNI but a low HDI. Nigeria is an oil-rich country but most of its wealth goes to a few wealthy families and multinational companies. In the case of Nigeria, increased wealth seems to make little difference to the majority of the people.

Country	GNI in US$ (PPP)	Life expectancy	Infant mortality rate	Maternal mortality rate	HDI
Bangladesh	4,560	73	27	176	0.608
India	7,680	69	32	174	0.640
Nepal	3,090	71	28	258	0.574
Pakistan	5,840	67	61	178	0.562
Sri Lanka	13,090	75	8	30	0.770
Ghana	4,650	63	36	319	0.592
Malawi	1,310	64	39	634	0.477
Mozambique	1,300	59	53	489	0.437
Nigeria	5,700	54	65	814	0.532
Sierra Leone	1,520	52	82	1,360	0.419
Tanzania	3,160	66	38	398	0.538
Uganda	1,970	60	35	343	0.516
Zambia	4,100	62	42	224	0.588

Selected South Asian countries

Selected sub-Saharan African countries

▲ **Figure 10** GNI and social development indicators for selected countries in South Asia and sub-Saharan Africa.

Life expectancy: the average number of years a person can expect to live when they are born.
Infant mortality rate: the number of children per thousand live births who die before their first birthday.
Maternal mortality rate: the estimated number of mothers per 100,000 who die in childbirth.

Activities

1. Explain why geographers shouldn't rely only on HDI when investigating patterns of development.
2. Using data from Figure 10, draw a scattergraph to investigate the relationship between GNI and life expectancy. Add a line of best fit and describe what your graph is telling you.

GNI per capita

An economic measure of development. The GNI of a country is the value of goods and services produced by people living in that country and by people overseas who are still citizens of that country. GNI per capita is that value divided by the population. Think of it as an average income. It is measured in US dollars.

GNI per capita (PPP)

A version of GNI that takes into account the cost of living. In some countries, everyday items such as food are cheaper than in others, so your income goes further. GNI (PPP) takes this into account.

GEOGRAPHICAL SKILLS

Testing relationships between sets of data

Is economic development related to improved health care and education? It is possible to answer this question by analysing data, using graphs. Before starting, it helps to make a prediction, or hypothesis, for example 'countries with lowest GNI have low HDI indicators'. Testing this hypothesis needs two sets of data that are related – known as bivariate data. In this case, the data are GNI per capita (PPP) and HDI. Find these columns of data in Figure 10. These data are known as variables. To test their relationship, you should draw a scattergraph. Follow these steps:

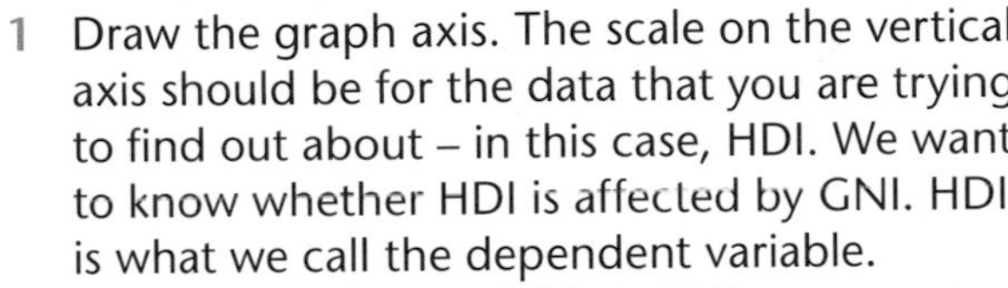

1 Draw the graph axis. The scale on the vertical axis should be for the data that you are trying to find out about – in this case, HDI. We want to know whether HDI is affected by GNI. HDI is what we call the dependent variable.
2 The independent variable is GNI – we want to test to see whether it affects HDI. This goes along the horizontal axis.
3 Label the axes 'HDI' (vertical) and 'GNI' (horizontal).
4 Plot the points for each country. These are called 'scatter points', and the graph is therefore called a scattergraph.
5 Your points should look like the one in Figure 11. There is a pattern to the plotted data – a line has been plotted to help you to identify this. This is known as a line of best fit. It does not join up points, but simply follows their general trend. There should be the same number of points on each side of the line of best fit.
6 In this case, the best fit line shows that as one indicator increases – (GNI) so does the other (HDI). The hypothesis has been proved correct!

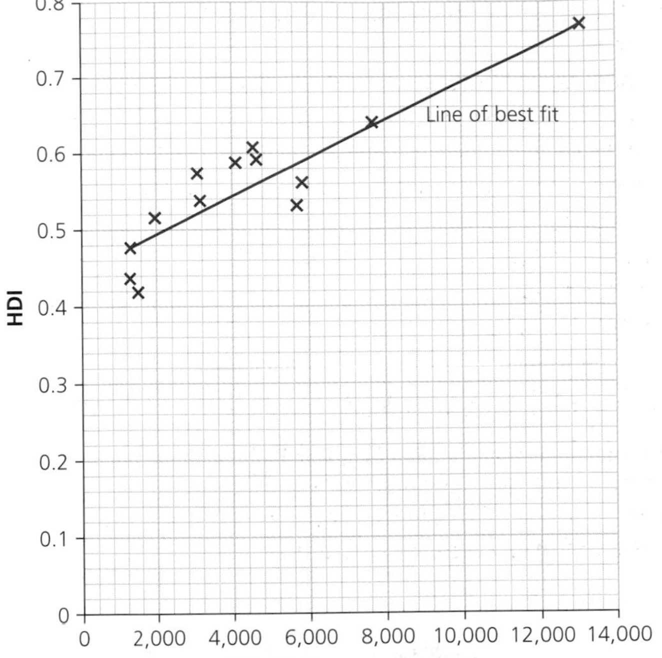

▲ **Figure 11** A scattergraph to test the relationship between HDI and GNI using data from Figure 10.

Enquiry

What should Nigeria do to reduce inequality?

- In pairs, devise hypotheses to test the relationship between any two other pairs of data shown in Figure 10.
- Draw a scattergraph with a best fit line to investigate this relationship.
- Describe what your graph is telling you.
- Explain how well the hypothesis that you devised has worked out.
- Based on what you have found out, write a 400-word report to the Nigerian government about why you think more money should be invested in health and education, and what benefits the government would get in return.

Chapter 2 Contemporary development issues

Population issues

Just over 3 billion people live in the regions of sub-Saharan Africa and South Asia. Sub-Saharan Africa has 48 countries, 24 of which are LICs like Malawi and the Gambia. South Asia has eight countries, including India, a very large NIC. Afghanistan and Nepal are the only LICs in South Asia.

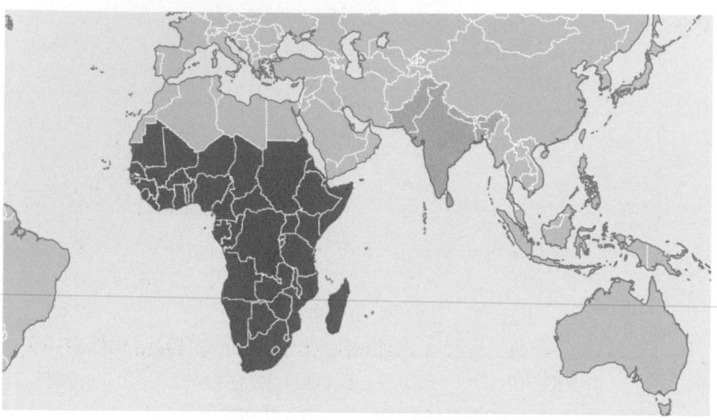

Population (million)	2019	2035	2050
Sub-Saharan Africa	1,065.3	1,579.9	2,148.4
South Asia	1,945.9	2,274.5	2,473.1
World	7,691.5	8,932.4	9,854.2

▲ **Figure 1** Projected population change in sub-Saharan Africa and South Asia (figures in millions).

Population structure and change

Populations change in size and in age/gender structure over time. LICs like Malawi tend to have higher **birth rates** and higher **death rates** than MICs or NICs like India or Bangladesh. Study Figure 2. Countries like Malawi, which have higher birth rates and higher death rates, have **youthful population** structures. This population pyramid has a very wide base and it narrows very quickly. This shows that Malawi has a large proportion of children and young adults, but only a few people live into older age. India's population has a different structure. The death rate in India is much lower than in Malawi, so a greater proportion of people live into older age. The birth rate is also lower than Malawi's and it is still falling, so India has a lot of young adults, but the proportion of children is much smaller than in an LIC.

A number of social, political and economic factors influence the way in which the birth and death rates change over time. **Social factors** include the levels of education and gender equality. Countries where women are treated fairly in society and which have good education systems have low birth rates.

Political factors include the amount of government spending on education and health care. LICs tend to have higher death rates from diseases such as malaria or diarrhoeal diseases, which could be preventable with greater spending on health care. In some countries, the government may promote family planning and encourage the use of birth control to bring down the birth rate.

Average household income is an example of an **economic factor** that can influence birth and death rates. In LICs, the average income is under US$1,025 a year. Parents rely on their children to support the family income, so larger families are common. Poor families may not have access to clean water, so disease is common and death rates are higher.

Birth rate

The number of births in each year for every 1,000 population.

Death rate

The number of births in each year for every 1,000 population.

Youthful population

A population with a high proportion of children because the birth rate is relatively high.

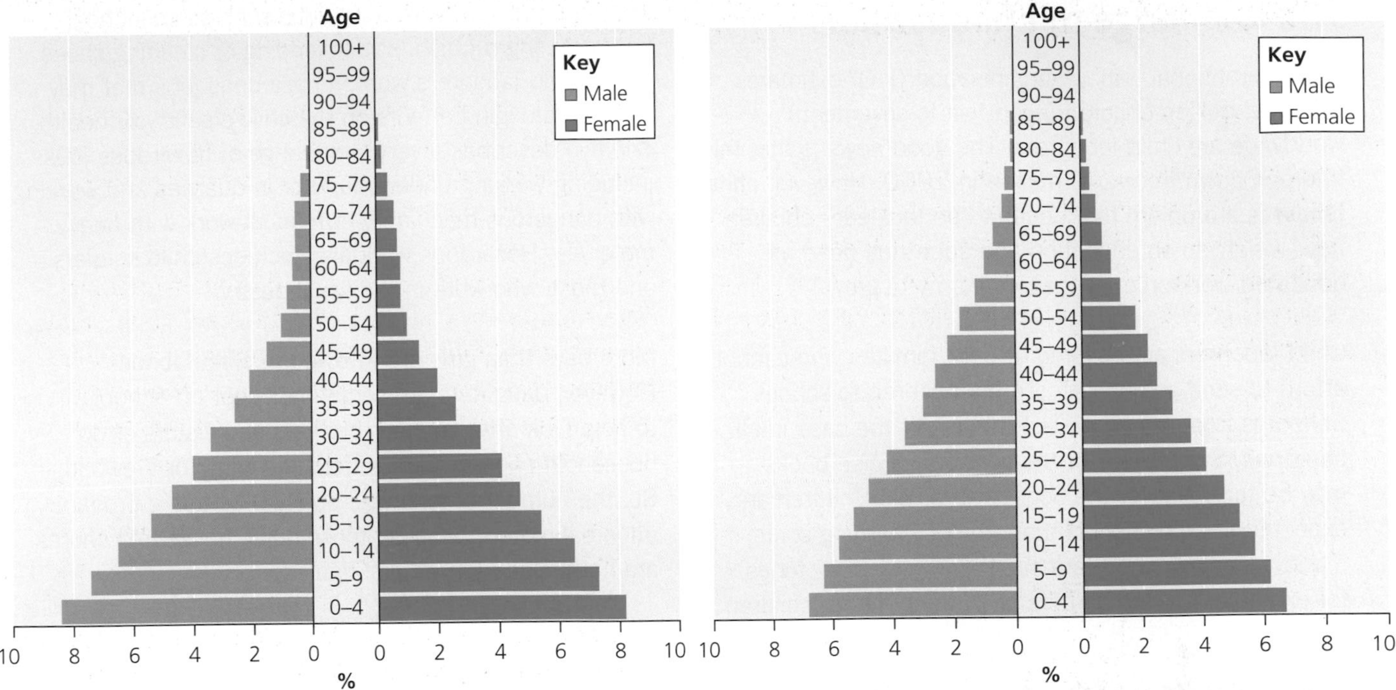

▲ **Figure 2** Population structure of Malawi (2019).

▲ **Figure 3** Projected population structure of Malawi (2039).

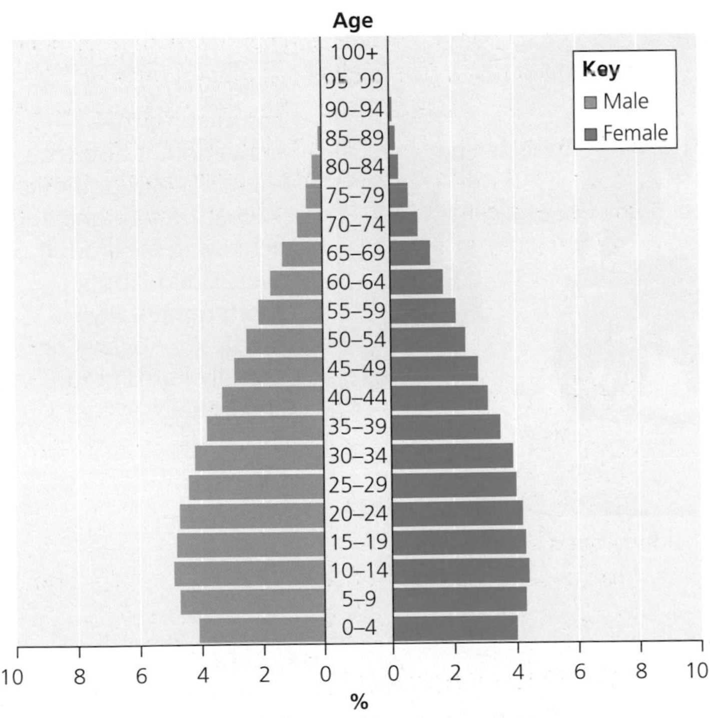

▲ **Figure 4** Population structure of India (2019).

Activities

1. Use the data in Figure 1 to:
   a) Draw a suitable graph.
   b) Calculate the percentage of world population in sub-Saharan Africa in each year. What does it suggest is happening to the birth rate and death rate in this region?
2. Use Figures 2 and 4 to compare the population pyramids of Malawi and India.
3. a) Use Figures 2 and 3 to describe how the population of Malawi is expected to change in the next two decades.
   b) Suggest possible reasons for this change. Think about social, political and economic factors.

The challenge of child labour

The International Labour Organisation (ILO) estimates that 152 million children (aged five to seventeen) worldwide are child labourers. The good news is that this figure is down from 246 million in 2000. However, **child labour** is a problem that needs to be tackled – children have a right to an education and countries need an educated workforce for their economy to grow.

Child labourers are children of poor families who cannot afford to send some or all of their children to school. School is free in the UK, but this is not the case in all countries. School fees or the cost of exercise books may be too great for the poorest families. Children are expected to help support the family by earning some money. Many work on farms. Others work in factories, for example, making carpets or clothing. Other children work in the informal sector – jobs such as collecting and recycling rubbish, street vending or cleaning shoes. Some child labourers work in hazardous jobs that may lead to damaging effects on the child's safety or health. The ILO describes a very wide range of hazardous jobs, including working underground or in quarries and working with dangerous machinery or manual work with heavy materials. Hazardous work also includes child soldiers and those who work in the sex industry.

More boys than girls are involved in child labour. However, daughters are sometimes kept off school to help look after younger brothers and sisters or do household chores, such as fetching water or firewood. So, the number of girls who do child labour is probably underestimated, because those doing household chores are not included in official figures.

▲ **Figure 5** Kavita, who is twelve years old, works along with her family at a brick kiln on the outskirts of Jammu, India.

Child labour

Work that robs children of their childhood. Because child labour prevents children from completing their education, it also robs them of their potential – what they might achieve in their adult lives. Child labour is often mentally, physically, socially or morally harmful to children.

Region	Child labour		Hazardous work	
	Millions	Per cent	Millions	Per cent
Africa	72.113	19.6	31.538	8.6
Arab states	1.162	2.9	0.616	1.5
Asia and the Pacific	62.077	7.4	28.469	3.4
Americas	10.735	5.3	6.553	3.2
Europe and Central Asia	5.534	4.1	5.349	4.0

◀ **Figure 6** Child labour and children working in hazardous jobs by region, 2016.

How should we tackle child labour?

The ILO works at an international scale to end child labour. They collect data so that targets can be set and progress can be monitored – like that shown in Figure 8. They cannot pass new laws banning child labour – this is the responsibility of individual governments – but they can advise governments on policies that work based on their experience of working with countries in which child labour is falling. The ILO recommends that governments look at two main areas:

- Improving access to education, so that all children are in school.
- Improving social security systems, so that the poorest and most disadvantaged members of society can rely on the state to give them support rather than relying solely on their children.

Official figures show that India has about 10 million child labourers, although the actual number could be much higher. About 43 per cent of these children do not attend school at all. Some children have to work to help pay off a family debt – it is a modern form of slavery. India is making slow progress at reducing child labour. For example, in rural parts of Tamil Nadu and Madhya Pradesh, trade unions are working with local leaders and employers to create child-labour-free villages.

152 million children in child labour. This is roughly equal to the entire population of the UK, Germany and Spain!

64.1 million girls involved in child labour

109 million child labourers work in farming

87.5 million boys involved in child labour

72.5 million children work in hazardous jobs

▲ **Figure 7** Child labour in numbers, 2016.

Activities

1 Explain why some poor parents do not send all of their children to primary school.
2 Study Figure 6. Which region has:
   a) the largest number of children in child labour and hazardous work?
   b) the largest percentage of children in child labour and hazardous work?
3 Use evidence from Figures 6 and 7 to create a poster that presents information about child labour in graphical form.
4 Describe the trend shown in Figure 8.
5 Explain why the ILO cannot prevent child labour in India.

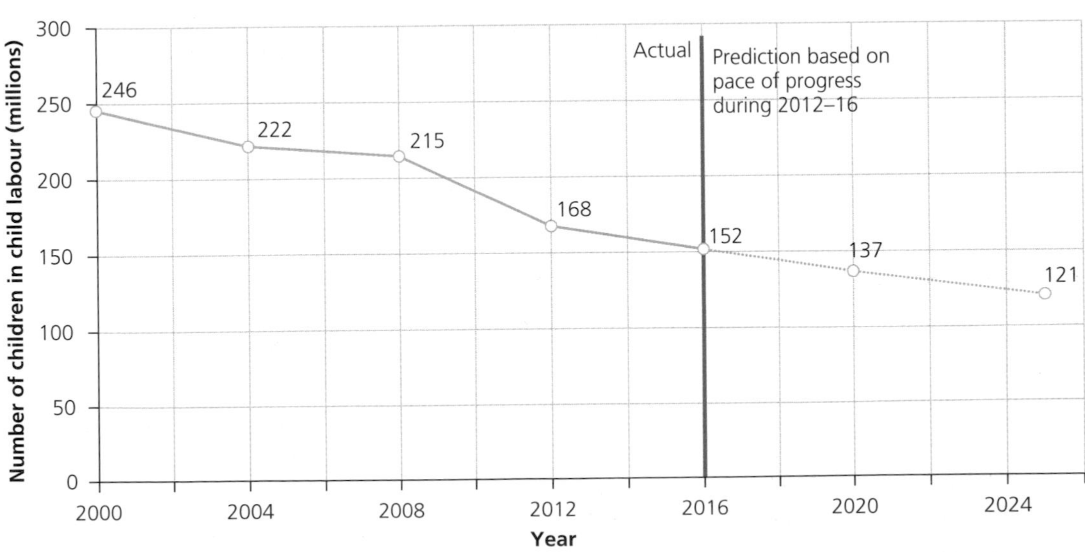

▲ **Figure 8** Actual and predicted trends in child labour.

Education of girls

Despite progress in many countries, millions of children still miss part or all of their education. UNESCO states that 16 million girls will never set foot in a classroom. Of the 750 million adults without basic literacy skills, about two thirds are women.

Large gender gaps still occur in schools of many countries in sub-Saharan Africa and South Asia. Many obstacles prevent girls from getting a full education. These include poverty, rural isolation, minority status of some ethnic groups, disability, early marriage and pregnancy, and traditional attitudes about the status and role of women. In countries such as India, where more women than men are uneducated, there are serious consequences, for example:

- The child of an uneducated mother is twice as likely to die before the age of one as a child whose mother had a full education.
- Women who are well educated tend to marry later and have smaller families.
- Education (especially at secondary and university level) empowers women, i.e. it gives them a higher status and better chances in life.

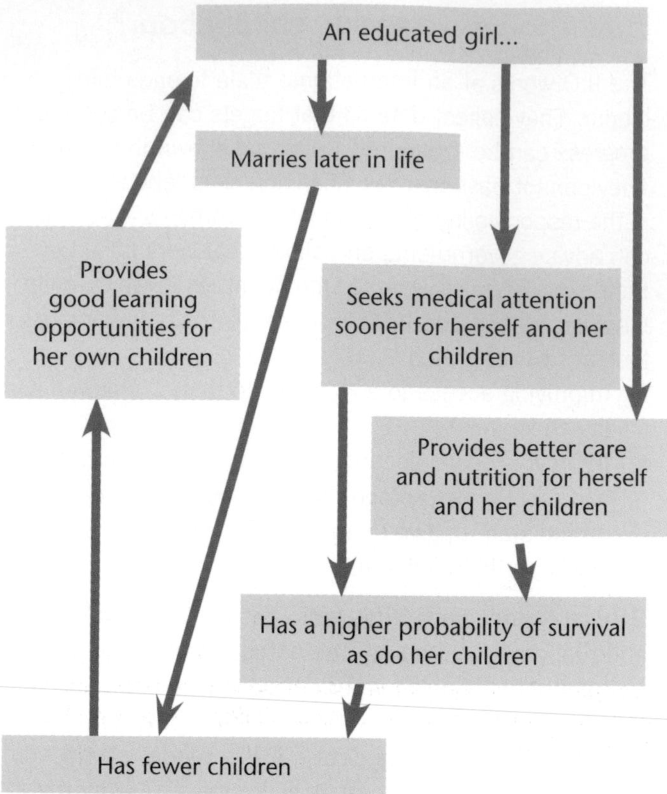

▲ **Figure 9** The advantages of improving education for girls.

Breaking free from child labour

Until a year ago, eight-year-old Laxmina was too busy working to even think of going to school. She earned about 30 rupees (less than a dollar) a day in return for delivering milk to nearby villages more prosperous than her own.

But all that has changed now. Laxmina has been attending an alternative learning centre (ALC), along with 40 other children in Uttar Pradesh. It is one among many such centres set up four years ago with UNICEF support to help educate children who have never been to school.

Over twenty per cent of India's working children are from Uttar Pradesh. Most of them work at odd jobs, in factories and in the carpet industry, for meagre wages. But their labour plays a key role in supplementing their families' meagre income. One of the main reasons for the high prevalence of child labour in these areas is the burden of debt, which forces families to send their children to work.

UNICEF addresses the issue of child labour through a combination of approaches. These include changing attitudes of parents, forming self-help groups, improving the quality of mainstream education, providing transitional schools to return children to learning levels appropriate to their age.

Education plays a crucial role in eliminating child labour. UNICEF's approach therefore focuses on motivating communities to send girls and boys (who have never been to school or who have dropped out) to ALCs.

The centres have been set up mostly in areas that do not have a school within a 1.5 km radius. Each ALC has around 40 students. The aim is to help children complete primary education – which normally takes five years – within three years. At the end of this period, the children are integrated into formal school.

The initiative, funded by IKEA (with about US$500,000), through UNICEF's German National Committee has covered around 650 villages in two districts of Uttar Pradesh. There are currently around 200 ALCs. These provide an education for more than 7,000 children, of whom 55 per cent are girls.

▲ **Figure 10** Text extract from UNICEF India website. It is estimated that 23 per cent of children in Uttar Pradesh are involved in child labour; about 70 per cent of the population can read and write, and 57 per cent of women are literate.

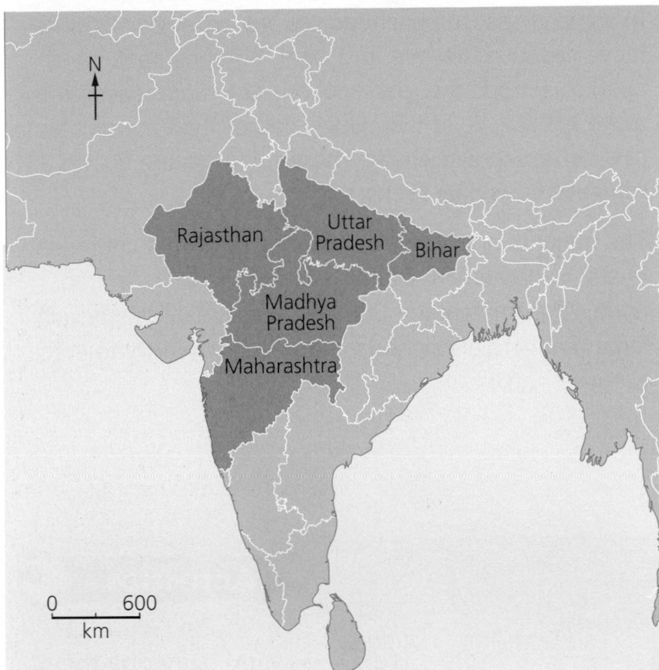

▲ **Figure 11** Child labour in India is highest in the five states of Bihar, Uttar Pradesh, Rajasthan, Madhya Pradesh and Maharashtra.

	Poverty (%)	Sex ratio	Female literacy (%)
Andhra Pradesh	9.20	993	59.1
Arunachal Pradesh	34.67	938	57.7
Assam	31.98	958	66.3
Bihar	33.74	918	51.5
Chhattisgarh	39.93	991	60.2
Goa	5.09	973	84.7
Gujarat	16.63	919	69.7
Haryana	11.16	879	65.9
Himachal Pradesh	8.06	972	75.9
Jammu and Kashmir	10.35	889	56.4
Jharkhand	36.96	948	55.4
Karnataka	20.91	973	68.1
Kerala	7.05	1,084	92.1
Madhya Pradesh	31.65	931	59.2
Maharashtra	17.35	929	75.9
Manipur	36.89	985	70.3
Meghalaya	11.87	989	72.9
Mizoram	20.40	976	89.3
Nagaland	18.88	931	76.1
Orissa	32.59	979	64.0
Punjab	8.26	895	70.7
Rajasthan	14.71	928	52.1
Sikkim	8.19	890	75.6
Tamil Nadu	11.28	996	73.4
Tripura	14.05	960	82.7
Uttar Pradesh	29.43	912	57.2
Uttarakhand	11.26	963	70.0
West Bengal	19.98	950	70.5

▲ **Figure 12** Selected data for Indian states; sex ratio is the number of women for every 1,000 men in the population; where men and women have equal health opportunities, this figure is close to 1,000 (data from 2011 census).

Activities

1 Study Figure 10.
   a) Describe how this project is helping to reduce child labour.
   b) Suggest the strengths and limitations of this project.
2 Use Figure 9 to extend the explanation of each of the following statements.
   An educated mother...
   - spots the early signs of ill health in her child so ...
   - understands the importance of a balanced diet so ...
   - recognises the importance of a full education for her daughter so ...
3 Use the data in Figure 12 to investigate the relationship between poverty and female literacy.
   a) Create a hypothesis.
   b) Draw a scattergraph to investigate your hypothesis (see page 267).
   c) What conclusions can you reach?

Enquiry

How closely are education and health issues linked?

Use the data in Figure 12 to investigate the relationship between female literacy and sex ratio in Indian states. If there is a relationship, how can you explain it?

Refugees

Most people become migrants because they believe that their move will end in a better job and quality of life. Such migrants are usually called **economic migrants**. They expect that their move will result in better pay, and better training and career opportunities. These benefits are described as pull factors.

However, the decision to migrate is also influenced by push factors – problems that are pushing people away from their existing homes. Push factors include:

- unemployment or underemployment and poverty
- natural hazards, political change, persecution or conflict.

In some cases, the push factors are extreme. Migrants move because they are in danger, perhaps because of conflict, terrorism or persecution. The main reason they move is because of this danger rather than the pull factor of a better job elsewhere. These migrants are known as **refugees** or **asylum seekers**.

The majority of the world's refugees are from countries in sub-Saharan Africa and Asia. The conflicts in Iraq/Syria, Afghanistan and Somalia have created especially large numbers of migrants – some of whom try to enter Europe.

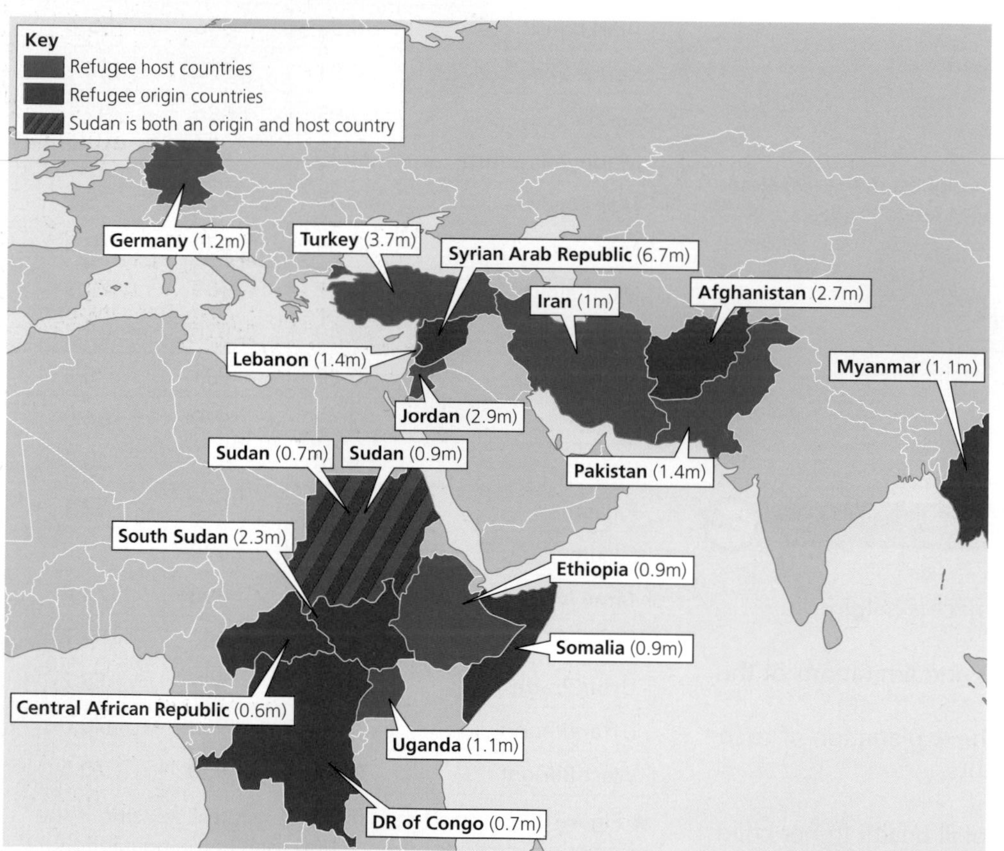

▲ **Figure 13** The number of people forced to move due to conflict (millions).

Activities 

1 Study Figure 13.
   a) Describe the pattern of refugee source countries and refugee host countries.
   b) Research some of the host countries. Are they rich or poor countries? Suggest how these host countries may be affected.

2 Describe an effective way that each number in Figure 15 could be represented in a form that would grab attention.

Country	Refugees (millions)
Syrian Arab Republic	6.7
Afghanistan	2.7
South Sudan	2.3
Myanmar	1.1
Somalia	0.9

▲ **Figure 14** The five main sources of refugees, 2018.

37,000 people were forced to leave their homes each day

84 per cent of the world's refugees are cared for by developing countries – not HICs

51 per cent of the world's refugees are children

▲ **Figure 15** Refugees in numbers, 2018.

The refugee crisis in Lebanon

Lebanon is a small country of 4.5 million people in the Middle East. A civil war has been fought in the neighbouring country of Syria since 2011. In 2018, there were approximately 1.5 million Syrian refugees living in Lebanon and 74% lacked legal status, so could be forced to return. Living conditions for the refugees are harsh. Some live in refugee camps – tent cities – while others are squatters, living in overcrowded buildings in Lebanon's cities.

Lebanon is a relatively poor country: GNI per person is US$9,500, and there is not enough work for the local population and the refugees. It is estimated that the refugee crisis costs Lebanon US$4.5 billion each year. Access to clean water and proper sanitation are major concerns. Lebanon has a dry, Mediterranean-style climate and clean water is scarce. Over half of refugees consider that local water supplies are unsafe.

▲ **Figure 16** A school for Syrian refugees in a garage under an apartment building in Minyara, Lebanon.

We are not living – we are surviving. I live in an overcrowded tenement block with hundreds of other refugees. I have no privacy. My husband made the dangerous journey by boat to Europe. He is now in Germany. One day, perhaps, we will join him.

Syrian refugee mother of three children

I no longer go to school. I do odd jobs here to try to support my family. I want to save US$1,000 – that's how much the people smugglers charge to get you into Europe by boat. I know that it's dangerous and that people drown, but there is no future here.

Refugee boy, aged 15

▲ **Figure 18** The views of a Syrian refugee and her son.

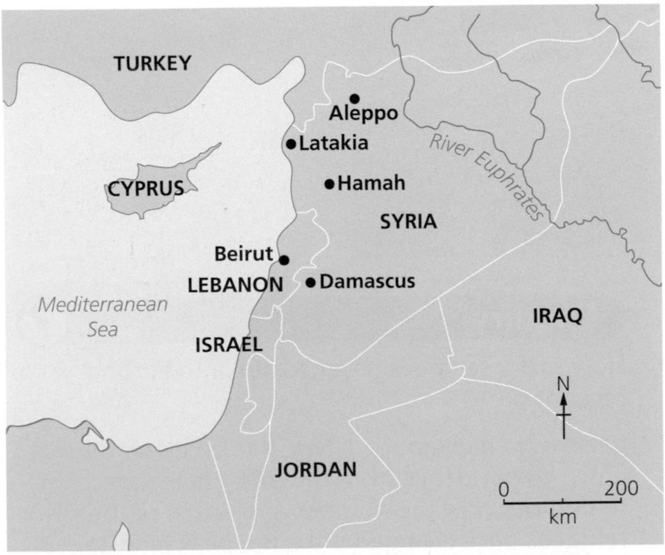

TURKEY
Aleppo
Latakia
CYPRUS
Hamah
SYRIA
River Euphrates
Beirut
LEBANON
Damascus
Mediterranean Sea
ISRAEL
IRAQ
JORDAN
N
0 200 km

▲ **Figure 17** The location of Syria and Lebanon.

Activities

3 Describe the location of Lebanon.
4 What is the ratio of refugees to local population in Lebanon?
5 a) Describe the school conditions in Figure 16.
  b) Describe one strength and one limitation of this UNICEF project.

Enquiry

What challenges face Lebanon and the Syrian refugees living there?

Use evidence from this page, and your own research, to sort the challenges facing refugees in Lebanon under these headings: economic, social and environmental.

Refugee movements into Europe

The Schengen Agreement (1995) enables the easy movement of people between most EU states. Passports do not usually have to be shown when people travel from one country to another within the **Schengen Zone**. However, Europe's leaders want to control the movement of people from non-EU countries – including refugees and economic migrants from African and Asian countries. So migration into Europe is controlled carefully. This has led to some migrants trying to enter Europe illegally, avoiding the border control agencies.

Most illegal migrants from Africa or Asia try to enter Europe by making a sea crossing across the Mediterranean into one of the countries in southern Europe. They usually pay people smugglers large amounts of money to transport them across on small, overcrowded fishing boats or inflatables. Many of these boats are not seaworthy and have no navigation equipment. The immigrants have no food and little water onboard, and the boats rarely have life rafts. In doing so, the migrants are put in great danger and many have drowned during attempted crossings.

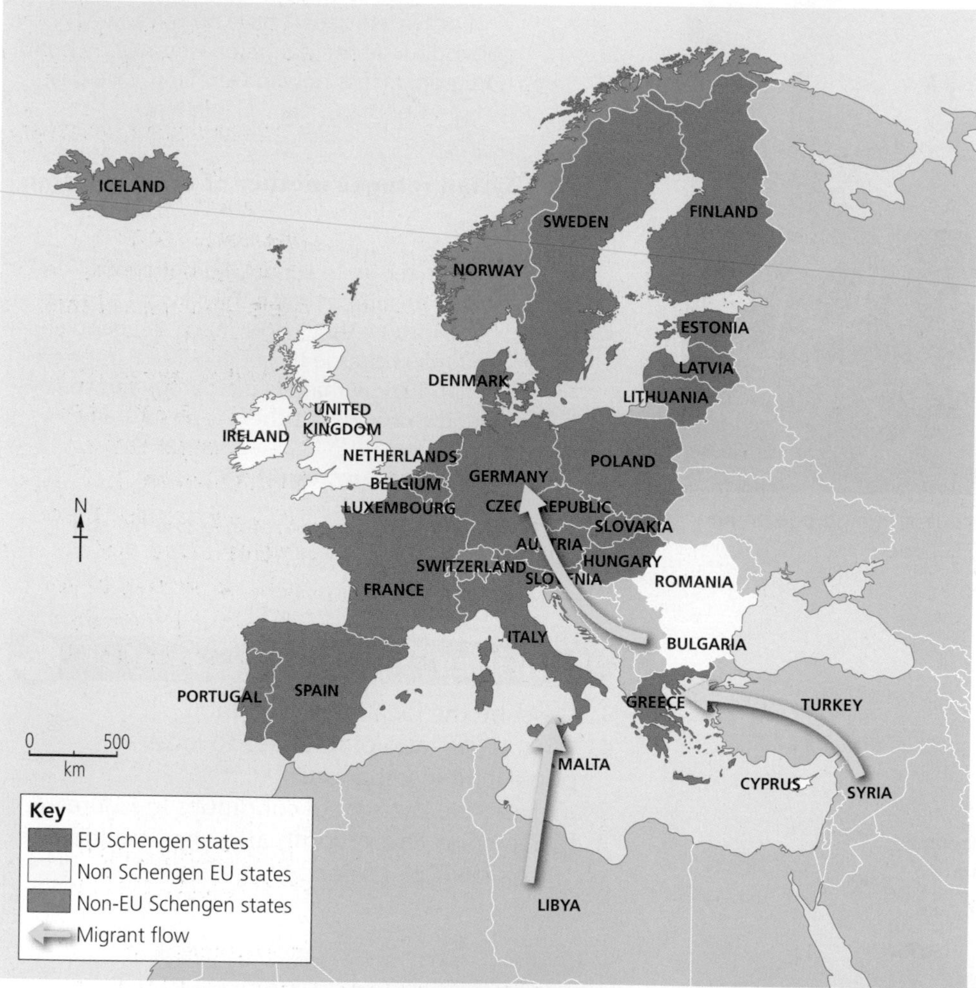

◀ **Figure 19** Migration routes into and across Europe.

Weblinks

Data about refugee movements across the Mediterranean from the UN Refugee Agency.

https://data2.unhcr.org/en/situations/mediterranean

Enquiry

How many refugees have crossed into Europe recently?

Use the weblink to the UNHCR to research the crisis. Use evidence from this and other sites to find out how the number and origin of the refugees have changed since the peak in 2015.

Crossing Europe

When illegal migrants manage to enter Europe, their problems do not necessarily end. Many want to travel across Europe to join family or friends who moved earlier. For example, some Iraqi refugees try to reach the UK as they have relatives already living here. But not every country in Europe is a member of the Schengen Zone. This creates blockages, where migrants are held at borders, unable to move. One such blockage is at Calais, in Northern France. Here, migrants and refugees wait for an opportunity to be smuggled on to a lorry that is crossing into the UK.

▲ **Figure 20** A shanty home occupied by refugees in Calais.

Tackling the refugee crisis

There are different national approaches to refugee movements within Europe. Some countries, including Germany and Sweden, have welcomed refugees and try to integrate them into society. They see the refugees as victims who need help – 44 per cent of refugees to Europe in 2019 were women and children. They also believe that refugees will help their economies – many Syrian refugees are highly educated.

Other countries, like Austria, have tried to limit the number of refugees entering the country to just 80 per day. They claim that, for security reasons, it would be difficult to take more.

It has also been necessary to get international agreement between several European countries during the crisis. Italian and Greek coastguards have worked extremely hard to save migrants who are in danger in the Mediterranean. The UK navy has also been involved. One solution to the crisis may be to seize and destroy the boats used by people smugglers.

Activities

1. Explain why the Schengen Zone has made the movement of refugees across most of Europe easier.
2. Study Figure 19.
   a) Describe two different routes used by illegal migrants and refugees entering Europe.
   b) Explain why routes into southern Europe are dangerous.
   c) Explain why many migrants take the longer route through Bulgaria.
   d) Explain why refugees in Greece found it difficult to get to Germany when Austria restricted entry at its border.
3. Use evidence from Figure 16 to suggest how living in the Jungle may affect quality of life for migrants.
4. Describe the potential opportunities and challenges created when refugees from Syria settle in European countries.
5. Explain why international agreements are necessary to help tackle the issue of refugee movement into and across Europe.

Infant mortality

Infant mortality rate is the number of children per 1,000 births who die before the age of 1 each year. The country with the highest infant mortality rate is the Central African Republic, which has 87 deaths for every 1,000 births. That means that, for every 11 births, one child will die. There are 26 countries in the world where the infant mortality rate is greater than 50 and 25 of these are in Africa (the other is Afghanistan).

Investigating the impact of malaria

Malaria is spread by mosquitoes carrying a parasite that infects a person when bitten. It is an entirely preventable disease. Malaria is a health threat in tropical regions of the world. Around 40 per cent of the world's population live in areas where malaria is endemic (i.e. malaria is constantly present) but 80 per cent of deaths occur in just 15 countries, and most of these are in sub-Saharan Africa. About 65 per cent of deaths occur in children under 5.

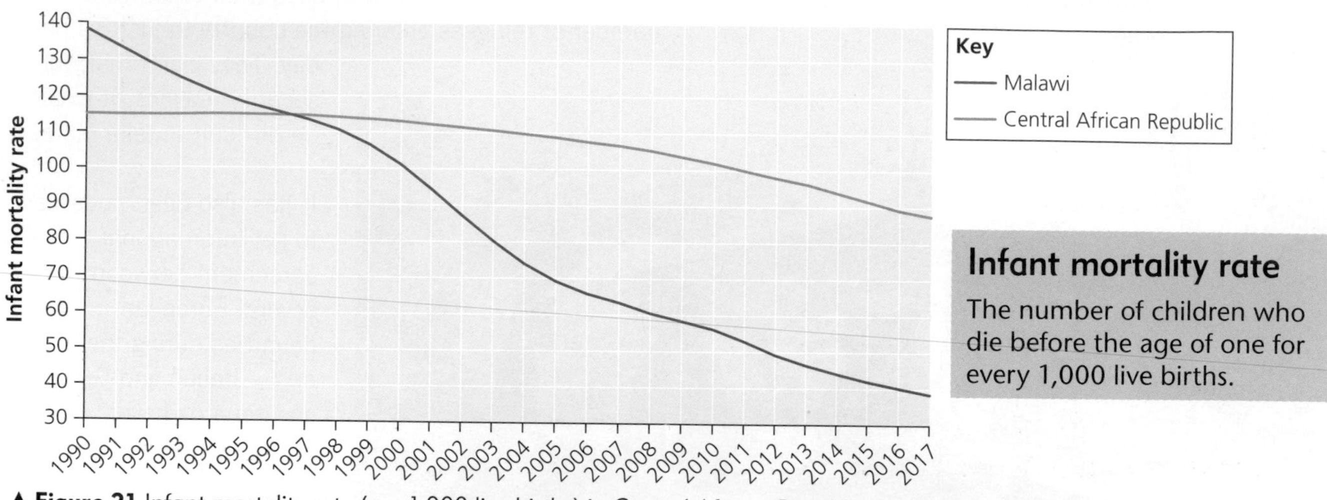

▲ **Figure 21** Infant mortality rate (per 1,000 live births) in Central African Republic and Malawi (1990–2017).

Infant mortality rate

The number of children who die before the age of one for every 1,000 live births.

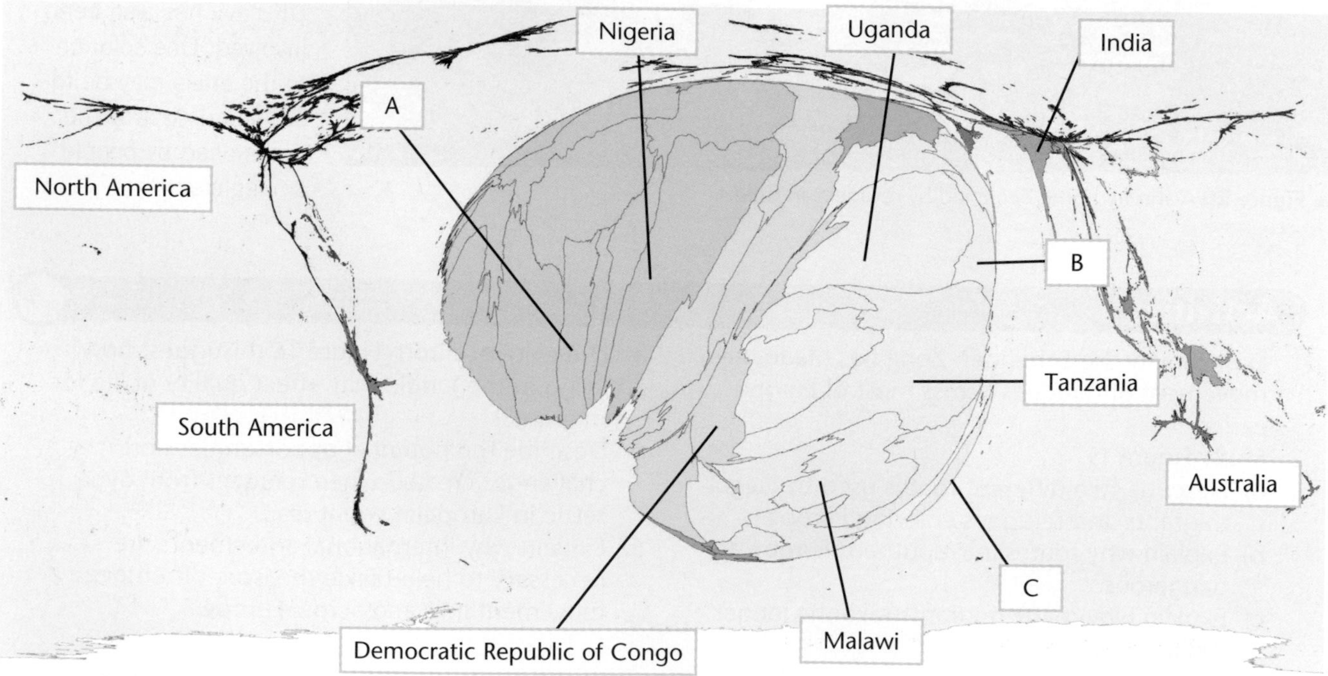

▲ **Figure 22** Deaths from malaria in 2016; countries are shown in proportion to the number of deaths from malaria.

Malaria in Malawi

Malaria is one of Malawi's most serious health problems. In Malawi, the risk from malaria varies across the country and also throughout the year. The shores of Lake Malawi provide ideal breeding conditions for mosquitoes, with warm temperatures and stagnant water. The highland areas are generally cooler and drier. Here, malaria is seasonal with cases reaching a peak during the rainy season when ditches and puddles quickly form and attract mosquitoes. Around 83 per cent of Malawi's population live in rural areas. Deaths from malaria are significantly higher in rural areas, such as Nkhata Bay, than they are in the cities of Blantyre and Lilongwe. Houses in rural areas are often constructed of mud bricks with thatched roofs. These homes offer little protection from mosquitoes.

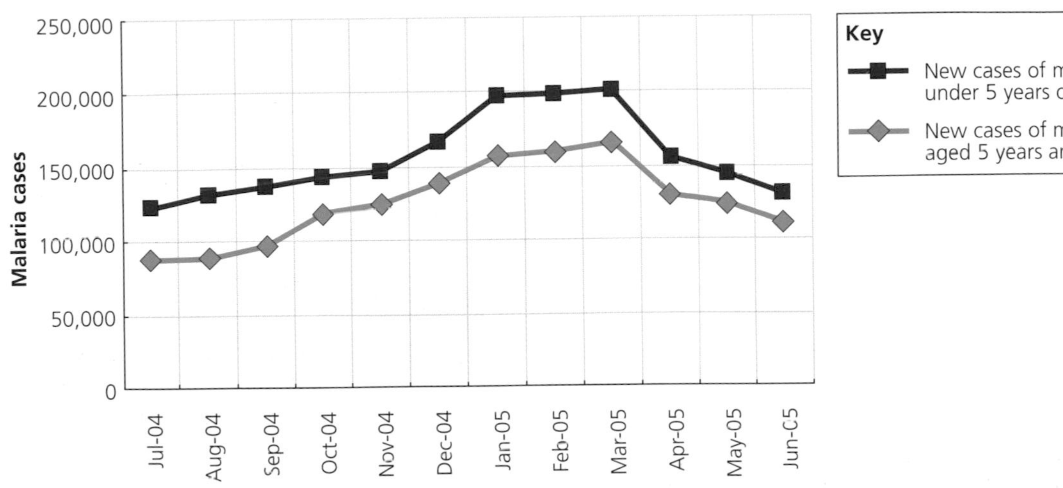

▲ **Figure 23** New malaria cases by month in Malawi.

	Nkhata Bay		
	Mean min. temperature (°C)	Mean max. temperature (°C)	Average precipitation (mm)
Jan	21	30	232
Feb	21	30	237
Mar	21	30	260
Apr	19	29	322
May	17	28	97
Jun	14	27	71
Jul	13	26	47
Aug	14	29	9
Sep	15	30	2
Oct	17	34	11
Nov	20	34	90
Dec	21	32	181

▲ **Figure 24** Climate for Nkhata Bay, 2004–13 averages.

Activities

1. Use Figure 21 to compare how infant mortality has changed in these two sub-Saharan African countries.
2. Use Figure 22 and an atlas to find out:
   a) which tropical region has very few cases of malaria
   b) the names of countries A, B and C
   c) which region of the world, after sub-Saharan Africa, has the next largest number of cases of malaria.
3. Study Figure 23. Describe the annual pattern of new malaria cases for children under five. Refer to figures in your answer.
4. Use Figure 24.
   a) Draw climate graphs for Nkhata Bay.
   b) Describe the annual pattern of rainfall.
5. Using evidence from Figures 23 and 24, explain why new cases of malaria are more common at certain times of the year.

Are some people at more risk from malaria than others?

The health risks of malaria vary widely for different groups of people. Not everyone who catches malaria will die. In fact, over time, continued infection from mosquito bites will lead to a person becoming immune. Prevalence rates in Malawi show that 60 per cent of babies and children under three years old had malaria compared with only 12 per cent for adult men. As children and adults age, they develop resistance to the disease and the number of deaths go down.

Some people are at greater risk of death from malaria than others. The groups at higher risk are children and pregnant women, especially those in their first pregnancy. Around 40 per cent of child deaths are from malaria. People with poor immune systems are also at high risk. It is estimated that 9 per cent of Malawi's population is living with the HIV virus (see page 283), which destroys a person's immune system. They are unable to fend off diseases and the death rate from malaria is higher than for people without HIV.

Pregnancy also leads to a slight decline in immune levels. This results in slightly more pregnant women dying from the disease than non-pregnant women.

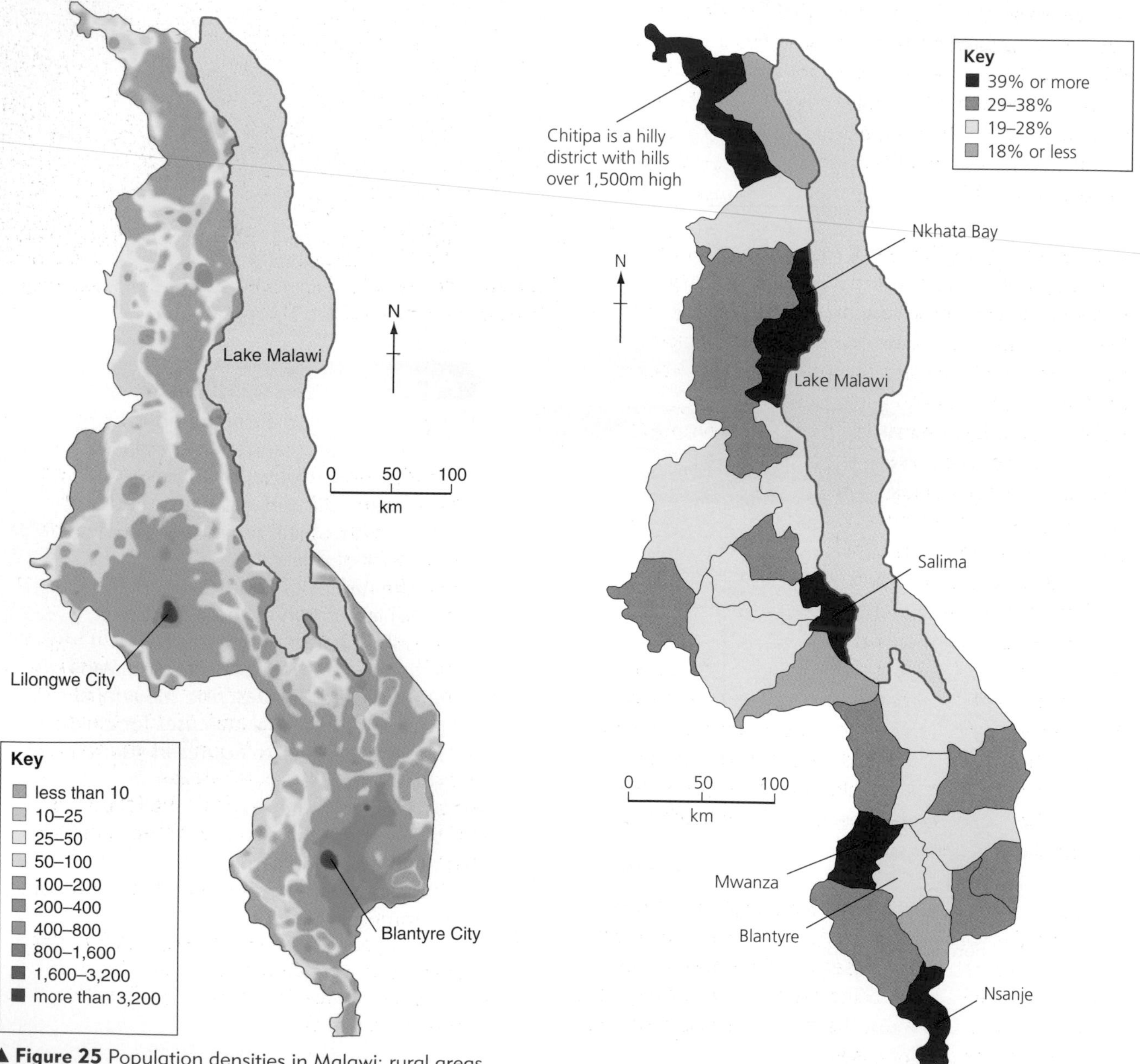

▲ **Figure 25** Population densities in Malawi; rural areas generally have lower population densities than urban areas.

▲ **Figure 26** Distribution of malaria in Malawi.

Combatting malaria

A number of strategies have been implemented in Malawi to combat the disease. One strategy has been to encourage people to use insecticide-treated bed nets (ITNs). This is a very effective way of reducing incidents of malaria and is relatively cheap. Each bed net only costs only £3. However, many people in rural Malawi are unable to afford this. In the last fifteen years, the Ministry of Health, and charities such as Nothing but Nets, have distributed bed nets across Malawi. In 1997, only eight per cent of homes in Malawi had bed nets. By 2004, the average was 50 per cent. By 2015, 80 per cent of households had an ITN. However, this doesn't mean that there are enough bed nets to protect everyone in those households. Only 33 per cent of households have one net for every two people. It's important to protect the people who are most vulnerable. The good news is that 66 per cent of children under five and 60 per cent of pregnant women sleep under an ITN.

Another strategy is the use of insecticides. Insecticides are sprayed in areas where mosquitoes are likely to come into contact with humans. The problem is that mosquitoes have developed resistance to some insecticides and they aren't as effective as they could be. Similarly, the mosquito parasite has become resistant to drugs. New drugs need to be developed urgently. Once a person starts to show signs of malaria, they need to take anti-malarial drugs as soon as possible to stand a better chance of fighting the disease. Unfortunately, not everyone in Malawi is able to access drugs when they need them. This is often because people in rural areas have to travel a long way to reach a doctor. Also, because the early symptoms of malaria are similar to many other conditions, people often don't realise until it's too late. Malaria has the biggest impact on poor people, as they cannot afford to buy nets and get the treatment they need. It can be a vicious circle: people are prevented from getting out of poverty because of the burden malaria has on their lives.

In 2019, the government of Malawi began to immunise young children against malaria. The new vaccine prevents malaria in 40 per cent of cases in babies aged between 5 and 17 months. The vaccine cannot prevent every case of malaria, but it can make the symptoms less severe in the other 60 per cent of cases.

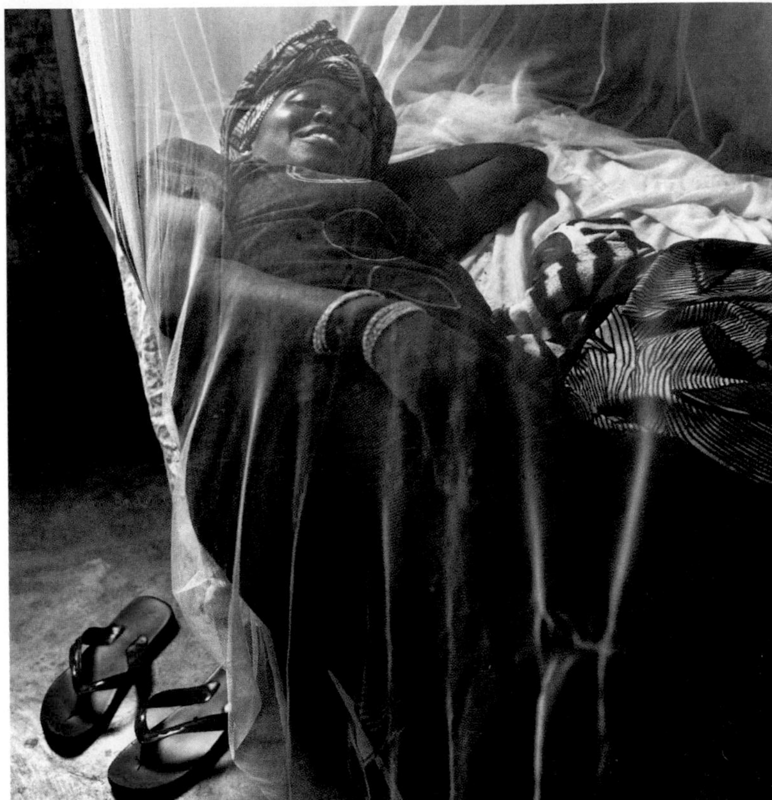

▲ **Figure 27** Deaths due to malaria can be prevented using insecticide-treated bed nets (ITNs).

Activities

1 a) Identify which three groups of people are most at risk from malaria.
  b) Explain why some people are more at risk than others.

2 a) Use Figure 26 to describe the distribution of malaria in Malawi.
  b) Use evidence from pages 278–280 to suggest different reasons for the high rates of malaria in two of the named regions on in Figure 26.
  c) Compare Figures 25 and 26. Use these maps to provide evidence that malaria is more common in rural areas.

3 Work in pairs. Discuss each of the following and then summarise your conclusions in two spider diagrams.
  a) Does malaria cause poverty or is it a result of poverty?
  b) Which groups should be targeted for ITNs?

4 Evaluate the success of the use of ITNs in Malawi. What can they achieve? What else needs to be done?

The challenges created by HIV

In 2017, an estimated 36.9 million people were living with HIV around the world. Sub-Saharan Africa is the region of the world that has been hardest hit by HIV, an incurable disease that attacks the immune system and which eventually leads to AIDS. HIV is a major cause of poverty in many communities. HIV most commonly infects people of working age. Deaths among the workforce not only cause distress for families, but also reduce the earning power of the family.

Rates of infection from HIV reached a peak in the late 1990s. Since then, in most countries of sub-Saharan Africa, the percentage of adults who are living with the virus has levelled out or fallen. This is because fewer people are becoming infected due to the success of education programmes. In Uganda, for example, HIV infection rates fell when the government introduced a programme of training for health care workers and education and counselling for the public.

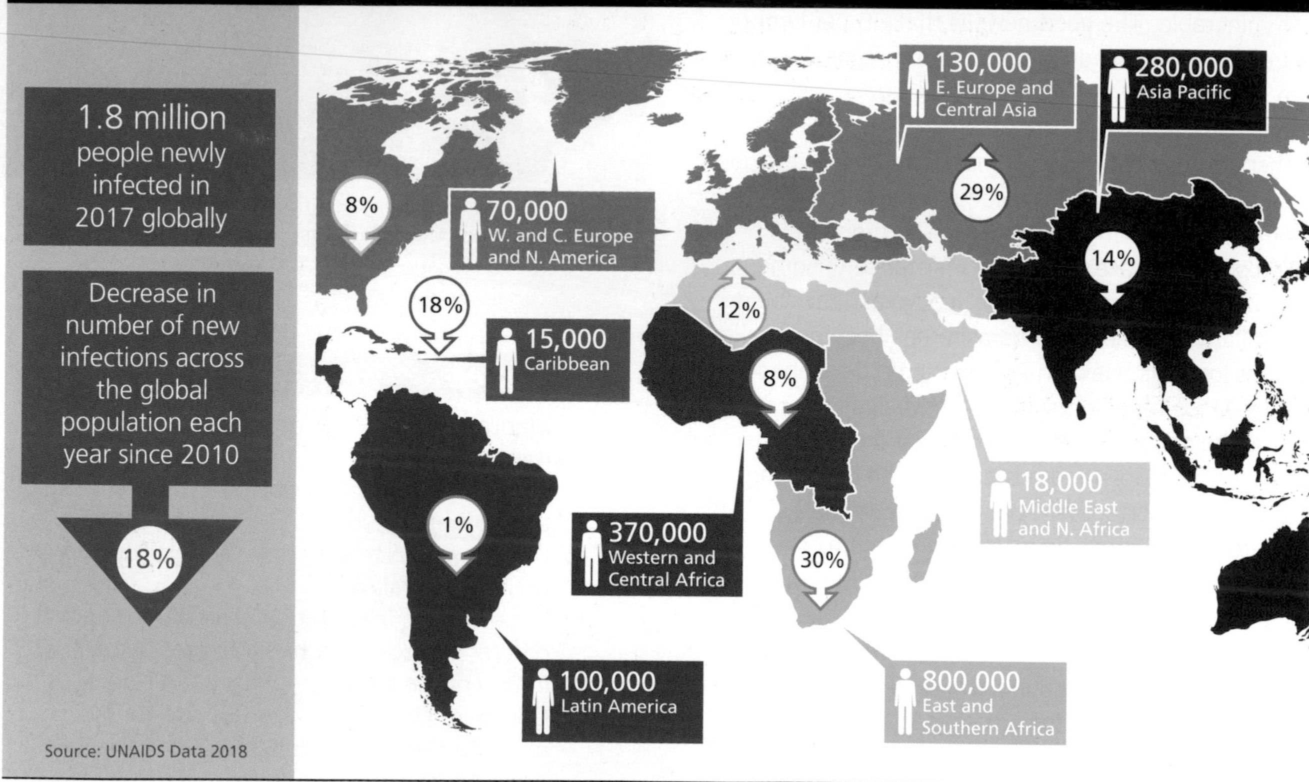

▲ **Figure 28** Number of new HIV infections in 2017 and change since 2010.

Activities

1 Use Figure 28 to describe:
   a) the global distribution of new HIV infections in 2017.
   b) how the percentage of new infections has changed since 2010.
2 Explain the links between HIV and poverty. Use Figure 31 to help you.
3 Explain why improving education may help reduce the prevalence of HIV.

HIV in Malawi

Around 1 million people are infected with HIV in Malawi, or 9.2 per cent of the population aged 15–49. This is one of the highest rates in the world. The government of Malawi has made huge efforts to combat the HIV epidemic. This is an example of **top-down** development. As a consequence, the number of new infections each year has fallen, from 98,000 in 2005 to 38,000 in 2018. One key to Malawi's success in reducing HIV infections has been the government introduction of HIV testing and counselling (HTC) services. Under this scheme, health workers can offer to test a patient for HIV when the patient visits a clinic, a mobile clinic or at national health events. In the first three months of this scheme in 2017, nearly 1 million people were tested. By comparison, 1.8 million were tested in the whole of 2014. Other elements of the government's plan to reduce HIV infection are:

- the free provision of condoms
- HIV education, including raising awareness via radio programmes
- preventing mother-to-child transmission by offering all pregnant women with HIV antiretroviral treatment for life.

Partly as a consequence of Malawi's efforts to reduce HIV infections, and its actions to prevent malaria, life expectancy in the country is now rising. In 2005, people lived, on average, to an age of 49. By 2018, life expectancy had risen to 63.

Top-down

An approach to development that is taken by governments or by international agreements between countries. In this type of development, key decisions are made by politicians or world leaders.

1 million live with HIV

13,000 deaths due to AIDS in 2018

15 percent of young women and 18 percent of young men report having sex before the age of 15

▲ **Figure 29** HIV/AIDS in Malawi in numbers, 2018.

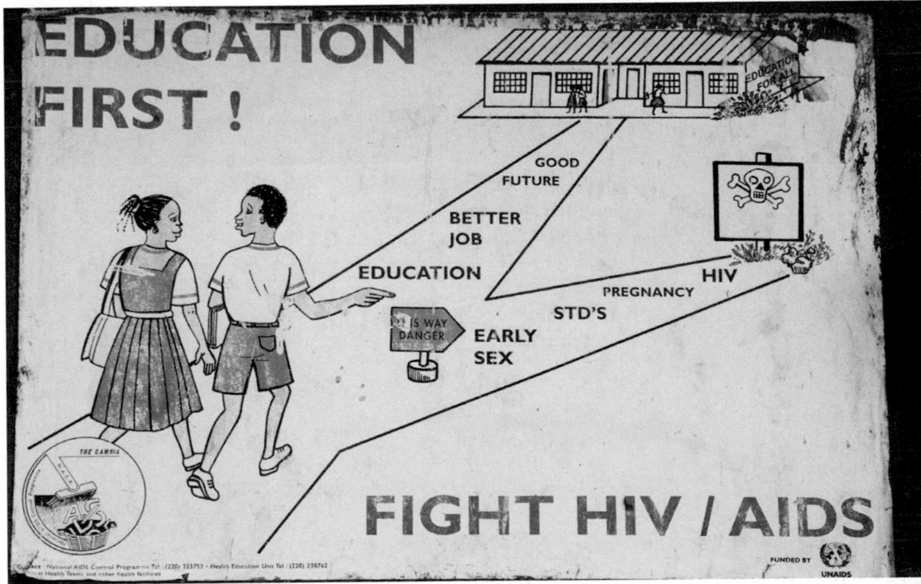

▲ **Figure 30** A poster in the West African country of the Gambia.

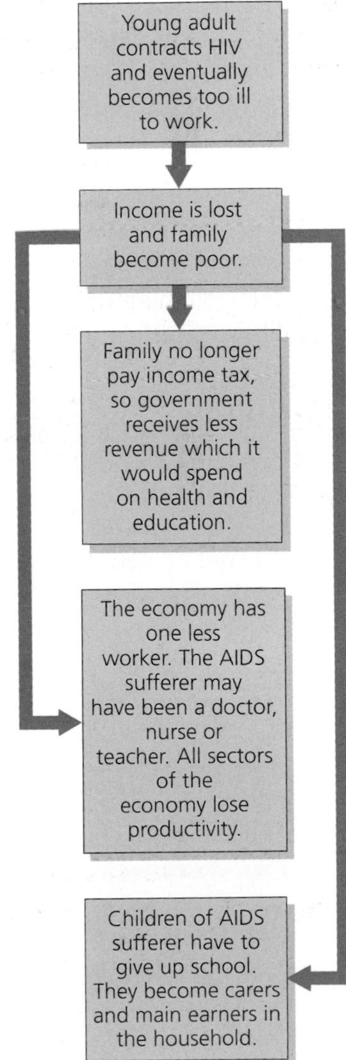

▲ **Figure 31** The social and economic consequences of HIV.

Different approaches to development

The countries of sub-Saharan Africa face a number of health-care problems. Issues such as controlling the spread of malaria and HIV can be dealt with at different scales:

- At the global scale. Organisations such as the World Health Organization (WHO) and United Nations design strategies to improve health and help governments to introduce these strategies. They also keep records of health data and set targets for improvement, such as the **sustainable development goals** (page 236).
- At the national scale. Governments use public money to improve health care. They can spend money on building new clinics, training health workers and improving health education.
- At a local scale. Communities and non-government organisations (which include charities such as Oxfam or WaterAid) introduce schemes to improve the health of local people, such as building new wells to provide clean water.

When decisions are made by national politicians or world organisations the development is top-down. When decisions are made by the people who live and work in their own communities, development follows a **bottom-up** approach (self-help).

Sustainable development goals

A set of aims proposed by the United Nations (UN) in 2015, applying to 17 different areas of life. The overall aim is to end poverty, protect the planet and ensure that all people enjoy peace and prosperity by 2030.

3,100 children die each year as a result of dirty water and poor toilets

5.6 million people don't have clean water

9.6 million people don't have a decent toilet

▲ **Figure 32** Malawi's water and sanitation in numbers.

Providing clean water in Malawi

Clean water and sanitation (decent toilets) are essential to good health. Dirty water can spread bacteria that cause diseases such as cholera and typhoid. UNICEF estimates that diarrhoea is the cause of 11.4 per cent of infant mortality and child mortality in Malawi. Only 26 per cent of primary schools in Malawi have good sanitation and only 4 per cent of schools have hand-washing facilities with soap.

Malawi's government has made providing clean water and sanitation a high priority. Even so, about one third of all Malawians do not have clean water and half of the population do not have a decent toilet. Local people can help solve this problem by getting involved in self-help projects run by charities such as WaterAid. WaterAid works with the Malawian government and local communities to provide clean water and sanitation. For example, they provide new water pumps in rural areas so that clean water can be pumped up from underground stores. They also train local people to install and maintain these pumps, so the scheme will be sustainable. Figure 34 describes one such project.

Bottom-up

An approach to development in which the local community is involved in decision making. Bottom-up (or self-help) approaches to development rely on local people being involved in starting and maintaining a development project.

	Urban	Rural	National
Percentage of households with access to basic drinking water	87	63	67
Percentage of households with access to basic sanitation	47	41	42

▲ **Figure 33** Access to clean water and sanitation in Malawi.

Dalia Soda is one of only three female pump mechanics in Salima District. Highly respected by co-workers and communities alike, she was already a keen volunteer in the region before she learned her trade.

'I knew Nzeremu Village before the borehole came,' Dalia explains. 'I was encouraging people to contribute something towards it. They were collecting water from the lake.'

After training from WaterAid, Dalia now keeps Nzeremu's and several other villages' boreholes working. The job has helped her build her own house and pay for all seven of her children's school fees.

Looking to the future, Dalia has started taking her 19-year-old son with her on jobs so that he can assist – one day, when she is too old to work, she hopes he can take over.

'I'd be very happy if he wanted to be a pump mechanic. I know it would be of service to a lot of the people in the community.'

▲ **Figure 34** Dalia Soda, pump mechanic.

Activities

1. Describe the main difference between top-down and bottom-up approaches to development.
2. Explain why improving water supply and sanitation is a key aim for Malawi's government.
3. Choose a suitable technique to represent the data in Figure 33.
4. Describe how the WaterAid scheme described in Figure 34 has helped the local community. Suggest how it has helped:
   a) health
   b) confidence and well-being.

Enquiry

'Bottom-up (or self-help) approaches to health development are more effective than top-down approaches.'

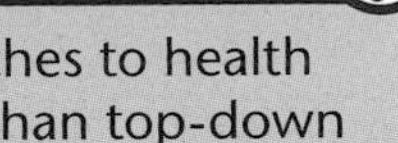

To what extent do you agree with this statement? Use evidence from pages 279–285 to support your answer. Think about how many people benefit from the schemes that are described on these pages and how sustainable they might be.

The link between health and water

Drinking or bathing in unclean water can cause diarrhoea, cholera, dysentery, typhoid and polio. WHO estimate that diarrhoea causes the deaths of 485,000 people each year. Many of these are children.

Clean water is essential to social and economic development. Safe water is needed to maintain water supplies, provide hygiene, grow food and supply industrial processes. Without sufficient safe water, infant mortality and maternal mortality rates increase. Having **water security** is an important aim for a country. It means having enough water to keep the population healthy and fed, and for the economy to develop sustainably – without damaging the prospects for future generations.

Water supply and polio in sub-Saharan Africa

The northeastern part of Nigeria is in the Sahel region of Africa. This region suffers from water insecurity. There are a number of reasons for this problem:

- The region has a hot, semi-arid climate with a long dry season.
- The rainy season is unreliable; in many years since 1965 there has been below-average rainfall.
- The Nigerian government has failed to provide enough piped water supply or sanitation to fast-growing cities, such as Kano.
- Poverty and drought have led to political instability and the rise of an extreme Islamist group called Boko Haram.

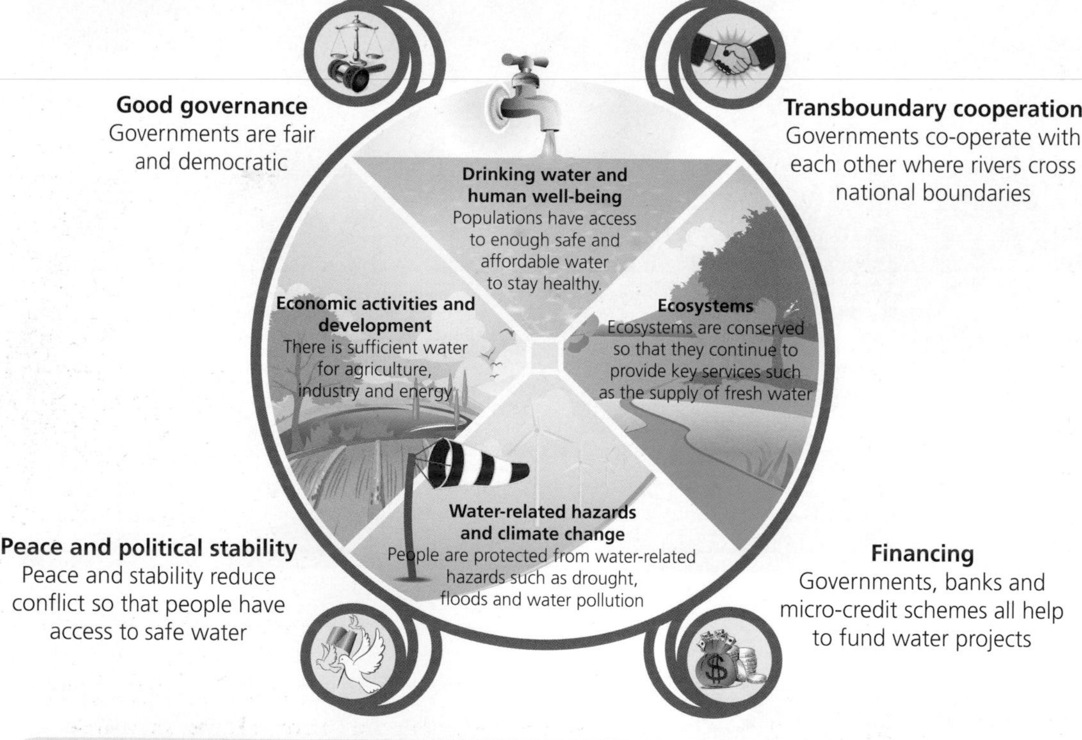

◀ **Figure 35** UN definition of water security (2013). The central part of the diagram describes the features of water security. The outer part of the diagram describes how water security can be achieved.

Activities

1. Study Figure 35. List the benefits of water security for people, the economy and environment. Record your answer in a Venn diagram.
2. Study the information about water insecurity in Kano State and Kano City.
   a) Make a list of the physical and human causes.
   b) Suggest at least four reasons why Kano is at huge risk of water insecurity.
   c) Use this example to explain why water security can only be achieved if political leaders work hard.
3. a) Use Figure 38 to draw a line graph showing population growth of Kano City.
   b) Describe the trend of your graph.
   c) Outline two main reasons for this trend.
4. a) Use Figure 36 to draw a line graph of polio cases in Nigeria.
   b) Explain the factors that have influenced the shape of this graph.

Water insecurity and health in Kano

Kano is Nigeria's third-largest city. Its population continues to grow quickly, partly because of rural to urban migration from the surrounding rural areas where low rainfall and poverty are push factors. Kano's water supply comes from the Tiga Dam (completed in 1974) and the Challawa Gorge Dam (completed in 1992). The Tiga Dam also supplies water to the Kano River Irrigation Project, which uses water to grow food for Kano. However, Kano has no sewerage system or sewage treatment plants. People living in high-density shanty towns in the city use pit latrines. Poor sanitation means that people are at risk of diseases such as cholera and polio. Polio attacks the nervous system and can cripple its victims. But polio can be eradicated by immunising young children.

Between 2003 and 2004, local Muslim leaders and the state government of Kano decided to oppose any future vaccination of children in Kano State. The state government now supports the vaccination programme, but some people have remained violently opposed to it. In February 2013, extremists opened fire on two polio clinics in Kano, killing nine health workers. The state government and UNICEF have led a huge programme to educate local people about the benefits of the polio vaccine. If no new cases of polio are reported in 2019, then Nigeria will have been free of the disease for three years. This means that polio will have been eradicated from Nigeria.

Year	Cases of polio reported
2001	56
2002	202
2003	355
2004	782
2005	830
2006	1,122
2007	285
2008	798
2009	388
2010	21
2011	62
2012	122
2013	53
2014	6
2015	0
2016	4
2017	0
2018	0

▲ **Figure 36** Polio cases in Nigeria.

▲ **Figure 37** Immunisation campaign worker Hassan Makama keeps count while Riskat Giwa, UNICEF officer of social mobilisation, vaccinates children against polio in Kano, Nigeria.

1970	1975	1980	1985	1990	1995	2000	2005	2010	2015	2020	2025	2030
0.54	0.86	1.36	1.86	2.06	2.34	2.60	2.90	3.22	3.59	4.17	5.11	6.20

▲ **Figure 38** Population of Kano (millions).

Water security

When there is enough clean water for healthy people and for the needs of the economy.

Enquiry

Can polio be eradicated? Focus on the problem of polio in Afghanistan and Pakistan. Can polio be eradicated in these countries? Justify your decision.

THEME 8

Environmental challenges

Chapter 1 Consumerism and its impact on the environment

What links the UK consumer to the global environment?

Food miles are a measure of the distance your weekly shopping has travelled before it arrives on your plate. Food grown in the UK is transported further now than it ever was 50 years ago. That is because of the growth of large supermarket chains and their complex distribution systems (see Figure 1). Food and other agricultural products now make up almost a third of goods carried on our roads – with lorries using diesel and emitting CO_2, a greenhouse gas.

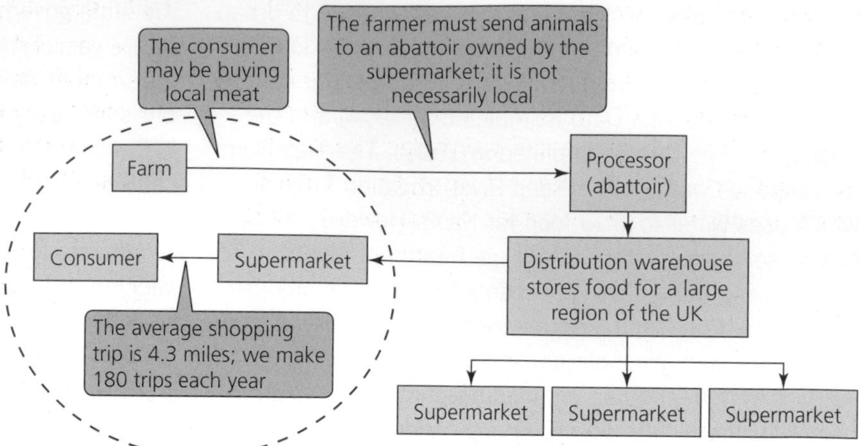

▲ **Figure 1** Food miles are increased by complex supply chains; you may be eating meat reared on a local farm, but it may still have travelled many food miles.

Much of our food is grown abroad and then imported into the UK, so the food miles for these products are even higher. We import half of our vegetables (most come from the EU) and 95 per cent of our fruit. Meats such as beef (from South America), lamb (from New Zealand) and chicken (from the EU) can be frozen and transported to the UK by container boat. Energy costs, and carbon emissions, for transporting this meat will be higher than buying locally produced meat.

Carbon emissions are even higher if goods are transported by air. Products such as soft fruit (for example raspberries and strawberries) and flowers need to be transported quickly by air to keep them fresh. Many can be grown in the UK, but only for a few months each year. They are imported from tropical countries (such as Kenya) or southern hemisphere countries (such as South Africa) so we can have them out of season.

Type of transport	Grams per tonne/km
Large container ships (over 35,000 tonnes)	3.0
Container ships (10,000–35,000 tonnes)	7.9
Trucks/lorries	80.0
Air freight	435.0

▲ **Figure 2** Carbon emissions from different types of transport.

Imports from	Value in Euros (billions)
South Africa	1.9
Turkey	1.7
Costa Rica	1.5
Chile	1.4
Colombia	1.0
Ecuador	1.0
Brazil	0.7

▲ **Figure 3** Imports of fresh fruit to the EU in 2018.

Enquiry

Should we only eat locally grown food that is in season?

Research the UK's fruit imports.

- Using Figure 3 to help you, which fruits do we import and from which countries?
- Can we reduce our demand for imported fruit? Which fruits are grown in the UK?
- Design a survey to investigate attitudes to this issue. What questions will you ask? What would be a representative sample for your survey?

So, should I eat locally produced food?

You would imagine that buying locally grown food is good – especially if you buy from an independent shop that does not have complex supply chains like those shown in Figure 1. You are supporting local farmers and the food miles should be shorter. However, some argue that the concept of food miles simplifies a complex subject and we should not ignore food grown abroad, like the Kenyan beans in Figure 4. The production of food uses all sorts of resources, including fertiliser, pesticides and water. Fertilisers and pesticides need to be manufactured, which uses energy, then supplied by lorry to retailers and to the farmer, using diesel, then applied to the fields with a tractor, using more diesel. Each of these processes emits greenhouse gases and has an impact on the environment. This way of thinking about products is called **life cycle analysis**. It is a more sophisticated way of thinking about the impacts of consumerism than food miles.

▲ **Figure 4** Should we buy these green beans that have been flown from Kenya?

▲ **Figure 5** Mr Pugh, the family-run butcher's shop, in Bishop's Castle; Mr Pugh buys all of his meat from farms less than 10 km from the shop; it is slaughtered in Leintwardine, 22 km from the shop.

Beans (in Kenya) are grown using manual labour – nothing is mechanised. They don't use tractors, they use cow muck as fertiliser; and they have low-tech irrigation systems in Kenya. They also provide employment to many people in the developing world. So you have to weigh that against the air miles used to get them to the supermarket.

▲ **Figure 6** Professor Gareth Edwards-Jones of Bangor University, an expert on African agriculture, talking about green beans grown in Kenya, quoted in 'How the myth of food miles hurts the planet' by Robin McKie and Caroline Davies, *Guardian*, 23 March 2008. Courtesy of Guardian News & Media Ltd.

Food miles

How far the food has been transported to get from producer to consumer. Food miles are a simple way to measure the impact that consumerism has on the environment. Life cycle analysis is a more sophisticated way.

Activities

1. Use Figure 1 to explain why meat bought from a supermarket is likely to have more food miles than meat bought from a local butcher.
2. Represent the data in Figure 2 using a suitable technique.
3. Study Figure 3.
   a) Which of these countries are:
      i) in the EU
      ii) tropical?
   b) Choose a suitable technique to represent this data.
4. Describe the strengths and limitations of the food miles concept.
5. Create a table that outlines the arguments for and against buying fruit and vegetables from abroad.

Consumerism and waste

We all consume stuff – food, water, clothes, electronics and energy. We don't often think about where it came from or its impacts on our environment when we finish with it – but we should. The population of Earth today uses resources equal to 1.6 planets. We call this our **ecological footprint**. As the population grows, and our desire for more consumer goods grows too, our footprint grows. By the 2030s the UN estimates we will need two Earths to support us. Increasing consumer demand can be linked to the destruction of rainforests (see pages 294–5) and the collapse of fish stocks (see pages 290–1), as well as increased greenhouse gas emissions.

Ecological footprint

A measure of the impact each of us has on the environment. It considers the amount of land we need, each year, to provide the food, water, energy and services each of us needs – including the amount of space and resources needed to get rid of our waste. Ecological footprint is measured in global hectares per person (GHA). We have enough on Earth for 2.1 GHA each. If we use more than that each year, our lifestyle is unsustainable because we are consuming resources faster than they can be replaced.

Country	Footprint per capita (GHA)	HDI
Australia	6.6	0.939
Germany	4.8	0.936
USA	8.1	0.924
UK	4.4	0.922
UAE	8.9	0.836
Mexico	2.6	0.774
Brazil	2.8	0.759
China	3.6	0.752
South Africa	3.2	0.699
Indonesia	1.7	0.694
Vietnam	2.1	0.694
India	1.2	0.644
Bangladesh	0.8	0.608
Ghana	2.0	0.592
Malawi	0.7	0.477
Gambia	1.0	0.460

▲ **Figure 7** Ecological footprint compared with Human Development Index (HDI) for selected countries (2018).

Consumer societies create huge quantities of waste – households throw away food and packaging every week. In the past, most of this was collected and burnt or dumped in a landfill site. However, we create so much waste that we are running out of space to dump it. Besides, useful materials such as card, paper, plastics, metals and glass can be recycled. This reduces demand for new resources and, in some cases, it's cheaper to recycle than create packaging from scratch. Aluminium drinks cans are a good example – it takes much less electricity to recycle a drinks can than to smelt aluminium from its ore.

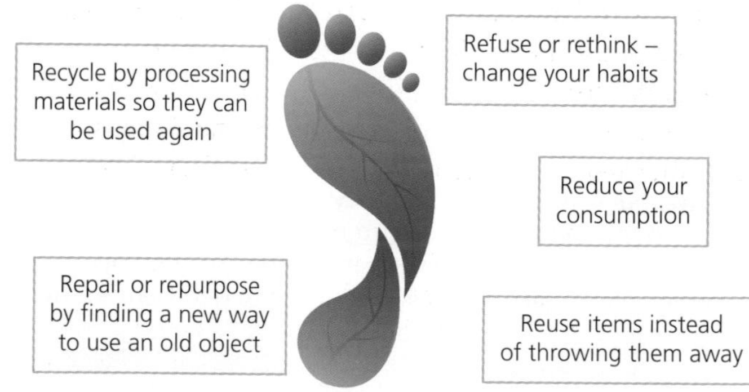

▲ **Figure 8** Ways to reduce our ecological footprint.

Activities

1 Study Figure 7.
   a) Design a hypothesis linking ecological footprint and HDI and make a prediction.
   b) Draw a scattergraph to test your hypothesis. Think carefully about which is the dependent variable (which should be plotted on the vertical axis). See page 267.
   c) What conclusions can you draw about the links between consumerism and development? Make sure you explain the reasons for the links.

2 Discuss the five Rs in Figure 8.
   a) Make a list of at least ten items around your home. Give an example of how each could be treated using the ideas in Figure 8.
   b) What should we do about waste? Is one of the five Rs better than the others? Explain your answer.

What happens to our e-waste?

What happens to your old phone, computer or tablet when you have finished with it? We describe this as **e-waste**. These items contain various metals, including copper (covered in plastic) and small amounts of gold (in computer chips), indium and palladium. Recycling these materials reduces the need to extract more raw materials from the Earth – so reducing our ecological footprint. However, they also contain some hazardous materials, such as mercury, lead and flame-retardant chemicals. They need to be recycled carefully or these materials can pollute the environment and threaten human health.

The UN estimates that 50 million tonnes of e-waste is created each year. Much comes from the rapidly growing economies of NICs such as India and China, as well as from European countries and other HICs. Environmental campaigners are concerned that some e-waste generated in Europe is exported to countries such as India, China, Ghana and Nigeria where recycling costs are much lower. This results in places like Agbogbloshie, a huge e-waste recycling dump in Accra, Ghana. It is estimated that between 50,000 and 80,000 people live and work in the informal settlement that has grown up next to the dump. Computers, monitors and phones are broken down and burnt – melting the plastics and exposing the metals for recycling. Smoke from the fires fills the air and is a health hazard. Tar from the molten plastics pollutes the stream that runs alongside the dump. The work is in the informal sector, so there are no health and safety rules.

▲ **Figure 10** Young men dismantle computer servers at Agbogbloshie, Accra.

E-waste

Electronic waste products, such as computers and mobile phones.

Activities

3 Describe the pattern shown in Figure 9. Consider whether each source and destination country is an HIC, NIC or LIC.

4 Describe the social and environmental consequences of e-waste recycling in informal settlements such as Agbogbloshie in Ghana.

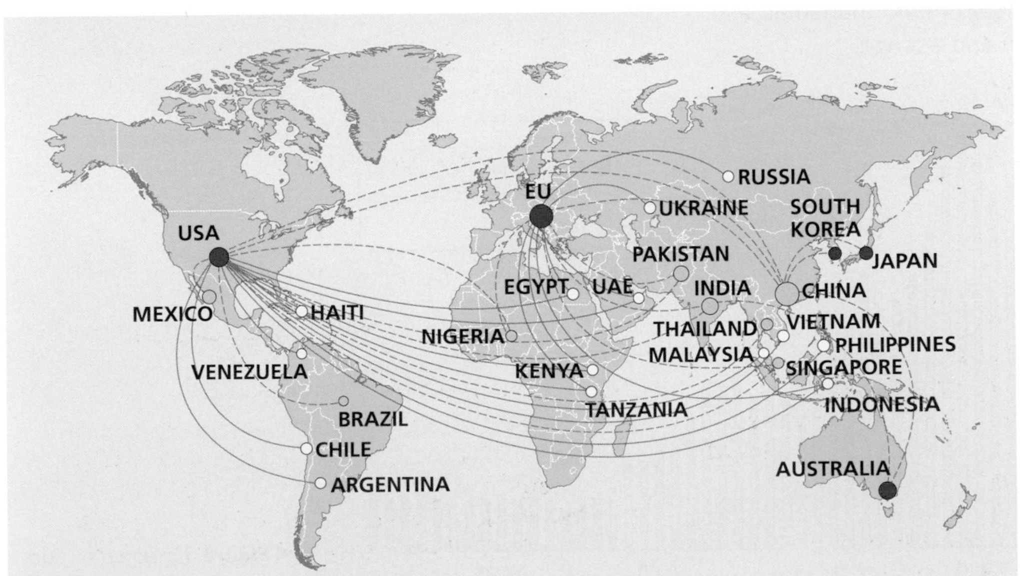

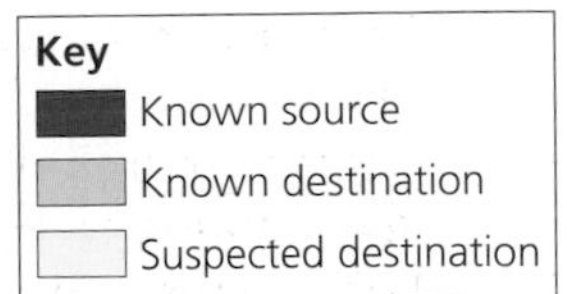

▲ **Figure 9** Known and suspected movements of e-waste.

Are there enough fish in the North Sea?

'Fish and chips' is a popular UK take-away food. Traditionally, the fish was cod or haddock caught in the North Sea – an example of a marine ecosystem. But in recent years, cod has been replaced with other fish such as coley or pollock as cod has been in short supply. Why has this happened?

Over-fishing is not sustainable because, when too many young fish are caught, there are not enough fish left in the sea to breed and replace the fish stocks. This is shown in Figures 12 and 14. It can take up to four years for cod to become mature enough to breed, so fish stocks have to be managed to give them a chance to survive. It is illegal in the UK to land a cod fish that is less than 35 cm long. These fish have not yet bred and if caught they must be thrown back into the sea.

▲ **Figure 11** Cod for sale at the market in Peterhead, Scotland. Peterhead is the UK's largest fishing port.

How can cod be protected?

The EU manages the fish stocks around Europe, including the North Sea, by imposing quotas. This is where a limit is put on the number of fish that can be caught by member states in certain sea areas. They also limit the number of days in a month when the fishermen can go out to fish. The quota system has angered UK fishermen. They have argued that the restrictions have been too severe. Quotas mean they catch less fish so they earn less money and some fishermen have gone out of business. There was also criticism of how quotas were allocated to different member states of the EU. Catching too many or small fish leads to hefty fines, so fish are sometimes thrown back into the sea, dead or alive. A minimum net size should mean that small fish can swim back out of the net and escape.

Activities

1 a) Describe the overall trend of Figure 12.
  b) Suggest how changes to the catch affected people working in the North Sea fishing industry.
  c) What proportion of the cod catch was discarded (thrown back) in 2008?

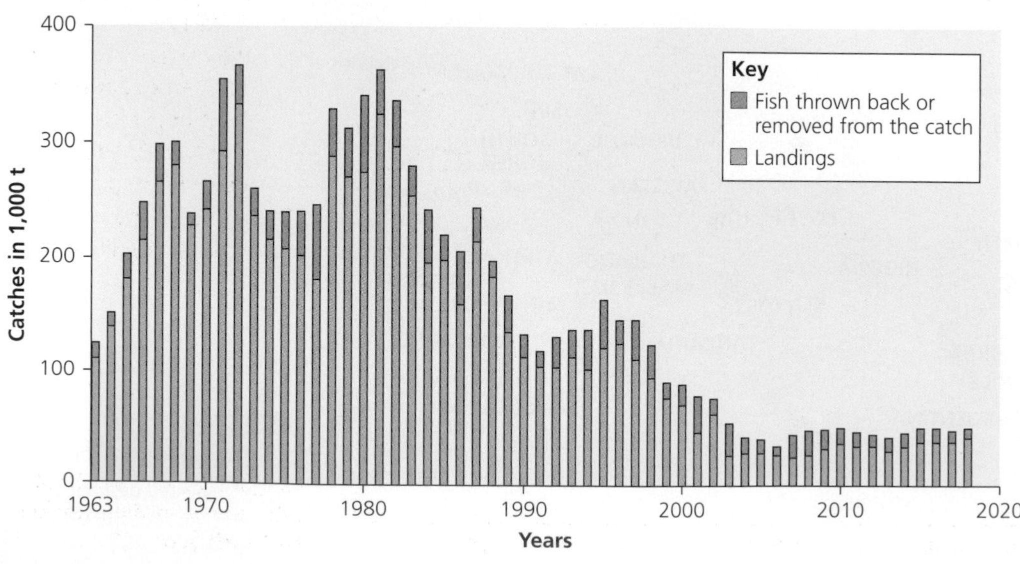

◀ **Figure 12** Catch of cod in the North Sea (in 1,000s of metric tonnes).

Has the cod been saved?

Figures show that stocks of North Sea cod began to increase after the very low level of 2004 (see Figure 14). Preventing over-fishing has allowed the cod to breed. The fishing industry is now much smaller than it was in the 1990s, with fewer smaller family businesses still working. So is it okay for consumers to eat cod again? Scientists warn that the cod is not yet out of danger.

> Quotas don't go far enough. We need to stop fishing over-exploited stocks altogether for one year. Then stocks would be restored to a fully sustainable level in four years.

Scientist

> We need to look at how we catch the fish. Trawlers use dragnets and catch more fish but damage the sea bed. Gillnets are vertical nets which catch less fish with less waste. We need to fish responsibly.

Environmental campaigner

> We need to protect our fishing industry. If we set quotas as low as the scientists would like, our fishing industry would collapse. We have set quotas low enough to conserve fish stocks but high enough to keep people in jobs.

Government minister

> I like cod, but we have a responsibility to conserve the world's resources. If I have to buy other sustainable types of fish for a while, so be it.

Consumer

> This town relies on the fishing industry. People work on the boats, in the fish market and in the boat repair yard. When the government limits the amount of fish we can catch then the whole town suffers.

Fish and chip shop owner in Peterhead

▲ **Figure 13** Views of fishing.

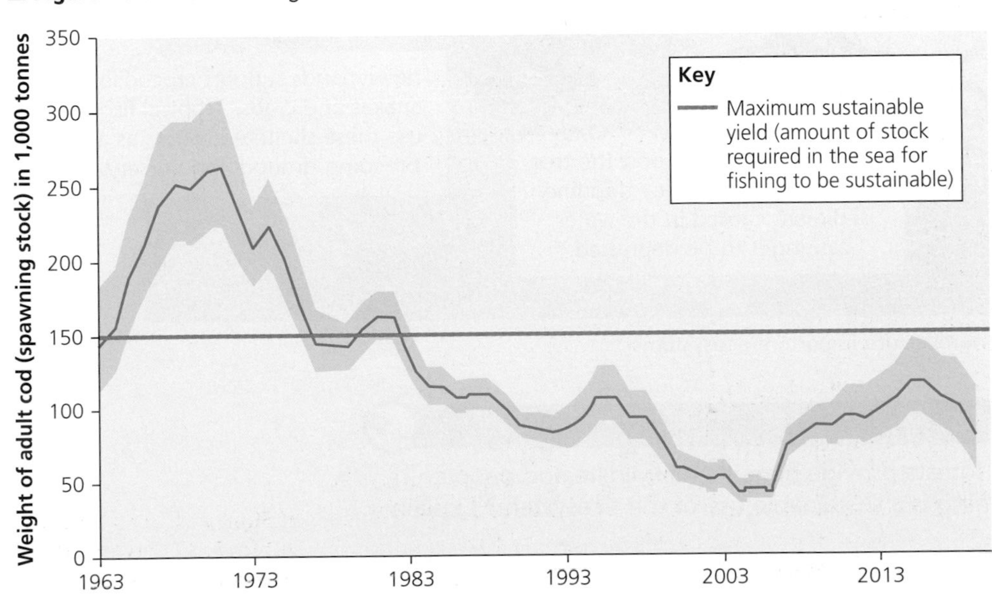

◀ **Figure 14** The weight of adult cod (spawning stock) in the North Sea.

Activities 

2 a) Compare the trend of Figure 14 to Figure 12.
b) At what date might scientists have asked for fishing quotas to be introduced?
c) Explain why fishermen and scientists might disagree about the meaning of the final three years of this trend line.

3 Discuss the views shown in Figure 13. Give one reason why it is important to:
a) protect the ecosystem of the North Sea
b) protect jobs in the fishing industry.

4 Suggest two different ways that UK consumers could help conserve the North Sea ecosystem and support the fishing industry.

Is shrimp farming sustainable?

Mangrove forests grow on tropical coastlines. The trees of the mangrove tolerate flooding by both fresh and salt water, so this is both a forest and a wetland, and it supports a very wide range of fish, insects and animals.

Mangroves help to absorb wave energy during storms so they help to protect coastal communities. Despite this, big business regards mangroves as useless wasteland. Trees are cut down and the swampy land redeveloped for tourism or **aquaculture**. Over 25 million hectares of mangrove forest are estimated to have been destroyed in the last 100 years. The fastest rates of destruction have been in Asia during the late 20th century. Globally, around 150,000 hectares of mangrove are destroyed every year, which is equal to about 1 per cent of the global total. Only 6.9 per cent of mangrove forests are protected. One of the most common reasons for the destruction of mangroves has been the rapid growth of shrimp (or prawn) farming. Mangroves are cleared away to make space for artificial ponds. These ponds are flooded with salt water and used to rear the shrimp.

Shrimp is low in fat and high in protein. Most consumers are in Europe, North America and Japan. Around 55 per cent of all shrimp we eat is farmed. The rest is caught from our oceans, largely from tropical waters such as the Bay of Bengal where large trawlers catch wild shrimp in nets. Over-fishing wild shrimp damages the food web of the marine ecosystem. A smaller quantity is caught in cold waters such as in the North Atlantic around the coast of Iceland.

Most shrimp aquaculture is in China, Thailand, Indonesia, Brazil, Ecuador and Bangladesh. Shrimp farming is carried out by both big business and many small farmers. Some argue that shrimp farming has helped poor farmers to diversify their income and reduce poverty. The global shrimp industry is thought to be worth US$12–15 billion.

Shrimp aquaculture is an intensive form of farming. It takes between three and six months to rear shrimp so farmers produce two to three crops a year. The shrimp are treated with pesticides and antibiotics. Organic waste, chemicals and antibiotics escape from the ponds polluting fresh groundwater supplies that local communities use for drinking.

▲ **Figure 15** Why mangroves are important ecosystems.

Activities

1. Describe how mangrove forests provide benefits for wildlife and people.
2. Do you think shrimp farming is a sustainable use of this ecosystem? Explain your point of view.

Activities

3 a) Use a suitable technique to represent the data in Figure 17.
b) Describe the trends shown by your graph.
c) What percentage of shrimp production came from aquaculture in:
i) 2008
ii) 2014?

4 Study the points of view in Figure 18.
a) What are the long-term benefits of shrimp farming and who gets these benefits?
b) What problems does shrimp farming create for:
i) people
ii) the environment?
c) Discuss what consumers in the UK can do to help ensure that ecosystems (either mangroves or other ecosystems) are used sustainably.

▲ **Figure 16** A shrimp farm in Bangladesh.

Shrimp production (1,000s of metric tonnes)							
	2008	2009	2010	2011	2012	2013	2014
Wild	3,217	3,269	3,263	3,442	3,568	3,541	3,591
Aquaculture	3,400	3,532	3,629	4,046	4,168	4,320	4,581

▲ **Figure 17** Global shrimp production (2008–14) in 1,000s of metric tonnes.

Aquaculture

A type of farming that involves the rearing of fish or shellfish for food.

Local people lose out because they can no longer use the timber or other resources available in the mangrove forest. Local fishermen have noticed a fall in the number of fish they catch. This may be because the mangroves are a nursery ground for young fish. Shrimp farming releases a lot of fertilisers and other chemicals into the environment. Fresh-water wells are polluted by these chemicals. These are problems that are likely to affect coastal communities for many years after the farms have been abandoned.

A local fisherman

The biggest consumers of shrimp are the USA, Canada, Japan and Europe. Perhaps consumers will be able to influence what happens to mangroves if we demand to know more about how our food is produced. Then we might decide to only buy shrimps or other fish that have been farmed sustainably.

Consumer in the UK

People make quick profits from shrimp farming. However, after a few years ponds are abandoned because of disease and pollution. In Asia there are approximately 250,000 hectares of abandoned, polluted ponds where healthy forests once grew. This boom–bust cycle is about to be repeated in Latin America, Africa and the Pacific where shrimp farming is growing in popularity.

Economics expert

▲ **Figure 18** Views on whether the use of mangroves for shrimp farming is sustainable.

Enquiry

'Mangroves are a swampy wasteland. There is nothing wrong in clearing them away if it means jobs are created and poverty reduced.'
To what extent do you agree with this statement? Justify your point of view.

How are tropical rainforests used for food production?

▲ **Figure 19** The Borneo orang-utan.

Borneo is an island in South East Asia, divided between Malaysia, Brunei and Indonesia. It has 1 per cent of the world's land area but 6 per cent of the world's species of wildlife. This makes Borneo an important **biodiversity hotspot**. The rainforest is thought to be 140 million years old and contains many thousands of species of plants and animals, some of which are now endangered, like the Borneo orang-utan.

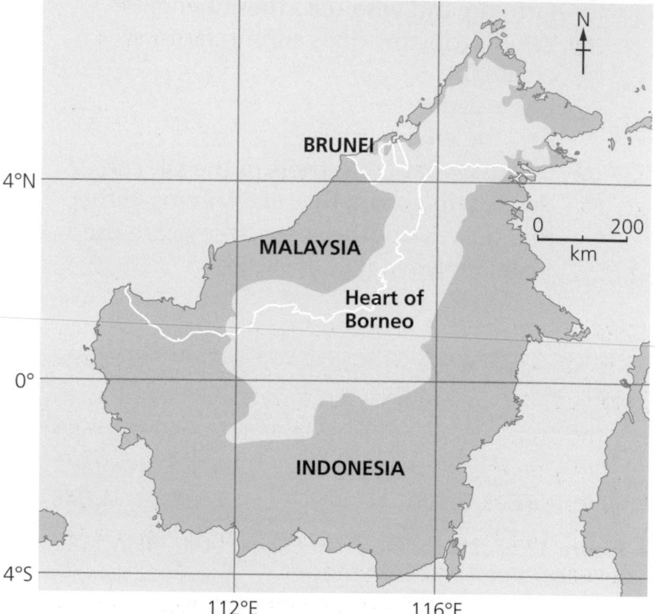

▲ **Figure 20** The island of Borneo and the 'Heart of Borneo'.

Global resources

Malaysia and Indonesia are newly industrialised countries (NICs). Their economies have grown as they have traded their natural resources and developed manufacturing industries. They have cashed in on the global demand for produce such as timber, fuel and food. Borneo's rainforests provide much needed timber, but at what cost? In 1985, 73.3 per cent of Borneo was covered in forest. By 2005, only 50.4 per cent was left. Many areas of rainforest have been cleared, destroying the homes of indigenous people, and the habitat of animals such as the orang-utan.

As the rainforest is cleared, it is replaced with crops such as oil palm. Global production of palm oil doubled between 1997 and 2008. Palm oil is a healthier alternative for other fats used in food and cooking. The desire to find greener fuels has led to oil palm being grown to produce biofuel. Oil palm plantations are a form of **monoculture** with a much lower biodiversity than the rainforest they replace.

Percentage	Use
71	Foods, e.g. margarine and processed foods such as cake, biscuits, chocolate
24	Consumer products, e.g. cosmetics and detergents
5	Fuel

▲ **Figure 21** How palm oil is used.

Activities

1 a) Represent the data in Figure 21.
b) Explain why palm oil production creates benefits for Malaysia and Indonesia.
c) Outline the environmental and social problems created by palm oil production.

Monoculture

When farms grow a single crop over large areas of land. Examples of monoculture in tropical regions include soybeans (soya) and palm oil.

Biodiversity hotspot

A region which has a particularly great variety of organisms. Some of these organisms are endemic, meaning they are only found in this particular location.

Is palm oil good for Borneo?

Palm oil has higher yields and lower production costs than other oilseed crops, such as soybean and sunflower. It requires less fertiliser and pesticides and can be grown on small farms as well as large plantations.

Country	Metric tonnes
Colombia	1,680,000
Indonesia	43,000,000
Malaysia	21,200,000
Nigeria	1,015,000
Thailand	3,000,000
Other	6,119,000

▲ **Figure 22** Estimated world palm oil production in 2019 (values in metric tonnes).

▲ **Figure 23** Plantations of oil palms. Sabah, Borneo Island, Malaysia.

What about the rainforest?

Clearing the land and putting in infrastructure, such as roads and electricity supply, provides jobs and incomes, but often these go to migrant workers, and not the local communities. Many indigenous people, who rely on the rainforest for their food and homes, have been displaced. Animals can become isolated living in the fragments of habitat that are still intact between the oil palm plantations. The rapid decline in areas of the rainforest, especially through illegal logging, is a threat to many species of plants and animals. The loss of the rainforest can even change local weather patterns.

In 2007, the three countries of Malaysia, Brunei and Indonesia, along with NGO support, declared to conserve and manage the forest resources that are still largely undamaged in the central part of Borneo. This is the 'Heart of Borneo' and you can see the location of this forest in Figure 20. NGOs such as WWF and Greenpeace have campaigned to raise public awareness of the deforestation. They have encouraged multinational companies to support sustainable use of land for palm oil production. In 2015, 18 per cent of palm oil was produced using sustainable methods. In 2015, Colgate-Palmolive and Procter & Gamble both pledged to responsibly source primary commodities such as palm oil that are used in their products.

▲ **Figure 24** Greenpeace Protect Paradise campaign.

Activities

2 Use Figure 22 to calculate the percentage of global palm oil production in:
   a) Malaysia
   b) Indonesia.

3 a) Use an atlas to locate the countries listed in Figure 22. Describe their distribution.
   b) Explain why most palm oil production is located in these countries.

4 a) Suggest why the oil palm plantations are not a suitable habitat for orang-utans.
   b) Suggest why tropical monocultures have much lower biodiversity than tropical rainforests.

How might climate change affect people and environments?

So far we have seen how our consumerism affects specific environments, such as the mangrove forests of South Asia. However, our ecological footprint also has a global impact: the emission of greenhouse gases. These gases cause climate change, a process we examined on pages 156–7. A warmer atmosphere is a more unstable one, meaning that wild weather events will become more common. A warmer atmosphere will mean that the air masses over oceans will have greater moisture content. For the UK, this is likely to mean milder and wetter weather with more frequent storms. Many scientists interviewed by the media during the winter storms of 2015–16 (pages 41–5) argued that the extreme rainfall was consistent with predictions of climate change. The 'Foresight Report' considered how climate change and growing populations might affect the UK by the year 2080. The main findings were:

- The number of people at high risk of flooding could rise from 1.5 million to 3.5 million.
- The economic cost of flood damage will rise. At the moment flooding costs the UK £1 billion a year. By 2080 it could cost as much as £27 billion.
- Towns and cities will be at risk of flash floods even if they are not built near a river. Drains that are supposed to carry rainwater away will not be able to cope with sudden downpours of rain. This kind of flooding could affect as many as 710,000 people.

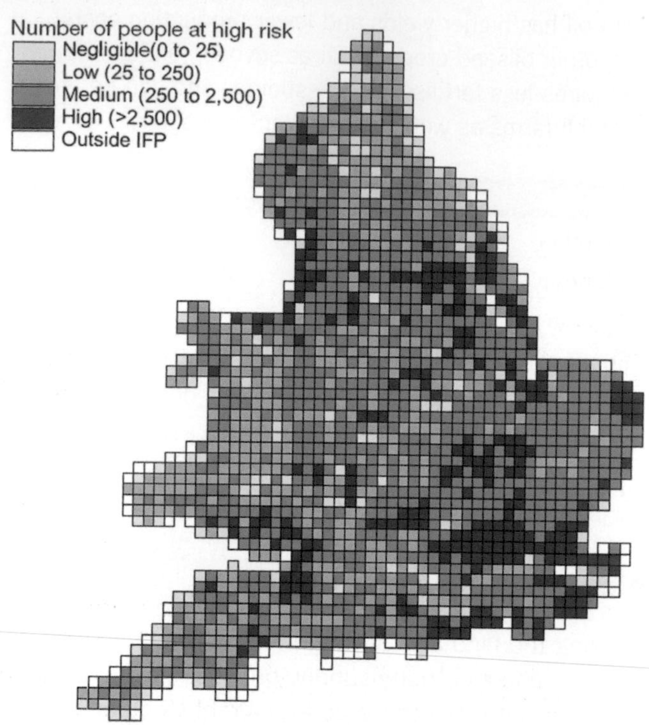

▲ **Figure 25** The number of people at risk of river and coastal floods in 2080, assuming that carbon dioxide emissions remain high.

Category 5: Over 250 kph. Complete failure of some smaller buildings. Failure of the roofs of large industrial buildings. Extensive coastal flooding damages the ground floor of many buildings.

Category 4: 211–250 kph. Complete destruction of the roofs of smaller buildings and more extensive damage to the walls. All signs and trees are blown down. Flooding of coastal areas 3 to 5 hours before the arrival of the storm may cut off escape routes.

Category 3: 178–210 kph. Severe damage to the roofs of small buildings. Some structural damage to walls. Mobile homes destroyed. Poorly constructed road signs destroyed. Large trees blown down.

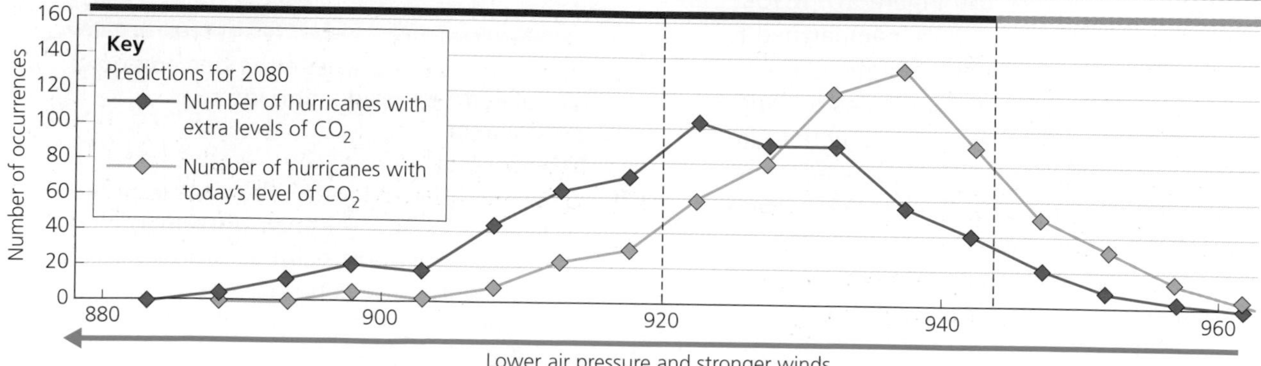

▲ **Figure 26** The National Oceanic and Atmospheric Administration (NOAA) has used computer models to predict the frequency and intensity of hurricanes (cyclones) in 2080.

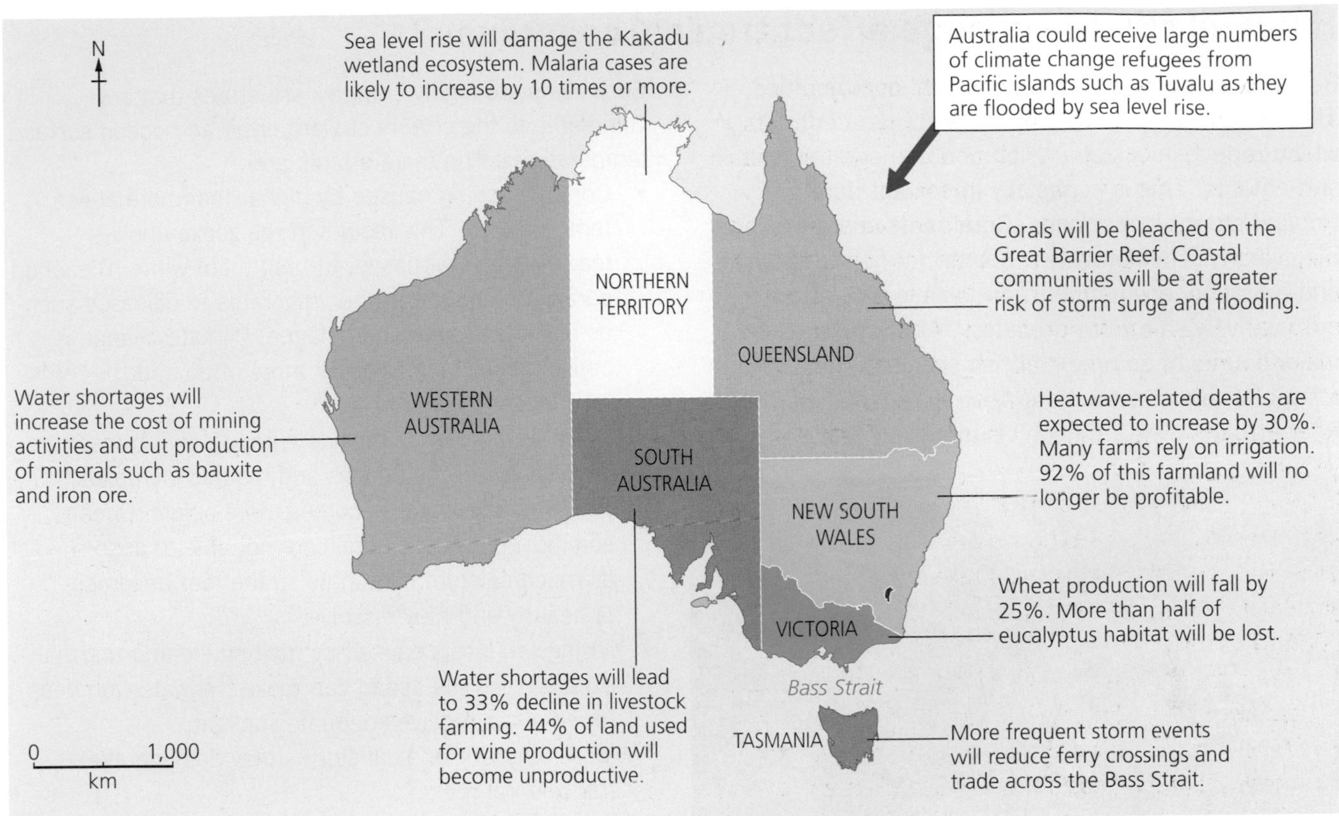

▲ **Figure 27** The predicted impacts of climate change on Australian states by 2100.

Australia is a vast country and the impacts of climate change are likely to be varied. The Australian Bureau of Meteorology warn that wild weather will be more common, with longer and more severe droughts leading to larger and more dangerous bush fires like the ones in 2019–2020. Frequent tropical storms and stronger cyclones (see pages 163–5) could affect Australia's northern coastlines. Climate change means that areas of extremely low pressure will develop more frequently, as shown in Figure 26. Australia's famous Great Barrier Reef, on the tropical Queensland coast could be affected by bleaching – a process described on page 300.

Activities

1. Use Figure 25 to describe the distribution of areas where there are high numbers of people at risk of flooding.
2. a) Describe how climate change will have negative impacts for people, the economy and environment of the UK. Use evidence from pages 41–5 to help you.
   b) Sort the impacts into those that are short-term (we can see examples now) and longer-term (we can expect to see them in the near future).
3. Study Figure 26. Describe how the frequency and magnitude of hurricanes (cyclones) is expected to change.
4. Study Figure 27. Select three impacts and explain how climate change may cause these issues.

Enquiry

How serious is the threat of climate change in Australia?

- Select five impacts of climate change in Australia and place them in rank order. The top-ranked one should be the most problematic/serious impact.
- Justify your ranking.
- Suggest who should be responsible for trying to fix the most serious of these problems.

How will climate change affect tropical coastlines?

Reefs provide **key services** to coastal communities. They reduce wave energy by up to 97 per cent, which reduces coastal erosion and damage to coastal settlements. This is especially important during tropical storms or cyclones. Coral reefs contain a huge biodiversity. They provide a habitat for fish to spawn and a nursery environment for juvenile fish where they are relatively safe from predators. As such, reefs are a major source of commercial fish species. They are also a popular tourist attraction. Australia's Great Barrier Reef attracts over 2 million visitors each year.

▲ **Figure 28** The bright colours of coral come from the zooxanthellae, which also provide them with nutrients and energy.

Coral reefs are highly complex structures that are vulnerable to the effects of warmer air and ocean surface temperatures. The main effects are:

- Coral bleaching caused by higher than normal sea temperatures. This results in the zooxanthellae leaving the coral tissue, turning them white. Bleached corals are unhealthy and vulnerable to diseases such as blackband and white plague. Climate change is causing bleaching to occur more often and the reefs take longer to recover.
- Much of the carbon dioxide entering our atmosphere is dissolved in the oceans and, as this increases, the pH of the water becomes more acidic. Ocean acidification means corals are not able to absorb as much calcium carbonate so the reef becomes unhealthy and may dissolve.
- Rising sea levels caused by melting ice and thermal expansion of the ocean can make the water too deep for corals to receive adequate sunlight.
- More frequent tropical storms may damage already fragile coral reefs.

Key services

The benefits provided to people by an ecosystem. Coral reefs, for example, absorb wave energy so protect coastlines during a storm. Forests store water and release it slowly, preventing floods.

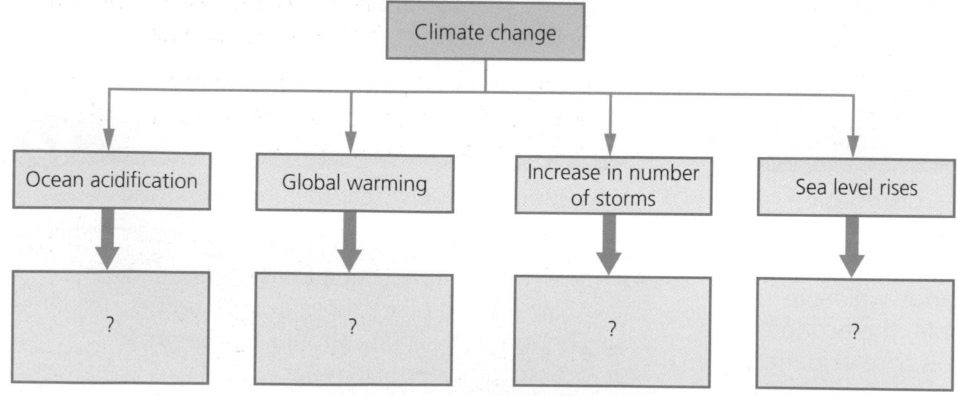

◀ **Figure 29** Impact of climate change on coral reefs.

Higher sea temperatures cause coral bleaching which weakens the structure

Deeper water leads to reduced amounts of sunlight and poorer water quality

Acidified water has lower pH levels which means less calcium carbonate is absorbed so coral skeletons dissolve faster

Delicate reef structures are damaged by heavy rain and run-off from the land

Activities

1. Describe three key services provided by coral reefs.
2. Make a copy of Figure 29. Add the labels (right) in their correct places to complete the diagram, summarising the effects of climate change on coral reefs.

How can we respond to climate change?

Since 1990 representatives of governments from around the world have met periodically to discuss the issue of climate change. The Kyoto Protocol (1997) is an international agreement that commits countries to targets to reduce greenhouse gas emissions. The latest version is the Paris Agreement (2015). The outcome of the Paris Agreement was a long-term aim to keep global warming to well below 2°C compared to pre-industrial levels.

However, the Paris Agreement recognises that countries can make cuts to emissions at different rates, depending on their level of economic development. Wealthy countries, such as Australia and the UK, should be able to make cuts by investing in new and renewable technologies. Some of these are described on pages 300–1. NICs such as India and China, however, have recently invested in coal-fired power stations. Even so, some parts of their populations do not have access to a reliable electricity supply. These countries want to continue to make economic progress. Doing so will mean further greenhouse gas emissions.

▲ **Figure 30** Solar farms in England, like this one, generate electricity using PV cells.

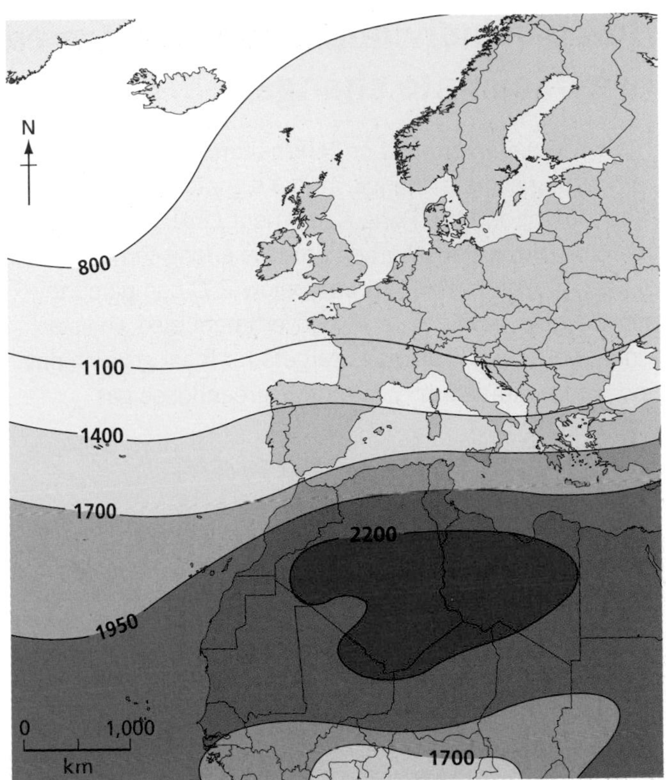

▲ **Figure 31** Patterns of solar energy across Europe and Africa (kilowatts/m²/year).

Activities

3 Use Figure 31 to describe the distribution of countries that have:
   a) between 1,100 and 1,400 kilowatts/m²/year
   b) more than 2,200 kilowatts/m²/year.

4 Using Figure 30, suggest the advantages and disadvantages of solar farms for the UK's economy and environment.

Enquiry

'European countries should adopt new technologies while NICs continue to emit greenhouse gases.'

To what extent do you agree with this statement? You should suggest:

- a range of different new technologies that could be used in Europe
- the reasons for and against allowing NICs to continue to emit greenhouse gases.

How can individuals and government reduce the risk of climate change?

The UK Government has an international role in combatting climate change. It is a signatory of the Doha Amendment and the Paris Agreement (2015). The outcome of the Paris Agreement was a long-term aim to keep global warming to well below 2°C compared to pre-industrial levels. The UK Government also works at a national scale. It provides targets for local government and works with industry to reduce greenhouse gas emissions by:

- investing in low-carbon energy sources;
- improving fuel standards in cars;
- increasing energy efficiency in new buildings.

Councils must publish a policy statement on climate change by law.

Figure 32 describes some of the ways in which Bristol City Council is trying to meet its own climate change targets.

Activities

1 Study Figure 32.
   a) Explain why ideas 5 and 7 will help Bristol meet its climate change target.
   b) Make a diamond nine diagram (like the one on page 83) and place the strategies from Figure 32 in the diagram, putting those that you think are essential at the top of the diagram.
   c) Justify your choice of the top three strategies.

2 Using Figure 32 for ideas, identify five ways in which individuals can reduce the risk of climate change.

	What will it do?	Why it helps
1	Warmer homes. £105 million will be spent fitting external wall insulation to homes (including blocks of flats).	£3.5 million could be saved each year on heating bills. Possible 5 per cent reduction in the amount of gas used. 17,900 tonnes of CO_2 saved each year.
2	Install district heating (or heat networks). These will duct spare heat between the university, the hospital and buildings in the city centre.	Heating buildings and hot water use are two of the main reasons for CO_2 emissions in large cities. District heating uses energy efficient boilers.
3	Lead by example. Improve the energy efficiency of council offices, the museum, library and two schools.	£500,000 energy saving and 2,000 tonnes of CO_2 saved each year.
4	High energy performance. Install efficient boilers in all new council buildings, including schools, care homes and council housing.	It is cheaper in the long term to install energy efficient technologies when buildings are new rather than trying to modify older buildings.
5	Solar photovoltaic programme. Two solar farms were constructed in 2014 at a cost of £35.9 million.	
6	Metro-bus scheme. Build 6 km of new roads, 18 km of new bus lanes and purchase 50 new hybrid vehicles.	Shift passengers from car to bus to reduce congestion and CO_2 emissions from commuter traffic.
7	Sustainable transport. Invest in 10 km of cycle lanes and promote cycling and walking for people aged 8–80.	
8	Land use planning. Locating new homes to reduce the need for commuting and allow the use of district heating.	Future proof new housing developments by making them sustainable.
9	Improve mass transit. Spend £90 million improving suburban train services.	Improve air quality and transport safety. Reduce congestion and CO_2 emissions.

▲ **Figure 32** How Bristol City Council hopes to meet its targets to reduce CO_2 emissions.

Collecting qualitative data

Some geographical data is easily quantified – meaning that it is easy to measure and record an actual number. Examples include pedestrian counts or the number of vacant shops in a high street. Other useful data is difficult to quantify: other people's opinions, for example, on their views about leading a low-carbon lifestyle. This is **qualitative data** and it can be collected in a number of ways:

- in a lengthy interview
- using a questionnaire
- with a quick survey such as a Likert scale.

If you are designing a questionnaire it is a good idea to have examples of both closed and open questions. Closed questions have set answers which you can tick.

For example: How do you get to school?

Walk [ ] Cycle [ ] Bus [ ] Car [ ]

An open question is where people can give their own unstructured answer, e.g. How do you think the school could reduce its GHG emissions?

A Likert scale is where people are asked to use a scale when responding to a question.

Activities

1. Show one way you would represent each of the data sets from the survey. Justify your choice.
2. What conclusions can you draw from the responses given?
3. Suggest two other questions you could ask to investigate individual attitudes to climate change.
4. Explain why data representation and data analysis are easier if closed rather than open questions are asked in questionnaires.
5. Suggest an open question that you might ask. Explain why this open question would reveal useful information about how individuals feel about climate change.

Qualitative data

Geographical evidence in the form of words, sounds or images.

Only offshore wind farms should be developed in future			
1	2	3	4
Agree	Slightly agree	Slightly disagree	Disagree

▲ **Figure 33** An example of a Likert scale.

Question, with responses measured on a Likert scale of 1–10 (1 = not at all seriously and 10 = very seriously)	Mean score
How seriously do you regard the threat of climate change?	7.5
How effective do you think that individuals can be in reducing the threat of climate change?	4.5

▲ **Figure 34** An example of a climate change enquiry.

A group of geography students were interested in investigating how people of different ages are responding to climate change. They set an enquiry question:

Are younger people more willing to change their lifestyles than older people?

As part of their enquiry, they used a Likert Survey with 100 people. They also asked the respondents to tick up to five things that they already do (as individuals) to help reduce the threat of climate change. The results are shown below.

Possible actions that individuals can take	Number of responses
Use low-energy bulbs	82
Choose energy-efficient goods	57
Improve insulation for the home	53
Recycle all plastics, glass, etc.	89
Lower the thermostat settings for heating	23
Walk or cycle to avoid using the car	2
Use public transport	17
Buy locally produced food	11

▲ **Figure 35** Possible actions that individuals can take.

Chapter 2 Management of ecosystems

Can the tourist industry reduce its impact on the environment?

In 2012 the number of tourists passed the 1 billion mark for the first time in history. We saw in Theme 6 that tourism can have huge impacts on the economies of many developing countries. Tourism can have negative, as well as positive, impacts on the people and communities who are visited, however. Travel emits greenhouse gases and tourism generates waste products as well as increasing demands on resources such as water, food and energy. So, how can these negative impacts of tourism be reduced?

Ideally, tourism should be sustainable, meaning that it should have long-lasting benefits. However, for tourism to be developed sustainably it needs to satisfy several conflicting needs. These are summarised in Figure 1.

Local people need to benefit. This may take the form of new jobs and better pay. Where poverty is widespread it should also provide better basic services such as clean water, sewage treatment systems and schools for local people.

The environment (including wildlife/ecosystems) should not be damaged so much by the growth of tourism that it cannot recover.

The growth of tourism should not create problems for future generations of local people. For example, if the development of tourism uses more clean water than can be replaced by natural processes then tourism is unsustainable.

The growth of tourism should not create so many problems that tourists soon stop coming because the environment has been spoilt.

▲ **Figure 1** The conflicting demands of sustainable tourism.

Ecotourism

Tourism that has very low environmental impacts. Ecotourism supports the conservation of habitats and wildlife. In some cases this means that money generated by tourism helps to fund conservation projects. In other cases, the tourists are volunteers who become involved in conservation while they are on holiday – sometimes known as **voluntourism**.

Ethical tourism and responsible travel

Ethical tourism and responsible travel are concepts used to describe sustainable forms of tourism. **Ethical tourism** means that local beliefs, traditions and customs are respected and conserved. Ethical tourism also aims to reduce the negative impacts of the industry on the environment. **Responsible travel** has a similar meaning. Local families benefit economically from jobs and services that are created by ethical tourism and responsible travel.

▼ **Figure 2** Tourism can have several negative impacts on the environment.

Activities

1. Outline at least three different environmental impacts of tourism. Use Figure 2 to help you.
2. Use Figure 4 to describe the distribution of conservation areas in Limpopo.
3. Explain what it is meant by the phrase 'diversify the rural economy'.

Wildlife tourism in Limpopo, South Africa

Limpopo Province, South Africa, has a semi-arid climate and largely unspoilt savanna ecosystem that is a major tourist attraction. Farming, ecotourism and game reserves are the largest employers in Limpopo. However, most jobs in these industries are low skilled and poorly paid. Unemployment is high – as much as 37 per cent in some rural areas. The climate makes farming difficult and incomes are low. There is an urgent need to diversify the rural economy: this means creating a variety of new jobs that are not necessarily connected to farming. One option is the creation of jobs in ecotourism.

District in Limpopo	Population	Population with piped water (%)	Households with electricity (%)
Mutale	91,870	27	8
Lephalale	115,767	67	70

▲ **Figure 3** Percentage of the population who have access to piped water and electricity in selected rural districts of Limpopo.

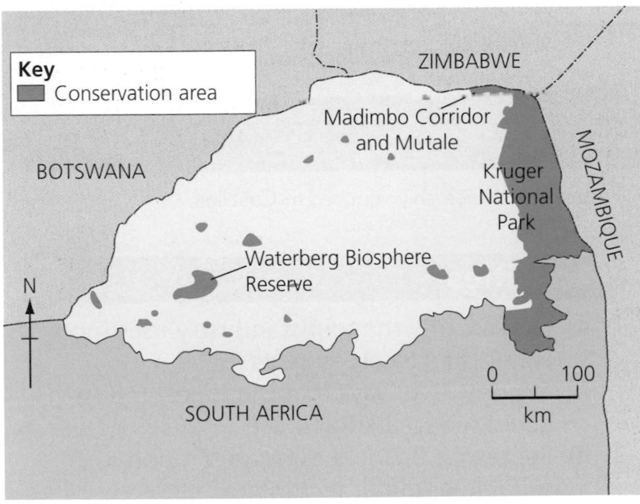

▲ **Figure 4** Conservation areas in Limpopo.

South African and foreign tourists already visit a number of sites in Limpopo, most notably the Kruger National Park in the east of the province. The Waterberg Biosphere Reserve in Lephalale is another popular destination. The reserve contains 75 mammal species (including the big five: lion, leopard, rhino, buffalo and elephant) and 300 species of bird. About 80,000 people live in the reserve. Tourists visiting Waterberg create jobs in a variety of ways. They pay to stay with local families or in luxury lodges that employ cleaners, cooks and bar staff. Local people are also employed to act as tour guides and wardens in the reserve.

Game ranching is one way that wealthy land owners are creating new incomes and managing wildlife. It involves the capture of wild animals, such as the buffalo in Figure 5, and transferring these animals to wildlife reserves as breeding stock where they can be protected from illegal poaching. In this way the population of these wild animals has gradually been rising.

A limited amount of hunting by tourists is then allowed. The main income comes from tourists who pay large sums of money to shoot antelope, buffalo and other wild animals. The meat is sold in South Africa, generating more income. This money is then used to pay for conservation projects such as breeding programmes, habitat conservation and anti-poaching patrols.

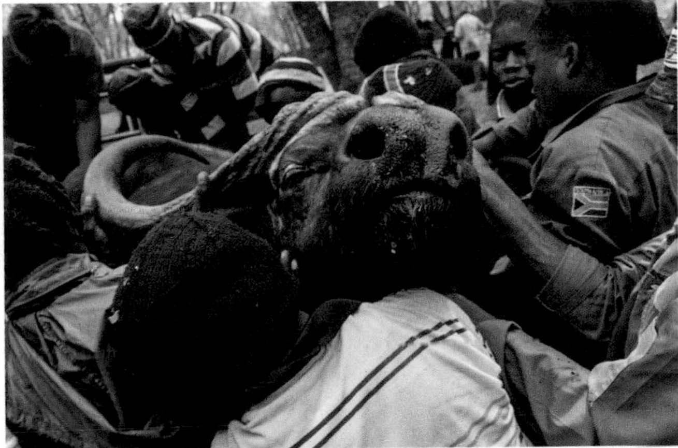

▲ **Figure 5** A darted buffalo is lifted on to a truck on a private farm near Lephalale; after medical checks it will be sold to a breeder.

Activities

4 Calculate the actual number of people in Mutale and Lephalale who do not have access to piped water and electricity.

5 Explain how a new ecotourism project in Mutale or Lephalale could:

a) improve standards of living for local people

b) conserve wildlife.

Enquiry

'All hunting in South Africa should be banned.'

To what extent do you agree with this statement?

- Discuss whether hunting by tourists can be considered to be a form of ethical tourism.
- Outline the arguments for and against the controlled hunting of wild animals in South Africa as part of an ecotourism project.

The impact of tourism on water supply

Tourism places a high demand on local water resources. Water is needed for food preparation, cleaning hotels, laundry of sheets and towels, and keeping hotel gardens watered and swimming pools full. A lot of water is also used to keep golf courses green – an especially big problem in hot, dry climates such as those of Mediterranean holiday destinations.

Data on water consumption by the tourist industry is not collected at a national scale, so those who research this issue need to base their work on estimates. A recent study by the Travel Foundation concluded that tourism uses more water than local residents – their findings are summarised in Figure 7.

Region	Amount of water used by tourists compared to local residents (%)
Mediterranean	150–200
Caribbean (except Jamaica)	150–200
Jamaica	400–1,000
North Africa (e.g. Tunisia)	400–1,000
Kenya	1,800–2,000
Sri Lanka	1,800–2,000

▲ **Figure 7** Water consumption by tourists compared to local residents; 200 per cent means that tourists use twice as much water as locals.

Is preventing the development of golf courses the answer?

Saint Lucia is a Caribbean island that depends heavily on mass tourism for income. It is estimated that a huge 33 per cent of its fresh water supplies are used by the tourist industry and a further 6 per cent are used to irrigate golf courses to keep them green.

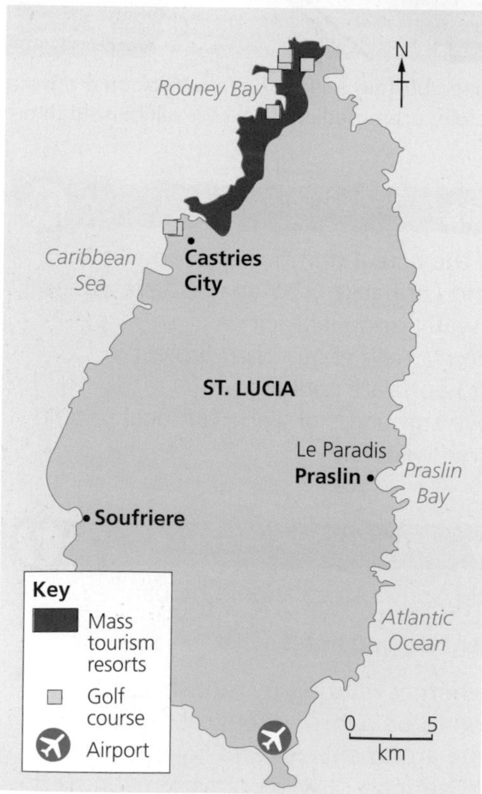

▲ **Figure 6** Location of Le Paradis and the location of other golf courses in Saint Lucia.

▲ **Figure 8** A cruise ship moored in Castries.

Activities

1 a) Explain why the tourist industry uses more water than local residents.
b) Suggest two ways that the tourist industry could reduce its water use.

2 a) Represent the data in Figure 7 using a suitable graph or pictogram.
b) Suggest why water consumption is so much higher in some tourist destinations.

3 Use Figure 6 to describe the:
a) distribution of golf courses in Saint Lucia
b) location of Le Paradis resort and golf club

4 Outline at least two ways in which the development of Le Paradis may not have been sustainable. Use Figure 1 on page 304 to help you.

5 Explain the limitations of cruise tourism for the economy of Saint Lucia (see pages 229 and 231 to help you).

6 Describe the benefits of saving the remaining mangrove in Praslin Bay (use Figure 15 on page 294 to help you).

The Atlantic coast of Saint Lucia is largely undeveloped for tourism. Developers bought 200 hectares of land in Praslin Bay to build a luxury resort and golf course called Le Paradis. Local campaigners protested about the destruction of mangrove forest and dry tropical rainforest but the development began. It was scheduled to be completed in 2007 but the developers ran out of money and the project has never been completed. In 2020, the unfinished resort was still for sale.

▲ **Figure 9** The remaining area of mangrove forest in Praslin Bay. Most mangrove was destroyed when the resort was built; the remaining area is under threat if the development is completed.

▲ **Figure 10** The abandoned Le Paradis complex in 2010.

Saint Lucia government spokesperson

We want the owner of the site to drop their asking price so that another developer can buy the project and complete it. The project includes a hotel, marina for small yachts and a luxury golf course. Completing the project will create much-needed jobs in the east of the island.

Local resident and forest worker

The planned golf course would destroy a dry tropical woodland habitat that is home to rare endemic birds of the thrush family. I work for the forest authorities and as an ecotourism guide in my spare time. If the golf course goes ahead Saint Lucia's biodiversity will be damaged further and I may get less work.

Local resident

I do not want the golf course to be completed. A few hundred families live in Praslin Bay. Our water comes from the stream and boreholes. The water table drops in the summer months and I am worried that our borehole will run dry if water is used to irrigate the golf course.

▲ **Figure 11** Conflicting attitudes towards the development of Le Paradis resort.

Enquiry

Do you think that Le Paradis resort and golf course should be completed? Justify your decision.

Sustainable management of the Great Barrier Reef Marine Park

The Great Barrier Reef Marine Park in northern Australia was established in 1975. It was the first coral reef ecosystem to be given UNESCO World Heritage Status in 1981. It is now considered to be one of the best managed marine ecosystems in the world. As well as being important for its biodiversity, protection is given to more than 70 Aboriginal and other groups whose traditions and culture are under threat from development. The area also has many places of historical significance including lighthouses and shipwrecks. The Marine Park attracts over 2 million tourists a year as well as 5 million recreational users.

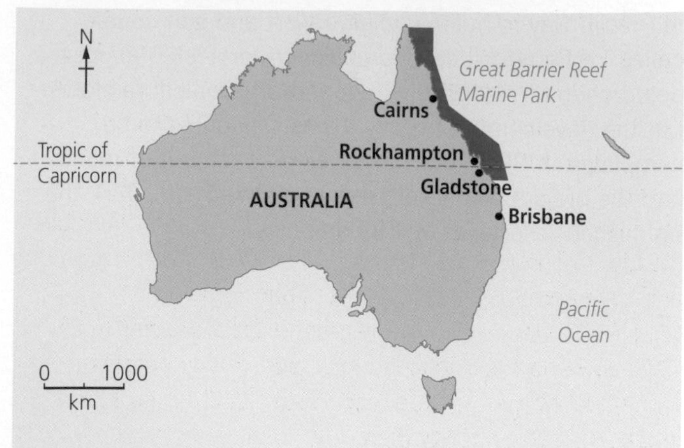

▲ **Figure 12** The location of the Great Barrier Reef Marine Park.

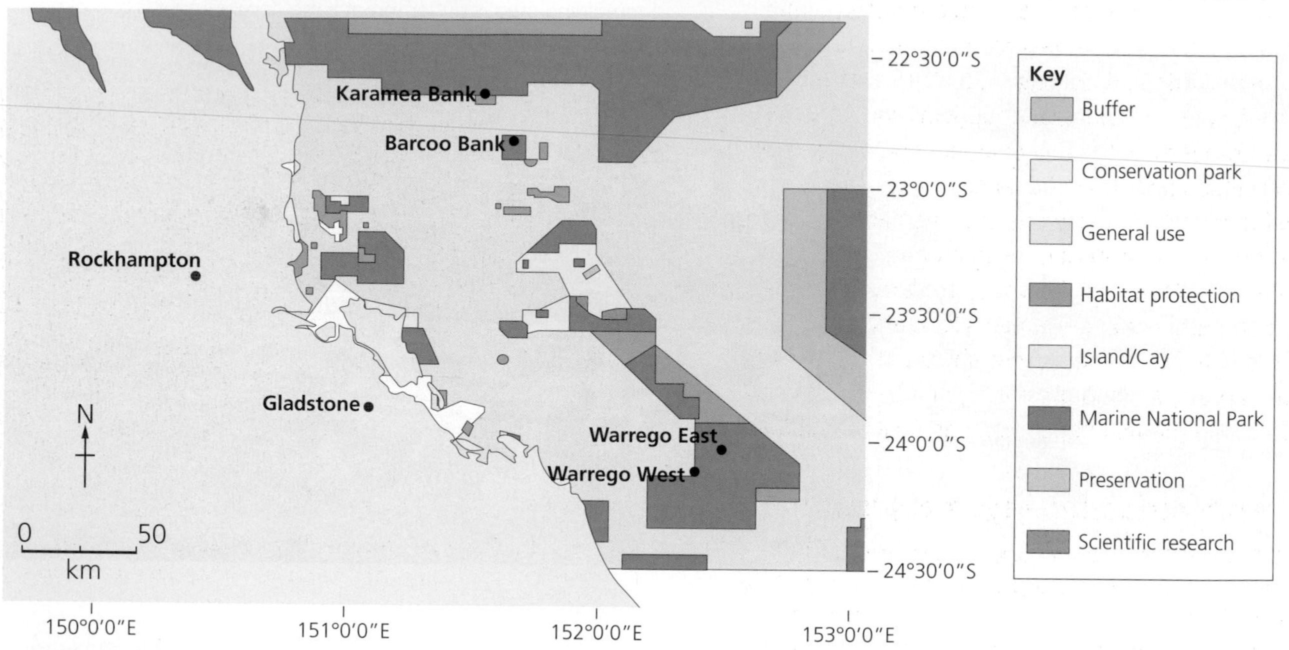

Zone	What is and isn't allowed
Preservation	No one can enter without written permission.
Marine National Park	Fishing and oyster collecting are not allowed. Boating, swimming, snorkelling and sailing are allowed.
Scientific research	For scientific study. Members of the public are usually not allowed.
Buffer	Some fishing is allowed. The public are allowed to enjoy the natural environment in this zone.
Conservation park	Only a limited amount of fishing is allowed.
Habitat protection	Sensitive and vulnerable habitats are protected from damaging activities.
General use	Trawling is not allowed. Crabbing, boating, diving, photography, line fishing and trawling are all allowed.

▲ **Figure 13** Zones of the Great Barrier Reef Marine Park to the east of Gladstone.

How is the park managed?

The Great Barrier Reef Marine Park has been divided into different **conservation management zones**. The level of protection increases from General Use Zone up to the most restrictive Preservation Zone. At honeypots like Cairns and the Whitsunday Islands, there are special management plans. Boat length and visitor group size are restricted to prevent overcrowding of these vulnerable locations. There is also a joint programme of education and enforcement with the Queensland Parks and Wildlife Service. Every five years the Great Barrier Reef Outlook Report is published and from this, a detailed management plan is produced. Reef 2050 is the latest long-term sustainability plan.

Zoning the reef protects the unique marine animals, plants and habitats, as well as threatened species like the green turtle and dugong. Industries that rely on the reef, such as fishing and tourism, can continue. This provides social and economic benefits to local communities and the national economy. Wider benefits include opportunities for recreational, cultural, educational and scientific research into coral reef ecosystems. Research from the latest report has shown that fish numbers and average size are increasing. Coral trout is now 50 per cent more abundant. Bigger fish mean more eggs, and increasing numbers in closed areas mean the fish population can spill into other zones.

▲ **Figure 14** The dugong, a large marine mammal sometimes known as the sea cow.

▲ **Figure 15** The green turtle is threatened by habitat loss, the wildlife trade, fishing for its meat and accidental drowning in fishing nets.

Conservation management zone

An area of a National Park or conservation area which is zoned to help protect the environment from damage by human activities. A zoning plan identifies which activities are permitted in each zone. The aim is to separate uses of the ecosystem that might conflict with each other.

Activities

1. Why do you think the Great Barrier Reef Marine Park was given World Heritage Status?
2. Explain why it needs to be managed.
3. a) Describe how the Marine Park zoning system works.
   b) Explain why the system is an example of sustainable management.
4. Describe the advantages and disadvantages of the zoning system for each of the following groups of people:
   a) fishermen
   b) research scientists
   c) divers and tourists.

Enquiry

Do you think that such a zoning system would work in a National Park in England and Wales such as Snowdonia or the Norfolk Broads?

Weigh up the benefits and problems of such a system. Support your answer by referring to a particular National Park in the UK.

Why should we restore ecosystems?

China has 53 million hectares of wetland. Wetland ecosystems are sometimes described as the kidneys of the Earth because they store and filter water. In this way they protect people from floods and help to sustain a clean supply of drinking water. But wetlands in China are vulnerable to damage by farming and urban growth. A survey in 2014 showed that 60 per cent of China's wetlands were in very poor or relatively poor condition. We saw on pages 186–7 how and why wetland and moorland ecosystems in the UK are now being restored. Similar projects are happening in other parts of the world, including in Heilongjiang Province in China. Here, a mixture of wetland restoration, tree planting and ecotourism projects has helped to improve biodiversity and increase local incomes.

53 per cent of the 8 million population are farmers.

10,090 hectares of trees have been planted.

6 bird reserves have been created.

39,769 hectares of woodland is managed.

3,441 hectares of farmland have been converted back to wetland.

▲ **Figure 18** Sanjiang Plain restoration project in numbers.

▲ **Figure 16** Tourists taking a boat trip on the Ussuri River in the Sanjiang wetlands.

▲ **Figure 17** The location of the Sanjiang wetlands.

The Sanjiang Plain, in Heilongjiang, is a huge wetland ecosystem of natural marshes, water meadows and forests. These habitats flood seasonally. They naturally support a wide biodiversity, especially of ducks and geese. Years of digging ditches and straightening rivers had destroyed 4.32 million hectares of wetlands – 80 per cent of this ecosystem's original extent.

Activities

1 Use Figure 17 to describe the location of Heilongjiang Province.
2 Describe the features that make the landscape of Figure 16 distinctive.

In 2005 a ten-year project began to restore the wetlands, increase biodiversity and increase local incomes through tourism. The project, which cost in excess of US$30 million, focused on the restoration of six large wetland bird reserves. Field ditches were blocked so that the water table would rise – recreating ponds and water meadows. Thousands of hectares of trees were planted around the edge of the reserves. This meant that there was less land available for farming, so the project has supported farmers to diversify into ecotourism. Some farmers now keep bees and sell the honey to tourists. Others are involved as guides for tourists on bird-watching holidays. Greenhouses were built so that farmers can earn more money from much smaller amounts of land, releasing land for conservation. Farmers involved in conservation can now earn up to 40 times more than before.

The project has run campaigns to raise public awareness of the value of conservation. Wetland conservation has been introduced to the curriculum in local schools. The reserves are monitored by patrols and CCTV cameras to try to prevent illegal farming, fishing and hunting.

Impacts of the Sanjiang wetlands project

Li Yuanwen was anxious and unhappy when local government authorities told him to quit farming a piece of land in the Zhenbaodao reserve in Heilongjiang Province in the northeast of the People's Republic of China (PRC).

For twelve years, he had been earning $3,300 to $6,500 a year growing corn, soybean and rice on a twenty-hectare plot he had developed on vacant land, even though it was in one of 24 officially designated nature reserves on the vast alluvial Sanjiang floodplain.

His fears proved to be unfounded. Yuanwen now works as a salaried custodian of a new ecotourism area in the Zhenbaodao reserve where visitors come to fish, boat, hike, camp and watch birds. His income, further increased by eco-friendly beekeeping and the sale of honey and mushrooms to tourists, is far greater than in his best years of farming.

Zhiang Liang had been growing corn on ten hectares of altered nature reserve for six years when his plot was incorporated into the restoration project. He now produces tomatoes in three greenhouses that occupy no more than 1,050 m^2 of space – less than one per cent of what he cultivated before – but his income has risen from $4,410 to $6,860 a year.

Official monitoring of birds in the six nature reserves has shown an increase in their numbers from 510,559 in 2008 to 683,612 in 2011. In the Qixinghe reserve, scientist Cui Shoubin, who has been tracking the egret population, has seen it rise from 600 in 2011 to more than 1,000 in 2014.

Shoubin is delighted that the nature reserves are being visited by more tourists and scientists. 'I'm convinced that what we're finding out in wetland management and the work we're doing on the habitats of rare migratory birds is important not just for the Sanjiang Plain and the PRC but for the whole world.'

▲ **Figure 19** Text extract from a Chinese website describing the Sanjiang wetlands project.

Activities

3 Explain why it was necessary to:
  a) diversify jobs so that incomes could be increased
  b) involve local communities in campaigns and education programmes.
4 Describe the main advantages of the Sanjiang Project for:
  a) local environments and wildlife
  b) local communities.
5 Suggest how the project may help:
  a) communities living downstream (pages 186–7 may help you)
  b) the global environment.

Weblink

www.wwt.org.uk – the website of the Wildfowl and Wetland Trust in the UK.

Enquiry

Who should conserve habitats in the UK?

- Use the weblink to the Wildfowl and Wetland Trust and follow links to the London Wetland Centre. This wetland habitat has been restored.
- How is the centre used for conservation and tourism?
- Who is responsible?

PRACTICE ACTIVITIES

Theme 1

1 **a)** Study the OS Map extract shown in Figure 1.1. It shows part of the Jurassic Coast on the south coast of England.

▼ **Figure 1.1** An OS map extract at a scale of 1:50,000.

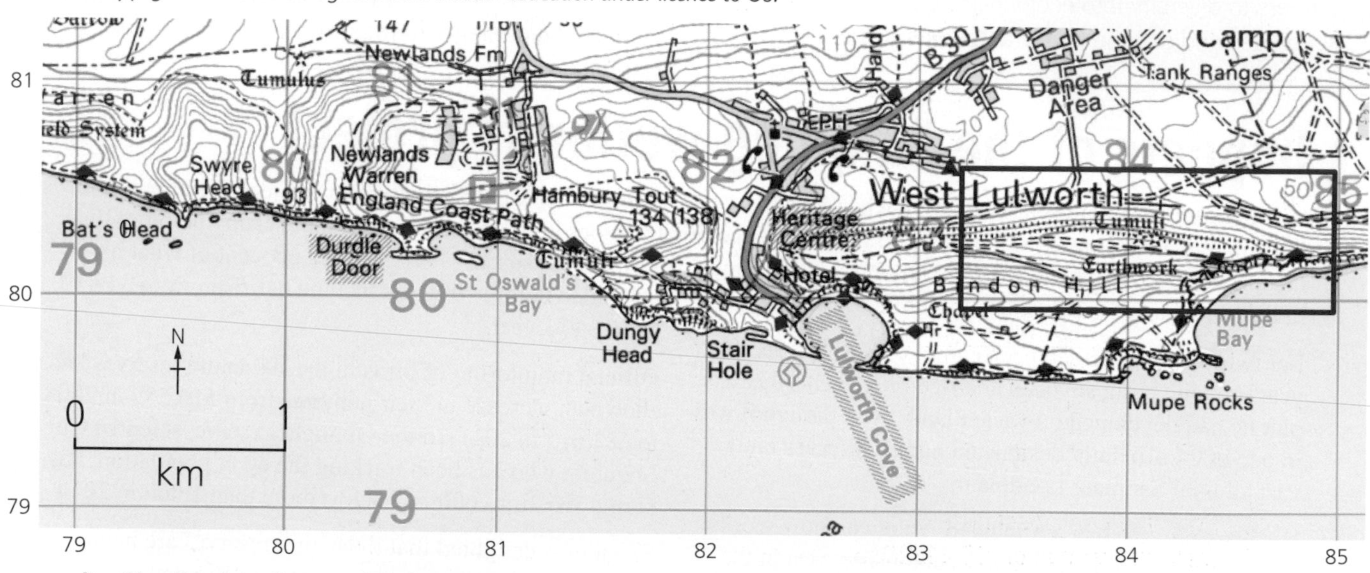

i) Give the four-figure grid reference for Lulworth Cove. Choose the correct answer below. [1]

A 7982 **B** 8280 **C** 8279

ii) Which of the following is the most accurate description of the relief inside the red box? [1]

A An area of rather flat land.

B A river valley with steep slopes.

C A long hill with steep slopes.

iii) Describe the location of Durdle Door using distance and direction from the Heritage Centre in West Lulworth. [2]

b) Durdle Door and Lulworth Cove are coastal landforms made by erosion. Durdle Door is a coastal arch.

i) Describe the process of abrasion at the coast. [2]

ii) Describe how erosion creates a coastal arch. [4]

iii) Give **two** reasons why the rate of coastal erosion depends on geology. [4]

iv) Explain why human management of the coast can lead to increased rates of erosion. [6]

c) Study Figure 1.2. This hydrograph shows discharge of the River Eden in Cumbria (December 2015). During December 2015 there were severe floods in Cumbria.

▼ **Figure 1.2** Hydrograph for the River Eden (1–14 December 2015), Cumbria.

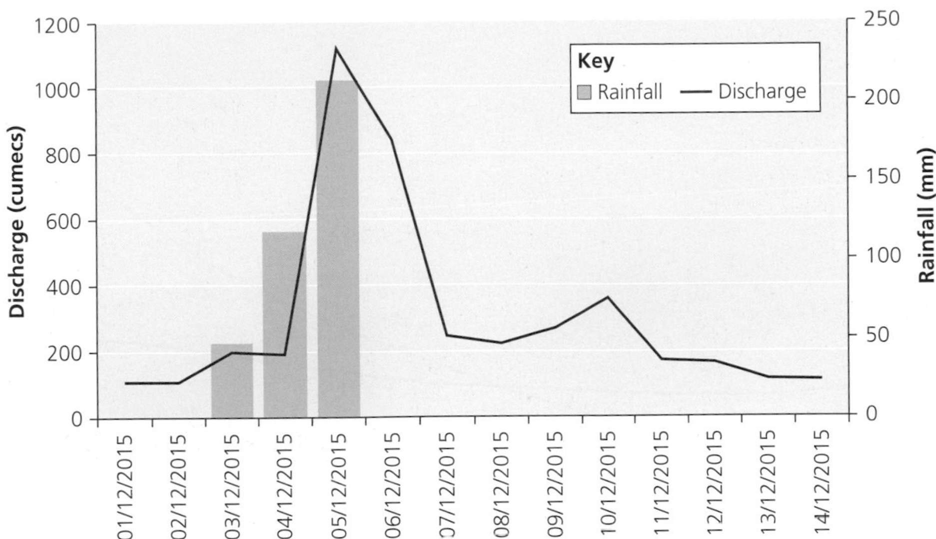

i) State the peak discharge. [1]

ii) State the rainfall on 05/12/2015. [1]

iii) Describe the relationship between rainfall and discharge between 01/12/2015 and 08/12/2015. [4]

d) Study Figures 1.3 and 1.4. They show two examples of engineering strategies to reduce river flooding in Cumbria. In Figure 1.3 the channel on the left was made by engineers over 100 years ago. The new channel on the right of Figure 1.3 has been made recently.

▼ **Figure 1.3** River Swindale, Cumbria.

▼ **Figure 1.4** River Eden, Carlisle, Cumbria.

To what extent are the strategies shown in Figures 1.3 and 1.4 likely to change the discharge of the river? Use evidence from both photos and also refer to the hydrograph, Figure 1.2, to support your answer. [8]

Theme 2

2 a) Study Figure 2.1. It shows population change in two global cities. Lagos is a global city in Nigeria, Africa.

▼ **Figure 2.1** Population change in Lagos, Nigeria, and London, UK.

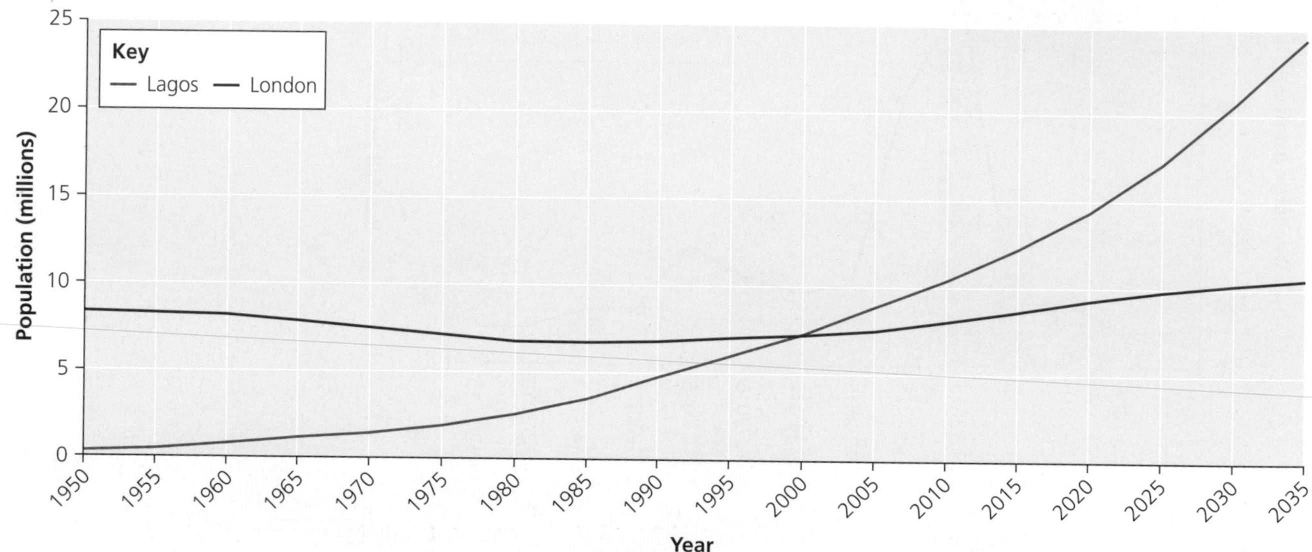

i) Which **three** of the following statements are true? [3]

A The population of London decreased by about 1.4 million between 1960 and 1985.

B The population of London increased from 3.5 million in 1985 to 8.5 million in 2015.

C The population of Lagos has increased faster than the population of London.

D Lagos had a population of 12.2 million in 2015.

E Lagos is expected to be three times larger than London in 2035.

ii) By how much is the population of Lagos expected to grow between 2015 and 2035? Show your working. [2]

b) Lagos is growing partly because of migration from other parts of Nigeria and partly because of natural population increase.

i) Describe **one** push factor for migration that is common in LICs or NICs. [2]

ii) Explain why the population of global cities in LICs/NICs is growing. [6]

c) Lagos has a large number of informal (or squatter) settlements. One of these informal settlements is called Makoko. Figure 2.2 shows the location of Makoko.

▼ **Figure 2.2** The city of Lagos, Nigeria.

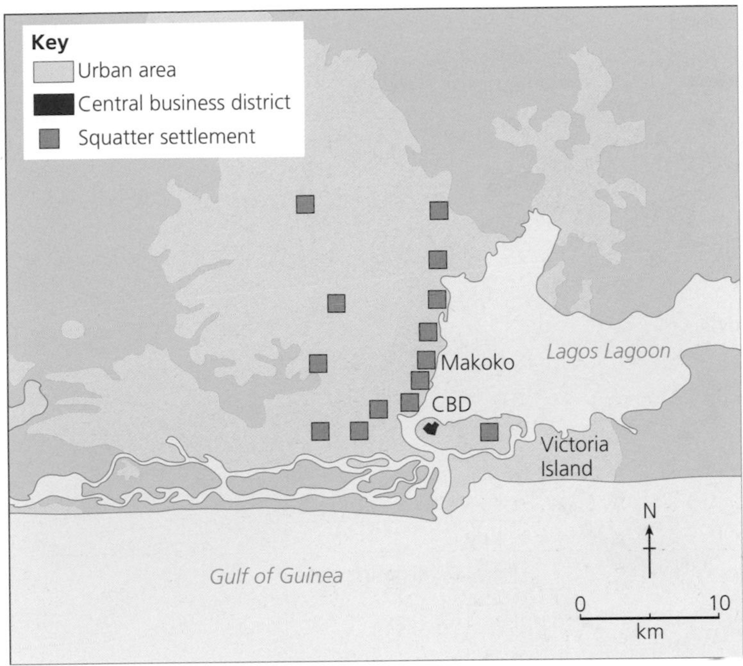

i) Describe the location of Makoko. [2]

ii) Describe the distribution of squatter settlements within Lagos. [3]

iii) Give **two** reasons why so many people in cities like Lagos work in the informal economy. [4]

d) Study Figures 2.3 and 2.4. Figure 2.3 shows Makoko. Figure 2.4 gives information about women in developing cities like Lagos.

▼ **Figure 2.3** The squatter settlement of Makoko, Lagos.

Analyse the challenges facing people who live in the squatter settlements of Lagos. Use evidence in Figures 2.2, 2.3 and 2.4 to support your answer. [8]

▼ **Figure 2.4** The state of women who live in developing cities like Lagos.

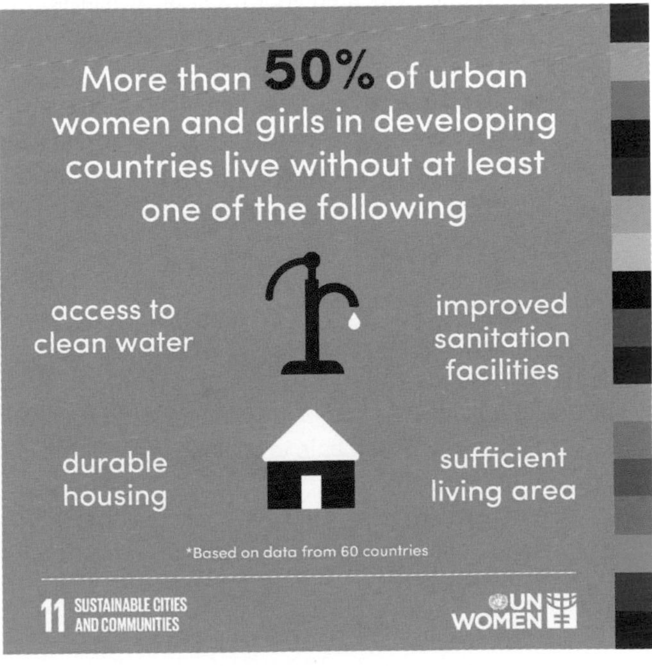

Theme 3

3 a) Study Figure 3.1.

▼ **Figure 3.1** The location of tectonic features in Iceland.

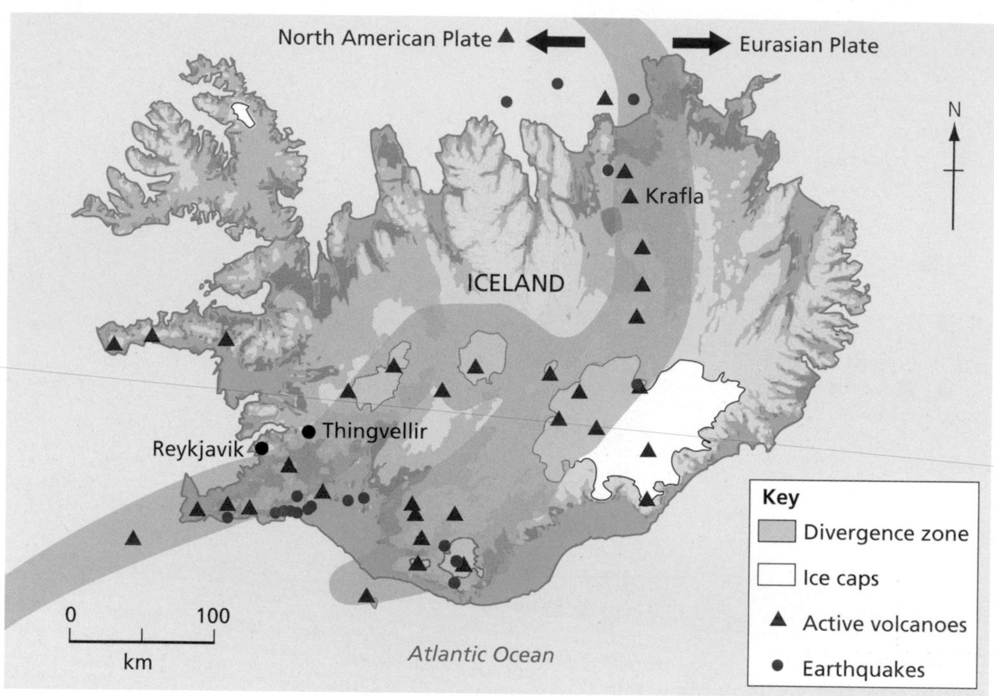

i) Which **two** statements are true? [2]

A Krafla is located 200 km from Reykjavik.
B Krafla is located 300 km from Reykjavik.
C Krafla is located southwest of Reykjavik.
D Krafla is located northeast of Reykjavik.

ii) What is the definition of a divergence zone?

A A plate margin where plates are sliding past each other.
B A plate margin where plates are moving away from one another.
C A plate margin where one plate is destroyed beneath another.

iii) Use Figure 3.1 to describe the location of the divergence zone within Iceland. [2]

iv) Use Figure 3.1 to describe the distribution of active volcanoes in Iceland. [3]

v) Thingvellir is located within a rift valley. Describe how a rift valley is formed. [4]

b) Iceland is located over a large plume of hot rocks in the mantle. This causes frequent volcanic eruptions. What is the correct name for this type of feature? [1]

A subduction zone **B** volcanic hot spot **C** magma chamber

c) Describe **one** smaller-scale feature of a volcanic landscape that you have studied. [4]

d) Study Figure 3.2. It shows groups of people who are vulnerable to the impacts of tectonic hazards such as volcanic eruptions and earthquakes.

▼ **Figure 3.2** Groups of people who are vulnerable to tectonic hazards.

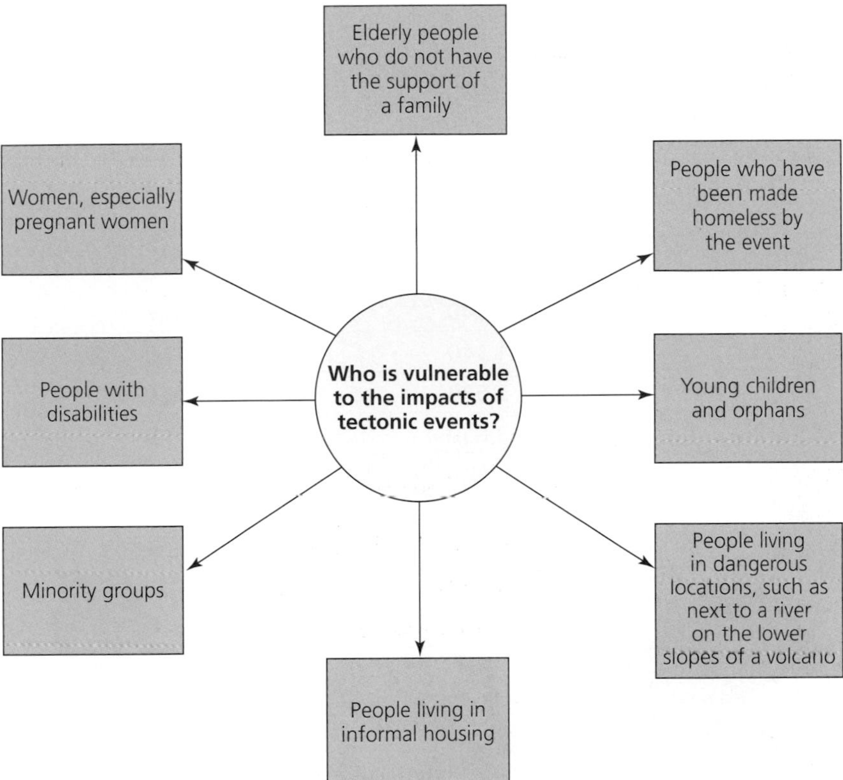

Explain why some groups of people are more vulnerable to the impacts of tectonic hazards. [6]

e) Study Figure 3.3. It shows information about several countries where volcanic eruptions are common. Figure 3.4 shows the impact of a volcanic eruption in Indonesia. It shows the impact of a lahar (volcanic mudflow). The lahar was caused by heavy rain after the eruption of Mount Merapi.

▼ **Figure 3.3** Data for selected countries that have active volcanoes.

Country	Number of active volcanoes	Cities with a population greater than 300,000	Gross national income (GNI) per person (US$)	Percentage of people living on less than US$1.90 per day
Chile	108	6	14,670	0.7
DR Congo	6	16	490	76.6
Iceland	35	0	60,740	0
Indonesia	147	32	3,840	5.7
Japan	118	32	41,340	0
Papua New Guinea	67	1	2,530	38.0
Philippines	53	30	3,850	7.8

▼ **Figure 3.4** The impact of a lahar.

Analyse the factors that make people vulnerable to the impacts of volcanic eruptions. Use evidence in Figures 3.3 and 3.4 to support your answer. [8]

Theme 4

4 a) Study Figure 4.1. There were severe coastal floods on the east coast of England on this date.

▼ **Figure 4.1** Air pressure over the UK on 5 December 2013.

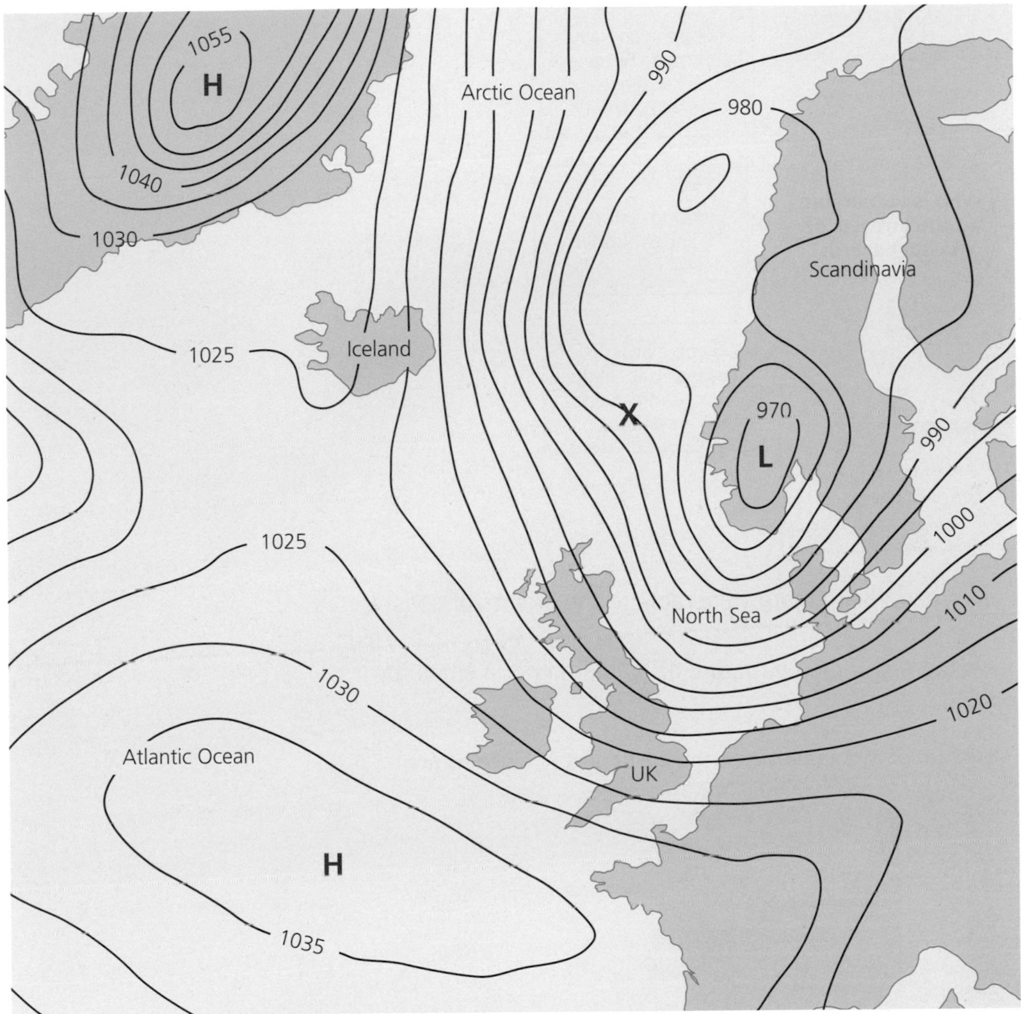

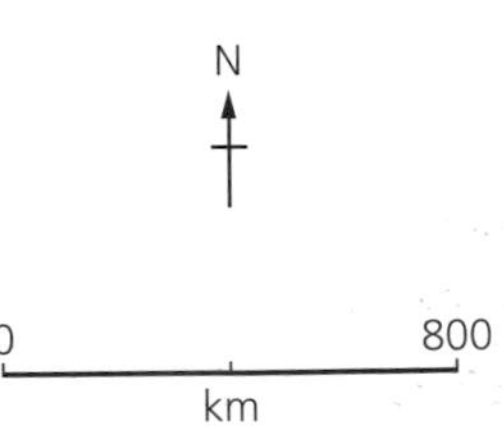

i) Describe the location of the low pressure (marked with the letter L). [2]

ii) Give the air pressure at point X. [1]

iii) Explain why low pressure can lead to coastal flooding. [4]

b) Study Figure 4.2. It shows groups of people who are vulnerable to coastal floods.

▼ **Figure 4.2** Groups of people who are vulnerable to coastal floods.

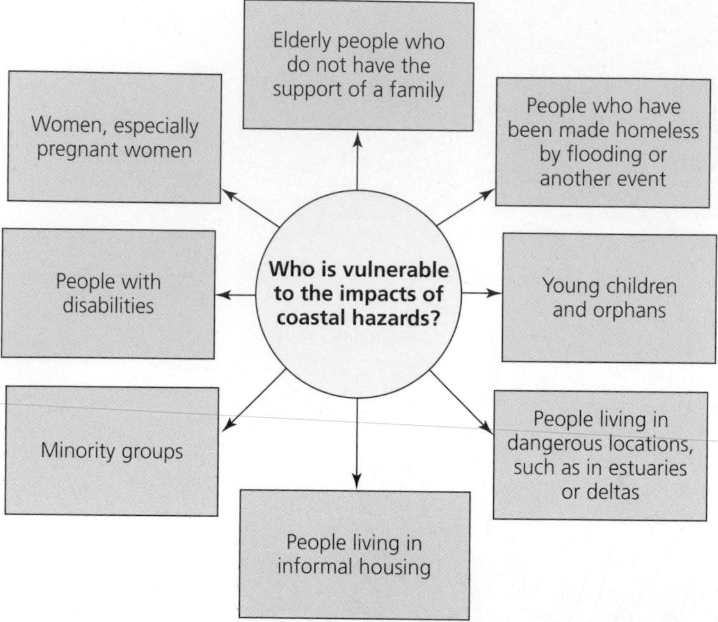

Explain why some groups of people are more vulnerable to the effects of coastal flooding. [6]

c) Study Figure 4.3. It shows the population structure of Mablethorpe, a small town on the east coast of England.

▼ **Figure 4.3** Population structure of Mablethorpe on the east coast of Lincolnshire.

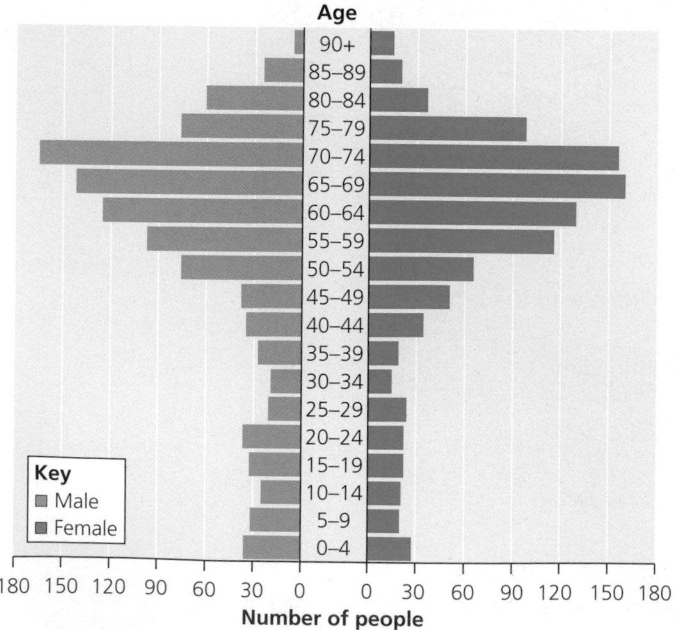

i) State the number of males aged 60–64. [1]

ii) Calculate the total number of people aged 70–74. [2]

d) Describe how emergency planning can be used to reduce the risk of coastal floods. [4]

e) Study Figure 4.4 and Figure 4.5.

▼ **Figure 4.4** Fetch in the North Sea.

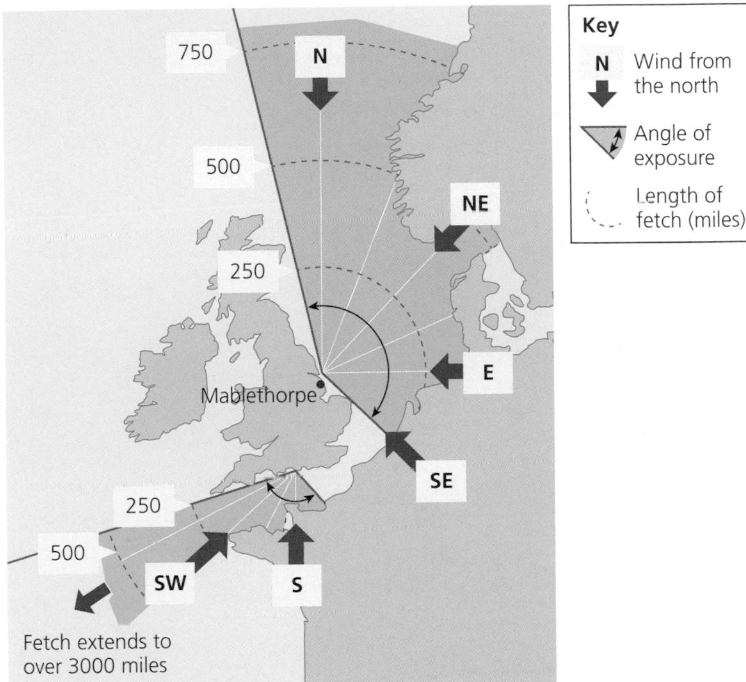

▼ **Figure 4.5** Caravan park at Mablethorpe.

Analyse the factors that make Mablethorpe vulnerable to flooding. Use evidence in Figures 4.3, 4.4 and 4.5 to support your answer. [8]

Theme 5

5 a) Study Figure 5.1.

▼ **Figure 5.1** Carbon dioxide emissions (metric tonnes per person) for three selected countries.

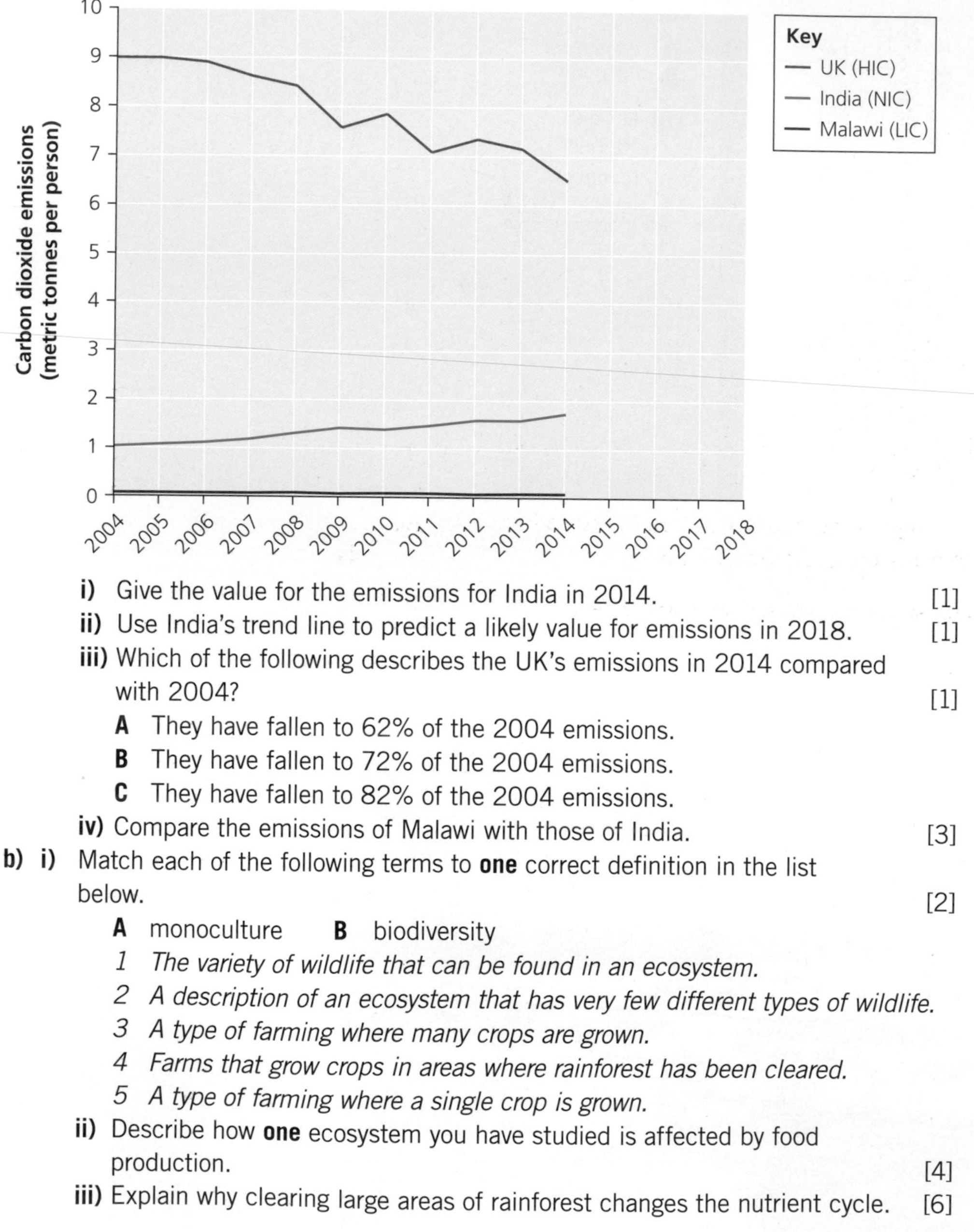

i) Give the value for the emissions for India in 2014. [1]

ii) Use India's trend line to predict a likely value for emissions in 2018. [1]

iii) Which of the following describes the UK's emissions in 2014 compared with 2004? [1]

A They have fallen to 62% of the 2004 emissions.

B They have fallen to 72% of the 2004 emissions.

C They have fallen to 82% of the 2004 emissions.

iv) Compare the emissions of Malawi with those of India. [3]

b) i) Match each of the following terms to **one** correct definition in the list below. [2]

A monoculture **B** biodiversity

1 The variety of wildlife that can be found in an ecosystem.

2 A description of an ecosystem that has very few different types of wildlife.

3 A type of farming where many crops are grown.

4 Farms that grow crops in areas where rainforest has been cleared.

5 A type of farming where a single crop is grown.

ii) Describe how **one** ecosystem you have studied is affected by food production. [4]

iii) Explain why clearing large areas of rainforest changes the nutrient cycle. [6]

c) In Brazil, fires are used to clear areas of forest before the land is used for farming.

▼ **Figure 5.2** Total number of forest fires in Brazil each year between 1 January and 29 August.

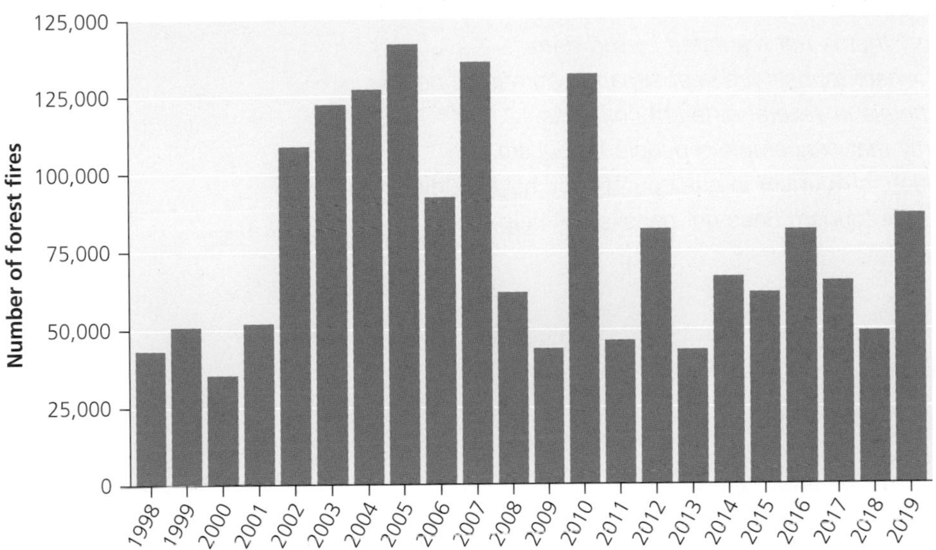

i) Describe the overall change in the number of forest fires in Brazil between 2008 and 2019. Use evidence in Figure 5.2 only. [4]

▼ **Figure 5.3** Fragment of rainforest surrounded by soybean plantations.

ii) Evaluate the environmental and social impacts of turning Brazil's forests into agricultural land for crops like soybean. Use evidence in Figures 5.2 and 5.3 to support your answer. [8]

Theme 6

6 a) i) Match each of the following terms to **one** correct definition in the list below. [2]

A enclave tourism **B** informal sector

1 Tourism that takes into account the needs of local people.

2 The part of the economy that is not regulated by the state.

3 A tourist development where tourists are kept separate from local people.

4 A business that owns hotels in several different countries.

5 The part of the economy that only employs people in tourism.

ii) Describe the positive impacts of tourism in one country you have studied. [4]

iii) Give **two** reasons why enclave tourism does not necessarily help the economic development of a LIC. [4]

b) Study Figure 6.1.

▼ Figure 6.1 Number of tourist arrivals each year to St Lucia (thousands).

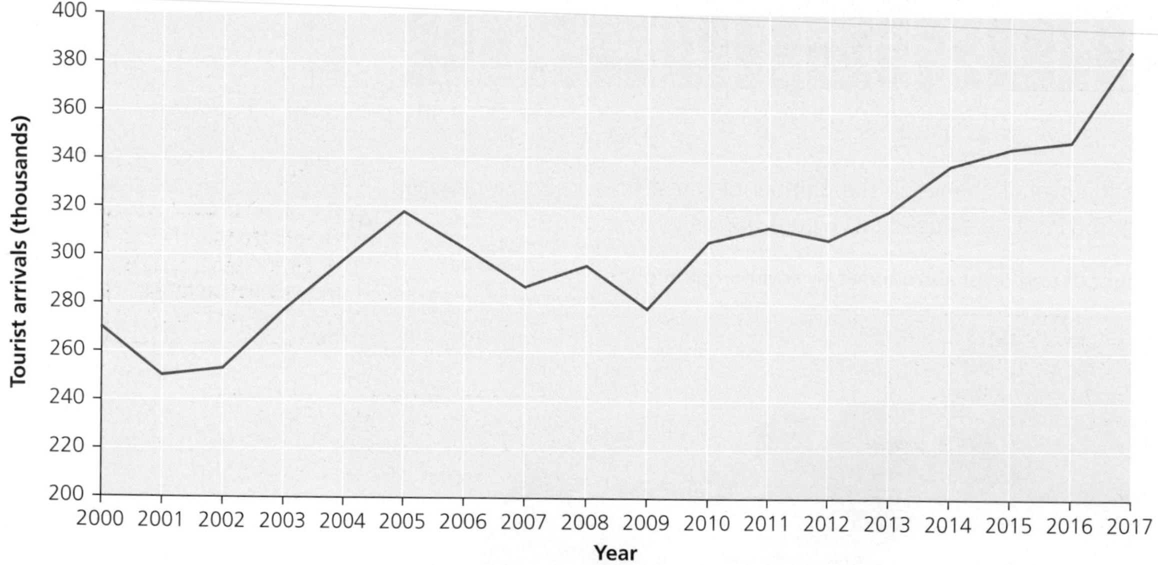

i) Describe how the number of tourist arrivals to St Lucia changed between 2000 and 2009. [2]

ii) Describe how the number of tourist arrivals to St Lucia changed between 2009 and 2017. [2]

iii) Calculate the percentage increase in tourist arrivals between 2000 and 2017. Show your working. [4]

c) Explain why countries use protectionist trade policies. [6]

d) Study Figure 6.2. It shows information about the use of India's groundwater supply.

▼ **Figure 6.2** India's groundwater supply.

State	Availability (billions cubic metres)	Consumption (billions cubic metres)
Punjab	20.3	34.9
Rajasthan	10.8	14.8
Delhi	0.3	0.4
Haryana	9.8	13.1

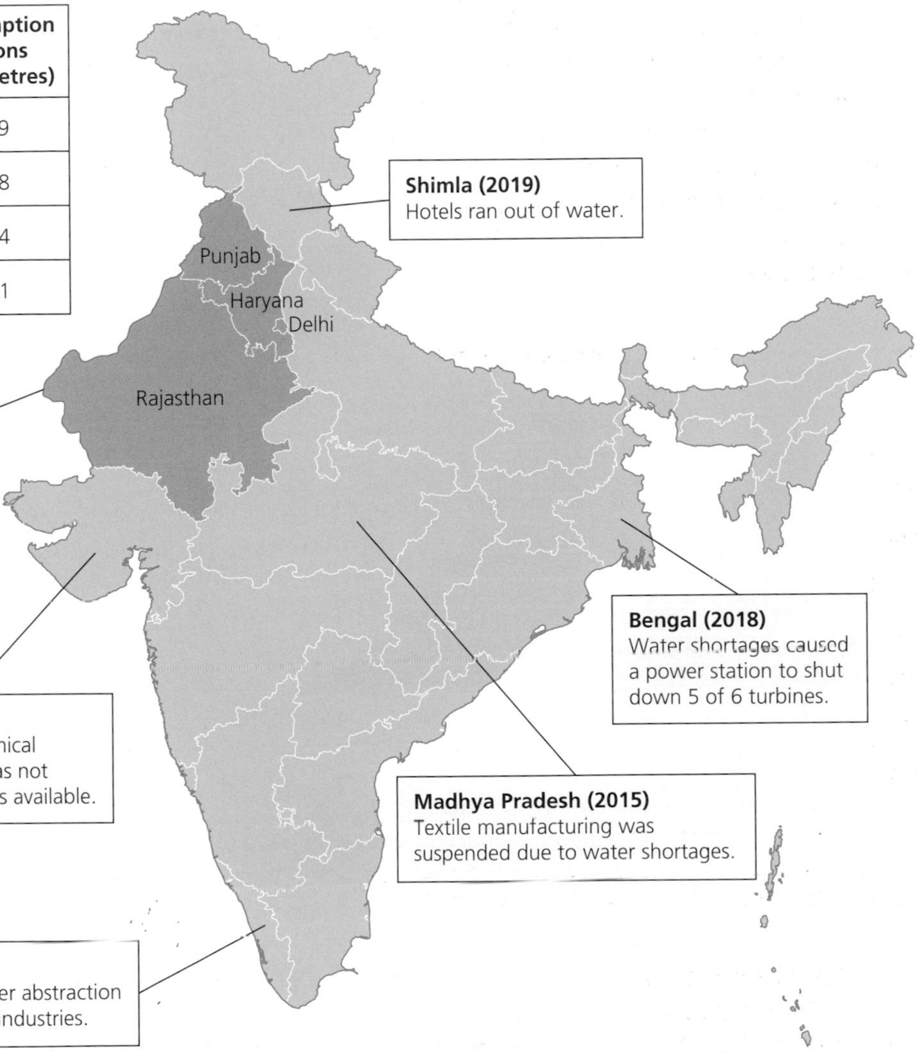

These four states use groundwater more quickly than it is recharged.

Punjab

Haryana

Delhi

Rajasthan

Shimla (2019)
Hotels ran out of water.

Bengal (2018)
Water shortages caused a power station to shut down 5 of 6 turbines.

Gujarat (2016)
A large petrochemical factory was shut as not enough water was available.

Madhya Pradesh (2015)
Textile manufacturing was suspended due to water shortages.

Kerala (2004)
Street protests over abstraction of water by local industries.

Analyse the social, environmental and economic consequences of over-abstraction of groundwater in India. Use evidence from Figures 6.2 and 6.3 to support your answer. [8]

▼ **Figure 6.3** The dried-up River Vaigai in Tamil Nadu, India.

Theme 7

7 a) Study Figure 7.1. It shows the population structure for the Gambia, which is an LIC in sub-Saharan Africa, and Sri Lanka, which is an MIC in South Asia.

▼ **Figure 7.1** Population pyramids for the Gambia (in sub-Saharan Africa) and Sri Lanka (in South Asia in 2017).

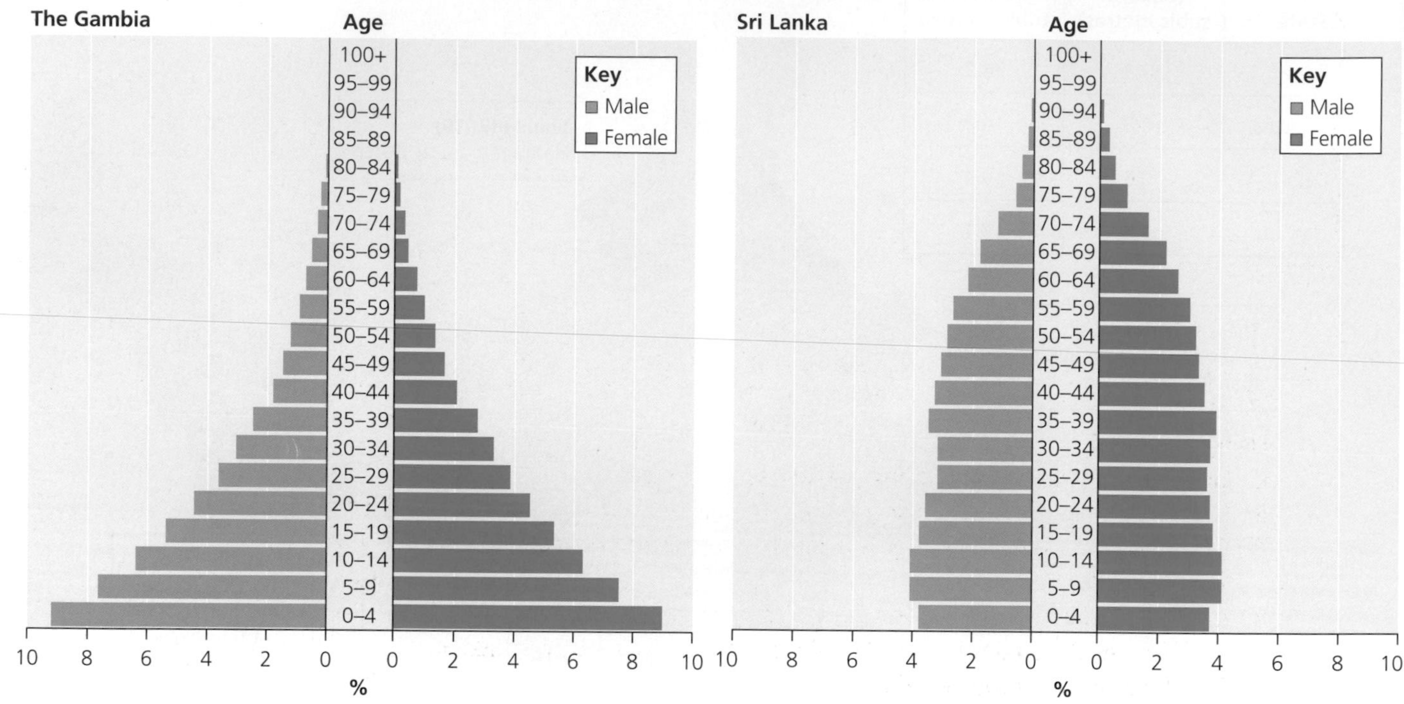

i) Which **three** of the following statements are true? [3]

A The Gambia has a larger proportion of elderly people than Sri Lanka.
B The Gambia has a larger proportion of children than Sri Lanka.
C 16.5% of Gambians are females under the age of 9.
D 8.4% of Sri Lankans are aged 35–39.
E Sri Lanka has more females than males aged 60–69.
F People are more likely to live into older age in the Gambia than in Sri Lanka.

ii) Give **two** reasons why LICs like the Gambia have higher birth rates than countries that have higher incomes, like Sri Lanka. [4]

b) Explain why many countries in sub-Saharan Africa have high rates of infant mortality. [6]

c) i) Give **two** reasons for child labour. [4]

ii) Describe how the issue of child labour is being tackled. [4]

d) Study Figure 7.2. It shows the number of children who do not have a primary school education in Pakistan, a country in South Asia.

▼ **Figure 7.2** Children missing primary school, Pakistan (millions).

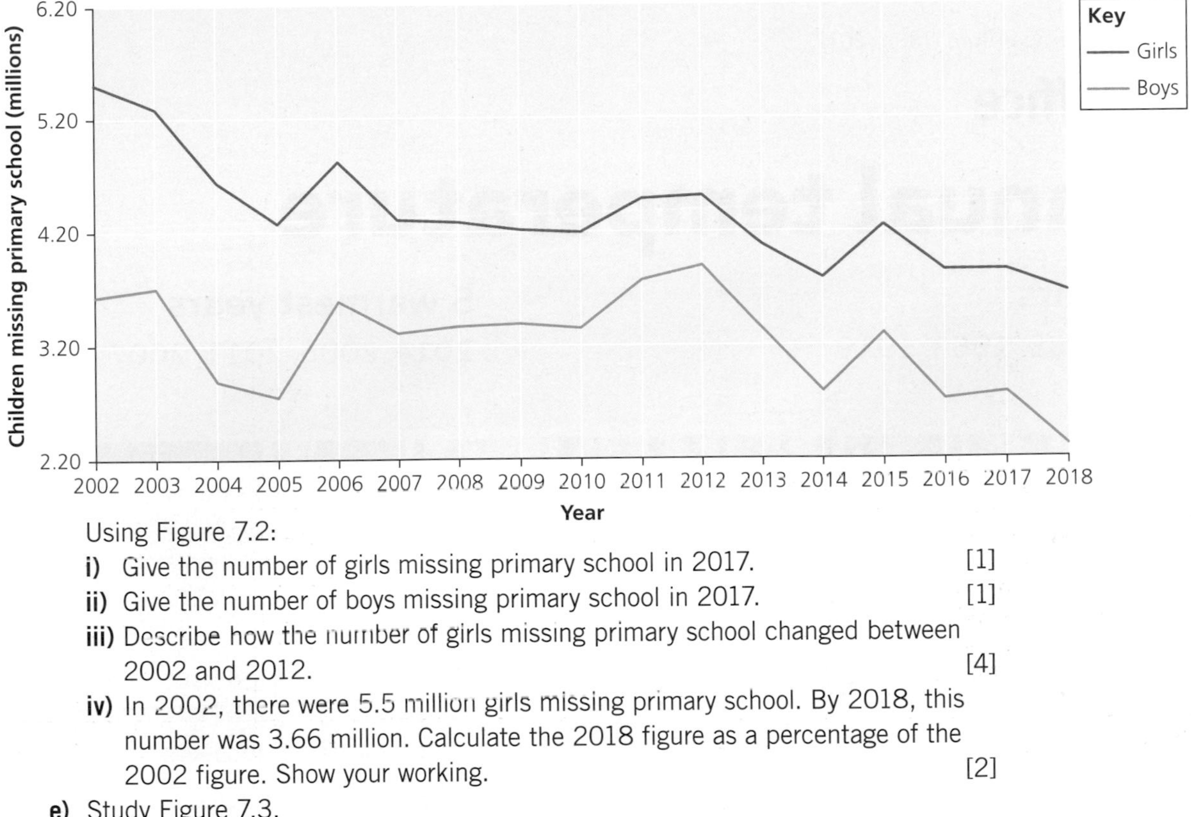

Using Figure 7.2:

i) Give the number of girls missing primary school in 2017. [1]

ii) Give the number of boys missing primary school in 2017. [1]

iii) Describe how the number of girls missing primary school changed between 2002 and 2012. [4]

iv) In 2002, there were 5.5 million girls missing primary school. By 2018, this number was 3.66 million. Calculate the 2018 figure as a percentage of the 2002 figure. Show your working. [2]

e) Study Figure 7.3.

▼ **Figure 7.3** A young worker in a textile factory, India.

Evaluate the impacts of child labour on the people and the economy of countries in South Asia. Use evidence from Figures 7.2 and 7.3 to support your answer. [8]

Theme 8

8 **a)** Study Figure 8.1. It represents UK annual temperatures: cooler years are blue, warmer years are red.

▼ **Figure 8.1** UK annual temperature (1884–2019).

UK annual temperature

5 coolest years

1892, 1888, 1885, 1963, 1919

5 warmest years

2014, 2006, 2011, 2007, 2017

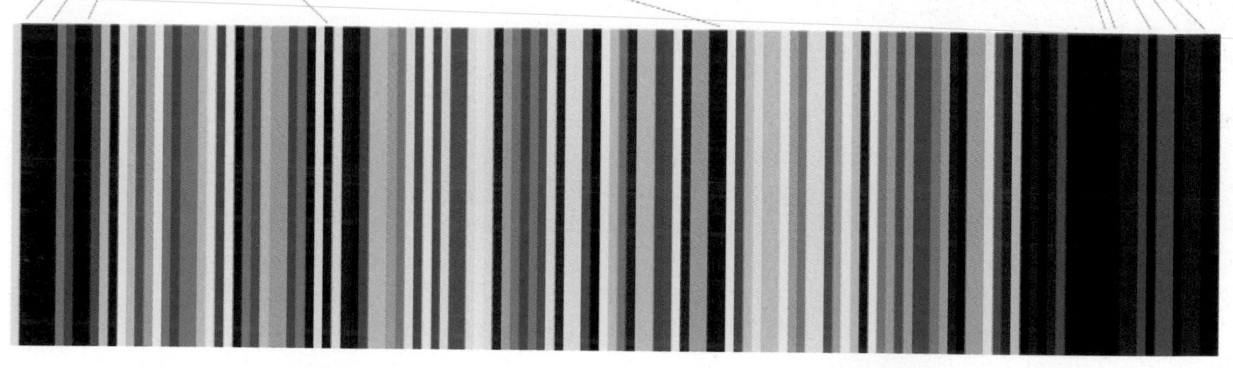

1884 1900 1920 1940 1960 1980 2000 2020

i) Calculate the median date for the five warmest years. Show your working. [2]

ii) Calculate the range for the five coolest years. [1]

iii) Identify **one** strength and **one** limitation of this method of showing these data. [4]

b) i) Match each of the following terms to **one** correct definition in the list below. [2]

A ecological footprint **B** consumerism

1 The amount of food needed to keep people healthy.

2 The part of the economy associated with selling and buying products.

3 The average amount of land needed by people to maintain their lifestyle.

4 The amount of water needed to grow the food we need.

5 The part of the economy associated with making electronic goods.

ii) Describe how agribusiness has had an impact on the environment in one example you have studied. [4]

iii) Explain why consumerism in HICs has led to the destruction of tropical rainforests. [6]

c) Recycling e-waste (waste from electronic products) is a growing issue. Study Figures 8.2 and 8.3.

▼ **Figure 8.2** Teenage boys in Accra, Ghana, burning cables from electronic equipment so that the copper can be recycled.

▼ **Figure 8.3** Regions that send and receive e-waste.

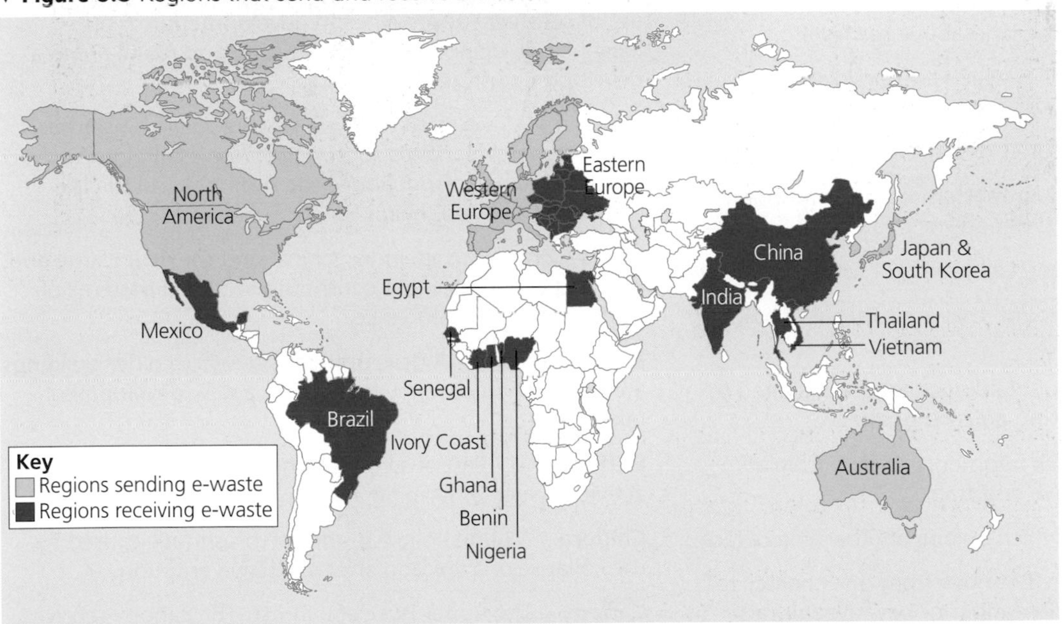

Norway	28.3
Switzerland	26.3
Iceland	25.9
Denmark	23.9
United Kingdom	23.4

Niger	0.4
Ethiopia	0.5
Afghanistan	0.6
Uganda	0.6
Nepal	0.8

The highest and lowest e-waste generating nations (kg per capita)

'E-waste should be recycled in the countries where it is produced.' To what extent do you agree with this statement? Use evidence from Figures 8.2 and 8.3 to support your answer. [8]

GLOSSARY

Abrasion – Erosion, or wearing away, of the landscape caused by friction. Abrasion occurs when rivers or waves are carrying sand or pebbles. The water uses these materials to do the work of erosion.

Abstraction – When water is taken from a river, reservoir or underground source to be used it is abstracted.

Aerosol – Tiny particles of dust, volcanic ash and gas in the atmosphere that can reflect the Sun's energy back into space.

Afforestation – The planting of forests.

Ageing population – A country which has a high proportion of people aged over 65 is said to have an ageing population.

Agribusinesses – Farming that is organised by large businesses – often by multinational companies.

Agro-forestry – A type of farming in which a mixture of crop, shrubs, fruit trees and nut trees are grown.

Air mass – A large parcel of air in the atmosphere. All parts of the air mass have similar temperature and moisture content at ground level.

Altitude – The height of the land above sea level.

Annual regime – The way in which a river's discharge varies throughout the year.

Anomaly – Evidence that does not fit the general trend.

Anticyclone – A high pressure system in the atmosphere associated with dry, settled periods of weather.

Aquaculture – The commercial farming of fish and shellfish.

Aquifers – Rocks in the ground that are capable of holding large quantities of water.

Aseismic – A description of buildings that are designed to withstand the shaking during an earthquake.

Ash cloud – Powdered rock fragments thrown from a volcano during an explosive eruption.

Aspect – The direction in which a slope or other feature faces.

Asylum seekers – People who move from one country to another because they are in danger or because they are persecuted because of their religion or political views.

Attrition – A type of erosion where rocks smash against each other making them smaller and more rounded.

Average life expectancy – The average age someone can expect to live to. Average life expectancy is usually calculated separately for men and women.

Backwash – The flow of water back into the sea after a wave has broken on a beach.

Bay – A body of shallow water that forms where softer rocks have been eroded by the sea. Headlands and bays exist because rocks erode at different rates.

Beach – A coastal feature that provides a natural defence against erosion and coastal flooding. A wide, thick beach absorbs wave energy. Building groynes makes a beach wider and thicker.

Bilateral aid – Financial support, or the gift of food, clothing or other emergency support, that is given directly from the government of one country to another.

Biodiversity – The variety of living things.

Biodiversity hotspot – A region with a particularly great variety of organisms. Central America (or Meso-America) is one such hotspot.

Biomes – Very large ecosystems e.g. tropical rainforests or deserts.

Birth rate – The number of children born in one year for every 1,000 people in a country's population.

Bivariate data – Two sets of numbers that are linked by some kind of relationship.

Bottom-up – An approach to development in which the local community is involved in decision making. Bottom-up (or self-help) approaches to development rely on local people being involved in starting and maintaining a development project.

Braided – A river pattern made when a shallow river has deposited gravel islands so that the river is split into several smaller channels. From above the river looks a little like platted (or braided) hair.

BRICs – Brazil, Russia, India, China (and Mexico). Large and growing economies that contribute to global patterns of trade and interdependence.

Brownfield site – A development site where older buildings are demolished or renovated before a new development takes place.

Buffer zone – Part of a conservation area in which some activities, such as farming and tourism, may be allowed.

Caldera – A huge hollow in the earth's surface caused by the collapse of a volcano after a massive eruption.

Canopy – The upper layer of a forest. The canopy receives most sunlight so contains many leaves, flowers and fruit.

Capacity – The ability of a group of people to withstand a problem such as a natural disaster. Capacity is the opposite of **vulnerability**.

Carbon sinks – Places where carbon is stored over very long periods of time, for example in fossil fuels.

Carrying capacity – The ability of a landscape or ecosystem to absorb the activity of people without any lasting damage. Some ecosystems have larger carrying capacities than others.

Central Business District (CBD) – The area of a town or city in which most shops and offices are clustered together.

Chawl – The name for a type of flat or tenement building found in many Indian cities.

Child labour – Work that robs children of their childhood. Because child labour prevents children from completing their education, it also robs them of their potential – what they might achieve in their adult lives. Child labour is often mentally, physically, socially or morally harmful to children.

Choropleth – A type of map that uses different colours or shades to represent data shown in areas of the map.

Cinder cone – A volcanic hill that is conical in shape. A cinder cone is formed by the eruption of red hot lava that is thrown from the vent. The globs of lava are full of gas bubbles. They solidify to form pebble-sized fragments of rock that have the texture of honeycomb.

Climate – Average patterns in weather conditions that are experienced over a long period of time (at least 40 years).

Commuting – The regular (daily or weekly) movement of workers from their home in one place to their work in a different place.

Conservation management zone – An area of a national park or conservation area which is zoned to help protect the environment from damage by human activities. A zoning plan identifies which activities are permitted in each zone.

Constructive plate margin – A boundary between two plates of the Earth's crust which are moving away from one another. The term **divergence zone** has the same meaning. See **destructive plate margin** for the opposite type of plate boundary.

Continental climate – A climate with hot summers and cold winters. Poland and western Russia have continental climates.

Convection – A process by which hot liquids or gases rise and then spread apart before cooling and sinking. Convection occurs in the atmosphere and also in the mantle where it may be one process that helps to drive plate movement.

Coral bleaching – A process which results in zooxanthellae being forced out of the tissue of corals, turning the colourful corals white. The process occurs when the temperature of sea water increases.

Coriolis Effect – The way in which the rotation of the Earth deflects the movement of objects such as airplanes or hurricanes.

Corrosion – The wearing away of the landscape by chemical processes such as solution.

Counter-urbanisation – The movement of people and businesses from large cities to smaller towns and rural areas.

Cumecs – An abbreviation of cubic metres per second – which is a measure of the **discharge** of a river.

Cyclone – A low pressure system in the atmosphere associated with unsettled weather, wind and rain.

Dalits – The lowest social group in Indian society, sometimes referred to as 'untouchables'. There are 200 million Dalits in India.

Death rate – The number of deaths in each year for every 1,000 population.

Dependency – When a country relies too heavily on one way of earning foreign income. For example, some Caribbean countries rely too much on money from tourism.

Dependent variable – A set of data whose values depend on another set of data which is the independent variable. The relationship may be tested in a **scattergraph**.

Depopulation – The net loss of people from an area due to outward migration and low birth rates.

Deposition – The laying down of material in the landscape. Deposition occurs when the force that was carrying the sediment is reduced.

Depression (weather) – A weather system associated with low air pressure. Depressions bring changeable weather that includes rain and windy conditions.

Deprivation – To lack features, such as employment or basic services, which are usually regarded as necessary for a reasonable standard of living.

Desertification – A process that turns areas of semi-arid grassland into desert. It is caused by drought, the removal of shrubs and trees and/or the use of unsuitable farming techniques.

Destructive plate margin – A boundary between two plates of the Earth's crust which are moving towards one another. See **constructive plate margin** for the opposite type of boundary.

Development aid – Help which is given to tackle poverty and improve quality of life over the long term to improve education or health care.

Direct employment – Jobs that are created within a business. For example, direct employment is created for baggage handlers if a new airport is constructed. See **indirect employment**.

Discharge – The amount of water flowing through a river channel or out of an aquifer. Discharge is measured in cubic metres per second (cumecs).

Divergence zones – A boundary between two plates of the Earth's crust which are moving away from one another. The term **constructive plate margin** has the same meaning.

Diversification – Where a much wider variety of new business opportunities and jobs are created in a region.

Drainage basin – The area a river collects its water from. This is also called a catchment area.

Drought – A long period of time with little precipitation.

Drought-resistant – Plants that are able to survive periods with below average rainfall.

Ecological footprint – The amount of land that is needed to support a person's lifestyle.

Economic migrants – People who move to another country in search of work. Economic migrants are sometimes encouraged to move by the host country in which case they are known as guest workers.

Ecosystem – A community of plants and animals and the environment in which they live. Ecosystems include both living parts (e.g. plants) and non-living parts (e.g. air and water).

Ecotourism – Small-scale tourist projects that create money for conservation as well as creating local jobs.

Embedded – The amount of water or energy that is required to make a product.

Emergency aid – Help that is given urgently after a natural disaster or a conflict to protect the lives of the survivors.

Empower – To create opportunities for a disadvantaged group in society so that they have the ability to help themselves.

Enclave tourism – Tourism based on 'all-inclusive' packages, which means that tourists pay one price and get all their food, drink and entertainment from the hotel.

Endemic – An endemic disease is one that is often found amongst people in a particular place. An endemic species (of plant or animal) is one that is found in a certain place.

Enivronmental refugee – Someone forced to flee their home because of a natural disaster such as coastal floods, drought or climate change.

Epicentre – The place on the surface of the Earth that experiences ground shaking from an earthquake first. The epicentre is directly above the focus.

Erosional processes – These result from the power of a flowing river or waves and wear away the bed and banks of the river channel or the shoreline. They include hydraulic action, abrasion, attrition and solution.

Escarpment – A long, steep-sided slope. The term is often shortened to scarp. **Rift valleys** usually have steep escarpments on either side.

Ethical tourism – Tourism that benefits local people and which respects their culture, environment, beliefs and traditions.

E-waste – Electronic waste products such as computers and mobile phones.

Exports – Goods that are sold to other countries.

Fallow period – A term used in agriculture to describe the period of time during which a piece of land is allowed to rest between crops.

Fertility rate – The average number of children born to each woman in a country. If the fertility rate is greater than two, the population will grow.

Fetch – The distance over which wind has blown to create waves on the sea. The greater the fetch, the larger the waves.

Floodplain – The flat area beside a river channel that is covered in water during a flood event.

Focus – The underground point from which energy spreads during an earthquake.

Fold mountains – Large mountain ranges that have been formed by folding as two tectonic plates collide into one another.

Food miles – How far the food has been transported to get from producer to consumer.

Foreign direct investment – An investment of money by a company in a development, such as a new factory, located in another country.

Fractures – Cracks in the Earth's crust or in rocks at the surface that are often formed by plate movement or volcanic activity.

Free trade – When countries trade without any limits to the amount of goods that can be exported and imported

Garden cities – New, planned urban areas (towns) that have village-like communities and plenty of space for private gardens and public open space.

Geochemical (monitoring) – Scientific research into the Earth's chemistry by measuring gases that escape from the ground. This research may be used to check volcanic activity before and during an eruption.

Geomagnetic (monitoring) – Scientific research into the Earth's magnetic field. This research may be used to check volcanic activity before and during an eruption.

Geyser – A geothermal feature of volcanic landscapes in which groundwater, heated by hot rocks, is ejected from the ground.

Glacials – Cold periods in Earth's history when glaciers have advanced and ice sheets increased in size.

Global city – A city that is well connected by the process of globalisation. For example, global cities are usually important transport hubs with major airports and ports. They often have headquarters for **multinational companies**.

Globalisation – Flows of people, ideas, money and goods are making an increasingly complex global web that links people and places from distant continents together.

Gorge – A steep-sided, narrow valley. Gorges are often found below a waterfall.

Graben – A valley with a wide, flat bottom that has been formed by downward faulting. The term **rift valley** has the same meaning.

Green belt – A government policy used to prevent the spread of cities into the countryside. It is very difficult to get planning permission for new homes in a green belt.

Greenfield site – A plot of land that has not been used before for building.

Greenhouse effect – A natural process that traps heat in the atmosphere. This process has been enhanced by human activity.

Greenhouse gases (GGs) – Gases such as carbon dioxide and methane that are able to trap heat in the atmosphere.

Gross national income (GNI) per capita – The average income in a country. It is also known as gross national product (GNP) per person.

Groundwater (store) – Water in the ground below the water table.

Groundwater flow – The flow of water through rocks.

Gulleys – Narrow, V-shaped channels cut by running water on steep slopes.

Hard engineering – Artificial structures such as sea walls or concrete river embankments.

Hazard map – A type of plan or map that shows the extent of a hazard such as flood risk or danger of volcanic eruption.

Hazard mapping – Plotting the predicted levels of risk of a natural hazard, such as a volcanic eruption, on to a map.

Hazard monitoring – In terms of volcanic activity, this involves using a variety of different scientific instruments to observe volcanoes. The data collected may give the scientists clues about whether or not a volcano is about to erupt. Scientists use these data to give local people enough time to evacuate.

Headland – A rocky outcrop that sticks out into the sea. Headlands are made of rocks that are resistant to erosion.

Heatwave – When temperatures are at least 4.5°C above mean temperature for two or more days.

Hierarchy – The rank order of places, from hamlets to towns to cities.

High-pressure system (anticyclone) – A large-scale weather system that occurs where air in the lower atmosphere is descending and pressing down on the Earth.

Highly skilled – Workers who have high levels of qualification or technical ability and training. See **unskilled workers**.

Honey pot site – A place of special interest that attracts many tourists and is often congested at peak times.

Host countries – Countries that receive investment from multinational companies.

Human Development Index (HDI) – A measure of development that takes into account a country's level of education, its wealth and average life expectancy.

Hydraulic action – Erosion caused when water and air are forced into gaps in rock or soil.

Hydrograph – A line graph that shows either the discharge or height of a river over time.

Hypothesis – An idea or theory that can be investigated.

Identity – The features (e.g. social, cultural or landscape) that give each place its distinctive character.

Impermeable – Soil or rock which does not allow water to pass through it, such as clay.

Import duty – A tax placed on goods brought into a country to make them more expensive.

Imports – The purchase of goods from another country.

Independent travel – When tourists make their own holiday arrangements rather than buying a package holiday from a travel agent.

Independent variable – A set of data whose values stand alone and are not altered by other sets of data. See **dependent variable.**

Indirect employment – Jobs that come as a result of the investment by a business but not within the business itself. For example, indirect employment is created for existing taxi drivers if a new airport is constructed.

Infant mortality rate (IMR) – The number of children who die before the age of one for every 1,000 that are born.

Infiltration – The movement of rain water or snow melt into the soil.

Informal economy (informal sector) – The part of the economy that is not fully controlled by government. People working in this sector are often self-employed or in small businesses. They do not pay tax but they do not receive any state benefits either.

Infrastructure – The basic structures and services needed by any society such as water supplies, sewage systems, roads or bridges.

Interglacials – Warmer periods in Earth's history when glaciers have retreated and ice sheets have decreased in size.

Interlocking spurs – A feature of V-shaped valleys where the river meanders from side to side so that the hillsides interlock rather like the teeth of a zip.

Intermediate technology – Technology that is appropriate for use in a developing country because it does not need expensive parts or high-tech repairs.

Intertidal zone – The part of the shoreline that is between high tide and low tide.

Intertropical convergence zone (ITCZ) – A broad band of atmosphere that circles the tropical latitudes. The ITCZ is characterised by low pressure, cloud and heavy rain.

Jet stream – A strong wind that circulates around the Earth.

Key services – The way in which ecosystems provide benefits for people. For example, mangrove forests act as coastal buffers, soaking up wave energy during a storm and reducing the risk of erosion and flooding.

Kite diagram – A special type of graph, in the shape of a kite, which is typically used to show changes in vegetation.

Knowledge economy – Jobs that require high levels of education or training.

Lag time – The time delay between a rain event and the maximum (peak) discharge.

Lagoon – A shallow pond of salt water

Lahar – An Indonesian word that describes a flood of water and volcanic ash or a mudslide down the slope of a volcano. Lahars are caused when rainwater mixes with loose volcanic ash. Lahars are very dangerous hazards.

Landlocked – A country that has no coastline and, therefore, has no sea ports. Many of the world's poorest countries are landlocked.

Landslide – The sudden collapse of a hillside under its own weight. Landslides are sometimes triggered by an earthquake or by erosion at the foot of the slope.

Land use zoning – Where land uses that have a low value, such as car parks or playing fields, are not protected by flood walls. These zones provide safe areas for water to be stored during a flood event so that water is kept away from more valuable land uses such as homes.

Lateral erosion – The process by which a river can cut sideways into its own river bank.

Latitude – The distance to the north or south of the Equator.

Lava field – A very large, rather flat landform that is created when a large lava flow solidifies. Lava fields are usually formed by the eruption of lava from a shield volcano or a fissure eruption.

Lava flows – Features formed as molten rock (lava) runs away from a volcanic vent. Lava flows often form large flat features known as **lava fields**.

Lava tube – A natural tunnel through solidified volcanic rocks that is formed when hot lava flows just below the surface of the ground.

Leakage – When money, spent by tourists, benefits companies in other countries rather than people working in the country that the tourists are visiting.

Life cycle analysis – A way of trying to determine the full impact of every stage of the manufacture and delivery of consumer goods.

Line of best fit – A line that represents the trend through the points plotted on a scattergraph.

Load – The sediment carried by a river.

Long-haul (flights) – Flights to distant places. Long-haul flights from the UK go to locations outside of Europe.

Longshore drift – A process by which beach material is moved along the coast.

Low elevation coastal zone (LECZ) – Flat, low-lying land close to the sea that could be at risk of coastal flooding or sea level rise.

Low-pressure system (cyclone) – A condition of the atmosphere that brings wet and windy weather to the UK. These systems begin when warm, moist air rises from the ground. This creates air pressure that is lower than normal at the centre of the system and air is drawn in to fill the space, which can generate strong wind.

Magma – Molten, or semi-molten, rock that is stored beneath the Earth's surface in, for example, a **magma chamber** beneath a volcano.

Magma chamber – A hollow or cavern beneath a volcano that contains hot, molten rock.

Magnitude – A description of the strength and scale of natural hazards such as a volcanic eruptions, tsunami or earthquakes.

Managed retreat – A strategy that is used to manage some coastal environments. Managed retreat (also known as managed realignment) means that defences are breached and the coastline is allowed to retreat inland.

Management zones – Within an area of conservation, such as a National Park, different activities will be permitted within area or zone.

Mangrove forests – A type of tropical forest that grows in coastal regions.

Mantle – A zone of hot rocks that lies beneath the solid rocks of the Earth's crust.

Manufactured goods – Items that have been made in a workshop or factory.

Maritime-climate – The climatic condition of land close to sea. The sea moderates temperatures meaning that there are only small variations in temperature.

Mass movement – When soil, rocks and stones slip, slide or slump down a slope. Some mass movement is very slow, for example, soil creep. Other mass movements are very rapid, for example, a rock fall.

Mass tourism – When very large numbers of tourists, who have bought a package holiday, visit a large resort.

Mass transit – A type of transport system that is able to move large numbers of people through a city, for example, an underground rail system.

Meander – A river landform. A sweeping curve or bend in the river's course.

Mega-city – An urban area (city) that has a population greater than 10 million people.

Mid-ocean ridge – Long chains of mountains that run down the centre of several oceans, including down the middle of the Atlantic Ocean. The ocean ridges are formed by plate movement.

Millennium Development Goal (MDG) – Millennium Development Goals. These are targets set by the United Nations to try to encourage and measure improvements to human development.

Moment magnitude scale (M_w) – A measure of the strength of an earthquake.

Monoculture – A type of agriculture (farming) in which only one crop is grown over very large areas of land.

Monsoon – A climate type experienced in South Asia in which a seasonal pattern of wind brings a distinct wet season.

Multilateral aid – Financial support that is given by a large number of different governments to a large organisation like the United Nations or the World Bank. This organisation then uses the aid to support countries that are in need of support.

Multinational companies (MNCs) – Large businesses such as Sony, Microsoft and McDonalds, which have branches in several countries. Multinational companies are also known as transnational companies.

Multiplier effect – An upward spiral of the economy and its benefits on employment. Positive multipliers are often triggered by a large investment, for example, the opening of a new factory.

Net immigration – When more people move into the region than leave it.

Net migration – The difference between the number of immigrants and emigrants in any one year.

Newly industrialised country (NIC) – Newly industrialised countries such as India, Thailand or Indonesia have a large percentage of the workforce working in the secondary (manufacturing) sector.

NIMBYism – Not In My Back Yard. People who object to a development because they live close by are said to be NIMBYs.

Nutrient cycle – The flow and storage of minerals within an ecosystem. Nutrients are stored in plants, soil and rocks. Nutrient flows occur through processes such as weathering and leaching.

Nutrient flows – The movement of minerals from one store to another.

Nutrient stores – A part of an ecosystem in which nutrients are kept.

Ocean acidification – A process by which the pH level of sea water decreases, making the water more acidic. Acidification is caused when extra CO_2 is absorbed into the sea.

Ocean currents – Predictable flows of water through the seas and oceans. Some currents are flows of relatively warm water, like the Gulf Stream. Other currents are relatively cold, like the Labrador.

Ocean trenches – Long, deep gorges in the sea bed that occur around the edges of some oceans including the Pacific Ocean. The ocean trenches are formed by plate movement.

Offshore bar – A feature on the sea bed formed by the deposition of sand.

Outsourcing – To get a product or service from a supplier that is outside the company.

Over-abstraction – When water is abstracted at a faster rate than it is recharged, leading to a store of water decreasing in size.

Overland flow – The flow of water across the ground surface.

Pavement dwellers – People who live in make-shift homes on the footpaths of some developing cities, especially in Indian cities.

Peak discharge – The maximum amount of water in a river during a flood.

Permeability – The ability of a rock to allow water to pass through it.

Permeable – A rock which allows water to pass through it, such as limestone.

Place – A geographical concept used to describe what makes somewhere special, unique or distinct.

Plates – Rigid sections of crust. The plates lie on top of the mantle. They are able to move relative to each other. The movement is slow, but the force generated by their movement creates earthquakes and volcanic hazards.

Plunge pool – The pool of water found at the base of a waterfall. Plunge pools are erosional features created by abrasion and hydraulic action of the plunging water.

Point bar – A river beach formed of sand and gravel that is deposited on the inside bends of a meander.

Porosity – The ability of a rock to store water in tiny air spaces (pores).

Porous – A rock which has many tiny gaps within it (pores) that allow it to store water, such as chalk and sandstone.

Postglacial rebound – An adjustment in the level of the Earth's crust. The crust was depressed by the mass of ice lying on it during glacial periods of the ice age. Since the end of the last glacial period the crust has been slowly rising back to its original level.

Poverty line – People who live below this amount of income are said to live in poverty.

Primary consumers – Animals that eat vegetation (**producers**) in the food chain. These animals may be eaten by **secondary consumers**.

Producers – Plants that are able to create starch from the sun's energy. Producers are at the bottom of the food chain.

Protectionist policies – Schemes that are used by countries to protect their own industries from cheap imports. For example, by placing a tariff on imports, the imported goods become more expensive, so people are more likely to buy the product that has been made in their own country.

Pull factors – Reasons that attract migrants to move to a new home.

Purchasing power parity (PPP) – A way of comparing the average wealth of a country by taking the cost of living in those countries into account.

Push factors – Reasons that force people to move away from their existing home.

Pyroclastic flows – A mixture of hot gas, ash and fragments of volcanic rock that fall, in a tumbling motion, down the slopes of a volcano during some explosive eruptions.

Quadrat – A piece of equipment used in fieldwork. Quadrats are square frames that come in several different sizes that are used during sampling.

Qualitative data – Information that is not numerical.

Quantitative data – Information that can be measured and recorded as numbers.

Quotas – Restrictions on the amount of particular goods that can be imported each year.

Rainwater harvesting – The collection and storage of rain water, for example from the roof of a house.

Raw materials – Materials such as timber, stone or crude oil that have not been processed or refined.

Re-afforestation – The planting of large areas with trees.

Recharge – Water that enters an aquifer and refills a groundwater store.

Refugees – People who are in danger and who leave their homes for their own safety. Refugees may move because of a natural disaster such as a volcanic eruption or because of conflict.

Reservoir – A body of water that is used as a supply of water.

Responsible travel – Tourism that benefits local people and respects their culture, environment, beliefs and traditions. This term has the same meaning as **ethical tourism**.

Retreat – The gradual backward movement of a landform due to the process of erosion. The coastline retreats due to the erosion of a cliff and a waterfall retreats towards the source of a river as it is eroded.

Re-urbanisation – The recent trend for the population of city centres to grow.

Rift valley – A steep-sided valley formed by the pulling apart (or rifting) of the Earth's crust during plate movement.

River landforms – Natural features of the Earth's surface associated with rivers. V-shaped valleys and waterfalls are made by erosional processes. Floodplains are made by deposition. River landforms vary in scale. Meanders, for example, are often large landforms. They contain smaller features, such as slip-off slopes.

Rock fall – The sudden collapse of rocks from a cliff or steep slope.

Rural diversification – The development of new businesses in the countryside. These new businesses are often in leisure or tourism rather than in farming.

Rural to urban migration – The movement of people from the countryside to towns and cities.

Sahel – The semi-arid region of North Africa to the south of the Sahara desert. Sahel means 'shore' in Arabic.

Scale – A geographical concept used to describe the size or area covered by a feature. Scale varies from small (or local) through to regional, national and global.

Scattergraph – A type of diagram that can be used to test whether or not there is a relationship between two sets of data. See **dependent variable** and **independent variable**.

Schengen Zone – The area within the EU in which border control has been abandoned allowing people to cross from one country to another without showing a passport.

Scoria – Small fragments of volcanic rock that are ejected during an eruption. Scoria is full of bubble-like holes.

Sea arches – Natural arch-shaped features in cliffs on the coastline that are formed by the erosion of a cave in a headland.

Secondary consumers – Animals that are higher up the food chain and that eat **primary consumers**.

Seismometers – Scientific instruments used to measure the strength and frequency of earth shaking or earthquakes.

Self-help – Improvement projects carried out by ordinary people rather than by businesses or governments. Compare this with **top-down development**.

Shield volcano – A large volcano that has gentle slopes. Some of the volcanoes in Iceland and Hawaii have this shape.

Shoreline Management Plan – The plan that details how a local authority will manage each stretch of coastline in the UK in the future.

Slab pull – A process by which the sinking of the crust at a destructive plate margin pulls the oceanic plate away from the constructive plate margin.

Slip-off slope – The gentle slope on a river beach (or **point bar**) that is formed by deposition of sediment on the inside bends of a meander.

Slumping – The gradual collapse of a hillside under its own weight. Slumping sometimes occurs where an **unconsolidated** rock glides over an impermeable rock type such as clay.

Small island developing states – A group of 58 countries with a combined population of 65 million people. Many SIDS are very small and some are located in remote and isolated parts of the world. Most SIDS are vulnerable to climate change and other natural disasters, such as earthquakes and tsunamis.

Social housing – Homes that are provided at an affordable price for rent. Social housing is either provided by the local council or by a housing association.

Soft engineering – Alternative method of reducing floods by planting trees or allowing areas to flood naturally.

Soil erosion – The loss of soil due to either wind or heavy rain. Gulley erosion is a major cause of soil erosion in countries that have a seasonal wet and dry climate.

Spatial – Patterns or geographical features that vary over two dimensions so that they can be shown on a map.

Sphere of influence – The area over which a geographical feature or event is able to create an impact. These impacts may be good or bad.

Spit – A coastal landform formed by the deposition of sediment in a low mound where the coastline changes direction, for example, at the mouth of a river.

Squatter homes (slums) – Homes where the householders have no legal rights to the land, i.e. they do not have legal housing tenure. Informal settlements are commonly known as shanty towns or squatter homes. They are referred to as slums in India.

Stacks – Natural features of an eroded cliff landscape. Stacks are formed by the collapse of a sea arch.

Standard of living – The level of wealth and comfort experienced by any group of people or individual.

Storm beach – A steeply sloping beach. Strong waves throw pebbles into a steep ridge at the top of the beach during a storm.

Storm surge – The rise in sea level that can cause coastal flooding during a storm or hurricane. The surge is due to a combination of two things. First, the low air pressure means that sea level can rise. Second, the strong winds can force a bulge of water on to the shoreline.

Stratovolcano – A large, steep-sided volcano formed from layers of solidified lava and ash that have been built up by many different eruptions.

Subduction – The process where one plate is destroyed as it is slowly pulled underneath another plate.

Subjective – Evidence that is personal and which varies depending on someone's point of view.

Subsidy – A payment that a country makes to its own farmers and businesses so that their goods can be sold at a lower price to consumers.

Subsistence – A type of economic activity where very little money is used. In subsistence farming the farmer only produces enough food to feed the family. There is very little surplus that can be sold for cash.

Surface stores – Places where water is found on the surface such as lakes and rivers.

Sustainable community – A community which is designed to have minimum impact on the environment. Such communities may make use of energy efficiency, renewable technologies and local services in order to reduce transport costs.

Sustainable development goals – A set of aims proposed by the United Nations (UN) in 2015, applying to 17 different areas of life. The overall aim is to end poverty, protect the planet and ensure that all people enjoy peace and prosperity by 2030.

Swash – The flow of water up the beach as a wave breaks on the shore.

Taiga – Natural forest ecosystems found in the cold climates of Northern Europe and America.

Tariff – A type of tax that may be charged on goods as they enter a country.

Telework – To work from home using technologies such as phones and the internet to communicate with a central workplace or office.

Throughflow – The downhill flow of water through soil.

Tiltmeter – A scientific instrument that is used to measure tilting of the ground. Tiltmeters can be used to measure small changes in ground shape that occur when the magma chamber beneath a volcano is filling with molten rock.

Top-down – When decisions about development are made by governments or officials rather than by ordinary people. Compare this to **self-help** schemes.

Tourist enclave – A tourist resort that is separated from local communities. Some tourist developments are designed so that tourists are discouraged from leaving the hotel or resort. In this way, the tourist spends more money with the company, and very little with local businesses.

Trade bloc – A group of countries that agree to buy and sell goods without any restrictions or tariffs (taxes). The European Union is one example.

Transect – A line along which data is collected. Transects usually cut across different geographical features.

Transport – The movement of material through the landscape.

Tributary – A smaller river which flows into a larger river channel.

Tropical rainforest – Large forest ecosystems (or biomes) that exist in the hot, wet climate found on either side of the Equator.

Tsunami – A series of large and powerful waves on the surface of a lake or the ocean. A tsunami may be caused when water is displaced by the movement of the ocean floor/lake bed during an earthquake.

Tundra – An ecosystem largely found in the Arctic region. The tundra is treeless because the growing season is short and the average monthly temperature is below 10°C.

Unconsolidated – A rock that is only loosely compacted and that is not properly glued together. Deposits left by melting of ice at the end of the last age are largely unconsolidated.

Unintended consequence – A side-effect of human activity that can, for example, affect coastal processes.

Unskilled workers – Workers who have low levels of qualification or little technical ability and training. See **highly skilled**.

Unstable – Warm air that is rising may be described as unstable. Unstable air causes clouds to build up and form rain.

Upland – A landscape that is hilly or mountainous. Upland landscapes contain large areas of open space with few field boundaries.

Urban heat island – When a city has temperatures that are warmer than in the surrounding rural area.

Urban microclimate – The small scale, local climate of a large city which is influenced by its buildings and traffic.

Urban sphere of influence – The influence over the surrounding area that every urban area has through its close links with nearby towns and countryside.

Urbanisation – The physical and human growth of towns and cities.

Urban–rural continuum – The range of human environments one can experience from the very densely built-up parts of our largest cities to the most remote, sparsely populated rural communities.

U-shaped valleys – Valleys that have steep slopes and a flat valley floor. U-shaped valleys are erosional features created by the movement of a glacier.

Variables – Sets of data. See **dependent variable** and **independent variable**.

Vertical erosion – When the force of water, that is wearing away the landscape, is concentrated downwards. Vertical erosion is common in steeply flowing streams and also in gulley erosion.

Viscosity – A measure of the stickiness of lava. Viscosity depends on factors such as the temperature of the lava and its chemical composition.

Volcanic Explosivity Index (VEI) – A measure of the size of a volcanic eruption that is based on the amount of ash ejected and the height of the ash cloud.

Volcanic hotspot – An area of the Earth's crust in which volcanic activity is concentrated. Some hotspots are on plate margins (such as Iceland) and other hotspots are in the middle of plates (such as Hawaaii).

Vulnerability – Exposure to a risk such as a natural disaster. Some groups of people in society are more vulnerable to risk than others. Vulnerability can be overcome by building **capacity**.

Water cycle – The continuous flow of water between the earth's surface and the atmosphere – also called the hydrological cycle.

Water footprint – The amount of water used to make an item of food or make a product such as an item of clothing.

Water security – When a society has enough water to ensure that everyone has clean water, sanitation and good health and the economy has enough water to grow food and make things.

Water stress – A lack of clean water that causes problems for people and the economy.

Waterfall – A landform created when a river plunges over a vertical or near vertical surface.

Wave-cut notch – A slot with overhanging rocks that has been cut into the bottom of a cliff by wave action.

Wave-cut platform – A coastal landform made of rocky shelf in front of a cliff. The wave-cut platform is caused by erosion and left by the retreat of the cliff.

Weather – Our day-to-day experience of temperature, wind, precipitation and sunshine.

Weathering – Processes that weaken joints in rocks. They include frost action, chemical reactions and the growth of plant roots.

Wholesale clearance – The demolition of a large quantity of old unfit housing and the redevelopment of new, better homes.

Wildlife corridor – Strips of habitat that allow wild animals to migrate from one ecosystem to another, for example hedgerows.

Youthful population – A population with a high proportion of children because the birth rate is relatively high.

Zooxanthellae – Algae that live symbiotically within the cells of other organisms. This relationship creates the coral polyps that we see in coral reefs.

INDEX

INDEX

PHOTO ACKNOWLEDGEMENTS

The publishers would like to thank the following for permission to reproduce copyright material:

p.3 *t* © Janet Baxter Photography; **p.5** © Jeff Morgan 13/Alamy Stock Photo; **p.6** © Realimage/Alamy Stock Photo; **p.18** © David Waters/Alamy Stock Photo; **p.19** © robertharding/Alamy Stock Photo; **p.22** © B&JPhotos/Alamy Stock Photo; **p.25** © Billy Stock/Alamy Stock Photo; **p.28** *t* © Ashley Cooper/Alamy Stock Photo; *m* © Anthony Collins/Alamy Stock Photo; **p.31** © PHILIP SMITH/Alamy Stock Photo; **p.32** © Peter Barritt/Alamy Stock Photo; **p.33** *t and m* © Glyn Owen; *b* © Tony Boydon/Alamy Stock Photo; **p.35** © robertharding/Alamy Stock Photo; **p.43** © Mark Runnacles/Getty Images; **p.44** © Andrew Calverley/Alamy Stock Photo; **p.45** © Cumbria Crack; **p.47** © SWNS/Alamy Stock Photo; **p.48** *m* © Paul Glendell/Alamy Stock Photo; *b* © JMF News/Alamy Stock Photo; **p.50** *t* © Avalon; **p.60** *b* © Richard Downs/Alamy Stock Photo; **p.69** © MJM/Alamy Stock Photo; **p.71** © Peter Macdiarmid/Getty Images News/Getty Images; **p.72** © Raf Makda/View Pictures/Rex Features; **p.76** © Andy Leeder; **p.77** © Andy Leeder; **p.78** © Rawiwan/stock.adobe.com; **p.80** © A.P.S. (UK)/Alamy Stock Photo; **p.83** © Paul Chambers/pcp(news)/Alamy Stock Photo **p.86** © ZUMA Press, Inc./Alamy Stock Photo; **p.88** *t* © Robert Gray/Alamy Stock Photo; *bl* © vdbvsl/Alamy Stock Photo; *br* © Jeffrey Blackler/Alamy Stock Photo; **p.90** © Paul Kennedy/Alamy Stock Photo; **p.91** © Dinodia Photos/Alamy Stock Photo; **p.94** © Dean Hoskins/Alamy Stock Photo; **p.95** © roger parkes/Alamy Stock Photo; **p.96** *t* © galit seligmann/Alamy Stock Photo; *b* © galit seligmann/Alamy Stock Photo; **p.97** © Saifee Burhani Upliftment Trust; **p.103** © Mikhail Riches/Tim Crocker 2019; **p.106** © Efrain Padro/Alamy Stock Photo; **p.109** © Warren Kovach/Alamy Stock Photo; **p.110** *t* © Martin Hertel/360cities.net via Getty Images; *b* © frans lemmens/Alamy Stock Photo; **p.113** © Manuel Ribeiro/Alamy Stock Photo; **p.116** © Jacqui Owen **p.117** © epa european pressphoto agency b.v./Alamy Stock Photo; **p.118** © Stocktrek Images, Inc./Alamy Stock Photo; **p.121** © Lianne Milton – Panos; **p.122** © CLARA PRIMA/AFP/Getty Images; **p.123** © Andrea Fitrianto (http://creativecommons.org/licenses/by-nc-sa/3.0/); **p.125** © ephotocorp/Alamy Stock Photo; **p.127** © Nripal Adhikary; **p.128** © NICOLAS ASFOURI/AFP/Getty Images; **p.129** © Philip Game/Alamy Stock Photo; **p.130** © nobleIMAGES/Alamy Stock Photo; **p.134** © Environment Agency; **p.142** © robertharding/Alamy Stock Photo; **p.146** © Stephen Pond/Stringer; **p.147** © Mark Richardson/Alamy Stock Photo; **p.149** © Irene Abdou/Alamy Stock Photo; **p.151** © Chris Wood/Alamy Stock Photo; **p.152** *bl* © Richard Smith/Alamy Stock Photo; *r* © Cartoonstock; **p.155** Chichester Canal © Joseph Mallord William Turner via Getty images; **p.156** © Sue Cunningham/Sue Cunningham Photographic/Alamy Stock Photo; **p.158** © Arctic Images/Alamy Stock Photo; **p.163** © Pacific Press Agency/Alamy Stock Photo; **p.165** *t* © NASA, *b* © NASA; **p.166** © Logan Bush/Shutterstock.com; **p.169** © Bob Kreisel/Alamy Stock Photo; **p.170** © Ashley Cooper/Alamy Stock Photo; **p.171** © Nasa; **p.174** © Toby Melville/Thomson Reuters; **p.179** *b* © FLPA/Tui De Roy/Minden Pictures; **p.181** © FLPA/Tui De Roy/Minden Pictures; **p.184** *l* © Brad Simmons/Beateworks/Corbis; *r* © Richard Becker/Alamy Stock Photo; **p.185** © Gerry Ellis/Minden Pictures/FLPA; **p.186** *t and b* © Montgomeryshire Wildlife Trust; **p.187** © M Williams/Alamy Stock Photo; **p.190** © Val Davis; **p.191** © Val Davis; **p.193** © Images of Africa Photobank/Alamy Stock Photo; **p.194** © Ken Edwards/Alamy Stock Photo; **p.198** © Photoshot License Ltd/Alamy Stock Photo; **p.199** © blickwinkel/Alamy Stock Photo; **p.202** © Ian Nellist/Alamy Stock Photo; **p.203** © Neil Cooper/Alamy Stock Photo; **p.205** © Nyani Quarmyne/Panos Pictures; **p.206** © Jacob Silberberg/Panos Pictures; **p.208** © Prisma Bildagentur AG/Alamy Stock Photo; **p.209** © Ullstein Bild/Getty Images; **p.211** © Colm Regan, Man eating the world that someone else is holding, cartoon, Thin Black Lines: Political Cartoons & Development Education (Teachers in Development Education 1988); **p.213** *l* © david pearson/Alamy Stock Photo; *r* © Ariadne Van Zandbergen/Alamy Stock Photo; **p.214** *tl* © Danita Delimont/Alamy Stock Photo; *tr* © Getty Images/Raveendran/AFT; *bl* © PETER PARKS/AFP/Getty Images; **p.215** Stuart Freedman/Panos Pictures; **p.219** © Janine Wiedel Photolibrary/Alamy Stock Photo; **p.221** © Thomas Imo/Photothek via Getty Images; **p.222** © Simon Stirrup/Alamy Stock Photo; **p.224** © Everett Collection Historical/Alamy Stock Photo; **p.225** © Ulrich Doering/Alamy Stock Photo; **p.232** © imageBROKER/Alamy Stock Photo; **p.233** © robertharding/Alamy Stock Photo; **p.234** © Jack Sullivan/Alamy Stock Photo; **p.239** © Joerg Boethling/Alamy Stock Photo; **p.241** © Simon Rawles/Alamy Stock Photo; **p.242** © Aerial-photos.com/Alamy Stock Photo; **p.243** *tl* © Ton Koene/VWPics/Alamy Stock Photo; *bl* © The International Women's Tribune Centre; **p.244** © Dieter Telemans/Panos Pictures; **p.249** © Friedrich Stark/Alamy Stock Photo; **p.251** © Desiree Martin/AFP/Getty Image; **p.253** © Shankar Mourya/Hindustan Times via Getty Images; **p.254** © Joerg Boethling/Alamy Stock Photo; **p.261** © Paul White - UK Industries/Alamy Stock Photo; **p.262** *t* © Panos Pictures/Christien Jaspars; *mr* © Colm Regan, Man thinking about a car and woman thinking about shoes, cartoon, Thin Black Lines: Political Cartoons & Development Education (Teachers in Development Education 1988); **p.270** © Pacific Press/LightRocket via Getty Images; **p.273** © epa european pressphoto agency b.v./Alamy Stock Photo; **p.275** © Tina Manley/Alamy Stock Photo; **p.277** © Horizons WWP/TRVL/Alamy Stock Photo; **p.281** © Panos Pictures/Alfredo Caliz; **p.283** © Jenny Matthews/Panos Pictures; **p.285** © WaterAid **p.287** © ZUMA Press, Inc./Alamy Stock Photo; **p.291** © Friedrich Stark/Alamy Stock Photo; **p.292** © Simon Price/Alamy Stock Photo; **p.294** © Jacqui Owen; **p.295** Photo taken by Caroline Russell during fieldwork funded by Lancaster Environment Centre; **p.296** © YAY Media AS/Alamy Stock Photo; **p.297** *t* © Age Fotostock/Alamy Stock Photo; *m* © Greenpeace; **p.300** © WaterFrame/Alamy Stock Photo; **p.301** © A.P.S. (UK)/Alamy Stock Photo; **p.304** © cartoonstock.com; **p.305** © STEFAN HEUNIS/AFP/Getty Images; **p.309** *t* © ImageBroker/Alamy Stock Photo; *b* © Steffen Binke/Alamy Stock Photo; **p.310** © Lintao Zhang/Getty Images; **p.313** *l* © RSPB, *r* Stuart Walker/Alamy Stock Photo; **p.315** Tayvay/Shutterstock; **p.318** Matthew Oldfield Editorial Photography/Alamy Stock Photo; **p.321** Mark Richardson/Alamy Stock Photo; **p.323** BrazilPhotos/Alamy Stock Photo; **p.325** Vincent Sufiyan/Alamy Stock Photo; **p.327** Paul Prescott/Alamy Stock Photo; **p.329** Friedrich Stark/Alamy Stock Photo.

The images on the following pages are copyright of Andy Owen: pp. 1, 2, 3 (*bl, br*), 9, 12, 13, 14, 15, 16, 20, 21, 24, 26 (*t, b*), 36, 37, 50 (*b*), 55, 56, 60 (*t*), 64, 65, 73, 82, 84, 85, 100, 102, 104, 105, 111, 112 (*tl, br*), 113, 132, 133, 136, 137, 139, 176, 178, 179 (*bl, tr*), 180, 183, 188, 201, 229, 230, 289 (*tl, tr*), 306, 307 (*tr, c*).